T0342408

Sustainable Aviation Technology and Operations

Aerospace Series

Visit www.wiley.com to view more titles in the Aerospace Series.

Sustainable Aviation Technology and Operations

Research and Innovation Perspectives

Edited by

Roberto Sabatini
Professor, Department of Aerospace Engineering
College of Engineering
Khalifa University of Science and Technology
Abu Dhabi, UAE

Honorary Professor, Aerospace Engineering and Aviation
School of Engineering, STEM College
RMIT University, Melbourne
Victoria, Australia

Alessandro Gardi
Assistant Professor, Department of Aerospace Engineering
College of Engineering
Khalifa University of Science and Technology
Abu Dhabi, UAE

Associate of RMIT University
Aerospace Engineering and Aviation, School of Engineering
STEM College, Melbourne
Victoria, Australia

This edition first published 2024

© 2024 John Wiley & Sons Ltd.

All rights reserved. No part of this publication may be reproduced, stored in a retrieval system, or transmitted, in any form or by any means, electronic, mechanical, photocopying, recording or otherwise, except as permitted by law. Advice on how to obtain permission to reuse material from this title is available at http://www.wiley.com/go/permissions.

The right of Roberto Sabatini and Alessandro Gardi to be identified as the editors of this work has been asserted in accordance with law.

Registered Offices
John Wiley & Sons, Inc., 111 River Street, Hoboken, NJ 07030, USA
John Wiley & Sons Ltd, The Atrium, Southern Gate, Chichester, West Sussex, PO19 8SQ, UK

For details of our global editorial offices, customer services, and more information about Wiley products visit us at www.wiley.com.

Wiley also publishes its books in a variety of electronic formats and by print-on-demand. Some content that appears in standard print versions of this book may not be available in other formats.

Trademarks: Wiley and the Wiley logo are trademarks or registered trademarks of John Wiley & Sons, Inc. and/or its affiliates in the United States and other countries and may not be used without written permission. All other trademarks are the property of their respective owners. John Wiley & Sons, Inc. is not associated with any product or vendor mentioned in this book.

Limit of Liability/Disclaimer of Warranty
While the publisher and authors have used their best efforts in preparing this work, they make no representations or warranties with respect to the accuracy or completeness of the contents of this work and specifically disclaim all warranties, including without limitation any implied warranties of merchantability or fitness for a particular purpose. No warranty may be created or extended by sales representatives, written sales materials or promotional statements for this work. The fact that an organization, website, or product is referred to in this work as a citation and/or potential source of further information does not mean that the publisher and authors endorse the information or services the organization, website, or product may provide or recommendations it may make. This work is sold with the understanding that the publisher is not engaged in rendering professional services. The advice and strategies contained herein may not be suitable for your situation. You should consult with a specialist where appropriate. Further, readers should be aware that websites listed in this work may have changed or disappeared between when this work was written and when it is read. Neither the publisher nor authors shall be liable for any loss of profit or any other commercial damages, including but not limited to special, incidental, consequential, or other damages.

Library of Congress Cataloging-in-Publication Data

Names: Sabatini, Roberto, editor. | Gardi, Alessandro, editor.
Title: Sustainable aviation technology and operations : research and
   innovation perspectives / Roberto Sabatini, Professor, Department of Aerospace
   Engineering, Khalifa University of Science and Technology, Abu Dhabi,
   UAE; Alessandro Gardi, Assistant Professor, Department of Aerospace
   Engineering, Khalifa University of Science and Technology, Abu Dhabi, UAE
Description: Hoboken, NJ, USA : Wiley, 2024. | Series: Aerospace series
Identifiers: LCCN 2020025457 (print) | LCCN 2020025458 (ebook) | ISBN
   9781118932582 (cloth) | ISBN 9781118932612 (adobe pdf) | ISBN
   9781118932605 (epub)
Subjects: LCSH: Aeronautics–Technological innovations. | Aerospace
   engineering. | Sustainable development.
Classification: LCC TL553 .S23 2024 (print) | LCC TL553 (ebook) | DDC
   629.13028/6–dc23
LC record available at https://lccn.loc.gov/2020025457
LC ebook record available at https://lccn.loc.gov/2020025458

Cover image: © ICHIRO/Getty Images
Cover design by Wiley

Set in 9.5/12.5pt STIXTwoText by Straive, Chennai, India

Contents

List of Contributors

Rafic Ajaj
Department of Aerospace Engineering
Khalifa University of Science and
Technology, Abu Dhabi, UAE

Martin Burston
School of Engineering, RMIT University
Melbourne, Victoria, Australia

Enda Crossin
University of Canterbury
Christchurch, New Zealand

Raj Das
School of Engineering
RMIT University
Melbourne, Victoria, Australia

Graham Dorrington
School of Engineering
RMIT University
Bundoora, Victoria, Australia

George Dulikravich
Florida International University
Miami, Florida, USA

Joel Galos
Department of Materials Engineering
California Polytechnic State University
San Luis Obispo, CA, USA

Alessandro Gardi
Department of Aerospace Engineering
Khalifa University of Science and
Technology, Abu Dhabi, UAE

Nikola Gavrilović
ISAE-SUPAERO
University of Toulouse
Toulouse, France

Kai Guan
RMIT University
Bundoora, Victoria, Australia

Rohan Kapoor
School of Engineering
RMIT University
Bundoora, Victoria, Australia

Trevor Kistan
Thales Australia
Melbourne, Victoria, Australia

Arun Kumar
School of Engineering
RMIT University
Bundoora, Victoria, Australia

Yixiang Lim
Agency for Science, Technology and
Research (ASTAR)
Singapore

Matthew Marino
School of Engineering
RMIT University
Bundoora, Victoria, Australia

Jean-Marc Moschetta
Jean-Marc Moschetta Aerodynamics
Energetics and Propulsion Department
ISAE-SUPAERO Toulouse, France

Vladimir Parezanović
Department of Aerospace Engineering
Khalifa University of Science and
Technology, Abu Dhabi, UAE

Nichakorn Pongsakornsathien
School of Engineering
RMIT University
Bundoora, Victoria, Australia

Kavindu Ranasinghe
Insitec Pty Ltd
Melbourne, Victoria, Australia

Boško Rašuo
Faculty of Mechanical Engineering
University of Belgrade
Belgrade, Serbia

Stephen Rondinelli
RMIT University
Melbourne, Victoria, Australia

Roberto Sabatini
Department of Aerospace Engineering
Khalifa University of Science and
Technology, Abu Dhabi, UAE

Jose Silva
School of Engineering
RMIT University
Melbourne, Victoria, Australia

Jacob Sliwinski
RMIT University
Bundoora, Victoria, Australia

Anthony Zanetti
RMIT University
Melbourne, Victoria, Australia

About the Editors

Roberto Sabatini is a Professor of Aerospace Engineering at Khalifa University of Science and Technology (UAE) and an Honorary Professor of Aerospace Engineering and Aviation at RMIT University (Australia). Previously, Prof. Sabatini was also affiliated with Cranfield University (UK), where he led the research team contributing to the European Union Clean Sky Joint Technology Initiative for Aeronautics and Air Transport – Systems for Green Operations Integrated Technology Demonstrator. Prof. Sabatini holds various academic qualifications in aerospace and geospatial engineering, including a PhD from Cranfield University and a PhD from the University of Nottingham. Additionally, he holds the licenses of private pilot, flight test engineer and remote pilot. Throughout his career, Prof. Sabatini led numerous research projects funded by national governments, international organizations and aerospace/defence industry partners. He has authored, co-authored, or edited several books, and has had more than 300 articles published in refereed international journals and conference proceedings. Since 2019, he has been listed by the Stanford University's ranking among the top 2% most cited scientists globally in the field of aerospace and aeronautics. Prof. Sabatini is a Fellow of the Royal Aeronautical Society (RAeS), the Royal Institute of Navigation (RIN), the Institution of Engineers Australia (IEAust), and the International Engineering and Technology Institute (IETI), as well as a Senior Member of the American Institute of Aeronautics and Astronautics (AIAA) and the Institute of Electrical and Electronics Engineers (IEEE). He was conferred prestigious national and international awards, including: Best-in-field National Scientist in Aviation and Aerospace Engineering – The Australian Annual Research Report (2021); Distinguished Leadership Award – Aviation/Aerospace Australia (2021); Scientist of the Year – Australian Defence Industry Awards (2019); Science Award – Sustainable Aviation Research Society (2016); and Arch T. Colwell Merit Award – Society of Automotive Engineering (2015). Since 2017, Prof. Sabatini has represented the Australian Government in several occasions at the ICAO Committee on Aviation Environmental Protection (CAEP) Impact and Science Group (ISG). More recently, he has also contributed to the activities of the Joint Authorities for Rulemaking in Unmanned Systems (JARUS), the ICAO Drone Enable initiative, the FAA NextGen Tech Talk program, and the NASA UAS Traffic Management (UTM) and Advanced Air Mobility (AAM) working groups. Currently, he serves as Distinguished Lecturer of the IEEE Aerospace & Electronic Systems Society (AESS), Chair of the AESS Avionics Systems Panel (ASP) and member-at-large of the AESS Board of Governors. Additionally, he is a founding Editor of the IEEE Press Series on Aeronautics and Astronautics Systems, Editor for Progress in Aerospace Sciences, and Associate Editor for Aerospace Science and Technology, Robotica, the Journal of Navigation, and the IEEE Transactions on Aerospace and Electronic Systems.

Alessandro Gardi is an Assistant Professor in Aerospace Engineering at Khalifa University of Science and Technology (UAE), with more than ten years of experience in aerospace systems research and education. He received his BSc and MSc degrees in Aerospace Engineering from Politecnico di Milano (Italy) and a PhD in the same field from RMIT University (Australia). His work focusses on avionics, air traffic management, and sustainable aviation technology for conventional and autonomous aerospace vehicles. In this domain, he specializes in multidisciplinary and multi-objective optimization with emphasis on optimal control methods and Artificial Intelligence (AI) techniques for air and space vehicle design and operations. Before joining Khalifa University, Dr Gardi was affiliated with Cranfield University (UK) as a member of the Systems for Green Operations Integrated Technology Demonstrator (SGO-ITD) of the European Union Clean Sky Joint Technology Initiative for Aeronautics and Air Transport, one of the largest programs addressing aviation sustainability globally. Successively, he was awarded a multi-year Thales research fellowship in Australia, during which he continued and extended his research work on sustainable and digital aviation technologies. More recently, Dr Gardi has worked on advancing systems and software engineering methodologies for the design of aerospace and defence human-machine systems, utilizing neurophysiological and system integrity monitoring, Internet of Things (IoT) technology and cyber-resilience functionalities to operate autonomously for extended periods of time even in degraded conditions. These contributions also resulted in him being conferred the 2020 Early Career Award by the IEEE Aerospace Electronics Systems Society (AESS), as well as in his appointment as member of the Joint Authorities for Rulemaking in Unmanned Systems (JARUS) Automation Working Group and of the AESS Avionics Systems Panel (ASP). To date, Dr. Gardi has been a senior investigator in more than ten research projects funded by industry and government partners, and has produced more than 150 refereed publications. In addition to his primary affiliation at Khalifa University, Dr. Gardi is an Associate of RMIT University and serves as editor and reviewer for several high-impact journals.

About the Companion Website

This book is accompanied by the following website:

www.wiley.com/go/sustainableaviation

This website includes color version of selected figures.

1

Sustainable Aviation: An Introduction

Roberto Sabatini and Alessandro Gardi

Department of Aerospace Engineering, Khalifa University of Science and Technology, Abu Dhabi, UAE
School of Engineering, RMIT University, Melbourne, Victoria, Australia

The aviation industry plays an important role in the global economy. Before the recent crisis caused by the Coronavirus Disease 2019 (COVID-19) pandemic, air transport alone contributed US$2.7 trillion to the world GDP (3.6%) and supported 65.5 million jobs globally [1]. For several decades, the sector has been on an almost uninterrupted exponential growth trajectory, which demonstrated a remarkable resilience to economic and geo-political crises. According to forecasts predating the COVID-19 pandemic, air traffic was expected to double approximately every 25 years [2]. It was also expected that without intervention, aviation would contribute about 6-10% of all human-induced climate change by 2050 [3], while half of all air traffic would take off, land, or transit through the Asia-Pacific region. In the period 2019–2020, the COVID-19 pandemic has led to a reduction in global passenger traffic in the order of 60% (2,703 million passengers) and the airlines experienced a loss of approximately US$372 billion of gross passenger operating revenues [4, 5]. The situation gradually improved in 2021 and 2022, with a recovery of about 11% and 31% in the number of passengers, reflected by revenue losses of about US$324 billion in 2021 and US$175 in 2022 (compared to 2019).

While sending this book to the press, COVID-19 travel restrictions have been removed in most regions and the latest reports of the International Civil Aviation Organization (ICAO) show that both domestic and international air travel are resuming pre-pandemic levels [5–7]. Factors that could contribute to accelerate further the aviation market recovery and growth include: (1) an increasing demand for commercial Unmanned Aircraft Systems (UAS) and Advanced Air Mobility (AAM) services; (2) technological advances in eco-friendly design solutions (i.e., aerospace vehicles, propulsion, digital flight systems and ground-based infrastructure); (3) uptake of sustainable aviation technologies and associated evolutions of legal frameworks, design/certification standards and operational procedures. In the longer term, the expansion of commercial aviation operations above Flight Level 6-0-0 (FL 600) and the introduction of point-to-point space transport could also contribute to a further evolution and expansion of the aviation sector [8, 9]. Factors that could hinder the growth of the aviation sector include airlines' bankruptcy, order cancellations, increased cyber threats, insufficient investment in aviation infrastructure,

Sustainable Aviation Technology and Operations: Research and Innovation Perspectives, First Edition.
Edited by Roberto Sabatini and Alessandro Gardi.
© 2024 John Wiley & Sons Ltd. Published 2024 by John Wiley & Sons Ltd.
Companion Website: www.wiley.com/go/sustainableaviation

increasing geopolitical tensions, escalation of conflicts, and global recession, many of which are being observed in the post pandemic era.

Over the years, the concomitance of several economic, technological and environmental factors has put the sector under intense and growing pressure. Key factors include the rising costs of operations and fuels; a spiking global competition in relation to the rapid liberalisation of the market and the proliferation of alternative forms of high-speed transport; increased air traffic; capacity bottlenecks at major airports; the need to reduce the environmental impact and achieve greater sustainability in airport and aircraft operations; as well as new regulations and processes to cater for new generation aircraft that are technologically more complex and have new maintenance requirements.

To ensure the aviation sector continues to play a vital role in supporting economic development and employment worldwide, the future air transportation system needs to become even more customer-orientated, time and cost-efficient, secure, and environmentally sustainable than it is today. One of the main priorities for the sector is the rapid uptake of digital technology and, in particular, Cyber-Physical Systems (CPS) that can support the introduction of higher levels of automation, increased airspace capacity, and significant advances in environmental sustainability of both passenger and cargo air transport operations. From the environmental sustainability perspective, over the past two decades, various countries have set unprecedented performance targets for future air transport, such as greenhouse gas emissions having to halve by 2020 (relative to 2000) and be completely offset by 2050 [10]. Adding to these demands are the rising fuel costs, which have increased fourfold in the past 20 years, impeding the profitability of both large airlines and smaller aviation companies.

1.1 Sustainability Fundamentals

Integrating Environmental Susitainability (ES) into business models and associated business functions is an open challenge faced by many industry sectors, including aviation. There is no universally accepted definition for ES while a thematic search of the existing literature[1] shows a prevailing emphasis on the responsible interaction with the environment to avoid depletion or degradation of natural resources and allow for long-term environmental quality both locally and globally. Until recently, businesses have not been held accountable for the cost of damages made to the environment and society. One possible approach is to quantify the environmental degradation caused by a sector and the required measures for restoring the pre-existing conditions. The damages and restoration costs include various sector-specific contributing factors. However, in most cases, such costs are associated air/land/sea pollution and noise. As proposed by [11], the following equation could be used to quantify the cost of environmental degradations caused by economic development activities:

$$ED_T = N \times G_N \times ED_G \tag{1.1}$$

where ED_T is the total environmental degradation (in dollars), N is the population (total number of people), G_N is the Gross National Product (GNP) per capita (in dollars) and ED_G is the environmental degradation per unit of GNP.

---

1 Thematic search on "Environmental Sustainability". Source: *Science Direct* (https://www.sciencedirect .com/topics/agricultural-and-biological-sciences/environmental-sustainability).

So, according to Eq. (1.1), an increase in population would require a proportional reduction of the environmental degradation per unit of GNP in order to maintain the overall environmental degradation at the same level. Similarly, a growth of the GNP per capita would require a commensurate reduction of the environmental degradation per unit of GNP. However, in practice, this equation finds a limited applicability as it does not capture the need for a balance between environmental impacts and the social benefits to be obtained by economic development [12]. Efforts to address these limitations of early quantitative approaches have placed emphasis on the concept of Sustainable Development (SD). The United Nation (UN) 1987 Bruntland Report[2] [13] describes SD as: *"Development that meets the needs of the present without compromising the ability of future generations to meet their own needs."*

The concepts of sustainability and SD have been subjects for extensive research and political debate form many years. What is sustainable can be illustrated using the so-called Triple Bottom Line (TBL) or the "Three Spheres of Sustainability" concept originally introduced by [14]. A modern reinterpretation of this concept is shown in Figure 1.1.

One of the advantages of the TBL approach is that it recognises the importance of delivering sustainable economic value to shareholders by focusing on the bottom line profit that is generated. It also considers that if an enterprise is to be sustainable, it also needs to evaluate its performance in terms of the corresponding environmental and social bottom lines [15]. Several variants of the TBL model have been proposed but essentially this remains a valid high-level reference still utilised in current research work addressing the development of SBM in the corporate environment. The concepts of corporate social

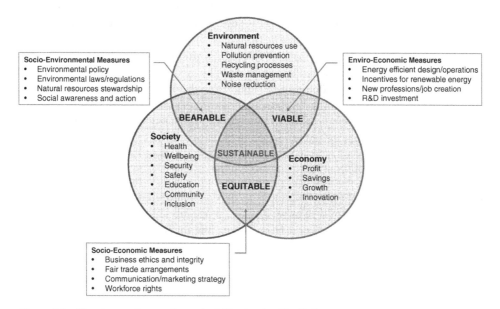

Socio-Environmental Measures
- Environmental policy
- Environmental laws/regulations
- Natural resources stewardship
- Social awareness and action

Environment
- Natural resources use
- Pollution prevention
- Recycling processes
- Waste management
- Noise reduction

Enviro-Economic Measures
- Energy efficient design/operations
- Incentives for renewable energy
- New professions/job creation
- R&D investment

BEARABLE   VIABLE

Society
- Health
- Wellbeing
- Security
- Safety
- Education
- Community
- Inclusion

SUSTAINABLE

EQUITABLE

Economy
- Profit
- Savings
- Growth
- Innovation

Socio-Economic Measures
- Business ethics and integrity
- Fair trade arrangements
- Communication/marketing strategy
- Workforce rights

Figure 1.1 The three spheres of sustainability. Inspired by [14].

2 In 1987, the World Commission on Environment and Development (WCED), published a report entitled "Our common future". The document came to be known as the "Brundtland Report" after the Commission's chairwoman, Gro Harlem Brundtland. It developed guiding principles for sustainable development and it is still adopted today as a key reference in the sector.

responsibility and environmental accountability have been widely discussed in the literature [16, 17]. The main function of the TBL approach is to make corporations aware of the environmental and social values they add or destroy in the world, in addition to the economic value they add [18–20].

Over the years, TBL has become a dominant approach in terms of corporate reporting [21, 22] and companies adopting TBL reporting have introduced significant changes to the way they do, or at least think about, business [23]. The three major criticisms of the TBL approach are in its measurement approach, its lack of integration across the three dimensions and its main function as a compliance mechanism rather than a basis for the development of SBM [24]. To tackle these limitations and the growing need for more specific approaches applicable to different industry sectors, researchers have proposed various approaches to SBM (or business models for sustainability). However, early attempts to develop and introduce SBM design methodologies where hindered by a strong focus on compliance (with existing laws and regulations) and responsible management (i.e., achieving some kind of perceived or measurable optimal balance in the socio-economical dimension). Almost invariably these early researchers concluded that more detailed investigations were needed to assess whether SBM could help developing integrative and competitive solutions by reducing negative and/or creating positive external effects for the natural environment and society [25–28].

These approaches limited the impact of this body of research and largely overlooked the huge transformative potential of SBM that introduce new mechanisms for commercial value creation and value capture both internally and externally to a particular enterprise. Recent research has addressed these limitations and developed more holistic approaches to SBM development. Geissdoerfer et al. (2016) defines a SBM as: *"A simplified representation of the elements, the interrelationship between these elements, and the interactions with its stakeholders that an organisational unit uses to create, deliver, capture, and exchange sustainable value."* The main idea pursued here is to radically modify the conventional approach to business modelling by embedding sustainability into the value chains of an organisation [29]. It is now a common view that the transition towards SBM requires the practitioners to look beyond the specific boundaries of an organisation, and it requires innovation activities to create sustainable values for the stakeholders [30].

Sustainable Development (SD) in aviation is typically mapped to the following fundamental concepts [31, 32]:

- The consumption of natural resources is managed at a rate which allows future generations to meet their needs as well as we do – i.e., usage rates of renewable (e.g., biofuels) should not exceed the rates of their regeneration, and the usage rates of non-renewable resources (e.g., petroleum fuel) should not exceed the development rate of their substitutes (e.g., biofuels).
- The growth of aviation supports a liveable environment for future generations – i.e., the rates of polluting emissions should not exceed the assimilative capacity of the environment and the aircraft noise exposure (perceived noise levels by the population and frequency of noise disturbance or awakening events) should not lead to a degraded health and quality of life.

As illustrated in Figure 1.2, the three fundamental components in sustainable aviation are the aircraft, the airport and the Air Traffic Management (ATM) systems.

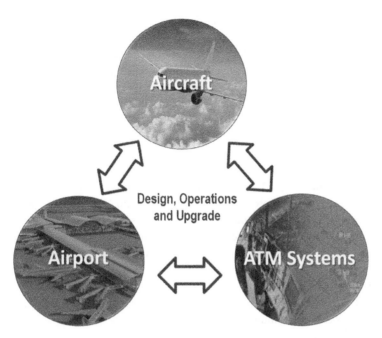

Figure 1.2 The three pillars of sustainable aviation research and innovation.

Designing/upgrading the aircraft to be more aerodynamically and operationally efficient entails advances in the following areas:

- **Propulsion and power:** targeting improvements in fuel efficiency, a transition to more sustainable energy management technologies, with associated reductions in gaseous and noise emissions.
- **Aerodynamics:** targeting drag reduction and consequential improvement of aerodynamic efficiency in various flight conditions, as well as reductions in airframe noise and wake turbulence.
- **Navigation and guidance:** leading to optimised flight paths for reductions in gaseous and noise emissions.
- **Computing, information and communication:** leading to more efficient management of on-board systems as well as more collaborative and higher levels of decision making, supporting more effective flight planning and operations.
- **Structural mechanics and materials:** targeting weight reduction across the aircraft, as well as lower impacts from the disposal processes.

ATM plays an important role is developing systems and procedures to support efficient use of airspace and networking between the various stakeholders. These solutions enhance the efficiency and effectiveness of flight operations by increasing the level of automation, improving the decision-making process and targeting the introduction of safety/security measures. The most promising technologies include [33]:

- **Communication, Navigation and Surveillance** (CNS) systems enabling 4-Dimensional Trajectory (4DT) based operations.
- **ATM systems supporting 4DT Planning, Negotiation and Validation** (4-PNV) with the Next Generation of Flight Management Systems (NG-FMS) on-board aircraft.

Airports also play a fundamental role in the SD of aviation. Designing/upgrading the airport infrastructure and operations to be more environmentally friendly, entails the adoption of various measures, such as: digital technology and multimodal transformation; operational procedures and restrictions [34]; land planning and management; financial measures (e.g., noise and atmospheric pollution charges); measuring and collecting data (on noise and pollutants); preventing/containing fuel and de-icing spillages; and managing the impact on wildlife [35].

Despite the existence of multiple interrelated socio-technical factors, the air transport literature discusses the topic of sustainability adopting a relatively narrow perspective and heavily focussing on reducing compliance costs or better utilising the existing airline/airport infrastructure to increase efficiency/quality of service and revenues. Other important sustainability factors (a tailored uptake of key aircraft/ATM technologies, airport "greening" and multimodal transformation, proper disposal/recycling of aircraft parts and consumables, etc.) have typically received less attention in the aviation political debate, despite the significant body of research published in the scientific and technical literature [12, 31, 33, 34, 36]. As a result of this, the regulatory initiatives led by ICAO and other national/international aviation authorities have been relatively limited in these sectors. Different models are used to describe the processes occurring in the atmosphere. Uncertainties in predictions can be attributed to [37]:

- The processes being modelled (missing or incorrect processes). Since our understanding of the atmospheric physics improves over time, these uncertainties can also reduce.
- Different factors influencing climate change. Uncertainties in aviation developments also make it difficult to predict the impact of aviation on climate beyond 5 to 10 years.

Factors considered in previous research include:

- Cost of air travel (and hence number of aircraft in operations);
- Economic activity and new market opportunities;
- Air transport liberalization and subsides;
- Improvements in aircraft fuel efficiency;
- Improvements in engine efficiency.

To reduce the impact of aviation on the environment, it is clearly necessary, first and foremost, to reduce the aircraft emissions. Newer aircraft have improved fuel efficiency, leading to reduced emissions. However, due to the growth of air traffic volume (expected to double every 20 years), these improvements are not sufficient to balance the environmental impact of aviation.

1.2 International Policy Framework

The establishment of an international policy framework within the UN allows technological improvements and operational changes to be implemented through policy documents, technical/operational standards, recommendations and economic measures, which are typically translated into legislation/regulations by national governments. This provides an opportunity for policy makers, scientists and industry to communicate and better assess the costs and benefits of implementing different measures. Additionally, the existence of an international framework provides assurance to producers and consumers that adopt new technologies and operational measures, allowing for a coordinated use of policy instruments to reduce environmental impacts and to increase the cost-effectiveness of the various mitigation/adaptation measures. The current policy framework includes:

- **UN Framework Convention on Climate Change (UNFCCC):** an international treaty addressing climate change, originally signed at the UN Conference on Environment and Development (UNCED) in 1992. The UNFCCC seeks for the stabilization of Greenhouse Gasses (GHG) concentrations in the atmosphere at a level that would prevent dangerous anthropogenic human-induced interference with the Earth's climate system. Such a level should be achieved within a timeframe sufficient to ensure ecosystems to adapt naturally to climate change, to ensure that food production is not threatened and to allow economic development to proceed in a sustainable manner.
- **Intergovernmental Panel on Climate Change (IPCC):** a scientific and intergovernmental body addressing human-induced climate change. The IPCC was originally established in 1988 by the World Meteorological Organisation (WMO) and the UN Environmental Programme (UNEP) and was later endorsed by the UN General Assembly. IPCC does not typically conduct its own original research but instead performs detailed reviews of the existing body of scientific knowledge, which are publicly disseminated in the form of comprehensive impact assessment reports.
- **International Civil Aviation Organization (ICAO):** a specialized agency of the UN responsible for harmonizing the international policies, standards and practices concerning aviation. The Committee on Aviation Environmental Protection (CAEP) is a technical committee of ICAO, responsible for assessing and formulating specific standards and recommendations related to aviation and the environment.

Cost-Benefit Analysis (CBA) has been widely adopted to assess the effects of (real or projected) environmental mitigation measures and can be a useful tool to guide policy decisions, but can be limited by uncertainties and/or incorrect assumptions introduced in the analysis [12, 34, 37]. More advanced econometric analysis techniques/tools have been introduced by CAEP and Elasticity of Demand (EOD) has been widely used by industry to assess the responsiveness of consumers to airfare increases (i.e., how cost increases due to new policies are passed to consumers and subsequently affect demand for aviation services).

Deregulation of the airline industry has become a predominant trend in various markets. Deregulation has resulted in cheaper flights and more competition in the industry. However, deregulation has also contributed to increases in traffic volume, fuel use, airport/airspace congestion and noise [2, 37, 38]. Fuel cost and consumption are important drivers for mitigation measures. Airlines have traditionally invested their profits into acquiring new technologies and more efficient aircraft for reducing operating costs through more efficient use of aircraft, optimal fleet mix and greater engine efficiency. However, it is observed that the rate of improvement achievable with presently known aerodynamic and power plant technologies will not allow offsetting the projected air traffic growth post-COVID. So, the current Research and Innovation (R&I) trends and opportunities identify digital aviation technologies as well as advances in energy production (in particular bio-fuels) as the main pathways to mitigating the environmental impact of aviation [12, 33, 39].

ICAO has been the main regulatory driver in modernising Communication, Navigation, Surveillance (CNS) for ATM and avionics systems but the focus, so far, has been almost exclusively on increasing efficiency (and safety) of the air transportation system. This, unfortunately, has not yet translated in successful worldwide cooperation efforts. Despite the ambitions targets set by large-scale regional R&I programs such as SESAR (Single European Sky ATM Research) and NextGen (Next Generation Air Transport Management), it appears that the impact of these US and EU initiatives has been hindered by a number of contributing factors and, so far, they have not delivered to their promises, [40]. The situation is even more fragmented in the Asia-Pacific region that, before COVID-19, was the fastest growing aviation market in the world [41].

Various potential economic instruments have been proposed over the years and many of them have been experimented or introduced in various nations. These instruments include:

- Fuel taxes and charges to promote fuel efficiency and reduce demand;
- Emissions charges aimed at encouraging the adoption of lower emitting technology;
- Emissions trading to encourage emissions reductions through market forces;
- Levies on empty aircraft seats to promote improvement in seat load factor;
- Levies on excessive traffic per destination served or type of equipment serving a destination;
- Levies on route length to reduce the number of flights exceeding the minimum distance;
- Subsidies or rebates to act as an incentive for polluters to change their behaviour, such as grants, soft loans, tax allowances or differentiation, and instruments similar to effluent, product, or administrative charges.

Other instruments identified included voluntary measures (e.g., carbon offsetting) and multi-modal transport (e.g., encouraging rail in place of air transport).

1.3 Sustainability Agenda

As discussed above, CAEP is a technical committee of ICAO that assists the nations in formulating new polices and adopting new Standards and Recommended Practices (SARPs) related to technologies/operations that reduce aircraft noise and GHG/noxious emissions, and more generally mitigate and keeps under control the aviation environmental impacts. CAEP undertakes specific studies as requested by ICAO. Its scope of activities encompasses the assessment of aircraft technology, operational improvement, market-based measures and alternative fuels. CAEP has the following high-level goals:

- To limit or reduce the number of people affected by significant aircraft noise;
- To limit or reduce the impact of aviation greenhouse gas emissions on the global climate;
- To limit or reduce the impact of aviation emissions on air quality and water/land contamination.

CAEP is composed by four permanent working groups and six dedicated task/support groups as illustrated in Figure 1.3. CAEP Working Group 1 (WG1) addresses aircraft noise technical issues. The main aim of WG1 is to keep international aircraft noise certification standards (Annex 16, Volume I) up-to-date and effective, while ensuring that the certification procedures are as simple and inexpensive as possible. WG2 addresses

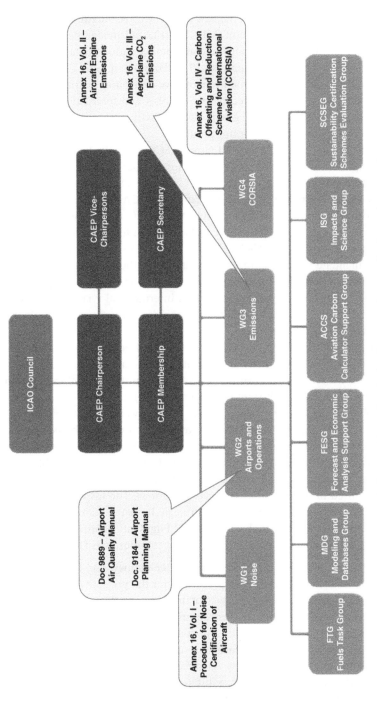

Figure 1.3 CAEP organisation chart. Source: ICAO (https://www.icao.int/environmental-protection/pages/caep.aspx).

aircraft noise and emissions issues linked to airports and operations. WG3 deals with aircraft performance and emission technical matters, including the updating of Annex 16 – Volume II and the development of the new aircraft CO_2 Standard, Annex 16 – Volume III. The Modelling and Databases Group (MDG) carries out modelling efforts to support the activities of the other CAEP groups and maintains various databases such as the movements, fleet and population databases. The Forecasting and Economic Analysis Support Group (FESG) has the important role of developing and maintaining the models and databases necessary to perform economic analysis and forecasting fleet growth. It provides support to the other working groups within CAEP and works with them on data issues that concern more than one working group.

The Aviation Carbon Calculator Support Group (ACCS) has the task of developing and updating an impartial, transparent methodology for computing the CO_2 emissions from passenger air travel. The Impacts and Science Group (ISG) is composed of academics, scientists and engineers responsible for informing the CAEP Secretariat on scientific findings (atmospheric pollution and noise) and the measures that the aviation industry should implement to limit the increase in global average temperature to less than 2°C above pre-industrial levels. The Global Market Based Measure Technical Task Force (GMTF) has a mandate to develop recommendations for the Monitoring, Reporting and Verification (MRV) system of international aviation emissions and for the quality of offset remits for use in a global market-based measure for international aviation. The Alternative Fuels Task Force (AFTF) assesses the potential emission reductions attainable from the use of alternative fuels in aviation.

Towards the end of the 1990's, the US and EU started addressing aviation SD as an integral part of their policy agendas and initiated large-scale R&I initiatives. The EU Advisory Council for Aviation R&I in Europe (ACARE) initially developed Vision 2020 and, successively FlightPath 2050, setting unprecedented emission reduction targets (both for gaseous pollutants and noise). In parallel, the Clean Sky (EU Framework 7) and Clean Sky 2 (Horizon 2020) programs were launched to address aircraft technology evolutions, while the ATM quota was assigned to SESAR. In the US, the NASA Environmentally Responsible Aviation (ERA) program addressed objectives similar to Clean Sky/Clean Sly 2 but with a much smaller budget and without progressing to the high Technical Readiness Level (TRL) required in the EU industry-driven programs. The ERA program completed its mandate in 2016 and was followed by the Strategic Implementation Plan (SIP), which is still ongoing and pursues similar objectives to EU FlightPath 2050 (Figure 1.4).

Some of the open questions that the global aviation community is facing are:

- Large uncertainties over future trends in traffic, technology, and therefore emissions, depending on the scenarios/assumptions selected for the projections. Key contributing factors include uncertainties about the pace of introduction of game-changing technologies and the impacts of the current infrastructure constraints ("bottlenecks") in limiting growth both in airport/airspace capacity and demand.
- The monetary impact of aviation emissions on the environment and the monetary benefits of mitigating those impacts. As already mentioned, different models and different scenarios/assumptions produce different results and there is no consensus on the appropriate level at which any environmental levy should be set.
- As the environmental benefits (reduction of gaseous and noise emissions) achievable with conventional aircraft/power plants configurations have reached a plateau, it is

	ACARE – SRA and SRIA (vs. 2000)		NASA – ERA (vs. 1998) and SIP (vs. 2005)					
Subsonic A/C Emissions	Vision 2020	FlightPath 2050	ERA 2015	ERA 2025	ERA 2035	SIP 2015–25	SIP 2025–35	SIP >2035
Fuel/CO$_2$	50% (38% 2015)	**75%**	50%	50%	60%	40–50%	50–60%	**60–80%**
NO$_X$	80% (----- 2015)	**90%**	75%	75%	80%	70–75% LTO* / 60–70% CRZ	80%	**>80%**
Noise	50% (37% 2015)	**65%**	32dB	42dB	71dB	22–32dB**	32–42dB	**42–52dB**

ACARE - Advisory Council for Aviation R&I in Europe, SRA - Strategic Research Agenda, SRIA - Strategic Research and Innovation Agenda, ERA - Environmentally Responsible Aviation, SIP - Strategic Implementation Plan

A/C - Aircraft, LTO - Landing and Take/Off, CRZ - Cruise, *Below CAEP6, **Below Chapter 4. All % reductions are in Passenger-km

Figure 1.4   Fuel, gaseous emissions and noise goals.

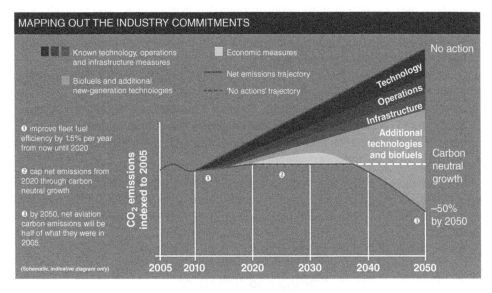

Figure 1.5   Carbon emission reduction goals and research drivers.

essential to investigate more radical approaches. These include advanced route/airspace optimisation techniques through the adoption of network-centric ATM and avionics technologies, innovative aircraft/engine design approaches and alternative aviation fuel, including biofuels. This concept is graphically illustrated in Figure 1.5. Although the illustration refers to CO$_2$ emission reduction goals, similar conclusions have been found for other GHG and noxious emissions [42].

Additionally, there is limited practical experience with emission taxes and trading schemes at a global level and there are uncertainties regarding the applicability of many economic and technical measures to countries not included in the UNFCCC.

Current strategies for ensuring aviation sustainability include regulating aircraft design/operations with environmentally-friendly policies (carbon tax/offsetting schemes, noise emission charges, replacing or ruling-out old fleet, etc.). However, in the long term, digital transformation initiatives are essential and will radically transform product and service lifecycle management processes both in the aerospace and aviation industries. Such initiatives will include:

- Adopting Multidisciplinary Design Optimisation and Multi-Objective Mission Optimisation (MDO/MOMO) tools to develop new CNS/ATM and Avionics (CNS+A) systems for eco-friendly flight operations (i.e., management of airspace, trajectory and mission) [33, 39, 43].
- Adopting MDO and other digital tools (e.g., artificial intelligence, robotic process automation and digital twins) for Design, Development, Test and Evaluation (DDT&E) and Maintenance, Repair and Overhaul (MRO) of more energy efficient and "cleaner" (i.e., less polluting) propulsive systems [39].
- Adopting MDO and other digital tools for DDT&E/MRO of lighter and more aerodynamically efficient manned/unmanned aircraft [39].
- Enabling the cost-effective introduction of alternative aviation fuels, especially third generation biofuels, by deploying the required CPS architectures (e.g., distributed sensor networks and AI-based health/quality monitoring) to improve crop quality, maximise fuel yield and minimise land take [44].
- Developing Intelligent Transport Systems (ITS) for multimodal airport transformation. These will include advanced digital solutions (e.g., sensor networks, user-apps, centralise/distributed traffic management, connected autonomous vehicle technologies) which aim to provide innovative services relating to various interconnected modes of transport and enable users to be better informed and make safer, more coordinated and "smarter" use of the transport network [12, 45].

However, there is a need to assess the impacts of various possible measures for encouraging the adoption of digital/sustainable aviation technologies, including the applicability to aviation of mature solutions and/or promising operational concepts developed in other sectors.

1.4 Emission Taxes, Trading and Offsetting

One of the earliest propositions brought forward at the onset of the global warming debate was a tax imposed on companies based on the amount of emissions they produce, specifically on GHGs such as CO_2 and was commonly known as "Carbon Tax" (CT). CT is simply a direct payment to government (collection body), based on the carbon content of the fuel being consumed. Given that the primary objective of the abatement policy is to lower carbon dioxide emissions, carbon taxes make sense economically and environmentally because they tax the externality (carbon) directly.

Under an Emission Trading (ET) system, the quantity of emissions is fixed (often called a "cap") and the right to emit becomes a tradable commodity. The cap (say 10,000 tons of carbon) is divided into transferable units (10,000 permits of 1 ton of carbon each). Permits are often referred to as "GHG units," "quotas" or "allowances." For compliance, actors participating in the system must hold a number of permits greater or equal to their actual emissions level. Once permits are allocated (by auction, sale or free allocation) to the actors participating in the system, they are then tradable. This enables emissions reductions to take place where least costly. Some key characteristics of ET schemes include:

- The emission levels are specified upfront, allowing more predictable estimates of emissions. This also allows for countries to agree upon specific emissions reduction levels, making international environmental agreements more negotiable.
- Emissions trading is more appealing to private industry, as firms can profit by selling their excess greenhouse gas allowances. Creating such a market for pollution could potentially drive emissions reductions below targets.
- Emissions trading is better equipped than taxes to deal with all six GHGs included in the Kyoto Protocol and sinks (e.g. trees which absorb and store carbon) in one comprehensive strategy. Each gas has a "greenhouse gas potential" (GWP, based on carbon dioxide). Thus, firms emitting more than one GHG have more flexibility in making reductions.
- Permits adjust automatically for inflation and external price shocks, while taxes do not. For example, the US has already experienced an extended period of stable greenhouse gas emissions levels from 1972 to 1985 because of high oil prices. Taxes would need to be designed to adjust for such external shocks.

Compared to ET, CT offer a broader scope for emissions reductions, extending to all carbon-based fuel consumption, including gasoline, home heating oil and aviation fuels.

- Compared to emissions trading, which involves significant transaction costs, taxes involve little transaction cost, over all stages of their lifetime.
- Taxes are not susceptible to speculative or hoarding behaviour by firms or nongovernmental organizations which may harm the market forces.
- Compared to emissions trading, which rely on the supply and demand of emission permits to control emissions, carbon taxes provide a permanent incentive to reduce emissions. Improvements in technology and operations might lead to reductions in the permit price, lowering the incentive to reduce emissions.
- Emissions trading proposals are highly complicated and technical, unlike taxes which are familiar instruments to policymakers. Ongoing costs are also low for tax systems because of the lack of monitoring and enforcement requirements.
- Emissions trading may prevent meaningful domestic reductions from taking place, as some countries might choose to buy emission permits. This rises significant equity issues among developed, developing and transitional economies.
- Carbon taxes earn revenue, which can be "recycled" back into the economy by reducing taxes on income, labour and/or capital investment. Permit systems have the potential to earn revenue, but only if permits are auctioned.

Carbon offsetting allows individuals and companies to reduce their carbon footprint by investing in environmental projects elsewhere. Credits are usually purchased and used by individuals or companies to cancel out or "offset" the emissions they generate during their day-to-day life or normal course of business (e.g., using air transport). Carbon offsets can be used to offset emissions voluntarily or to meet regulatory requirements. Carbon offsetting projects may include:

- Reducing the cost differential of renewable energy such as wind, solar, hydroelectric power or biofuel, thereby increasing its commercial viability;
- Combustion or containment of methane generated by landfills, industrial waste or farm animals – converting methane to CO_2;
- Increasing the energy efficiency of buildings, vehicles or power plants;
- Reforestation initiatives.

In 2009, the Airports Council International Europe (ACI Europe) introduced a carbon management initiative for airports, called the *Airport Carbon Accreditation* program, which allows airports to be recognised (through accreditation) for their efforts in managing and reducing their carbon emissions. Airports can be accredited to one of four levels in the program [46]:

- Level 1: Mapping, requiring carbon footprint measurement;
- Level 2: Reduction, requiring a carbon management plan to be in place;
- Level 3: Optimisation, requiring airports to engage stakeholders (airlines, catering, air traffic control, ground services, rail, etc.) to reduce the airport's carbon footprint;
- Level 3+: Neutrality, requiring airports to neutralise any residual emissions through carbon offsetting.

The accreditation requires airports to verify their activities (e.g., carbon monitoring and management processes) by a group of independent verifiers. The carbon footprint of an airport is verified in accordance with the ISO 14064 standard (Greenhouse Gas Accounting), which requires specific supporting evidence.

1.5 ATM and Avionics Systems

In the last two decades, a number of major ATM modernisation initiatives such as the Single European Sky ATM Research (SESAR) and the Next Generation Air Transportation System (NextGen), were launched around the globe to cope with the rapid growth of air traffic and mitigate the growing congestion and inefficiency issues. These initiatives support an evolution of the ATM system into a highly integrated network where civil, military, and remotely piloted aircraft will continuously and dynamically share the common airspace in a highly automated and collaborative decision-making environment. To meet the goals of enhanced flight safety, environmental performance, and efficiency while simultaneously accommodating the predicted traffic growth, several key policy directions have been identified by various governments internationally [47]: robust and integrated planning, adoption of advanced technology, international harmonisation of ATM systems, enhanced regional aviation safety, and environmental impact mitigation. In this context one key strategic priority for countries is to plan, develop, and implement a new ATM platform that meets the future needs of both civil and military aviation while enhancing ATM business competitiveness by addressing service capability, continuity, and environmental sustainability [48]. With air traffic expected to grow more substantially within the lifespan of the new transport aircraft, along with the introduction of new concepts to improve airspace organisation and airport operations, these major aviation renovation programmes around the world will play a critical role in the successful transition to new technologies and operational standards. Research is therefore needed to develop a new ATM regulatory framework and new systems for dynamic airspace management (DAM), free-flight and intent-based operations. This also encompasses the development of innovative methods and algorithms for the dynamic allocation of civil/military airspace resources and of CNS+A technologies enabling the unrestricted access of remotely piloted aerial systems (RPASs) to all classes of airspace.

Ground-based automatic dependent surveillance broadcast (ADS-B) currently provides wide area surveillance coverage, including those vast regions of the planet that are not under primary or secondary surveillance radar (SSR) coverage. A receiver autonomous integrity monitoring (RAIM) system enables controllers to anticipate and plan for a reversion to procedural separation if a GPS outage is predicted. For areas that are under radar surveillance (major air corridors and terminal manoeuvring areas) sensor-fused radar and ADS-B data have proved to be superior to radar data alone, particularly for tracking manoeuvring aircraft. Space-based ADS-B promises to expand the benefits of ADS-B to oceanic airspace and addresses the low reporting rate of automatic dependent surveillance contract (ADS-C). Optimised ATM procedures such as tailored arrivals [49] and the Green RNP project [50] have been trialled or already implemented. A growing number of airport/airline slot management and Air Traffic Flow Management (ATFM) centres around the world have contributed to optimises the allocation of airport and air traffic control (ATC) slots, while traffic management initiatives, such as ground delays programmes, tackle critical congestion situations, thereby simultaneously reducing fuel consumption, noise and gaseous emissions. Collaborative decision making (CDM) procedures improve common situational awareness and permit pre-tactical slot swapping. Current initiatives include user preferred routes (UPRs) and the extension of national CDM and ATFM operations to support long-range ATFM strategies for entire world regions. Conducting ATFM across national borders will improve its effectiveness, particularly for commercial airline companies. For example, delay can be absorbed en-route or allocated as ground delay if congestion is anticipated at the destination airport several FIRs away. Achieving this in the Asia-Pacific region without a single regulatory authority, like Eurocontrol or the Federal Aviation Administration (FAA), is one of the issues to be addressed but the benefits are evident. Early regional CDM trials between Bangkok and Singapore have proved promising [51], and it is clear that interoperability and harmonisation of standards will be key factors in moving forward.

In line with the ICAO's ASBU implementation timelines, new high-integrity and safety-critical CNS+A systems will be developed and deployed for strategic, tactical, and emergency ATM operations, and in particular:

- Civil/military dual-use CNS+A technologies, including a secure and reliable network infrastructure and airborne datalink for information sharing and CDM, network-centric

ATM technologies for strategic and tactical ATFM, DAM and real-time four-dimensional trajectory (4DT) operability.

- CNS+A technologies for RPAS, reliably meeting the required communication, navigation, and surveillance performance (RCP, RNP, and RSP) standards for unrestricted access of RPAS to airspace (non-segregated operations). In this perspective, essential steps are the adoption of fused cooperative/non-cooperative surveillance systems, beyond line-of-sight (BLOS) communication systems, high-integrity navigation systems and integrated avionics architectures.
- Satellite-based CNS systems, such as multi-constellation global navigation satellite systems (GNSS) and space-based datalink and ADS-B, for improved coverage of remote and oceanic airspace, precision approach, and auto-land.
- Airport ATM systems, mainly consisting of safety nets for ground and air traffic operations, remote tower systems (RTSs) and new standardised air traffic control operator (ATCO) work positions. In particular, the advanced surface movement guidance and control system (A-SMGCS) will also provide runway incursion and excursion detection and alerting similar to the airport movement area safety system (AMASS) and runway awareness and advisory system (RAAS) developed in Europe and the US.

A network-centric communication approach is required to allow greater sharing of ATM information, such as weather, airport operational status, flight data, airspace status and restrictions. Key network building blocks include high-integrity, high-throughput and secure avionics data-links for dual civil/military usage as well as a system wide information management (SWIM) system. Web service technologies for mobile, internet-based access will also be included to flexibly expand the number of participants in the CDM processes. Business intelligence and big data will also be implemented as part of SWIM for enhanced data mining. The implementation of enterprise-wide data warehouses by ANSPs will enable ATM to move beyond post-event reporting and to mine years of historical data to determine underlying traffic flow patterns and emission levels so as to derive enhanced models to address them. Automated air traffic flow management (A-ATFM) systems will enhance the continuous balancing of air-traffic demand with capacity to ensure the safe and efficient utilisation of airspace resources. Automated dynamic airspace management (ADAM) will enable the seamless optimal allocation of airspace resources. Real-time multi-objective 4DT optimisation and negotiation/validation algorithms, implemented in the next generation of ground-based and airborne CNS+A systems, will promote a continuous reduction in environmental impacts, which will be particularly significant in severe congestion and weather conditions. To enhance the operational efficiency at both regional and global levels, it is essential to address the interoperability of the ATM regulatory framework evolutions within and across regions, preferably taking the move from the European/US frameworks (being defined by SESAR and NextGen). This will likely contribute to the global ICAO initiatives in this domain, such as the Aviation System Block Upgrades (ASBUs). From a technological perspective, interoperability is also required at various levels, including signal-in-space (SIS), system level and human-machine interface and interaction (HMI^2).

SESAR [10] has defined three phases of ATM system development in an evolutionary roadmap, represented in Figure 1.6. These phases are:

Figure 1.6 Evolutionary roadmap for ATM Operations.

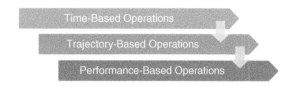

- Time-based operations, for which ATM strategic and tactical actions (including ATFM) are aimed at optimal traffic synchronisation.
- Trajectory-Based Operations (TBO), focussing on a further-evolved predictability, flexibility, and environment sustainability of air traffic, unleashing additional capacity.
- Performance-based operations, for which all the available CNS performance is exploited to establish a high-performance, network-centric, collaborative, integrated, and seamless ATM system, supporting high-density operations.

TBO are based on the adoption of 4DT defining the aircraft's flight path in three spatial dimensions (i.e. latitude, longitude, and altitude), and in time from origin to destination [52] and of the associated precise estimation and correction of current and predicted traffic positions. Each aircraft follows a 4DT, which is determined via a CDM process involving novel systems, such as the next generation flight management system (NG-FMS), and evolving tactically from the original reference business trajectory. Increased efficiency and higher throughput are obtained in a CNS+A context by actively managing 4DT. In the PBO context, the next generation air traffic management (NG-ATM) services will be matched to the performance capability of aircraft. Airlines deploying PBO-capable equipment will benefit from higher scheduling priority and easier access to congested areas. These regulations will impose requirements in terms of system performance rather than in terms of specific technology or equipment.

1.6 Lightweight Structures and Materials

The continuous push to reduce weight and enhance mechanical properties of aerostructures has led to significant advances in aircraft design and lifecycle management processes, as demonstrated by contemporary airliners such as the Boeing 787 "Dreamliner" and the Airbus A350 Extra Wide-Body (XWB) aircraft. These aircraft deliver substantial improvements when compared to previous generation airliners, largely through the selective use of new advanced materials in various parts of the airframe and propulsive components [53, 54]. In particular, the adoption of carbon fibre composites and other hybrid materials has facilitated the implementation of much lighter aircraft designs while improving the overall mechanical properties of aerostructures [55].

Lighter aircraft translates into reduced thrust and fuel consumption, with associated enhancements in payload capacity, range and endurance. Additionally, the lower thrust requirements allow for the integration of smaller, lighter and quieter engines, thereby leading to noise reduction and further fuel savings. Open research challenges and opportunities include methods for fatigue life assessment, maintenance and testing of composite structures (e.g., new composite repair technologies using hybrid material systems and new nanotechnologies to improve adhesive bonding processes, thermal properties and lightning protection); sandwich structure optimisation to prevent buckling and wrinkling of the soft core; net-like anisogrid structures to increase torsional rigidity and thereby avoid flutter; using natural fibres and binders (e.g., bamboo, cork, resins and latexes) to enhance certain mechanical properties and recyclability of composite materials; and reducing the production, assembly and operating costs of composites, which are significantly higher than aluminium alloys. The global carbon fibre market is projected to grow from $2.33 billion in 2021 to $4.08 billion in 2028 at a compound annual growth rate (CAGR) of 8.3% in the forecast period 2021-2028 [56]. The global demand for carbon fibre is very large and rapidly growing. Currently, the production and maintenance of modern airliners (e.g., Airbus A380, Airbus A350, Boeing 787 and Boeing 777), requires approximately 15,000 tons of carbon fibre per annum. Thus, with the current rates of production (significantly impacted by COVID-19), it is projected that there will be a shortage of supply to meet demand in the near future.

1.7 Advanced Aerodynamic Configurations

In terms of design, aircraft have now for quite a long time been following a conventional configuration which includes a central fuselage and a main wing, plus horizontal and vertical tailplanes. This configuration presents a few practical advantages but is rather far from the theoretical efficiency limits, as it relies on the functional separation between payload-carrying and lift-producing elements. Key limitations of this approach include:

- the fuselage producing substantial drag but insignificant lift, which weighs heavily against aerodynamic efficiency (i.e., the ratio between lift and drag of the entire aircraft in representative operational conditions);
- the concentration of very significant shear stresses and bending moments in small sections of the wing root, which then have to be reinforced adequately, adding substantial structural weight;
- natural tendency to develop large tip vortexes, which result in an energy-dissipating and operationally hazardous turbulent wake.

Advanced aircraft configurations attempt to enhance the aerodynamic efficiency of the aircraft in representative operational flight conditions, compared to conventional designs. Various solutions have been proposed throughout the years, including hybrid wing-bodies (e.g., blended wing-body, flying wing), box-wing aircraft and advanced morphing aircraft technologies. Some of the key gains in such technologies include respectively a 30% increase in aerodynamic efficiency or a 40% reduction in induced drag. Despite the relatively high confidence in these theoretical efficiency gains and some successful operational experience in the defence sector, the actual adoption of these advanced concepts in the civil transport domain has been encumbered by the limited maturity of certain technologies and a luke-warm attitude by major aircraft manufacturers, which adopted a more risk-averse approach financially, resulting in further evolutions of the conventional configuration. More recently, the diminishing returns associated with further investments in conventional aerodynamic technologies is eliciting a more courageous attitude in embracing the new configurations.

For instance, Airbus has unveiled that the most advanced hydrogen aircraft concept being investigated for marketization in just a couple of decades is a blended wing-body. The general and business aviation sectors also appear more interested in experimenting with some more advanced configurations and it is expected that this will also contribute to the large transport aircraft sector's willingness to develop and marketize the new concepts.

1.8 Advanced Propulsion Concepts

Aircraft engines are the source of the entirety of the adverse atmospheric emissions of the aircraft, and have therefore been at the focus of major aviation modernisation initiatives. The current standards revolve around air-breathing hydrocarbon-based combustion technology, either in the form of turbine-based engines or less commonly of reciprocating piston engines (typically for small aircraft). The limited thermal efficiency (i.e., transformation of chemical energy in thermodynamic energy) and propulsive efficiency (transformation of thermodynamic energy in mechanical energy) jointly concur to yield a very low overall efficiency for these engine technologies [42].

Technological advances are being pursued to further increase the bypass ratio (which promotes propulsive efficiency), reducing the weight of all components, the thermofluidic efficiency of compressors and turbines, the thermal resistance of materials and the introduction of less environmentally-harmful fuels and of electric aircraft technologies. In the near term, further evolutions of the conventional turbofan engine such as geared turbofan and 3-spool turbofan will dominate the scene, but these are already affected by diminished investment returns, hence are expected to give the way to turboelectric propulsion concepts and to boundary-layer ingestion engines.

1.9 Alternative Aviation Fuels

The International Air Transport Association (IATA) is targeting a 50% reduction in carbon dioxide emissions by 2050, compared to 2005 levels [57]. Sustainable Aviation Fuels (SAF) are expected to be one of the main tools in the mix of technologies which will allow aviation to attain the IATA 2050 and other more ambitious carbon neutrality targets, as the benefits associated with the introduction of new aircraft technologies are largely insufficient when taking the magnitude of the global growth of air transport into consideration. More effective actions to reduce fuel burn and carbon emissions are required in both the short and the long term.

For the last few years, in addition to effective fuel-burn reduction schemes, the aviation industry has been working on 'drop-in' SAF blends that are compatible with current powerplants and fully comply with international standards. The activity therefore mainly involved aircraft retrofits and operational strategies (ground and air) that had to be approved and certified.

The planned scale-up of SAF involves production from biomass (including land plants and algae) and other carbon feedstock sources (e.g. bio-waste) in order to increase the blending with Jet A1 progressively [58, 59]. As part of this initiative, the most efficient feedstock sources and harvesting methods are identified. The methods for feedstock collection at selected hubs, storage, and pre-processing are also determined. The processing and refining of feedstock to produce certification-compliant SAF, blending of SAF with Jet A1 and its commercial distribution via a fully integrated and secure supply chain is one of the major challenges in implementing this green initiative. Economic viability analysis for identifying environmental auditing methods and overall cost-risk analysis will be performed. This initiative additionally seeks to address the certification, government excise rebates, and regulatory legislation issues. Eventually, end-user operational monitoring will be implemented, including actual greenhouse emissions.

For the long term, a radical shift away from the petroleum-derived fuels in aviation will be essential for attaining carbon-neutrality in the fuel lifecycle, and will be highly desirable due to the predicted price trends and strategic reliance on imports. To allow this radical change, an evolution of specifications and logistics is required. The future alternative global fuel supply chain strategies (especially within the Asia-Pacific region) as well as the feasibility of manageable aircraft systems and airport infrastructure changes in both the civil and military aviation sectors will be a major factor in the long-term and enduring success of SAF [60].

1.10 Systems Engineering Evolutions

In the last decades, airlines have been transitioning towards performance-based contracting that guarantees minimum performance levels, including operating costs and thus shifting risks from the airline to the aircraft manufacturer. This requires research and innovation initiatives focussed on reducing costs in aircraft manufacturing and operations through improved maintenance (e.g. system health monitoring, diagnosis and prognosis) and upgrades (e.g. avionics hardware and software). Ensuring and sustaining structural and systems integrity of ageing aircraft is a major challenge for the aviation industry. Research on the design, development, implementation, and certification of technological and non-technological solutions is necessary to address the associated issues. In this regard, a multi-scale approach has to be employed, as the proper way to address the unique and specific demands of the aviation sector (from general aviation to commercial airlines).

Over the past decade, many commercial airline companies have significantly evolved their fleets and services, with an expansion of long-range operations and the associated rise of global scale hubs, such as in the Middle East. On the other hand, there has been a worldwide proliferation of low-cost carriers (LCCs), posing new challenges to the conventional airline business model. Inevitably, an increasing competition from LCCs employing more fuel-efficient aircraft is forcing the aviation industry as a whole to identify cost-effective solutions for servicing ageing aircraft, which form a substantial percentage of conventional airline fleets. The inclusion of commercial-off-the-shelf (COTS) components has become common practice for extending service life. However, the introduction of COTS in aviation poses several challenges for obsolescence management and adds complexities to the configuration and certification processes. Research is therefore needed to make technology insertion less onerous by developing open architectures, common interfaces, backward compatibility, and harmonisation methods for integrating old and new system components.

As mentioned above, the extensive adoption of composites and lightweight hybrid materials on the latest generation of airliners (e.g. Boeing 787 Dreamliner and Airbus 350 XWB utilise substantially advanced fibre-reinforced composites) and military aircraft also poses new challenges in terms of logistics supportability. Thus, systems engineering research is addressing the cost-effective management of safety standards, including non-destructive inspection and testing of composite components for continuing airworthiness, economic composite repair processes, and training/skilling-up of the aircraft maintenance workforce. To reflect these changes, current initiatives for green aviation are investigating the adoption of advanced techniques and models for aircraft through-life support. The development of rapid non-destructive inspection and testing techniques that enables the fast characterisation of structural damages and their impact on structural integrity is the major factor driving new aircraft design and development through-life support. Additionally, training requirements and associated standards for the next generation of aerospace/aviation professionals are constantly evolving to reflect the adoption of new cost-effective manufacturing and repair/maintenance processes for composites aircraft, modular architectures, and civil/military aircraft data networks. Additionally, new Integrated Vehicle Health Management (IVHM) solutions are being investigated to improve logistic supportability and COTS components repair/replacement strategies. Aircraft mid-life update, reliable COTS components insertion, and the evolution of current structural and system integrity monitoring, diagnosis and prognosis approaches are identified as the potential solutions for extending both new and ageing aircraft service life.

1.11 Airport Evolutions

Due to the rapid increase in the number of air passengers internationally, several airports worldwide are under significant pressure to increase their efficiency and capacity, while also improving airport safety, security, and environmental sustainability. Figure 1.7 shows the trends in passenger traffic at the ten busiest airports worldwide. Some new entrants have evolved considerably over the past two decades as major "global connector" airlines have consolidated their business models and LCCs have become important players in the industry. Environmental regulations and international rules have greatly shifted emphasis and recent airport technologies, such as new types of aircraft, satellite-based ATM, improved security controls, and extensive adoption of information technology, are being introduced at an increasing rate. From a business perspective, the organisational and financing characteristics of airports are also changing rapidly, stimulated to a large extent by airline deregulation and technological changes.

The traditional model that places airport management in the hands of a central bureaucracy within the national governments does not meet the needs of large airports in a fast-changing industry. The emerging airport business models focus on the concept of the airport authority, which is usually a corporate entity owned by government or private investors or a combination of the two, which acts as an autonomous and flexible airport operator. To reflect these technological and business changes, advances in the strategies and models for planning, designing, and managing airports are being investigated. New aircraft types and ATM systems are requiring substantial airside development and

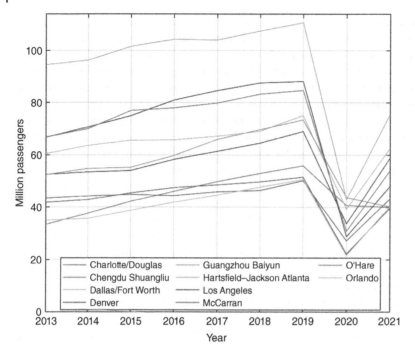

Figure 1.7 Passenger traffic in the world's busiest airports. Source: various, collected on Wikipedia "List of Busiest Airports by Passenger Traffic".

redevelopment initiatives, which involve aprons, fuelling systems, taxiways, holding areas, de-icing bays, runways, etc. Novel airport systems and procedures are being introduced to reduce noise and gaseous emissions, such as minimised auxiliary power unit (APU) usage, engine test, noise barriers, etc. Landside development and redevelopment initiatives are addressing renewable energy supply and security upgrades for the terminal building, greener vehicular ground transportation systems, improved airport ground access systems, and the transformation of large airports into multimodal transport nodes.

Advances in airport operations, including the efficient management of congestion and queues, implementation of airport collaborative decision making (A-CDM), Re-categorisation (RECAT) of wake turbulence and prediction initiatives, and traffic demand peak prediction and analysis methods are all being promoted. The coordination between airline operations centres (AOCs) and ATM systems, integrated departure and arrival management (DMAN/AMAN), and pre-departure sequencing and clearance are several key initiatives undertaken for the management of congestion in airports. Additionally, as part of improved airport operations, airline/operator business models and tools, including alternate slot trading mechanisms, are being identified.

1.12 Safety and Security Provisions

Over the past decade, global aviation security targets have increased significantly. External factors, such as the outbreak of various pandemics (including COVID-19), terrorist attacks, refugee crises and diplomatic clashes have greatly disrupted the aviation market at various times and have led to significant evolutions in airport security [61]. To account for the technological advances and process enhancements being introduced across all entities (airlines, airports, aircraft and ATM), an evolution of the cyber and physical security infrastructure in the aviation sector is required. The planned evolution of policies, objectives and planning is the baseline upon which a cyber-physical Security Management System (SEMS) will have to be developed. Safety assurance procedures, safety training and promotion, as well as risk management strategies are other key factors affecting the development of a SEMS. To further enhance security, additional airworthiness provisions and safety targets for both manned and unmanned aircraft are being investigated. Enhanced airport and ATM cyber and physical infrastructure will also provide substantial improvements to the overall air transportation system [62]. In particular, within the scope of terminal security, enhancements are being studied for security checkpoints/gates and passenger screening, waiting and transit areas, and security systems for baggage screening and reconciliation. Enhanced security measures will be developed for airside operations as well, including aircraft monitoring, aircrew/apron personnel movements, passenger boarding and detection of security threats. Important technology advancements are expected to address the security aspects of ATM facilities. In fact, the progressive adoption of greater automation, networking and data-links in ATM exposes the entire air transport system to new threats associated with unlawful interference as well as criminal and terrorist acts. The necessary technology-driven and process-driven regulatory framework evolutions for cyber-physical security (vehicles and infrastructure) are currently being explored [63].

References

1 Global Aviation Industry High-Level Group (2019). *Aviation Benefits Report 2019*, Airport Council International (ACI), Civil Air Navigation Services Organisation (CANSO), International Air Transport Association (IATA), International Civil Aviation Organisation (ICAO) and International Coordinating Council of Aerospace Industries Association (ICCAIA).

2 ICAO 2019, *Annual Report of the Council to the Assembly*, report n., International Civil Aviation Organization (ICAO).

3 Lee, D.S., Pitari, G., Grewe, V. et al. (2010). Transport impacts on atmosphere and climate: Aviation. *Atmospheric Environment* 44 (37): 4678–4734.

4 Hasegawa, T., Chen, S., and Duong, L. (2021). *Effects of Novel Coronavirus (COVID-19) on Civil Aviation: Economic Impact Analysis*. Montreal, Canada: ICAO Economic Development - Air Transport Bureau.

5 ICAO (2023). Effects of Novel Coronavirus (COVID-19) on Civil Aviation: Economic Impact Analysis. International Civil Aviation Organization (ICAO), Economic Development - Air Transport Bureau. Montreal, QC, Canada. https://www.icao.int/sustainability/Documents/Covid-19/ICAO_coronavirus_Econ_Impact.pdf (accessed 29 May 2023).

6 ICAO (2023). ICAO forecasts complete and sustainable recovery and growth of air passenger demand in 2023. International Civil Aviation Organization (ICAO). Montreal,

QC, Canada. https://www.icao.int/Newsroom/Pages/ICAO-forecasts-complete-and-sustainable-recovery-and-growth-of-air-passenger-demand-in-2023.aspx (accessed 29 May 2023).

7 ICAO (2023). Post-COVID-19 Forecasts Scenarios - Appendix A: Traffic Forecasts. International Civil Aviation Organization (ICAO). Montreal, QC, Canada. https://www.icao.int/sustainability/Documents/Post-COVID-19%20forecasts%20scenarios%20tables.pdf (accessed 29 May 2023).

8 Hilton, S., Sabatini, R., Gardi, A. et al. (2019). Space traffic management: towards safe and unsegregated space transport operations. *Progress in Aerospace Sciences* 105: 98–125.

9 FAA (2020). *Upper Class E Traffic Management (ETM) Concept of Operations v1.0*. USA: United States Federal Aviation Administration (FAA), Washington, DC.

10 *Strategic Research & Innovation Agenda (SRIA)*, vol. 1, Advisory Council for Aviation Research and Innovation in Europe (ACARE), 2012.

11 Hooper, P.D. and Gibbs, D. (1995). Cleaner technology: a means of an end or an end to a means. *Greener Management International* 3: 28–40.

12 Janić, M. (2007). *The Sustainability of Air Transportation: a Quantitative Analysis and Assessment*. Ashgate Publishing, Ltd.

13 United Nations (1987). Our common future. Report of the World Commission on Environment and Development (WCED). https://sustainabledevelopment.un.org/content/documents/5987our-common-future.pdf (accessed 27 February 2023).

14 Rodriguez, S., Roman, M., Sturhahn, S., and Terry, E. (2002). Sustainability Assessment and Reporting for the University of Michigan's Ann Arbor Campus. Center for Sustainable Systems. University of Michigan. Report No. CSS02-04.

15 Ingaldi, M. (2013). Sustainability as an element of environmental management in companies. *Production Engineering Archives* 7 (2): 29–32.

16 Ekwueme, C.M., Egbunike, C.F., Abdullahi, T.O., and Chukwuemeka, E.O. (2012). Benefits of triple bottom line disclosures on corporate performance: an exploratory study of corporate stakeholders. *African Journal of Hospitality, Tourism and Leisure* 2 (2): 1–15.

17 Henriques, A. and Richardson, J. (2004). *The Triple Bottom Line: Does it all Add Up?* London, UK: Earthscan. ISBN: 9781844070152.

18 Berger, I., Cunningham, P., and Drumwright, M. (2007). Mainstreaming corporate social responsibility: developing markets for virtue. *California Management Review* 49: 132–157.

19 Elkington, J. (1997). *Cannibals with Forks: The Triple Bottom Line of 21st Century Business*. USA: New Society Publishers.

20 Morland, M.P. (2006). Triple bottom line reporting as social grammar: integrating corporate social responsibility and corporate codes of conduct. *Business Ethics: A European Review* 15: 352–364.

21 Robins, F. (2006). The Challenge of TBL: A Responsibility to Whom? *Business and Society Review* 111: 1–14.

22 Savitz, A. and Weber, K. (2006). *The Triple Bottom Line: How Today's Best Run Companies are Achieving Economic, Social, and Environmental Success – and How You Can Too*. San Francisco: Wiley.

23 Kimmett, P. and Boyd, T. (2004). An institutional understanding of triple bottom line evaluations and the use of social and environmental metrics. In: *Proceedings of the 10th*

Annual Pacific Rim Real Estate Society (PRRES) Conference (ed. Y.S. Ming). Bangkok, Thailand (January 2004).

24 Sridhar, K. and Jones, G. (2013). The three fundamental criticisms of the Triple Bottom Line approach: an empirical study to link sustainability reports in companies based in the Asia-Pacific region and TBL shortcomings. *Asian Journal of Business Ethics* 2: 91–111.

25 Boons, F.A.A. and Lüdeke-Freund, F. (2013). Business models for sustainable innovation: state-of-the-art and steps towards a research agenda. *Journal of Cleaner Production* 45: 9–19.

26 Hansen, E.G., Große-Dunker, F., and Reichwald, R. (2009). Sustainability innovation cube. A framework to evaluate sustainability-oriented innovations. *International Journal of Innovation Management* 13: 683–713.

27 Schaltegger, S., Lüdeke-Freund, F., and Hansen, E.G. (2012). Business cases for sustainability: The role of business model innovation for corporate sustainability. *International Journal of Innovation & Sustainable Development* 6: 95–119.

28 Stubbs, W. and Cocklin, C. (2008). Conceptualizing a "sustainability business model". *Organization & Environment* 21: 103–127.

29 Geissdoerfer, M., Vladimirova, D., and Evans, S. (2018). Sustainable business model innovation: A review. *Journal of Cleaner Production* 198: 401–416.

30 Kennedy, S. and Bocken, N. (2020). Innovating business models for sustainability: an essential practice for responsible managers. In: *The Research Handbook of Responsible Management* (ed. O. Laasch, D. Jamali, E. Freeman, and R. Suddaby). Cheltenham: Edward Elgar. ISBN: 9781788971966.

31 Budd, L., Griggs, S., and Howarth, D. (ed.) (2013). *Sustainable Aviation Futures*, vol. 4. Transport and Sustainability: Emerald Group Publishing Ltd, Bingley, UK.

32 Gössling, S. and Upham, P. (2009). *Climate change and aviation: Issues, challenges and solutions*. London, UK: Earthscan.

33 Sabatini, R., Gardi, A., Ramasamy, S. et al. (2015). Modern Avionics and ATM Systems for Green Operations. In: *Encyclopedia of Aerospace Engineering* (ed. R. Blockley and W. Shyy). Wiley.

34 Janić, M. (2011). *Greening Airports: Advanced Technology and Operations*. Springer.

35 De Neufville, R., Odoni, A., Belobaba, P. et al. (2013). *Airport Systems - Planning, Design and Management*, 2e. New York, USA: McGraw-Hill.

36 Sabatini, R., Moore, T., and Ramasamy, S. (2017). Global navigation satellite systems performance analysis and augmentation strategies in aviation. *Progress in Aerospace Sciences* 95: 45–98.

37 Penner, J.E., Lister, D.H., Griggs, D.J., et al. 1999, *Aviation and the Global Atmosphere - A Special Report of IPCC Working Groups I and III in Collaboration with the Scientific Assessment Panel to the Montreal Protocol on Substances that Deplete the Ozone Layer*, report n. 9780521664042, United Nations Intergovernmental Panel on Climate Change (UN IPCC).

38 Blum, C., Aguilera, M.J.B., Roli, A. et al. (ed.) (2008). *Hybrid Metaheuristics: An Emerging Approach to Optimization*. Studies in Computational Intelligence, Springer.

39 Sabatini, R. 2017,' Integrated Air Traffic Management and Avionics Systems (CNS+A) for Environmentally Sustainable Aviation - Keynote paper', ICAO Committee on Aviation Environmental Protection (CAEP) Working Groups 2/3 Workshop, Sydney, Australia.

40 US DoT (2021). *NextGen benefits have not kept pace with initial projections, but opportunities remain to improve future modernization efforts*, report n. AV2021023, US Department of Transportation (DoT), Office of Inspector General.

41 Kistan, T., Gardi, A., Sabatini, R. et al. (2017). An evolutionary outlook of air traffic flow management techniques. *Progress in Aerospace Sciences* 88: 15–42.

42 Ranasinghe, K., Guan, K., Gardi, A. et al. (2019). Review of advanced low-emission technologies for sustainable aviation. *Energy* 188.

43 Gardi, A., Sabatini, R., and Ramasamy, S. (2016). Multi-objective optimisation of aircraft flight trajectories in the ATM and avionics context. *Progress in Aerospace Sciences* 83: 1–36.

44 Fahey, T., Pham, H., Gardi, A. et al. (2021). Active and passive electro-optical sensors for health assessment in food crops. *Sensors (Switzerland)* 21 (1): 1–40.

45 Bijjahalli, S. and Sabatini, R. (2021). A High-Integrity and Low-Cost Navigation System for Autonomous Vehicles. *IEEE Transactions on Intelligent Transportation Systems* 22 (1): 356–369.

46 Airport Carbon Accreditaion Website, URL: https://www.airportcarbonaccreditation .org/.

47 *Report of the Third Meeting of the Asia/Pacific Air Traffic Flow Management Steering Group (ATFM/SG/3) Singapore*, The International Civil Aviation Organization (ICAO), 10–14 March 2014.

48 CANSO, Bangkok-Singapore CDM Journal, Bangkok – Singapore Collaborative Decision Making (CDM), January 2013.

49 Sharwood S, "Data Analysis Takes Off", Article: Government Technology Review, Issue 10 Dec 2011-Jan 2012, pp 20–22,

50 *National Climate Assessment*, United States Global Change Joint Research Program, ed. 2014.

51 *Feasibility Study of Australian Feedstock and Production Capacity to Produce Sustainable Aviation Fuel*, Qantas Public Report, June 2013.

52 *Current Market Outlook, 2013–2032*, The Boeing Company, 2013.

53 Cikovic, A. and Damarodis, T. (2012). *The Boeing 787's role in new sustainability in the commercial aircraft industry*. University of Pittsburgh.

54 Kechidi, M. (2013). From'aircraft manufacturer'to'architect–integrator': Airbus's industrial organisation model. *International Journal of Technology and Globalisation* 7 (1): 8–22.

55 Immarigeon, J., Holt, R., Koul, A. et al. (1995). Lightweight materials for aircraft applications. *Materials Characterization* 35 (1): 41–67.

56 Fortune Business Insights, "The global carbon fiber market is projected to grow from $2.33 billion in 2021 to $4.08 billion in 2028 at a CAGR of 8.3% in forecast period, 2021-2028", URL: https://www.fortunebusinessinsights.com/industry-reports/carbon-fiber-market-101719

57 Australian Initiative for Sustainable Aviation Fuels (AISAF), Content available at: http://aisaf.org.au

58 *Aviation Security*, Australian Government – Department of Infrastructure and Regional Development.

59 *Aviation Safety Regulation Review*, Australian Government – Department of Infrastructure and Regional Development.

60 *Feasibility study of Australian Feedstock and Production Capacity to Produce Sustainable Aviation Fuel*, Qantas Airways Ltd, 2013.

61 SESAR (2015). European ATM Master Plan - The Roadmap for Delivering High Performing Aviation for Europe. Single European Sky ATM Research (SESAR) Joint Undertaking, Belgium. Available online at: https://ec.europa.eu/transport/sites/transport/files/modes/air/sesar/doc/eu-atm-master-plan-2015.pdf

62 Sabatini, R., Roy, A., Blasch, E. et al. (2020). Avionics Systems Panel Research and Innovation Perspectives. *IEEE Aerospace and Electronic Systems Magazine* 35 (12): 58–72.

63 Blasch, E., Sabatini, R., Roy, A., Kramer, K.A., Andrew, G., Schmidt, G.T., Insaurralde, C.C., and Fasano, G. (2019). Cyber Awareness Trends in Avionics. Proceediungs of the 38th IEEE/AIAA Digital Avionics Systems Conference (DASC 2019). San Diego, CA, USA, September 2019.

Section I
Aviation Sustainability Fundamentals

2

Climate Impacts of Aviation

Yixiang Lim[1], Alessandro Gardi[2,3], and Roberto Sabatini[2,3]

[1] *Agency for Science, Technology and Research (ASTAR), Singapore*
[2] *Department of Aerospace Engineering, Khalifa University of Science and Technology, Abu Dhabi, UAE*
[3] *School of Engineering, RMIT University, Melbourne, Victoria, Australia*

2.1 Introduction to Climate Change

Global warming is generally defined as *the rising average temperature of the Earth's atmosphere and oceans.* Scientists are more than 90% certain that a significant portion of the current global warming trend is caused by increasing concentrations of greenhouse gases produced by human activities (e.g. burning fossil fuel, deforestation). Such greenhouse effect is conceptually illustrated in Figure 2.1 and is the main subject of this chapter. Despite the growing body of evidence that human activities are having a prominent role, global warming is a highly controversial issue. The possible environmental impacts of global warming include:

- Sea levels rise due to melting of ice caps.
- Change in amount and pattern of precipitation (rain, snow, hail).
- Probable expansion of subtropical deserts.
- More frequent extreme-weather events (e.g. heat waves, droughts, heavy rainfall).
- Exceeding limits for adaption of many biological systems (plants, animals) leading to extinctions.

Aviation is a small contributor to global warming (estimated at about 2–3% of total global warming [1]; although without major changes to aircraft and their operation this may rise to 15% by 2050. Other contributors include power generation, industrial, transport, and agricultural activities. The greenhouse effect is one of the processes contributing to global warming and is caused when heat from the Earth is trapped by the atmosphere instead of being radiated into space. Greenhouse gases are responsible for absorbing this heat and contributing to the warming of the Earth (Figure 2.2).

Aviation's contribution to climate change is primarily due to aircraft emissions. The three key categories of aircraft emissions are:

- Radiatively active substances, which absorb or reflect radiation, causing either a warming or cooling effect. These include greenhouse gases like H_2O, CO_2 or particulate matter like sulphates.

Sustainable Aviation Technology and Operations: Research and Innovation Perspectives, First Edition.
Edited by Roberto Sabatini and Alessandro Gardi.
© 2024 John Wiley & Sons Ltd. Published 2024 by John Wiley & Sons Ltd.
Companion Website: www.wiley.com/go/sustainableaviation

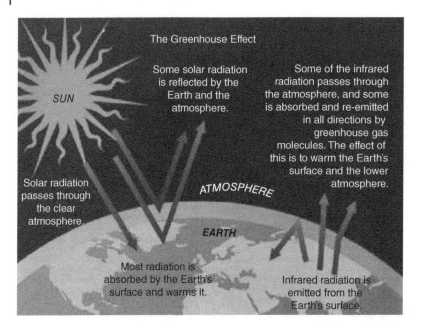

Figure 2.1 Synthesis of radiative forcing potential for common aviation emissions.

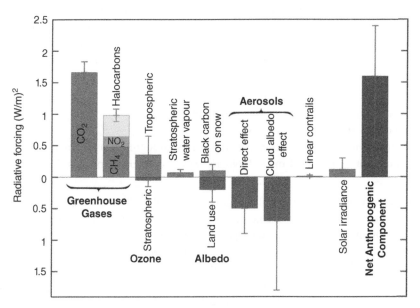

Figure 2.2 Radiative forcing related to various environmental impacts of aviation. Source: [6] and other references.

- Substances which modify the concentration of radiatively active substances (e.g. NOx emissions which lead to the formation of O_3, or the destruction of CH_4).
- Substances which trigger the generation of additional clouds or contrails (e.g. water vapour and particulates that trigger the formation of contrails).

Even though H_2O is the most abundant greenhouse gas (36% to 70%, compared to CO_2 with 9% to 26%), CO_2 has become the focus of environmental studies due to the longer lifetime (H_2O at 6 to 12 months; CO_2 with one hundred years) and an increasing CO_2 imbalance (more CO_2 being produced than can be absorbed by plants).

2.2 Climate Change Metrics

This section introduces the key metrics used for quantifying and assessing climate change. Table 2.1 lists the most frequently adopted metrics, categorised by time horizon and physical quantity considered. Radiative forcing (RF) has been the traditional measure and is discussed upfront (Section 2.2.1), but newer metrics, such as global warming potential (GWP) (Section 2.2.3) and global temperature change potential (GTP) (Section 2.2.4), have been developed in the past decade.

2.2.1 Radiative Forcing

Radiative forcing (RF) is defined by Ramaswamy et al. [2] as the net radiative flux change induced at the tropopause, expressed in Wm^{-2}. A negative net RF indicates that radiative energy is leaving the Earth's atmosphere, resulting in a cooling effect, while a positive net RF indicates that radiative energy is entering the earth's atmosphere, resulting in a warming effect. RF can be used to approximate the relationship between the change in global mean surface temperature ΔT_s based on the equation:

$$\Delta T_s \approx \lambda \cdot RF \tag{2.1}$$

where λ is the climate sensitivity parameter. The value of λ is dependant on the atmospheric general circulation model and ranges from 0.4 to 1.2 K $(W\,m^{-2})^{-1}$ [3].

RF is an instantaneous measure of the balance of energy within the Earth's atmosphere, but also has a temporal, backward-looking component because it includes the RF of long-lived substances (such as CO_2), which have accumulated in the atmosphere over

Table 2.1 Classification of global warning metrics based on the time horizon and on the physical quantity monitored.

Time horizon	Quantity measured		
	Energy flux	Temperature	Ratio
Forward-looking	RF	$\lambda \times RF$	RFI
Backward-looking	AGWP	AGTP	GWP, GTP

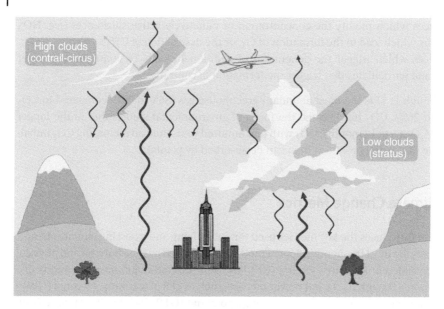

Figure 2.3 Radiative forcing from high and low clouds.

time. RF does not capture the future effect of such substances and therefore does not provide the total climate response of these long-lived substances. Figure 2.3 illustrates the RF potential of various substances related to aviation emissions.

2.2.2 Radiative Forcing Index

The radiative forcing index (RFI) is the ratio between the total annual aviation-related RF divided by the RF due to CO_2 in the same year, given by:

$$RFI = \frac{RF_{total}}{RF_{CO_2}} \tag{2.2}$$

The RFI allows the comparison of different sectors or activities with respect to their associated fossil fuel use and its impact on the climate. As an illustration, the RFI of aviation ranges between 2.2 and 3.4 while the baseline RFI (i.e. based on all changes in radiatively active gases from 1750 to the present) is about 1.3 [4]. Similar to RF, RFI does not capture the future effects of long-lived substances.

2.2.3 Global Warming Potential

GWP is a forward-looking metric that predicts the climate impact of an emission over a given timeframe. It is defined as the ratio between the time-integrated RF of an emission pulse to the time-integrated RF of a CO_2 pulse. The time-integration allows the effects of the emission's lifetime to be captured. The absolute global warming potential (AGWP) of species x is given (in W m^{-2} yr kg^{-1}) by [5]:

$$AGWP^x(H) = \int_0^H A_x \cdot exp\left(-\frac{t}{\alpha_x}\right) dt = A_x \alpha_x \left[1 - exp\left(-\frac{H}{\alpha_x}\right)\right] \tag{2.3}$$

where H is the time horizon, A_x is the specific radiative forcing (in W m^{-2} kg^{-1}) of emission x and α_x is the adjustment time, used for characterising the decay of x. Standard time horizons have been defined as 20, 100, and 500 years. The GWP of species x can subsequently be derived by taking the ratio of $AGWP^x$ with $AGWP^{CO_2}$, given by:

$$GWP^x(H) = \frac{AGWP^x(H)}{AGWP^{CO_2}(H)} \tag{2.4}$$

By definition, CO_2 has a GWP of one regardless of the time period used. The GWP of other species can vary depending on their lifetimes. Methane, for example, has a GWP of 86 (i.e. an AGWP 86 times that of CO_2) over 20 years and 34 (i.e. an AGWP 34 times that of CO_2) over 100 years [6].

2.2.4 Global Temperature Change Potential

GTP is also a forward looking metric, predicting the change in temperature as a result of a particular emission. It is defined as the ratio between the change in global mean surface temperature at a chosen point in time in response to an emission pulse, relative to that of carbon dioxide. The absolute global temperature change potential (AGTP) of species x at time H is given (in K kg^{-1}) by [5]:

$$AGTP^x(H) = \frac{A_x}{C(\tau^{-1} - \alpha_x^{-1})} \left[exp\left(-\frac{H}{\alpha_x}\right) - exp\left(-\frac{H}{\tau}\right) \right] \tag{2.5}$$

where τ is the time-scale of the climate response and C is the heat capacity of the climate system. τ and C are related to the climate sensitivity parameter λ by:

$$\tau = C\lambda \tag{2.6}$$

Similar to the GWP, GTP can be obtained by taking the ratio between $AGTP^x$ and $AGTP^{CO_2}$, given by:

$$GTP^x(H) = \frac{AGTP^x(H)}{AGTP^{CO_2}(H)} \tag{2.7}$$

GTP^{CO_2} is always one, while the GTP of other species varies in the same way as the GWP – the GTP of methane is 70 (i.e. an AGTP 70 times that of CO_2) over 20 years and 11 (i.e. an AGTP 11 times that of CO_2) over 100 years (Table 2.1).

2.3 CO$_2$ Emissions and their Impact on the Climate

CO_2 emissions are modelled as a function of the aircraft fuel flow, the relations of which can be found in the Base of Aircraft Data (BADA) [7]. The fuel flow FF is given in kg/min as the maximum thrust under nominal and idle conditions.

$$FF = max\left[\tau T_{max} C_{f1}\left(1 + \frac{v_{tas}}{C_{f2}}\right), C_{f3}\left(1 - \frac{H_p}{C_{f4}}\right) \right] \tag{2.8}$$

where τ is the throttle, v_{tas} is the true airspeed by (kts) and H_p is the geopotential pressure (ft). The aircraft-specific parameters T_{max}, C_{f1}, C_{f2}, C_{f3}, C_{f4} can be found in

BADA. The CO_2 emissions (kg) can then be approximated as a function of the fuel flow and elapsed time:

$$CO_2 = 3.16 \int FF(t)dt \tag{2.9}$$

The RF associated with gaseous emissions can be obtained through the simplified expressions found in Ramaswamy et al. [8]. Expressions are provided for CO_2, CH_4, N_2O, CFC-11, and CFC-12. Three expressions are provided for CO_2. The first expression is an updated version of IPCC [9] and is given as:

$$RF_{CO_2} = \alpha ln \left(\frac{C}{C_0} \right), \quad \alpha = 5.35 \tag{2.10}$$

The second expression is based on Shi [10] and is given as:

$$RF_{CO_2} = \alpha ln \left(\frac{C}{C_0} \right) + \beta(\sqrt{C} - \sqrt{C_0}), \quad \alpha = 4.841, \beta = 0.0906 \tag{2.11}$$

The third expression is based on Granier et al. [11] and is given as:

$$RF_{CO_2} = \alpha[g(C) - g(C_0)], \quad \alpha = 3.35 \tag{2.12}$$

$$g(C) = ln(1 + 1.2C + 0.005C^2 + 1.4 \times 10^{-6}C^3) \tag{2.13}$$

where C is CO_2 in parts per million (ppm). The conversion from kg to ppm is given by:

$$1kg \, CO_2 = 4.69 \times 10^{-12} ppm \, CO_2 \tag{2.14}$$

This allows A_{CO_2}, the radiative forcing from 1 kg of CO_2, to be obtained. As the decay of CO_2 occurs over many timescales, an extension of Eq. (2.3) is used to compute its AGWP, given by [12]:

$$AGWP^{CO_2}(H) = \int_0^H A_{CO_2} \cdot \left[a_0 + \sum_{i=1}^{n} a_i \cdot exp \left(-\frac{t}{\tau_i} \right) \right] dt$$

$$= A_{CO_2} \left\{ a_0 \cdot H + \sum_{i=1}^{n} a_i \cdot \tau_i \cdot \left[1 - exp \left(-\frac{H}{\tau_i} \right) \right] \right\} \tag{2.15}$$

where n is the number of timescales considered, a_0 is the fraction of total CO_2 that remains permanently in the atmosphere and a_i is the fraction of total CO_2 remaining in the atmosphere for a given timescale of τ_i years. Three timescales are given, such that $a_0 = 0.2173$, $a_1 = 0.2240$, $a_2 = 0.2824$, $a_3 = 0.2763$, $\tau_1 = 394.4$, $\tau_2 = 36.54$, and $\tau_3 = 4.304$.

2.4 Contrails and their Impact on the Climate

Contrails, or condensation trails, are formed from the condensation of water vapour in the plume of a jet engine and can be the precursor to those ice clouds, called cirrus, found at higher altitudes where ice super-saturation is present. The formation of these contrails is thermodynamically triggered by the engine exhaust, which contains heat, moisture, soot, and other forms of aerosols.

Wingtip contrails – which are usually short lived and will not be discussed here – form on aircraft wingtips and are triggered aerodynamically, when the local drop in pressure and temperature on the upper surface of the wing causes condensation of ambient air. As opposed to wingtip contrails, exhaust-triggered contrails may persist for long periods of time and have been shown to have an adverse effect on global warming through RF. Research done on the evolution of this type of contrail has shown three distinct phases in its lifecycle: the jet, vortex, and dispersion phase.

2.4.1 Jet Phase

In the jet phase, the contrail is catalysed when the hot and humid aircraft exhaust plume cools to ambient conditions. The state of the exhaust plume can be characterised by the mixing diagram, as seen in Figure 2.4.

The vapour partial pressure is depicted on the y-axis as the amount of vapour present in the air. The maximum amount of vapour, or the saturation point, is a function of ambient temperature, T, and is represented by the solid and dashed lines, corresponding to saturation points for water, $p_{w,\,sat}(T)$, and ice, $p_{i,\,sat}(T)$, respectively. $p_{w,\,sat}(T)$, and $p_{i,\,sat}(T)$ are obtained based on empirical relations by Sonntag [13], given by:

$$p_{w,sat}(T) = 100\,exp\left[\frac{-6096.9385}{T} + 16.635794 - 0.02711193T\right.$$

$$\left. + 1.673952e^{-5}T^2 + 2.433502\,ln(T)\right] \tag{2.16}$$

$$p_{i,sat}(T) = 100\,exp\left[\frac{-6024.5282}{T} + 24.7219 + 0.10613868T\right.$$

$$\left. - 1.3198825e^{-5}T^2 - 0.49382577ln(T)\right] \tag{2.17}$$

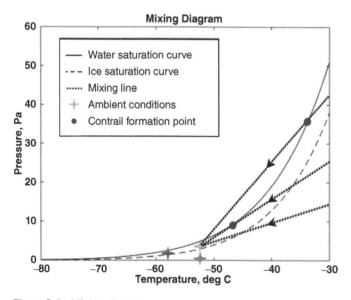

Figure 2.4 Mixing diagram.

where T is given in degrees kelvin. Saturation occurs when the ambient partial pressure of water, $p_{H_2O,amb}$, reaches either of these points and is conveniently expressed by taking the ratio of the two values. This ratio is termed the relative humidity (RH) and is given by:

$$RH_w(T) = p_{H_2O,amb}/p_{w,sat}(T) \qquad (2.18)$$

$$RH_i(T) = p_{H_2O,amb}/p_{i,sat}(T) \qquad (2.19)$$

where the subscripts w and i denote the relative humidities with respect to water and ice respectively. Note from Figure 2.4 that for the same temperature, the saturation point of ice occurs at a lower vapour partial pressure as compared to the saturation point of water. Note also that:

$$RH_i = RH_w \frac{p_{w,sat}(T)}{p_{i,sat}(T)} \qquad (2.20)$$

The exhaust plume, upon exiting the aircraft, quickly cools to ambient temperature as depicted by the dot-dashed lines in Figure 2.4. The different dot-dashed lines represent different initial exhaust conditions reaching the same ambient state (represented by the green star in Figure 2.4), which are typically around $-50\,°C$ and subsaturated with respect to water at cruise levels of around 33 000 ft. The slope G of the mixing line is a property of the aircraft characteristics – namely its propulsive efficiency η, EI_{H_2O} and specific heat of fuel Q – and ambient pressure. Larger values of G will result in a steeper slope, leading to condensation of the exhaust plume and the formation of contrails as it crosses the water saturation curve. G is determined using the Schmidt-Appleman Criterion, given by:

$$G = \frac{EI_{H2O} \cdot c_p \cdot p}{\epsilon \cdot Q \cdot (1 - \eta)} \qquad (2.21)$$

where G is the slope of the mixing line, in Pa K^{-1} describing the mixing of the exhaust plume with ambient air. The engine characteristics are EI_{H_2O}, the emission index of water vapour; Q, the heat per mass of fuel; and η, the propulsive efficiency. p is the ambient pressure, while c_p is the specific heat capacity of air, and ϵ is the ratio of the molecular masses of water and air.

The criterion for contrail formation is that the mixing line crosses the water saturation curve (represented by the red circle in Figure 2.4). Physically, this means that ambient air has sufficient excess vapour such that condensation occurs. If ambient conditions are also subsaturated with respect to ice (represented by the blue star in Figure 2.4), the contrails formed will quickly dissipate as the plume continues losing vapour pressure. However, if ambient conditions are ice-supersaturated (represented by the green and purple stars in Figure 2.4), contrails can persist. This is because the condensed water droplets freeze rapidly, providing ice nucleation sites that are important for the formation of ice. Ice crystals have a highly organised structure and cannot form easily, even on hydrophilic particles (this in turn implies that ice super-saturation is a metastable state).

For a constant RH (which indicates that the ambient partial pressure of vapour decreases in proportion to the saturation curves), lower ambient temperatures (represented by the purple star in Figure 2.4) can facilitate the formation of contrails, since the mixing line is more likely to cross the water saturation curve when cooling to such temperatures. Also, as the partial water pressure is lower at lower temperatures, the moisture content in the exhaust plume (denoted as EI_{H_2O}) contributes more significantly to the ambient water content. Finally, because ice saturation points are so low at low temperatures, high ice super-saturations can be observed at low temperatures for a given RH_w. This is why contrails tend to form and persist at high altitudes and low temperatures and are considered a precursor to cirrus clouds.

To determine the critical temperature for contrail formation (i.e. the temperatures corresponding to the stars in Figure 2.4), we use the criterion given in Schumann [14]. The temperature at the threshold point T_{LM} is first determined by:

$$T_{LM} = -46.46 + 9.43 \, ln(G - 0.053) + 0.72[ln(G - 0.053)]^2 \tag{2.22}$$

where T_{LM} is given in degrees Celsius, where the subscript 'L' denotes the threshold temperature for liquid saturation. The critical temperature at ambient conditions can then be found using curve fitting, where T and T_{LM} are given in degrees Celsius. This is given by:

$$T_{LC} = T_{LM} - (1 - RH_w)\frac{p_{w,sat}(T_{LM})}{G} - \Delta T_C \tag{2.23}$$

$$\Delta T_C = F_1 RH_w[W - F_2(1 - W)] \tag{2.24}$$

$$W = (1 - RH_w^2)^{x_2} \tag{2.25}$$

$$F_1 = x_1 + x_3 \, ln(G), \quad F_2 = \left[\frac{1}{4} - \left(RH_w - \frac{1}{2}\right)^2\right]^4 \tag{2.26}$$

where $x_1 = 5.686$, $x_2 = 0.3840$, $x_3 = 0.6594$, and $p_{w,sat}(T_{LM})$, RH_w and G are defined in Eqs. (2.16), (2.18), and (2.21) respectively. The criterion for contrail formation is thus

$$dT = T - T_{LC} < 0 \tag{2.27}$$

where T is again the ambient temperature, but this time given in degrees Celsius.

2.4.2 Vortex Phase

In the first few minutes of formation, the contrail transitions to the vortex phase as the exhaust plume gets caught in the downwash vortex generated by the aircraft. Large eddy simulations by Lewellen and Lewellen [15] suggest that the vortex phase occurs in the first three minutes and causes a sinking of the plume. As the plume sinks to a lower altitude, ambient pressure increases. The plume undergoes compression, which leads to adiabatic warming. Sublimation might occur, leading to a decrease in the ice particle count.

The parametric model developed by Schumann [14] can be used to estimate the initial depth and width of the contrail. This is done by first computing the maximum downward displacement Δz_w and then scaling this to obtain the initial contrail depth. The initial contrail width is then parameterized from the depth, dilution, and fuel flow. The parameters that determine the initial size of the contrail are:

Wake vortex separation $b_0 = \pi S_a/4$
Initial circulation $\Gamma_0 = 4M_a g/(\pi S_a \rho_{air} V_a)$
Effective time scale $t_0 = 2\pi b_0^2/\Gamma_0$

Initial velocity scale $w_0 = \Gamma_0/(2\pi b_0)$

Normalised dissipation rate $\varepsilon^* = (\varepsilon b_0)^{1/3}/w_0$.

These are dependent on the inputs

Wing span S_a

Aircraft mass M_a

True air speed V_a

Air density ρ_{air}

Brunt Vaisaila frequency N_{BV}

Turbulent kinetic energy dissipation rate ε.

Schumann's parameterisation distinguishes between strongly and weakly stably stratified conditions. If $N_{BV}t_0 \geq 0.8$, the maximum sinking Δz_w is given as:

$$\Delta z_w = 1.49\frac{w_0}{N_{BV}} \tag{2.28}$$

else, with $\varepsilon^* \leq 0.36$:

$$\frac{\Delta z_w}{b_0} = 7.68\,(1 - 4.07\,\varepsilon^* + 5.67\,\varepsilon^{*2})(0.79 - N_{BV}t_0) + 1.88 \tag{2.29}$$

The initial downward displacement Δz_1 is given by:

$$\Delta z_1 = C_{z1}\Delta z_w, \quad C_{z1} = 0.25 \tag{2.30}$$

where the subscript '1' denotes the end of the vortex phase. The initial depth D_1 is given by:

$$D_1 = C_{D0}\Delta z_w, C_{D0} = 0.5 \tag{2.31}$$

and the initial width is given by:

$$B_1 = N_{dil}(t_0)m_F/[(\pi/4)\rho D_1] \tag{2.32}$$

with the m_F being the fuel flow in $kg\,m^{-1}$ and the dilution N_{dil} of the wake vortex formation being:

$$N_{dil}(t) \approx 7000(t/t_s)^{0.8} \tag{2.33}$$

with $t_s = 1$ second. Adiabatic heating causes sublimation of the contrail plume, leading to a decrease in the mass mixing ratio of ice I and the ice particle number N, given by:

$$I_1 = I_0 - \Delta I_{ad} \tag{2.34}$$

$$I_0 = \frac{EI_{H_2O}m_F}{(\pi/4)\rho D_1 B_1} + q_0 - q_s(p_0, T_0) \tag{2.35}$$

$$\Delta I_{ad} = \frac{R_0}{R_1}\left[\frac{p_{ice}(T_0 + \Delta T_{ad})}{p_1} - \frac{p_{ice}(T_0)}{p_0}\right] \tag{2.36}$$

Here, q_0 is the specific humidity of ambient conditions at the start of the vortex phase and q_s is the specific humidity at saturation point. p_1 is the ambient pressure at the end of the vortex phase. R_0 and R_1 are the specific heat capacities of air and water respectively. ΔT_{ad} is the increase in temperature due to adiabatic warming and is given by:

$$\Delta T_{ad} = T_0(R_0/c_p)(p_1 - p_0)/p_0 \tag{2.37}$$

The ice number N_1 at the end of the vortex phase is determined by the soot index EI_{soot} and a survival factor f based on the ice mass ratio.

$$N_0 = EI_{soot} m_F \tag{2.38}$$

$$N_1 = f_{surv} N_0, \ f_{surv} = \frac{I_1}{I_0} \tag{2.39}$$

f represents the fraction of ice particles remaining after the vortex phase and is primarily dependent on the wake size, temperature, and relative humidity. Schumann [14] gives values of f that lie between 0.7 and 1 but lower values, corresponding to around 0.25, were found to be more common for most flight conditions for large aircraft. These are in good agreement with results by Unterstrasser et al. [16]. As the strength of the vortex (and subsequently the extent of adiabatic heating) is related to the aircraft mass, larger aircraft will have lower values of f for similar ambient conditions. Lewellen and Lewellen [15] concluded that the higher fuel flow of large aircraft (hence larger soot emissions and a larger number of ice particles in the jet phase) could be offset by the smaller survival factor, and that a B737 could give rise to contrails with similar persistence as a B747.

2.4.3 Dispersion Phase

Contrails that survive the vortex phase then undergo the dispersion phase, in which they are advected by wind and spread by wind shear, which is the first derivative of wind with respect to the vertical and lateral directions. The growth of contrails is largely dependent on ambient conditions, which affect the size and shape of the contrail plume as well as its ice particle properties. The state of each contrail segment along the flight path evolves with time and is described by the state vector $X(Position, Ambient, Plume, Particle)$. The position at time t is given by:

$$X.Position = (x(t), y(t), z(t), t) \tag{2.40}$$

where x, y and z are the longitude, latitude, and altitude of the contrail. The advection of the contrail over time is described by:

$$x(t + \Delta t) = x(t) + U(t)\Delta t \tag{2.41}$$

$$y(t + \Delta t) = y(t) + V(t)\Delta t \tag{2.42}$$

$$z(t + \Delta t) = z(t) + W(t)\Delta t \tag{2.43}$$

with the absolute distances converted to geodetic coordinates based on approximations provided by Veness [17]. The ambient conditions are given by:

$$X.Ambient = (p, T, \rho_{air}, q_a, q_s) \tag{2.44}$$

which are pressure, temperature, density, and ambient and saturation specific humidity respectively, obtained from the weather forecast data, and are functions of $X. Position$. The plume parameters evolve according to the contrail model from Schumann [14] and are given by:

$$X.Plume = (\sigma, B, D, D_{eff}, A, L, M, I, N, n, M_{H2O}, D_H, D_V) \tag{2.45}$$

Respectively, these are the:

Covariance matrix σ
Contrail width B
Contrail depth D
Effective depth D_{eff}
Contrail area A
Contrail length L
Air mass M
Ice mass mixing ratio I
Ice particle number N
Ice concentration n
Water mass M_{H_2O}
Horizontal diffusivity D_H
Vertical diffusivity D_V.

While the contrail is assumed to have uniform particle properties, in reality contrails typically possess a distribution of ice particle sizes, with the heavier particles sedimenting to the lower region of the contrail. The covariance matrix σ describes the shape of the contrail (assumed to be a Gaussian plume) with terms $\begin{bmatrix} \sigma_{yy} & \sigma_{yz} \\ \sigma_{yz} & \sigma_{zz} \end{bmatrix}$. The terms are initially given as:

$$\sigma_{yy}(t = t_0) = B^2/8 \tag{2.46}$$

$$\sigma_{zz}(t = t_0) = D^2/8 \tag{2.47}$$

$$\sigma_{yz}(t = t_0) = 0 \tag{2.48}$$

and evolve with time as:

$$\sigma_{yy}(t + \Delta t) = \left[\frac{2}{3} S^2 D_V \Delta t^3 + (S^2 \sigma_{zz}(t) + 2 D_S S) \Delta t^2 + 2(D_H + S \sigma_{yz}(t)) \Delta t + \sigma_{yy}(t) \right]$$
$$\cdot \left[\frac{L(t)}{L(t + \Delta t)} \right]^2 \tag{2.49}$$

$$\sigma_{zz}(t + \Delta t) = 2 D_V \Delta t + \sigma_{zz}(t) \tag{2.50}$$

$$\sigma_{yz}(t + \Delta t) = [S D_V \Delta t^2 + (2 D_S + S \sigma_{zz}(t)) \Delta t + \sigma_{yz}(t)] \cdot \left[\frac{L(t)}{L(t + \Delta t)} \right] \tag{2.51}$$

with the shear diffusivity D_S set to 0 and the vertical shear of the plume normal velocity S is taken to be the total shear $S_T = \sqrt{\left(\frac{\partial U}{\partial z}\right)^2 + \left(\frac{\partial V}{\partial z}\right)^2}$. The shape and size of the contrail are based on the covariance matrix:

$$B(t + \Delta t) = \sqrt{8 \sigma_{yy}(t + \Delta t)} \tag{2.52}$$

$$D(t + \Delta t) = \sqrt{8 \sigma_{zz}(t + \Delta t)} \tag{2.53}$$

$$A(t + \Delta t) = 2\pi \left\{ \frac{1}{3} S^2 D_V^2 (\Delta t)^4 + \frac{2}{3} S^2 D_V \sigma_{zz}(t)(\Delta t)^3 \right.$$
$$+ [2S\sigma_{zz}(t)(D_V - D_S) - 4(D_H D_V - D_S^2)](\Delta t)^2$$
$$\left. + [2\sigma_{zz}(t)(D_V + D_H) - 4D_S \sigma_{yz}(t)]\Delta t + \sigma_{yy}(t)\sigma_{zz}(t) - \sigma_{yz}^2(t) \right\}^{1/2} \quad (2.54)$$

$$D_{eff}(t + \Delta t) = \frac{A(t + \Delta t)}{B(t + \Delta t)} \quad (2.55)$$

and the length $L(t + \Delta t)$ is computed from the positions of the end-points of the contrail $x(t + \Delta t)$ and $y(t + \Delta t)$ in its position vector. The mass properties of the plume can then be calculated as:

$$M(t + \Delta t) = \rho A L|_{t+\Delta t} \quad (2.56)$$

$$I(t + \Delta t) = \frac{M(t) \cdot [I(t) + q_s(t) + \Delta M \cdot q_a]}{M(t + \Delta t)} - q_s(t + \Delta t) \quad (2.57)$$

$$M_{H2O}(t + \Delta t) = M(t + \Delta t) \cdot (I + q)|_{t+\Delta t} \quad (2.58)$$

where $\Delta M = M(t + \Delta t) - M(t)$ and $q_a = \frac{q_a(t+\Delta t) + q_a(t)}{2}$. The particle properties are given by:

$$X.Particle = (V_T, r_{eff}) \quad (2.59)$$

which respectively describe the terminal velocity (following Rogers [18]) and effective radius of the ice particles. These are given by:

$$r_{eff} = \left(\frac{\rho_{air} I}{n \rho_{ice} 4\pi/3} \right)^{1/3} \quad (2.60)$$

$$V_T = \begin{cases} k_1 r_{eff}^2, & r_{eff} < 40\,\mu m \\ k_2 r_{eff}, & 40\Delta m < r_{eff} < 600\,\mu m \\ k_3 \sqrt{r_{eff}}, & r_{eff} > 600\,\mu m \end{cases} \quad (2.61)$$

$$k_1 = 1.9e^8, \quad k_2 = 8e^3, \quad k_3 = 2.2e^2 \sqrt{\frac{\rho_0}{\rho}} \quad (2.62)$$

Particle loss due to turbulence and aggregation are modelled to determine the evolution of the ice number and concentration, with the adjustable parameters E_A and E_T set to 2:

$$(dN/dt)_{agg} = -E_A 8\pi r_{eff}^2 V_T N^2 / A \quad (2.63)$$

$$(dN/dt)_{turb} = -E_T \left(\frac{D_H}{max(B, D)^2} + \frac{D_V}{D_{eff}^2} \right) N \quad (2.64)$$

$$\alpha = -\frac{1}{N^2}(\partial N/\partial t)_{agg}, \quad \beta = -\frac{1}{N}(\partial N/\partial t)_{turb} \quad (2.65)$$

$$N(t + \Delta t) = \frac{N(t)\beta exp(-\beta\Delta t)}{\beta + N(t)\alpha[1 - exp(-\beta\Delta t)]} \frac{L(t)}{L(t + \Delta t)} \quad (2.66)$$

$$n = N/A \quad (2.67)$$

Finally, the horizontal and vertical diffusivities are updated as:

$$D_H = C_H(D^2 S_T), \quad C_H = 0.1 \quad (2.68)$$

$$D_V = \frac{C_V}{N_{BV}} w'_n + f_t(V_T D_{eff}), \quad C_V = 0.2, \quad w'_n = 0.1, \quad f_t = 0.1 \tag{2.69}$$

If contrails form in ice super-saturated regions, they will continue to grow as ambient water vapour becomes entrained within the contrail as it spreads. As the contrail ages, its particle concentration decreases while its particle size increases (i.e. the ice crystals become larger and more spread out in the contrail) eventually approaching the properties of cirrus clouds, with typical concentrations of around 10^1 cm^{-3} and effective radii of around 10^1 μm.

Contrails disperse via sublimation of its ice particles, which can occur when the contrail drifts into an ice sub-saturated region. This can occur from wind advection (laterally) or by sedimentation (vertically). The latter occurs when larger and heavier particles sink to the bottom of the contrail cirrus. Over sufficient time, the entire contrail will sink or ice particles that are large enough will form fallstreaks; these have been observed and recorded via lidar by Atlas et al. [19]. The fallstreak/contrail then sublimates in the warm temperatures at lower altitudes. The time integration of the contrail model ends when the ice concentration n falls below 10^3 m^{-3} (or $1\,l^{-1}$), the ice mass ratio I falls below 10^{-8} (or 10^{-2} mg kg^{-1}), or when the time exceeds a given threshold, set to five hours in this case.

2.4.4 Radiative Forcing

As illustrated in Figure 2.3, the net RF from clouds and contrails can be attributed to two competing effects; cooling, which occurs when the cloud/contrail reflects incoming short-wave radiation from the sun (largely determined by cloud/contrail albedo); and warming, which occurs the cloud/contrail absorbs and re-emits outgoing longwave radiation from the Earth (largely determined by cloud/contrail temperature).

The optical depth (a measure of absorptivity) of contrail-cirrus is much lower than the water (stratus) clouds found at lower altitudes because of the differences in particle and cloud thicknesses. Contrail cirrus clouds contain ice particles, which are less tightly packed than water molecules, and are also much smaller in depth than the water clouds. Contrail optical depth typically ranges from $10^{-3} < \tau < 1$ [20], whereas those of water clouds can reach up to $\tau > 10^2$. As a result, contrail cirrus has a lower albedo than low altitude clouds and has a reduced cooling effect as compared to low altitude stratus clouds. Additionally, as contrail cirrus is colder than stratus clouds, it readily absorbs the Earth's outgoing radiation, leading to an increased warming effect as compared to stratus clouds.

The RF model follows Schumann et al. [21]. The parameterisation scheme is based on a number of parameters that model the particle habit (i.e. the spherical, hollow, rosette, plate) as well as a number of independent properties. The contrail habit depends mainly on atmospheric conditions and may comprise a combination of different types, which can be accounted for with a weighted sum of each habit. The longwave radiation is positive and dependent on the following independent properties: the outgoing longwave radiation (OLR, Wm^{-2}), the atmospheric temperature (T, K), the optical depths of the contrail and its overhead cirrus at 550 nm (τ and τ_c), and the effective particle radius (r_{eff}, μm).

The contrail optical depth can be computed from Schumann [14] as follows:

$$\tau = \beta D_{eff} \tag{2.70}$$

$$\beta = 3Q_{ext}\rho I/(4\rho_{ice}r_{eff}) \tag{2.71}$$

$$D_{eff} = A/B \tag{2.72}$$

$$Q_{ext} = 2 - (4/\rho_\lambda) \frac{sin(\rho_\lambda) - [1 - cos(\rho_\lambda)]}{\rho_\lambda} \tag{2.73}$$

$$\rho_\lambda = 4\pi r_{eff} \ (\kappa - 1)/\lambda, \quad \kappa = 1.31, \quad \lambda = 550 \, nm \tag{2.74}$$

The optical depth of overhead cirrus is assumed to be $\tau_c = 0$ (i.e. zero cloud cover). The zenith angle θ is approximated with ESRL's equations [22], given by:

$$\gamma = \frac{2\pi}{365} * \left(day \, of \, year - 1 + \frac{hour - 12}{24} \right) \tag{2.75}$$

$$\begin{aligned} eqtime = 299.18[7.5e^{-5} + 1.868e^{-3}cos(\gamma) - 3.2077e^{-2}sin(\gamma) \\ - 1.4615e^{-2}cos(2\gamma) - 0.40849e^{-2}sin(2\gamma)] \end{aligned} \tag{2.76}$$

$$\begin{aligned} decl = [6.918e^{-3} - 3.99912e^{-1}cos(\gamma) + 7.026e^{-2}sin(\gamma) - 6.758e^{-3}cos(2\gamma) \\ + 9.07e^{-4}sin(2\gamma) - 2.70e^{-3}cos(3\gamma) + 1.48e^{-3}sin(3\gamma)] \end{aligned} \tag{2.77}$$

$$time \, offset = eqtime - 4 * long + 60 * timezone \tag{2.78}$$

$$tst = hour * 60 + min + \frac{sec}{60} + time \, offset \tag{2.79}$$

$$ha = \left(\frac{tst}{4} \right) - 180 \tag{2.80}$$

$$\mu = cos(\theta) = sin(lat)sin(decl) + cos(lat)cos(decl)cos(ha) \tag{2.81}$$

The longwave RF, RF_{LW}, is then given by:

$$RF_{LW} = [OLR - k_T(T - T_0)] \cdot \{1 - exp[-\delta_\tau F_{LW}(r_{eff})\tau]\}E_{LW}(\tau_c) \tag{2.82}$$

$$F_{LW}(r_{eff}) = 1 - exp(-\delta_{lr} r_{eff}) \tag{2.83}$$

$$E_{LW}(\tau_c) = exp(-\delta_{lc}\tau_c) \tag{2.84}$$

The shortwave radiation, RF_{SW}, is negative and depends on the following independent properties: τ, τ_c, r_{eff}, the cosine of the solar zenith angle $\mu = cos(\theta)$, and the effective albedo (A_{eff}). This is given by:

$$RF_{SW} = -SDR(t_A - A_{eff})^2 \cdot \alpha_c(\Delta, \tau, r_{eff}) \cdot E_{SW}(\mu, \tau_c) \tag{2.85}$$

$$\alpha_c(\Delta, \tau, r_{eff}) = R_C(\tau_{eff}) \cdot [C_\mu + A_\mu \cdot R'_c(\tau') \cdot F_\mu(\mu)] \tag{2.86}$$

$$\tau' = \tau F_{SW}(r_{eff}), \quad \tau_{eff} = \tau'/\mu \tag{2.87}$$

$$F_{SW}(r_{eff}) = 1 - F_r[1 - exp(-\delta_{sr} r_{eff})] \tag{2.88}$$

$$R_C(\tau_{eff}) = 1 - exp(-\Gamma\tau_{eff}), \quad R'_C(\tau_{eff}) = 1 - exp(-\gamma\tau_{eff}) \tag{2.89}$$

$$F_\Delta(\Delta) = \frac{(1-\mu)^{B_\mu}}{(1/2)^{B_\mu}} - 1 \tag{2.90}$$

$$E_{SW}(\mu, \tau_c) = exp(-\delta_{sc}\tau_c - \delta'_{sc}\tau_{c,eff}) \tag{2.91}$$

$$\tau_{c,eff} = \tau_c/\mu \tag{2.92}$$

The net instantaneous RF is simply the sum of the long and shortwave components:

$$RF_{net}(t) = RF_{LW}(t) + RF_{SW}(t) \tag{2.93}$$

2.5 Global Warming Forecasts

Different models are used to describe the processes in the atmosphere. Uncertainties in predictions can be attributed to:

- Different processes being modelled (missing or incorrect processes). As our understanding of the atmospheric processes improves over time, these uncertainties can also decrease.

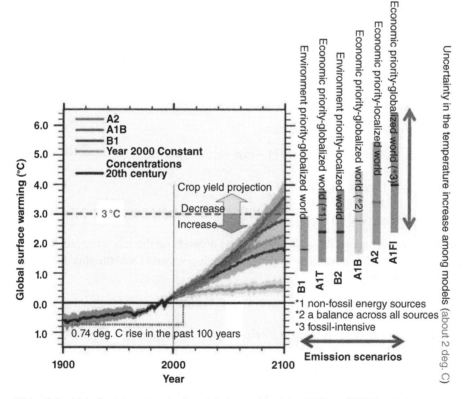

Figure 2.5 Main forecast scenarios for global warming. Adapted from IPCC Fourth Assessment Report: Climate Change 2007, United Nations Environmental Protection (UNEP), Intergovernmental Panel on Climate Change (IPCC), Geneve, Switzerland, 2007.

- Different factors influencing climate change. Uncertainties in aviation developments also make it difficult to predict the impact of aviation on the climate beyond 5–10 years. Factors include:
  - Cost of air travel (and hence number of aircraft).
  - Economic activity.
  - New market opportunities.
  - Air transport liberalisation.
  - Improvements in aircraft fuel efficiency.
  - Improvements in engine efficiency.

Figure 2.5 outlines the main forecasts scenario that were considered in the IPCC study.

References

1 Lee, D.S., Pitari, G., Grewe, V. et al. (2010). Transport impacts on atmosphere and climate: aviation. *Atmospheric Environment* 44: 4678–4734.

2 Ramaswamy, V., Boucher, O., Haigh, J. et al. (2001). Radiative forcing of climate. *Climate Change* 349.

3 Shine, K.P., Cook, J., Highwood, E.J., and Joshi, M.M. (2003). An alternative to radiative forcing for estimating the relative importance of climate change mechanisms. *Geophysical Research Letters* 30.

4 Fuglestvedt, J.S., Berntsen, T.K., Godal, O. et al. (2003). Metrics of climate change: assessing radiative forcing and emission indices. *Climatic Change* 58: 267–331.

5 Shine, K.P., Berntsen, T.K., Fuglestvedt, J.S. et al. (2007). Comparing the climate effect of emissions of short-and long-lived climate agents. *Philosophical Transactions of the Royal Society of London A: Mathematical, Physical and Engineering Sciences* 365: 1903–1914.

6 Myhre, G., Shindell, D., Bréon, F.-M. et al. (2013). Anthropogenic and natural radiative forcing. *Climate Change* 423: 658–740.

7 Eurocontrol (2013). *User Manual for the Base of Aircraft Data (BADA) Revision 3.11*. Paris, France: Eurocontrol.

8 Ramaswamy, V., Boucher, O., Haigh, J. et al. (2001). *Radiative Forcing of Climate Change*. Richland, WA: Pacific Northwest National Laboratory (PNNL).

9 IPCC. (1990). Climate change: the IPCC scientific assessment, Cambridge, MA.

10 Shi, G. (1992). Radiative forcing and greenhouse effect due to the atmospheric trace gases. *Science in China Series B-Chemistry, Life Sciences & Earth Sciences*.

11 Granier, C., Shine, K. P., Daniel, J. S. et al. (1998). Climate Effects of Ozone and Halocarbon Changes. WMO Report No. 44.

12 Joos, F., Roth, R., Fuglestvedt, J. et al. (2013). Carbon dioxide and climate impulse response functions for the computation of greenhouse gas metrics: a multi-model analysis. *Atmospheric Chemistry and Physics* 13: 2793–2825.

13 Sonntag, D. (1994). Advancements in the field of hygrometry. *Meteorologische Zeitschrift* 3: 51–66.

14 Schumann, U. (2012). A contrail cirrus prediction model. *Geoscientific Model Development* 5: 543–580.

15 Lewellen, D.C. and Lewellen, W.S. (2001). The effects of aircraft wake dynamics on contrail development. *Journal of the Atmospheric Sciences* 58: 390–406.

16 Unterstrasser, S., Gierens, K., and Spichtinger, P. (2008). The evolution of contrail microphysics in the vortex phase. *Meteorologische Zeitschrift* 17: 145–156.

17 Veness, C.. (2015, 25/3/2015). Movable type scripts. http://www.movable-type.co.uk/scripts/latlong.html (accessed 28 November 2020).

18 Rogers, R.R. (1976). *A Short Course in Cloud Physics: A*. Wheaton & Co.

19 Atlas, D., Wang, Z., and Duda, D.P. (2006). Contrails to cirrus—morphology, microphysics, and radiative properties. *Journal of Applied Meteorology and Climatology* 45: 5–19.

20 Kärcher, B., Burkhardt, U., Unterstrasser, S., and Minnis, P. (2009). Factors controlling contrail cirrus optical depth. *Atmospheric Chemistry and Physics* 9: 6229–6254.

21 Schumann, U., Mayer, B., Graf, K., and Mannstein, H. (2012). A parametric radiative forcing model for contrail cirrus. *Journal of Applied Meteorology and Climatology* 51: 1391–1406.

22 E. Team. (2015). ESRL global monitoring division – GRAD group. *2015(7/5/2015)*. http://www.esrl.noaa.gov/gmd/grad/solcalc/sollinks.html (accessed 28 November 2020).

3

Noise Pollution and Other Environmental and Health Impacts of Aviation

Alessandro Gardi[1,2], Rohan Kapoor[3], Yixiang Lim[4], and Roberto Sabatini[1,2]

[1]*Department of Aerospace Engineering, Khalifa University of Science and Technology, Abu Dhabi, UAE*
[2]*School of Engineering, RMIT University, Melbourne, Victoria, Australia*
[3]*School of Engineering RMIT University, Bundoora, Victoria, Australia*
[4]*Agency for Science, Technology and Research (ASTAR), Singapore*

3.1 Introduction

The operation of aircraft and of various other aviation-related vehicles and equipment produces several environmental and health impacts, ranging from gaseous and particulate emissions to noise. In this chapter, we perform a comprehensive review of these environmental effects, with the exclusion of climate impacts such as greenhouse gases and contrails that were already covered in detail separately. This chapter therefore primarily focusses on exhaust pollutants and noise as well as some of their well-known health impacts, both physical and psychological.

Key references used include the milestone report by the Intergovernmental Panel on Climate Change (IPCC) [1], two key updates [2, 3], and three relevant books [4–6], to which the reader should refer for further details. Some of the fundamental aspects introduced in this chapter are the focus of following sections of the book, such as the modelling of pollutant emissions and the fundamental aspects and technology evolutions of aeroengines, so the reader is referred to these for further details.

The review highlights the limitations of current mainstream modelling, assessment, and control/management approaches in relation to Urban Air Mobility (UAM) vehicles and other emerging concepts, especially in terms of noise impacts [7]. In particular, while reducing noise at source has been identified as paramount to enable scaling up of UAM and drone delivery operations in the near future, the impact of sound propagation and the psychoacoustic effects on the human listener are equally important for the same objective. For instance, optimizing the aircraft trajectory for minimizing the impact of noise on communities can significantly contribute to paving the way for the introduction of full scale UAM operations. Fully developed surrogate models will be highly instrumental to estimate the emissions and their propagation to the lower atmosphere and the planet surface, without relying on elaborate computational models or the need to generate and process large datasets.

Sustainable Aviation Technology and Operations: Research and Innovation Perspectives, First Edition.
Edited by Roberto Sabatini and Alessandro Gardi.
© 2024 John Wiley & Sons Ltd. Published 2024 by John Wiley & Sons Ltd.
Companion Website: www.wiley.com/go/sustainableaviation

3.2 Atmospheric Pollutants

Pollution generally refers to the artificial variation of the composition of a certain element (→ 'air pollution') and can take different forms, the most important of which are noxious gasses, liquid droplets, and particulate matter (PM). Engines and auxiliary power units (APU) are the largest source of atmospheric pollution from aircraft. Aviation relies on air-breathing internal combustion engines either of reciprocating piston type or of the turbine family. The exhaust emissions of all these propulsion systems depend mostly on the composition of fuel burned and the characteristics of the adopted combustion process. In terms of fuels, although the use of diesel, unleaded gasoline, and biofuels is on a steady growth trend, *Jet-A1* or *AvTur* (aviation turbine fuel, mostly equivalent to the military *JP8*) and *AvGas* (leaded aviation gasoline fuel) are still the most widely adopted hydrocarbon fuels for turbine and piston engines respectively. In chemical terms, these fuels of fossil origin are a mixture of a large variety of aliphatic and aromatic hydrocarbons, which originate from the distillation and cracking of crude oil. These hydrocarbon molecules have different structures and a variable number of carbon and hydrogen atoms, so it is preferable to refer to them using the generic molecular formula $C_x H_y$. Their balanced (stochiometric) combustion is expressed as:

$$C_x H_y + \left(x + \frac{y}{4}\right) O_2 \rightarrow x\, CO_2 + \frac{y}{2} H_2 O$$

For instance, adopting the average formula of $C_{12}H_{23}$, the reaction above is balanced as follows:

$$C_{12}H_{23} + 17.75\, O_2 \rightarrow 12\, CO_2 + 11.5\, H_2 O$$

In air-breathing engines, the hydrocarbon fuels are injected in the combustion chamber together with atmospheric air. When the oxygen available is greater than the amount required for stoichiometric combustion, the mixture is defined as lean, while conversely the mixture is termed rich when the amount of oxygen supplied is lower than the required quantity. In practice, neither the fuels nor the atmospheric air participating in the combustion are homogeneous in their composition and thermodynamic properties. Moreover, the thermodynamic conditions of the combustion process itself are not steady and uniform, due to the throttling of engine pressure ratios, external ambient conditions, and local disuniformities in the combustion chamber (i.e. poor fuel-oxidizer mixing, turbulence, stagnation regions, flame dynamics, etc.). Because of these factors, the local mixture ratio has significant spatiotemporal variations, facilitating the occurrence of incomplete combustion, which is responsible for the generation of several pollutants.

As a result, conventional aeroengines produce a wide variety of gaseous and condensed species as exhausts. Carbon dioxide (CO_2) and water vapour (H_2O) are the intended products of the ideal combustion of hydrocarbons with oxygen in stoichiometric conditions, so they are expectedly the largest components by mass and volume. Neither of these are directly harmful to living species, but they have a significant role in affecting the greenhouse effect, as already discussed in Chapter 2. Other species found in aeroengine exhausts are either products of incomplete combustion (e.g. CO, UHC, soot) or combustion by-products (e.g. NO_X, SO_X) and are almost entirely noxious to living beings and the environment. A brief introduction is provided in the following subsections, before the analytical discussion.

3.2.1 Carbon Species

Carbon monoxide (CO) and volatile organic compounds (VOC) are the products of incomplete combustion of the hydrocarbon fuel with atmospheric oxygen. Such incomplete combustion is due to a variety of factors, including insufficient combustion temperature or pressure, agglomeration or local stagnation due to poor mixing or excessive turbulence, temporary/local blowout of the combustion front (flame). At low engine throttle setting, the temperature in the combustion chamber can be significantly lower than nominal design conditions, leading to widespread incomplete combustion conditions and the associated production of these emissions. Low engine temperatures are a result of low engine power settings (e.g. idling or taxiing when the aircraft is on the ground). The mass conservation law applied across balanced chemical reaction implies that all exhaust emissions are a proportion of the mass of fuel consumed, so that the increase of VOC and unburned hydrocarbon (UHC) species in general results entirely from fuel not participating in the combustion, whereas the formation of CO is the result of partial oxidation of atomic carbon, which does not result in the formation of CO_2.

CO is an asphyxiating gas for all animal beings, while VOCs have a variety of health and environmental effects when either inhaled or ingested, including local irritation, toxicity particularly for the central nervous system (resulting in temporary or chronic loss of equilibrium, memory or motor coordination), and cancerogenic effects.

3.2.2 Nitrogen Oxides and Ammonia

Nitrogen oxides (NO_X) are a family of stable molecular species appearing in gaseous form in standard atmospheric conditions. They are the most significant undesired aeroengine emission by mass and are produced as a result of the chemical reaction of nitrogen and oxygen, both widely available in ambient air entering the engine. The main chemical reactions involved are the following:

$$N_2 + O_2 \rightarrow 2\,NO$$

$$2\,NO + O_2 \rightarrow 2\,NO_2$$

To spontaneously occur, these reactions require rather high temperatures and pressures, causing O_2 radicalisation (cf. 'Thermal NO_X formation'). These conditions are increasingly common in contemporary internal combustion engines both in aviation and other sectors, since increases of pressure ratios and combustion temperatures are associated with higher thermal and combustion efficiency.

When emitted at high altitudes where high concentrations of ozone (O_3) are available, some NO_X species react chemically with the ambient air and cause the depletion of the ozone layer. For instance, ozone is destroyed by reaction with NO as follows:

$$O_3 + NO \rightarrow NO_2 + O_2$$

As well documented, the ozone layer protects the earth from harmful ultraviolet (UV) radiation from the sun, and thus the emission of NO_X at high altitude results in a change in the radiative forcing due to an increased amount of incoming solar UV radiation reaching the lower atmosphere and the planet surface. In addition to the local concentration of oxygen and ozone, this reaction depends on the intensity of solar radiation at certain wavelengths, thus constituting a photostationary equilibrium. When NO_X is emitted at low atmospheric altitudes where O_3 is scarce, the balance of the above chemical reaction favours the creation of ozone, which becomes a greenhouse gas in the troposphere. In addition to these adverse climate impacts, NO_X pose health risks to living beings when inhaled or ingested.

The thermal NO_X formation process is maximised in proximity of the stoichiometric combustion, which an aeroengine is designed to achieve at/near maximum throttle.

Ammonia, conversely, is not a common exhaust by aeroengines burning conventional fuels but can nonetheless occur occasionally as a result of high temperatures followed by rapid cooling. The amount of ammonia is also expected to increase with the switch to more hydrogen-rich fuels and, clearly, become a common exhaust in ammonia-based fuels. Therefore, its emission will require attention as high local concentrations can be caustic to most living beings.

3.2.3 Particulate Matter

A variety of phenomena occurring in association with the combustion process in real atmospheric conditions can result in liquid droplets and solid particles being released by internal combustion engines as part of their exhausts.

Due to their very large variety, these particles are classified by size, with the smaller ones (10^{-9} to 10^{-6} m) being classified as aerosols, while larger ones (10^{-6} to 10^{-2} m) are categorized as dust. Aerosols do not encounter significant barriers when inhaled, so they can penetrate very deeply in human and animal respiratory systems, posing significant danger over time. Conversely, larger particulates such as droplets, sand, and ash would be naturally trapped but are still an important concern for various reasons, the most important being their capability to stimulate agglomeration by other substances including acidic/corrosive, toxic, and caustic species, which are more easily transported to the lower atmosphere and the planet surface.

3.2.4 Sulphur Oxides

Sulphur oxides are present as combustion products wherever sulphur is present in the fuel supplied to the engine. Fuels of fossil origin commonly contain sulphur unless it is removed during the refining process, so nowadays sulphur oxides are still a common aviation emission globally. Among the various members of this family, SO_2 is the most chemically stable so less environmentally hazardous, while other members such as SO_3, S_2O_2, and SO are less stable and usually more alarming. All of them can undergo a reaction with atmospheric oxygen and water vapour resulting in the formation of sulphuric acid (H_2SO_4) and other acids, which are the main causes of acid fog and rain.

3.3 Noise Pollution

The noise produced by aircraft during their operation represents an ecological, economic, and social problem, which significantly affects those communities in the proximity of airports and densely populated areas [8, 9]. The three main sources of aircraft noise are propulsive, aerodynamic, and other mechanical, thermochemical, and fluid dynamic processes. The nature and magnitude of noise generated by these sources depend significantly upon the aerostructural and propulsive configuration of the aircraft. Generally, the atmosphere acts as a low pass filter for the noise propagation spectrum, due to thermo-fluid dynamics and molecular absorption processes. The noise perceived by the human listener was found to depend on a number of factors, including the individual listener's cultural and socio-economic background as well as their psychological and physical wellbeing. The effects on affected humans vary from no effect or minor annoyance, to permanent physical and mental health effects [10]. Generally, noise generated from departing aircraft is greater than that of arriving aircraft due to the high throttle setting, but the minimum aircraft and receiver distance is much lower during approach in a conventional glide slope of 3° [11]. On departure, the noise level experienced on the ground from a particular aircraft is influenced by the aircraft type, size, standard instrument departure (SID) used, aircraft settings, climb rate, and the meteorological conditions. The relationship between the acoustic characteristics of the primary and secondary aircraft noise source and the flight mode parameters must be established for evaluating the noise levels. Planning, evaluation, and development (design and redesign) of airports are dependent on the noise contours determined around the airport. The different aircraft types, flight procedures, and propulsive systems contribute to the intricacy of the aircraft noise contour assessment process. A large dataset is generally used for evaluation purposes and, as a result, the complexity of the noise model to be adopted also increases. The measurement of aircraft noise at airports may involve several metrics including A-weighted sound exposure level (SEL/LAE) and effective perceived noise level (EPNL/LEPN) ([12]).

There is a significant amount of literature focusing on the noise of manned aviation, typically passenger aircraft, due to its huge impact not only on the people near airports who hear these aircraft during take-off and landing, but also on the passengers who are exposed to the aircraft sound for several hours at a time. In this regard, there is plenty of work detailing noise reduction strategies for jet and turbofan engine aircraft and airframe noise [13, 14]. Since the 1970s, the International Civil Aviation Organization (ICAO) has introduced various noise standards to reduce aircraft noise at source within its Balanced Approach to Aircraft Noise Management. As part of this initiative, noise standards have been introduced for light as well as large propeller aircraft, jet and supersonic aircraft, as well as helicopters and tiltrotors.

However, standards for multirotor UA as well as passenger ferrying 'flying taxis' are yet to be developed. Traditionally, propellers are characterized by axial flow (as in the case of a fixed-wing aircraft), whereas rotors most often operate in sideslip conditions (as in the case of a helicopter). However, most UA have multiple rotors aligned along the vertical axis, with the added provision to fly as a fixed wing aircraft provided in some. Unmanned aircraft

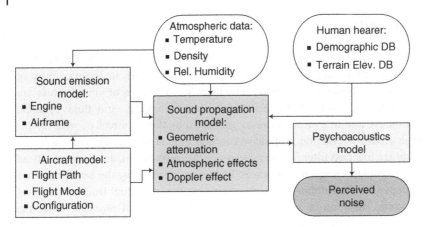

Figure 3.1 Noise modelling flow chart. From [7].

have proven their worth for tasks labelled as 'dull, dirty, or dangerous' for manned aircraft. Comparisons between a small single-engine, manned aircraft and a small UA show that the UA, on average, has larger endurance, lower fuel consumption and emissions, and has noise levels 6–9 dB lower [15]. However, with the increasing number of UA anticipated to fly at lower altitudes (<40 m AGL) in dense urban communities [16–18], the aggregate noise pollution for extended time periods poses grave health and wellbeing risks (2016).

The noise problem for UAM encompasses four fundamental aspects: the sound source itself, the UA, the flight path it takes, the atmospheric propagation, and the receiver – the human listener. As shown in Figure 3.1, there are more than just these elements and their relations can be complex. In order to determine a flight path which minimizes noise exposure to humans, multiple aspects must be considered. Firstly, there is the physics-based sound propagation model that can be applied to the acoustic output from the UA and which considers the path it takes through the atmosphere. Secondly, this sound is heard by a listener, who effectively filters the sound relative to other noise they hear at the time, as well as by their own perception and response to the noise (psychoacoustics). This human response can supplement the propagation model to predict the equivalent sound exposure on the ground as effectively perceived by humans. Finally, these models and their numerical results can be used to calculate a more optimal flight path which would cause the smallest acoustic disturbance. All aspects of this problem must be considered for effective noise mitigation. The focus of this work is on the third part of this problem, the sound propagation model, only considering the input of the sound source and flight path as a first step.

Current noise models for aircraft have evolved over the years, with acoustic databases developed, which feed into the noise engine algorithms. The modelling algorithms can be divided into three different types, namely the closest point of approach (CPA), segmentation, and simulation. The stages of model development can be categorized as shown in Figure 3.2, with advanced models being currently employed. The aircraft database consists of acoustic and performance data based on noise-power-distance (NPD) relationships, which are fed into the noise engine consisting of physics-based sound emission and propagation models. The noise engine is also fed data from the air traffic, radar, airport, and other operational data. All the models and data are processed to generate noise contours.

Figure 3.2 Noise model development.
Source: adapted from [82].

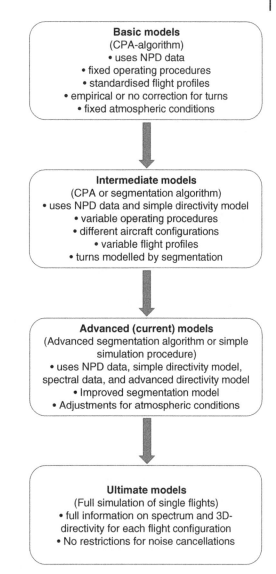

Noise standards are developed based on the noise limits set as a function of maximum take-off mass (MTOM) for larger aircraft. The noise evaluation measure for certification of aircraft is the sound exposure level (SEL) with 90% confidence limits [19]. The reference points for the SEL are take-off, overflight, and approach phases of the flight path. The measured noise levels are adjusted to the reference profile in order to calculate the effective perceived noise level (EPNL), which is the time integral of perceived noise level (PNL) over the noise event duration. Atmospheric conditions, both steady and unsteady, need to be considered for increased accuracy of noise models of low-flying aircraft. Additionally, physics-based noise models also need to account for platform dynamics and sound reflecting surfaces, such as buildings and ground surfaces. Several of these factors are overlooked or they are assumed to follow rather simplistic idealized models. The surrogate sound propagation models in this paper attempt to tackle some of the variables which are generally either neglected or assumed to be constant.

The aviation research and development community has begun to develop noise emission standards for UA. The basic noise emission standard EN ISO 3744:2010 is supplemented with the installation and mounting conditions for a UA (2021). The test operating conditions are defined where the noise tests are carried out for the hovering manoeuvre of the unmanned aircraft under MTOM. The A-weighted surface time-averaged SPL is determined multiple times until the difference of at least two of the measured values does not differ by more than 1 dB.

3.4 Sound Propagation

Fundamentally, sound can be defined as mechanical energy transmitted by pressure waves that require a material medium to propagate. These mechanical waves, are termed *acoustic waves* and microscopically involve rapid *to* and *from* displacements or vibrations of molecules in the medium. The acoustic energy is therefore transmitted from the source to its surroundings thanks to the propagation of these waves. Sound propagates as energy is transferred from one vibrating particle to the next as progressive waves, with particle displacements taking place in the same direction as the movement of the wave. The velocity of sound in a medium (c_m) varies with the bulk modulus (B) and density (ρ) of the medium as shown in Eq. (3.1). Sound travels faster in a medium with high bulk modulus or stiffness, like solids, as compared to a medium with lower bulk modulus, like fluids.

$$c_m = \sqrt{\frac{B}{\rho}} \tag{3.1}$$

The region of travel of acoustic waves is referred to as the *sound field*, while the space where the acoustic waves propagate freely without reflection, termed *free progressive waves*, is called a *free field*. The *to* and *from* displacement of the air particles about their mean position produces a local compression followed by a local rarefaction and so on. The instantaneous value of the fluctuating pressure disturbance on the ambient pressure is referred to as the *sound pressure* and is denoted by p'. It requires a time period T to complete one cycle of compression and rarefaction, where *sound pressure* at any given time t is therefore approximately equal to:

$$p'(t) \approx p'(t + T) \tag{3.2}$$

Humans can normally hear frequencies in the range of 20 Hz to 20 kHz. Frequencies below 20 Hz are referred to as infrasound, while frequencies greater than 20 kHz are called ultrasound.

The sound pressure motion is a simple harmonic motion, hence can be described in the form of a sinusoid. Sound pressure is a function of both time and distance from the source and can be described as:

$$p' = f(r, t) \tag{3.3}$$

where r is the distance from the centre of the acoustic source. For a point source, the variation of the sound pressure with time and distance can be expressed as:

$$p' = \frac{L_p}{r} \cos \omega \left(t - \frac{r}{c} \right)$$

(3.4)

where L_p is the intensity of the sound at source and ω is the angular frequency in radians per second. As the sound travels with speed c, the time taken by sound to travel a distance of r is given by r/c. The ration L_p/r is the amplitude of the local sound pressure. Since the magnitude of the sound pressure varies continuously with time, *effective sound pressure* (p_e) magnitude is usually given by the root mean square value of the instantaneous sound pressures over one period, as per the following equation [20]:

$$p_e = \left[\frac{1}{T} \int_0^T [p'(t)]^2 dt \right]^{1/2}$$

(3.5)

Merging Eqs. (3.4) and (3.5), p_e can be written as:

$$p_e = \left[\frac{1}{T} \int_0^T \left[\frac{A}{r} \cos \omega (t - r/c) \right]^2 dt \right]^{1/2} = \frac{L_p}{r\sqrt{2}}$$

(3.6)

Hence, the ratio of the effective sound pressure to the local sound pressure is given by the factor $\sqrt{2}$ and is known as the crest factor of a sound signal. The attenuation rate of sound waves varies with frequency, with higher frequencies attenuating at a faster rate. Attenuation can occur either due to reflection/scattering at interfaces or absorption [21]. However, higher frequencies, having short wavelengths, reflect strongly from small objects. Reflection from surfaces causes interference with the incident sound wave, which could be constructive or destructive. Interference depends upon the frequency of sound as well as the difference between the path length of direct and reflection paths [22–24]. Furthermore, the speed of sound in air varies with temperature, pressure, humidity, and wind, thereby affecting the propagation of sound. The generic equation for sound propagation can be given by:

$$L_p(r) = L_w + \Sigma_i A_i$$

(3.7)

where:

$L_p(r)$: The sound pressure level at distance r from the source [dB];
L_w: The sound power level of the source [dB];
A_i: The combination of modifying factors that either attenuate or enhance the transmission of the sound energy as it propagates from source to receiver.

Sound can be heard around corners and behind walls as well, due to bending of sound waves around obstacles thanks to the diffraction phenomenon. Acoustic sources have both far-field and near-field regions. Wavefronts produced by the sound source in near-field are not parallel and the intensity of the wave oscillates with the range and angle between sources. The sound pressure varies with distance as a complex function of the radiation characteristics of the source. However, in the far-field, wavefronts are nearly parallel, with intensity varying only with range to a centroid between sound sources, in accordance with the inverse squared rule. The near-field distance r_{nf} is given by:

$$r_{nf} = \frac{D^2}{\lambda}$$

(3.8)

where D is the equivalent aperture of the transmitter given by:

$$D = \frac{3.2}{k \sin\left(\frac{\theta_{3dB}}{2}\right)} \tag{3.9}$$

where k is the wave number and θ_{3dB} is the half power beam angle. The wavefront for a sound source radiating equally in all directions is a sphere of radius r, whose intensity I from the source of power W is given by:

$$I = \frac{W}{4\pi r^2} \tag{3.10}$$

As the intensity of sound that is audible to humans varies from 10^{-12} to $100\,\text{watt/m}^2$, a logarithmic scale is used for it. The *sound intensity level* (SIL), expressed in dB, corresponding to a sound intensity I is defined by:

$$SIL = 10\log\frac{I}{I_0} \tag{3.11}$$

where I_0 is the reference sound intensity of $10^{-12}\,\text{watt/m}^2$, representing the minimum intensity perceptible by human ear. Similarly, *sound pressure level* (SPL) in decibels is defined as follows:

$$SPL = 10\log\frac{p_e^2}{p_{e_0}^2} \tag{3.12}$$

where p_{e_0} is the reference pressure of $2\times10^{-5}\,\text{N/m}^2$. Analogous to Ohm's law, in free field, the effective pressure and the sound intensity in the direction of propagation are related as follows:

$$I = \frac{p_e^2}{\rho c} \tag{3.13}$$

where ρ is the density of air. Combining Eqs. (3.11)–(3.13), SIL can be shown to be related to SPL as follows:

$$SPL = SIL + 10\log\frac{I_0}{p_{e_0}^2}\rho c \tag{3.14}$$

$$SPL = SIL + 0.2 \tag{3.15}$$

Even when the prevailing air pressure and temperature conditions significantly deviate from sea-level standard values, the difference between the SPL and SIL is less than 1 dB. Hence, for all practical purposes, the SPL is numerically equal to SIL in a free sound field.

In case of two or more independent sound sources, their decibel values cannot be added algebraically, as they are logarithmic quantities. Consider two independent pure tone sound sources emitting frequencies ω_1 and ω_2 with amplitudes A/r_1 and B/r_2. The total sound pressure $p'_m(t)$ at any given point at time t is given by:

$$p'_m(t) = p'_1(t) + p'_2(t) \tag{3.16}$$

where

$$p'_1(t) = \frac{A}{r_1}\cos\omega_1(t_1 - r_1/c) \tag{3.17}$$

$$p'_2(t) = \frac{B}{r_2}\cos\omega_2(t_2 - r_2/c) \tag{3.18}$$

The resultant effective sound pressure is given by:

$$p_{e_m}^2 = \frac{1}{T}\int_0^T \left[\frac{A}{r_1}\cos\omega_1(t-r_1/c)\right]^2 dt + \frac{1}{T}\int_0^T \left[\frac{B}{r_2}\cos\omega_2(t-r_2/c)\right]^2 dt$$

$$+ \frac{1}{T}\int_0^T \left[\frac{2AB}{r_1 r_2}\cos\omega_1(t-r_1/c)\cos\omega_2(t-r_2/c)\right] dt \tag{3.19}$$

Equation (3.19) can be simplified as:

$$p_{e_m}^2 = \left[\frac{A}{r_1\sqrt{2}}\right]^2 + \left[\frac{B}{r_2\sqrt{2}}\right]^2 \tag{3.20}$$

or,

$$p_{e_m}^2 = p_{e_1}^2 + p_{e_2}^2 \tag{3.21}$$

The resultant SPL for two independent sound sources becomes:

$$SPL_m = 10\log\left[\frac{p_{e_1}^2 + p_{e_2}^2}{p_{e_0}^2}\right] \tag{3.22}$$

Equation (3.22) can also be written in terms of individual sound pressure levels as:

$$SPL_m = 10\log\left[10^{SPL_1/10} + 10^{SPL_2/10}\right] \tag{3.23}$$

Equation (3.23) can be generalised for N sound sources as:

$$SPL_m = 10\log\sum_{i=1}^{N}10^{SPL_i/10} \tag{3.24}$$

For two independent sound sources emitting the same frequency, the resultant SPL depends on the phase difference between the two sound waves.

$$SPL_m = 10\log\left[\frac{p_{e_1}^2 + p_{e_2}^2 + 2p_{e_1}p_{e_2}\cos\frac{\omega\Delta r}{c}}{p_{e_0}^2}\right] \tag{3.25}$$

where $\Delta r = r_2 - r_1$. If the two sound sources are in phase, $\cos\omega\Delta r/c = 1$, giving:

$$SPL_m = 10\log\left[\frac{(p_{e_1} + p_{e_2})^2}{p_{e_0}^2}\right] \tag{3.26}$$

If the two sounds are out of phase, $\cos\omega\Delta r/c = -1$, thus giving:

$$SPL_m = 10\log\left[\frac{(p_{e_1} - p_{e_2})^2}{p_{e_0}^2}\right] \tag{3.27}$$

Assuming a point source of sound in an unbounded homogenous atmosphere, the propagation of sound is affected by just two attenuating effects. While the first attenuation effect is geometric, which is solely dependent on the distance from the sound source, the second attenuating effect is the atmospheric absorption. Sound propagates due to the oscillation of air molecules about their mean position; with a higher frequency of sound leading to a higher rate of oscillation. This vibration of the air molecules leads to loss of energy through two dissipative mechanisms. While one of the mechanisms comprises frictional losses, which include both viscous action and heat conduction, the other mechanism involves the interaction of water vapour with the resonance of oxygen and nitrogen molecules. Hence, there are heat conduction, shear viscosity, and molecular relaxation losses [25].

3.4.1 Sound Attenuation in the Atmosphere

In practical situations, the propagation of sound in the atmosphere is affected by additional factors, such as ground effects, attenuation due to finite barriers and buildings, reflections, wind and temperature gradient effects, and atmospheric turbulence. The most notable atmospheric sound attenuation factors are discussed in detail in the following subsections 3.4.1.1–3.4.1.5.

3.4.1.1 Geometrical Divergence
Geometrical divergence refers to the spherical spreading in the free field from a point sound source. The attenuating effect is the geometric attenuation, which results from the spreading of the radiated sound energy over a sphere of increasing diameter as the wavefront propagates away from the source. Equation (3.28) shows the relationship between the sound power level of the source, L_w and sound pressure level, $L_p(r)$, at a distance r from that source. The variation of sound pressure level with distance from the source is shown in Eq. (3.7). Unlike atmospheric attenuation, geometric attenuation is independent of the frequency of the propagating sound wave. From Eq. (3.29), it can be inferred that sound intensity or sound pressure level, L_p decreases by 6 dB per doubling of distance away from the source.

$$L_p(r) = L_w + 10\log\left[\frac{1}{4\pi r^2}\right] \tag{3.28}$$

$$I = \frac{W}{4\pi r^2} \tag{3.29}$$

The geometrical divergence, in dB, is given by:

$$A_{div} = \left[20\log\left(\frac{d}{d_0}\right) + 11\right] \tag{3.30}$$

where d is the distance from the sound source to receiver (m) and d_0 is the reference distance which is 1 m from an omnidirectional point sound source.

3.4.1.2 Atmospheric Absorption
Air absorption becomes significant at higher frequencies and at long ranges, thereby acting as a low-pass filter at long ranges. The pressure of a planar sound wave at a distance x from a point of pressure P_0 is given by:

$$P = P_0 e^{\frac{-\alpha x}{2}} \tag{3.31}$$

The attenuation coefficient, α, for air absorption depends on frequency, humidity, temperature and pressure, with its value being calculated using Eqs. (3.32) to (3.12) [26].

$$\alpha = f^2\left[\left[\frac{1.84 \times 10^{-11}}{\left(\frac{T_0}{T}\right)^{1/2}\frac{P_s}{P_0}}\right] + \left(\frac{T_0}{T}\right)^{2.5} \times \left(\frac{0.10680e^{-3352/T}f_{r,N}}{f^2+f_{r,N}^2} + \frac{0.01278e^{-2239.1/T}f_{r,O}}{f^2+f_{r,O}^2}\right)\right] \tag{3.32}$$

where f is the frequency, T is the absolute temperature of the atmosphere in kelvins, T_0 is the reference value of T (293.15 K) and $f_{r,N}$ and $f_{r,O}$ are relaxation frequencies associated with the vibration of nitrogen and oxygen molecules, respectively and are given by:

$$f_{r,N} = \frac{p_s}{p_0}\left(\frac{T_0}{T}\right)^{1/2}\left(9 + 280He^{-4.17\left[(T_0/T)^{\frac{1}{3}}-1\right]}\right) \tag{3.33}$$

$$f_{r,O} = \frac{p_s}{p_0}\left(24.0 + 4.04 \times 10^4 H \frac{0.02 + H}{0.391 + H}\right) \tag{3.34}$$

where p_s is local atmospheric pressure, p_0 is the reference atmospheric pressure (101 325 Pa) and H is the percentage molar concentration of water vapour in the atmosphere which is given by:

$$H = \frac{\rho_{sat} r_h p_0}{p_s} \tag{3.35}$$

where r_h is the relative humidity and ρ_{sat} is given by:

$$\rho_{sat} = 10^{C_{sat}} \tag{3.36}$$

where C_{sat} is given by:

$$C_{sat} = -6.8346\left(\frac{T_0}{T}\right)^{1.261} + 4.6151 \tag{3.37}$$

Similarly, ρ_{sat} can also be written as [27, 28]:

$$\rho_{sat} = 1322.8\left(\frac{r_h}{T}\right)^{\left[\frac{25.22(T-273.15)}{T} - 5.31\ln\left(\frac{T}{273.15}\right)\right]} \tag{3.38}$$

Figure 3.3 shows the variation of absorption coefficient α with the frequency of sound at 29.15 K, one atmospheric pressure, 20% relative humidity, and H being 4.7×10^{-3}. As there are two relaxation frequencies associated with oxygen and nitrogen, the frequency dependence of the attenuation coefficient for sound in the air has three distinct regions. At very low frequencies, where the sound frequency is much lower than that associated with nitrogen molecules, the attenuation is dominated by vibrational relaxation of nitrogen molecules (α_1). The frequency dependence is quadratic with an apparent bulk viscosity associated with the nitrogen relaxation. In the intermediate region, the frequency is substantially larger than that associated with nitrogen relaxation, but still substantially less than that associated with oxygen relaxation, with quadratic frequency dependence, smaller coefficient, and apparent bulk viscosity that is associated with oxygen relaxation (α_2). In the higher frequency region, there is quadratic dependence again, although with an even smaller coefficient and with the intrinsic bulk viscosity associated with molecular rotation (α_3) [26].

Having calculated the absorption coefficient for a given temperature, pressure, relative humidity, and percentage molar concentration of water vapor in the atmosphere, the attenuation of sound due to atmospheric absorption, during propagation through a distance d(m) is given by:

$$A_{atm} = \alpha d/1000 \tag{3.39}$$

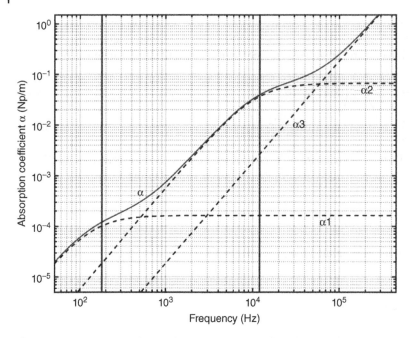

Figure 3.3 Attenuation of sound in air as a function of frequency (log-log plot). From [7].

3.4.1.3 Ground Effect

Adding the effect of a bounding ground plane to the sound propagation model allows for sound to propagate directly from source to receiver as well as through a secondary propagation path resulting from a reflection off the ground plane. This secondary propagation path can result in interference effects between the direct and reflected waves at the receiver. The interference effect can be constructive or destructive, depending on the relative amplitudes and phase of the direct and reflected waves. The relationship between the direct and reflected waves depends on a variety of factors including the difference between the direct and reflected path lengths, which is a function of source and receiver separation distance (d_p) as well as their height above the ground (h_s and h_r), the wavelength of sound and reflective properties of the ground which can cause variations in the phase and amplitude of the reflected sound wave.

Based on the acoustic properties of ground surfaces, they are classified into three types depending on the ground factor (G). Hard ground, which has a G of 0, includes concrete, paving, water, ice, and all other low porosity ground surfaces. On the other hand, porous ground, which has a G of 1, consists of grass, trees, foliage, and all other ground surfaces which are suitable for the growth of vegetation. Surfaces which are a combination of both hard and porous grounds, i.e. having a value of G ranging from 0 to 1, where G represents the fraction of the ground surface that is porous, are known as mixed ground. The total ground attenuation is obtained by summing up the attenuation A_s for the source region specified by the ground factor G_s, A_m for the middle region specified by the ground factor G_m and A_r for the receiver region specified by the ground factor G_r, as shown in Eq. (3.40).

$$A_{gr} = A_s + A_m + A_r \tag{3.40}$$

3.4.1.4 Screening

An object shall be considered as a screening obstacle if it meets the following requirements:

- The object has a surface density of at least $10 \, \text{kg/m}^2$;
- The surface of the object is closed without cracks or gaps.

The horizontal dimension of the object normal to the source-receiver line $(l_i + l_r)$ is larger than the acoustic wavelength λ at the nominal midband frequency for the octave band of interest, i.e. $l_i + l_r > \lambda$. The barrier diffraction could be either single diffraction in case of thin barriers or double diffraction in thick barriers. In case of more than two barriers, the barrier attenuation can be approximated to be a case of double diffraction, by choosing the two most effective barriers, neglecting the effect of the others. The effect diffraction (in dB) for downward sound propagation over the top edge and the vertical edge, respectively, is given by Eqs. (3.41) and (3.42) [29]. D_z is the barrier attenuation for each octave band and A_{gr} is the ground attenuation in the absence of the barrier, as described in Section 3.4.1.3.

$$A_{bar} = D_Z - A_{gr} > 0 \tag{3.41}$$

$$A_{bar} = D_Z > 0 \tag{3.42}$$

3.4.1.5 Wind and Temperature Gradient Effects

The bending of sound also occurs due to temperature and wind gradients in the atmosphere. Because of uneven heating of the Earth's surface, the atmosphere is constantly in motion. The turbulent flow of air across the rough solid surface of the Earth generates a boundary layer. The lower part of the meteorological boundary layer, called the surface layer, extends over 50–100 m in typical daytime conditions [26]. Turbulent fluxes vary by less than 10% of their magnitude in the surface layer, but the wind speed and temperature gradients are the largest. Turbulence can be modelled as a series of moving eddies with a distribution of sizes. Various turbulence models like Gaussian, Von Kármán, and Kolmogorov are used in atmospheric acoustics [30–32]. It has been shown that turbulence effects decrease with an increase in elevation of sound sources from the ground [33].

As the temperature decreases with height, i.e. a negative temperature gradient, in the absence of wind, this causes sound waves to bend, or refract, upwards due to the top of the wavefront travelling slower than the bottom. Such conditions usually exist during daytime on sunny days, leading to formation of shadow zones.

Wind velocity either adds or subtracts from the velocity of sound, depending upon whether the source is upwind or downwind of the receiver, height above ground, and temperature inversions. Sound is deflected towards regions of lower velocity, so with a typical wind profile of increasing wind speed with height, there is an upward bending of the sound with a shadow region upwind and a bending towards the ground in a downward direction. Wind effects tend to dominate over temperature effects when both are present. In the presence of both temperature and velocity gradients, the bending of sound in the direction of the wind is counteracted by a negative temperature gradient, while the shadow formation intensifies in the upwind direction. The general relationship between the speed of sound profile c(z), the temperature profile T(z) and wind speed profile u(z) in the direction of sound propagation, for a height z, is given by [4]:

$$c(z) = c(0) \sqrt{\frac{T(z) + 273.15}{273.15}} + u(z) \tag{3.43}$$

3.4.1.6 Other Sound Attenuation Factors

Various standardised techniques for measuring the attenuation of sound outdoors due to atmospheric absorption effects have been developed in ISO 9613-1 and ISO 9613-2 [29, 34]. This includes attenuation of sound due to miscellaneous effects like foliage, housing, or industrial sites. The analytical models presented rely on the values of various atmospheric parameters like temperature, pressure, relative humidity, wind speed, and time of the day. In addition, fog and precipitation can also affect the attenuation of sound [35]. Experiments show that precipitation affects the temperature variation, hence indirectly affecting sound attenuation outdoors [33]. Thus, to summarize, the attenuation of sound in the atmosphere can be given by:

$$A = A_{div} + A_{atm} + A_{gr} + A_{bar} + A_{misc} \tag{3.44}$$

where A_{misc} is the sound attenuation due to other miscellaneous effects like wind and temperature gradient effects, precipitation, foliage, and housing or industrial sites. Additionally, the speed of sound in air also varies with temperature, relative humidity (h), carbon dioxide content (h_c), and barometric pressure. A general equation for the same is given by [36, 37]:

$$\frac{c}{c_0} = a_0 + a_1 t + a_2 t^2 + a_3 h_c + a_4 h_c t + a_5 h_c t^2 + a_6 h + a_7 h t + a_8 h t^2$$
$$+ a_9 h t^3 + a_{10} h_c^2 + a_{11} h^2 + a_{12} h t h_c \tag{3.45}$$

where c and c_0 are the sound speed and the reference dry-air sound speed, respectively and a_0–a_{12} are coefficient constants given in Table 3.1. Sound speed can be deduced by multiplying Eq. (3.2) by the corresponding reference dry-air sound speed c_0. For a real gas at standard pressure (101.325 kPa), the dry-air sound speed can be approximated to be 331.29 m/s, with an uncertainty of approximately 200 ppm [38], which encompasses sound speeds from 331.224 to 331.356 m/s [39].

Table 3.1 Ranging parameters. Adapted from [37].

Coefficient constant	Value
a_0	1.000100
a_1	1.8286×10^{-3}
a_2	-1.6925×10^{-6}
a_3	-3.1066×10^{-3}
a_4	-7.9762×10^{-6}
a_5	3.4000×10^{-9}
a_6	8.9180×10^{-4}
a_7	7.7893×10^{-5}
a_8	1.3795×10^{-6}
a_9	9.5330×10^{-8}
a_{10}	1.2990×10^{-5}
a_{11}	4.8016×10^{-5}
a_{12}	-1.4660×10^{-6}

3.5 Noise Management for Traditional Aircraft

Aircraft noise guidelines have been developed by national or international aviation organisations including the ICAO guidance in its Circular 205, the Society of Automotive Engineers (SAEs) Committee A-21 and the European Union recommended use of the European Civil Aviation Conference – Conférence Européenne de l'Aviation Civile (ECAC-CEAC) Doc. 29 [40–42]. In order to model aircraft noise generation and prop-agation, a simplified model assuming progressive sound wave travelling in a plane is considered. The rate of transfer of energy per unit cross-sectional area for the sound wave is defined as sound intensity (SI) and the total intensity produced by several sources can be assumed to follow the superposition of effect principle and therefore is given by:

$$SI = S\,I_1 + SI_2 + SI_3 + \dots \tag{3.46}$$

The generation and propagation of aircraft noise through the atmosphere is paramount for trajectory optimisation. The optimiser needs to include a mathematical model of aircraft noise to allow the minimisation of perceived noise on the ground. In order to generate an optimal trajectory, the integrated noise model (INM) from the Federal Aviation Administration (FAA) is generally used as a reference model with respect to noise abatement aspects. Noise is calculated based on interpolation of data specified in the noise-power-distance (NPD) table containing empirical measurements for each aircraft type under reference conditions. The INM model uses a grid-based approach and a number of metrics, including exposure-based, maximum noise level, and time-based, are adopted [43]. In order to criti-cally evaluate the noise levels, the Aviation Environmental Design Tool (AEDT) developed by the FAA, serves as a multipurpose framework integrating the INM and the Model for Assessing Global Exposure to the Noise of Transport Aircraft (MAGENTA), a global noise model [44]. The overall sound level L is given by:

$$L = 10\,log_{10}\left(\sum_{i=1}^{n} 10^{L_i}\right) \tag{3.47}$$

where L is the sum of n noise levels L_i all at the same frequency. The sensitivity due to perturbation on the sound level dL_i on the sound levels L_i of n contributes is given by [11]:

$$L + dL = 10\,log_{10}\left(\sum_{i=1}^{n} 10^{(L_i+dL_i)/10}\right) \tag{3.48}$$

The concept of noise-radius is adopted generally by considering the complexity of aircraft noise modelling based on aircraft engine type, thrust setting, and atmospheric conditions [4]. The optimisation of aircraft vertical trajectory with minimum noise impact using the analytical jet noise model has proved to be effective [45]. The methodology adopted requires demographic data at each observer location and hence a demographic distribution database (D^3) is used in conjunction with the noise model. The D^3 model aids in estimating the population in a user-defined grid on a global or local scale. Additionally, the Digital Terrain Elevation Database (DTED) is part of the trajectory optimisation framework taking into account geographic information. The availability of D^3 and DTED are specif-ically important for assessing the environmental impact of aircraft noise in the terminal manoeuvring area (TMA). Four-dimensional (4D) trajectory optimisation is generally performed to avoid the densely populated areas in and around the airports, taking into account the topographical conditions, metrological data, and trajectory constraints. The constraints on the trajectory can be ATM operational, airspace, airline, flight parameters, and/or aircraft dynamics-based constraints. Several studies have been carried out for optimising the aircraft trajectory based on several cost functions, such as the number of sleep disturbances resulting in reduction of noise annoyance on specific regions around an airport [22–24, 46–50]. Reduced noise procedures implemented are noise abatement departure procedures (NADPs), including NADP1, NADP2 and its associated variations, as well as ICAO-A procedure [11]. Steep trajectory approach, spiral trajectories, and touch-down displacement principles have been proposed and trailed to reduce noise while landing. Generally, the optimisation or reduced noise is not harmonious with the cost function for minimising other environmental emissions. Reduction in noise by increasing time results in higher fuel consumption and, as a consequence, higher emissions [51, 52].

The impact of perceived noise on the population is commonly described by dose-response relationships. The awakening index (AI) defined by the percentage of persons woken at a specific location as a function of perceived indoor SEL is given by:

$$AI = 0.0087.(SEL_{indoor} - 30 \text{ dB})^{1.79} \tag{3.49}$$

The indoor SEL is estimated from the outdoor SEL by subtracting 20.5 dB (average loss due to the sound insulation of a typical house).

The number of awakenings for a specific noise station is calculated by multiplying the percentage of awakenings of the station i with the population P_i assigned to that receiver.

$$n_{awakenings} = \sum_{i=1}^{n} \%awakenings_i.P_i \tag{3.50}$$

The European Directive 2002/30/EC defines the concept of noise management as a balanced approach wherein international aviation organisations, governments, and aircraft manufacturers, as well as airliner and airport operators focus on the following key areas [53–55]:

- Reduction of noise at the source through aircraft technology improvements.
- Compatible land-use planning.
- Increased adoption of noise abatement procedures.
- Optimisation of the aircraft engine setting and trajectories to avoid the noise sensitive areas (NSAs).
- Introduction of operating restrictions.

A typical aircraft noise mitigation scenario is considered in Figure 3.1 in which probable 4D intents are obtained such that densely populated areas are identified with D^3 and DTED.

The ICAO Committee on Aviation Environmental Protection (CAEP) has agreed to set new standards for noise reduction worldwide (ICAO 2013). According to the FAA, the most beneficial area of future noise reduction is developing technology to reduce the source of the noise. It has an ongoing programme called the Continuous Lower Energy, Emission and Noise (CLEEN) programme. CLEEN is implemented to advance the development of technologies to further reduce noise from aircraft (FAA 2014). With CLEEN, mature, environmentally friendly technologies are being developed. These technologies support the achievement of next generation goals to reduce aircraft noise and the impact of noise in and around airports. The major aim of the CLEEN programme is to help and develop certifiable aircraft technology that reduces noise levels by 32 dB cumulative, relative to the ICAO noise standards (FAA 2014).

3.6 Noise Management for Drones and Advanced Air Mobility

This section delves into addressing the gap of a lack of noise emission standards for unmanned aircraft (UA). The aircraft noise at source is discussed for conventional aircraft and the lack of sufficient data to support noise certification for UAM operations is also highlighted. The human noise perception, along with the psychoacoustic effects, is also briefly discussed. The section also discusses the flight path dependencies on the perceived noise along with trajectory optimization for reduced noise impacts. Some recent advances in noise certification standards for UA and their integration in the flight information management systems are discussed before concluding with a proposed future pathway for UA noise measurement and certification standards, while also discussing the reduction of aircraft noise at source.

3.6.1 Emissions at Source and Certification Considerations

The noise certification procedures for conventional aircraft, including rotorcraft, are not applicable to UA and other non-conventional aircraft. Currently, there is insufficient data to support noise certification for UAM operations [56]. So, although the noise levels may be lower than those produced by conventional aircraft, UAM operations can be characterized by their novel operating mode, frequency, and other time-varying factors, which are distinctly different from conventional air traffic, and thus can be even more annoying and may get compounded by fear and privacy concerns.

Noise reduction technologies are being developed along with noise metrics and low-noise operational procedures by regulatory bodies in collaboration with the industry and academia [57]. As most of the proposed advanced air mobility vehicles and drones will have some form of vertical lift and forward flight capability powered by distributed electric or hybrid propulsion systems, the noise characteristics of these aircraft are expected to be different from existing helicopters and conventional general aviation aircraft. Additionally, as UA operate out of vertiports located in urban areas, their operations are in closer proximity to the noise receptors which can cause serious noise concerns. These noise concerns can seriously affect the community's acceptance of large-scale UA operations, which can ultimately determine the success or failure of this emerging industry.

Researchers are studying the urban environment [58, 59] and acoustic signature of small-scale rotors employed in the UA. The relationship between SPL and distance, altitude, as well as orientation of the UA is studied with respect to the receptor [60]. Correlation-based techniques have also been applied to quantify modulation in the acoustic field of a small-scale rotor [61]. Bispectral analysis was used to reveal the correlation

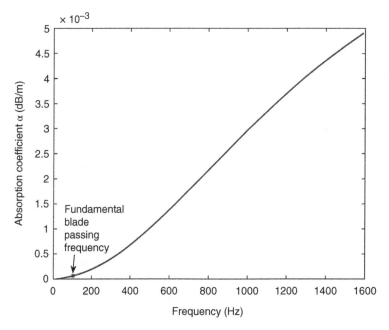

Figure 3.4   Sound absorption coefficient variation with frequency. From [7].

between the modulation strength-parameters and the degree of time-variation of the amplitude, with the blade passing frequency (BPF) being the modulating signal and the higher-frequency noise being the carrier signal. The variation of the sound absorption coefficient (α) with frequency for a typical small UA propeller is shown in Figure 3.4.

As most of the air taxi designs are multirotors having vertical take-off and landing (VTOL) or ultra-short takeoff and landing (uSTOL) capability, some lessons can be learnt from the noise emission from helicopters. The main source of noise for a helicopter is from its rotor blades. Blades produce several types of sound, ranging from air displacement or thickness noise, to loading noise from mostly lift and drag forces acting on the air flowing around the blade. Additionally, interactions with turbulent inflows of air and aerodynamic shocks on the blade surface also produces noise. A lot of air taxi designs incorporate ducted rotors, which can provide a barrier for sound.

3.6.2   Human Perception

Community acceptance of UA will hinge on the impact they make in day-to-day life, and one crucial aspect of this is noise. The masking of UA noise by traffic noise from a nearby highway or railway traffic could be an effective tool for low acoustic-annoyance path planning. Additionally, the perception of aircraft noise is key and there are many factors which can influence this perception, the major one being the context within which the sound is heard. For example, the noise of an ambulance, while annoying and disruptive, is understood to be made for a very important reason (saving lives). Conversely, neighbours playing loud music very late at night is equally disruptive but often not afforded the same allowances. The impact of perceived noise on the population is commonly described by dose-response relationships. An awakening index (AI) defined by the percentage of persons awakened at a specific location as a function of perceived indoor SEL is given by:

$$AI = 0.0087(\text{SEL}_{indoor} - 30dB)^{1.79} \tag{3.51}$$

The indoor SEL is estimated from the outdoor SEL by subtracting 20.5 dB (average loss due to the sound insulation of a typical house). The number of awakenings for a specific noise station is calculated by multiplying the percentage of awakenings of the station i with the population P_i assigned to that receiver.

$$n_{awakenings} = \sum_{i=1}^{n} \%awakenings_i \times P_i \tag{3.52}$$

A change of 5–6 dB in noise exposure can result in "sporadic complaints" becoming "widespread" (1971). In some cases, complaints are due to the nature or character of the sound rather than their decibel level. The frequency (pitch) of the sound is an important characteristic when considering annoyance as a response to noise in addition to the duration of the sound. Non-acoustic effects have also been shown to have a significant impact on perceived annoyance from the same sound. Moderating effects such as the listener's attitude towards the sound source or the activity the listener is undertaking whilst being exposed to the noise can in some cases have a larger impact on annoyance than sound-exposure alone [62].

Attempts to approximate the human response to noise in objective measurements have been made with varying degrees of accuracy. One of the most common examples is the A-weighting for the decibel scale. A purely decibel scale, while common, is considered inappropriate in some circumstances because there is no weighting given to the context of the sound. A single number cannot adequately capture the character of the sound [63, 64].

When it comes to specifically rating aircraft noise, energy-based descriptors have been proposed, as all sound sources in an environment can be integrated, although this approach fails to consider if the subjective meaning of each individual sound is different [65] (see the example of the ambulance). Helicopters are a good example of the character of the sound being distinctly different due to the phenomenon known as 'blade slap'. There is evidence to suggest that this acoustic characteristic ('thumping') is more annoying despite a lower overall SPL. However, it is an ongoing discussion as to how to accurately capture this additional level of irritation in noise metrics. Several authors argue that helicopter noise should be rated no differently to fixed-wing aircraft noise [66], while others argue for some kind of adjustment factor [67], either overall or type-dependent [68, 69].

3.6.3 Flight Path Considerations

The take-off and landing phases are the two flight regimes during which the noise generated by an aircraft has the largest impact on ground-based listeners, both due to the closer proximity to the ground (reducing the geometric attenuation), but also because of operational specificities such as high thrust, deployment of high-lift devices and landing gear, and thrust reversers (increasing the noise generated at source). For this reason, a significant body of research has historically addressed the modelling, quantification, and mitigation of take-off/landing noise impacts by conventional aircraft [4, 5].

The advent of UAM introduces a greater challenge as these new vehicles and services are expected to operate much more closely to densely inhabited areas (as compared to most metropolitan airports). Moreover, a much broader range of flight phases will be executed in closer proximity to the ground, with some flights occurring below the skyline and thus likely reverberating on building surfaces.

Effective noise emission and propagation models support two alternative mitigation processes:

1. The design of arrival and departure flight paths to minimize the overall noise exposure. This process requires adopting static digital terrain elevation databases (DTED) and digital demographic distribution (D3) models and to determine the paths which minimize the exposure levels, while still being safely and practically flyable for the aircraft.
2. The definition of land use planning initiatives to limit the amount of population potentially affected, or the extent of their exposure. This process requires assuming a predefined flight path and to generate detailed 'footprint'/contour maps, based on which an optimal land use planning can be implemented.

Both of these processes have been commonly adopted for many decades, but one major limitation is the fact that the solutions developed are 'static' in nature, thus not considering weather and other factors which may introduce spatiotemporal anomalies in the propagation phenomena, thus potentially introducing more temporarily optimal solutions.

Several studies have targeted noise exposure mitigation by trajectory optimization. Early studies in this subject area looked at the design/redesign of conventional arrival and departure procedures [70, 71]. More recent works in this domain [45, 72–78] opened the path for initiatives exploiting advanced navigation and guidance systems such as required navigation performance (RNP) and steep global navigation satellite systems (GNSS) arrival and approach procedures.

As mentioned in the previous paragraph, it is essential to model the aircraft's operational capabilities to ensure that the optimized trajectory is ultimately flyable (and safely so). This requires implementing some aircraft performance or dynamics models, which can range from steady-state performance to full kinematic or dynamic models [79].

Ongoing research is adopting both the simplistic empirical models as well as relatively complex and high-dimensional computational fluid dynamics (CFD) simulations. In either case, the atmospheric and weather conditions are very rarely considered and when they are treated in a typically static manner, these studies fail to capture the significant effects of temperature, wind, air density and other atmospheric properties on the propagation of noise generated from aircraft at various levels.

The optimizer needs to include a mathematical model of aircraft noise to allow the minimization of perceived noise on the ground. In order to critically evaluate the noise levels, the Aviation Environmental Design Tool (AEDT) developed by the Federal Aviation Administration (FAA), serves as a multi-purpose framework integrating the Integrated Noise Model (INM) and the Model for Assessing Global Exposure to the Noise of Transport Aircraft (MAGENTA), a global noise model [44].

The concept of noise-radius is adopted generally by considering the complexity of aircraft noise modelling based on aircraft engine type, thrust setting, and atmospheric conditions [4]. The optimisation of aircraft vertical trajectory with minimum noise impact using analytical jet noise models has proved to be effective [45]. The methodology adopted requires demographic data at each observer location and hence a D3 is used in conjunction with the noise model. The D3 model aids in estimating the population in a user defined grid in a global or local scale. Additionally, the DTED is also part of the trajectory optimization framework considering availability of geographic information. The availability of D3 and DTED are specifically important for assessing environmental impact of aircraft noise in the terminal manoeuvering area (TMA). Trajectory optimization is performed to avoid the densely populated areas in and around the airports, considering the topographical conditions, metrological data, and trajectory constraints. The constraints on the trajectory can be ATM operational, airspace, airline, flight parameters, and/or aircraft dynamics-based constraints. Several studies have been carried out for optimizing the aircraft trajectory based on a number of cost functions, such as number of sleep disturbances resulting in reduction of noise annoyance on specific regions around an airport [22–24, 46–50]. Reduced noise procedures implemented are Noise Abatement Departure Procedures (NADP), including NADP1, NADP2, and its associated variations, such as the ICAO-A procedure [11]. Steep trajectory approach, spiral trajectories, and touch-down displacement principles have been proposed and trialled to reduce noise while landing. Generally, the optimisation or reduced noise is not harmonious with the cost function for minimising other environmental emissions. Reduction in noise by increasing flight time results in higher fuel consumption and, as a consequence, higher emissions [51, 52].

Currently, it is generally accepted to apply the existing noise and emissions standards for manned aircraft to UA [15]. Also, air navigation service providers (ANSPs) like Airservices Australia have already defined system requirements for both users and various services for a flight information management system (FIMS) for UAM aircraft operations [80]. Among the system requirements for services, noise management service aims to ensure suitable management of noise impact on the community. The noise management service hosted by FIMS shall incorporate a noise threshold map (NTM), which will establish a basic grid over a geographic area and prescribe the maximum number of UA operations across different times of the day, based on various dBA level bands.

A library of acoustic parameters, such as nominal dB output and frequency range for standard operations, shall be incorporated for different UA. All operations are consolidated into a dynamic cumulative noise map (CNM), which will record both planned noise impact (PNI – before flight) and actual noise impact (ANI – after flight). A free field acoustic propagation model shall be developed, which, along with the UA acoustic parameters, will be used to calculate the A-weighted sound level (dBA) at ground level along the operational intent volume (OIV) centreline, where performance operational intent volume (POIV) is the subset of OIV. The ANI shall be determined through the combination of UA track data, the library of acoustic parameters, and the acoustic propagation model, and will be captured in the CNM. Besides, community inputs to the noise management service shall be allowed along with provision for archival of noise data.

However, since the UA will be flying at low altitudes (<40 m AGL) in case of drone deliveries and near buildings in case of 'flying taxis', there is a requirement for separate noise and emission standards for UA and small manned aircraft for urban air mobility. As the UAM framework development is currently underway, there is a strong anticipation of VTOL or uSTOL aircraft to provide UAM services in the early 2020s [81]. However, among other constraints, noise, has been identified as one of the main factors impacting the implementation of UAM systems.

3.7 Conclusions

This section addressed the several environmental impacts of conventional and emerging air mobility vehicles beyond the climate impacts, which were already addressed in Chapter 2 of this book. The main aspects covered include: other atmospheric pollutants, the aircraft noise problem, current noise emission, propagation, and psychoacoustic factors as they apply to manned aircraft, as well as the current work being done to develop suitable approaches to the modelling and mitigation of UA noise. Considering the unique challenges that would emerge with the widespread adoption of urban air mobility (UAM) as well as drone delivery operations, this chapter highlights the advantages of surrogate noise propagation models that consider a broader array of environmental and operational factors as compared to legacy aircraft noise models, and could potentially inform future noise certification standards for low-flying aircraft in urban environments. The physics-based sound propagation models presented in this chapter take into account the atmospheric effects and platform dynamics, and in future will include multipath effects without being computationally intensive, unlike the detailed CFD models or the current noise certification standards, which either neglect atmospheric effects altogether or assume constant values. Fully developed surrogate sound propagation models could either inform the design of future low-flying aircraft based on their anticipated acoustic signature or aid in the path planning of an existing aircraft design for minimal noise impact. Updated sound propagation and noise models can inform the design, development, testing, and evaluation of new noise certification as well as regulation procedures to be implemented for low-flying aircraft in densely populated areas. Since sound certification is carried out for flight conditions imposed by aviation authorities, sound mitigation efforts also need to focus on sound emission at source. Future noise certification standards for low-flying aircraft would need to consider the design improvements currently being undertaken to decrease sound emission at source. This would include improvements in aircraft and propulsion design as well as optimizing the flight trajectory of the UAs in order to mitigate their noise impacts on the ground, while ensuring safe separation from other manned and unmanned aircraft. The surrogate noise models could inform the flight path trajectories as well as regulate the number of aircraft in a given airspace, based on the noise threshold set for that airspace at that particular time of the day.

References

1 Penner J.E., Lister D.H., Griggs D.J. et al. (1999). Aviation and the Global Atmosphere – A Special Report of IPCC Working Groups I and III in Collaboration with the Scientific Assessment Panel to the Montreal Protocol on Substances that Deplete the Ozone Layer, United Nations Intergovernmental Panel on Climate Change (UN IPCC).

2 Lee, D.S., Fahey, D.W., Skowron, A. et al. (2021). The contribution of global aviation to anthropogenic climate forcing for 2000 to 2018. *Atmospheric Environment* 244: 117834.

3 Lee, D.S., Pitari, G., Grewe, V. et al. (2010). Transport impacts on atmosphere and climate: aviation. *Atmospheric Environment* 44 (37): 4678–4734.

4 Zaporozhets, O., Tokarev, V., and Attenborough, K. (2011). *Aircraft Noise: Assessment, Prediction and Control*. CRC Press.

5 Ruijgrok, G.J.J. (2007). *Elements of Aviation Acoustics*. Delft, Netherlands: VSSD.

6 Ruijgrok, G.J.J. and van Paassen, D.M. (2012). *Elements of Aircraft Pollution*. Delft, Netherlands: VSSD.

7 Kapoor, R., Kloet, N., Gardi, A. et al. (2021). Sound propagation modelling for manned and unmanned aircraft noise assessment and mitigation: a review. *Atmosphere* 12 (11): 1424.

8 Upham, P. (2003). *Towards Sustainable Aviation*. Earthscan.

9 Whitelegg, J. (2000). *AVIATION: The Social, Economic and Environmental Impact of Flying*. London: Ashden Trust.

10 Franssen, E., Van Wiechen, C., Nagelkerke, N., and Lebret, E. (2004). Aircraft noise around a large international airport and its impact on general health and medication use. *Occupational and Environmental Medicine* 61: 405–413.

11 Filippone, A. (2014). Aircraft noise prediction. *Progress in Aerospace Sciences* 68: 27–63.

12 Müller, G. and Möser, M. (2013). *Handbook of Engineering Acoustics*. Berlin Heidelberg: Springer.

13 Casalino, D., Diozzi, F., Sannino, R., and Paonessa, A. (2008). Aircraft noise reduction technologies: a bibliographic review. *Aerospace Science and Technology* 12 (1): 1–17.

14 Lighthill, M.J. (1963). Jet noise. *AIAA Journal* 1 (7): 1507–1517.

15 Organization I. C. A (2011). Unmanned Aircraft Systems (UAS) 978-92-9231-751-5 International Civil Aviation Organization.

16 Elbanhawi, M., Mohamed, A., Clothier, R. et al. (2017). Enabling technologies for autonomous MAV operations. *Progress in Aerospace Sciences* 91: 27–52.

17 Mohamed, A., Carrese, R., Fletcher, D., and Watkins, S. (2015). Scale-resolving simulation to predict the updraught regions over buildings for MAV orographic lift soaring. *Journal of Wind Engineering and Industrial Aerodynamics*, 140: 34–48.

18 Watkins, S., Burry, J., Mohamed, A. et al. (2020). Ten questions concerning the use of drones in urban environments. *Building and Environment* 167: 106458.

19 Organisation I. C. A (2017). Annex 16 - Environmental Protection – Volume I - Aircraft Noise. International Civil Aviation Organisation.

20 Ruijgrok, G.J. (1993). *Elements of Aviation Acoustics*. Delft University Press.

21 Zagzebski, J.A. (1996). *Essentials of Ultrasound Physics*. Mosby.

22 Navaratne, R., Tessaro, M., Gu, W. et al. (2012). Generic framework for multi-disciplinary trajectory optimization of aircraft and power plant integrated systems. *Journal of Aeronautics and Aerospace Engineering* 2: 1–14.

23 Pisani, D., Zammit-Mangion, D., and Sabatini, R. (2013). City-pair trajectory optimization in the presence of winds using the GATAC framework. *AIAA Guidance, Navigation and Control Conference 2013 (GNC 2013)*, Boston, MA, USA.

24 Sammut, M., Zammit-Mangion, D., and Sabatini, R. (2012). Optimization of fuel consumption in climb trajectories using genetic algorithm techniques. *AIAA Guidance, Navigation and Control Conference 2012 (GNC 2012)*, Minneapolis, MN, USA.

25 Bass, H., Sutherland, L., Zuckerwar, A. et al. (1995). Atmospheric absorption of sound: further developments. *The Journal of the Acoustical Society of America* 97 (1): 680–683.

26 Attenborough, K. (2014). Sound propagation in the atmosphere. In: *Springer Handbook of Acoustics* (ed. T.D. Rossing), pp. 117–155. Springer, New York, NY, USA.

27 Kneizys F. X., Shettle E., Abreu L., Chetwynd J., et al. (1988). Users guide to LOWTRAN 7, AIR FORCE GEOPHYSICS LAB HANSCOM AFB MA.

28 Sabatini, R. and Richardson, M. (2010). *Airborne Laser Systems Testing and Analysis*. The Research and Technology Organisation.

29 Norma I. (1996). 9613–2: Acoustic attenuation of sound during propagation outdoors–general methods of calculation. International Standard.

30 Daigle, G., Piercy, J., and Embleton, T. (1983). Line-of-sight propagation through atmospheric turbulence near the ground. *The Journal of the Acoustical Society of America* 74 (5): 1505–1513.

31 Juvé, D., Blanc-Benon, P. and Chevret, P. (1994). Sound propagation through a turbulent atmosphere: Influence of the turbulence model. *Proceedings of the Sixth International Symposium on Long Range Sound Propagation*, 270–282.

32 Von Karman, T. (1948). Progress in the statistical theory of turbulence. *Proceedings of the National Academy of Sciences* 34 (11): 530–539.

33 Johnson, M.A., Raspet, R., and Bobak, M.T. (1987). A turbulence model for sound propagation from an elevated source above level ground. *The Journal of the Acoustical Society of America* 81 (3): 638–646.

34 Normalización O.I. (1993). ISO 9613-1: 1993, Acoustics: attenuation of sound during propagation outdors. Caldulation of the absorption of sound by the atmosphere. International Organization for Standardization.

35 Wiener, F.M. and Keast, D.N. (1959). Experimental study of the propagation of sound over ground. *The Journal of the Acoustical Society of America* 31 (6): 724–733.

36 Wong, G.S. (1990). Approximate equations for some acoustical and thermodynamic properties of standard air. *Journal of the Acoustical Society of Japan (E)* 11 (3): 145–155.

37 Wong, G.S. (2014). Microphones and their calibration. In: *Springer Handbook of Acoustics*. Springer.

38 Wong, G.S. (1986). Speed of sound in standard air. *The Journal of the Acoustical Society of America* 79 (5): 1359–1366.

39 Wong, G.S., Embleton, T.F., and Ehrlich, S.L. (1995). AIP Handbook of Condenser Microphones (Theory, Calibration, and Measurements). Chapter 4: Primary pressure calibration by reciprocity. *American Institute of Physics*, College Park, MA, USA.

40 ECAC.CEAC (1997). Doc 29, Report on Standard Method of Computing Noise Contours around Civil Airports.

41 ICAO (1988). *Recommended Method for Computing Noise Contours around Airports, Circular 205*. Montreal, Quebec, Canada: International Civil Aviation Organization.

42 Quindry, T. L. (1976). Standards on noise measurements, rating schemes and definitions: A compilation. US Department of Commerce National Bureau of Standards (NBS) Special Publication 385, 1st ed., Washington, DC, USA.

43 Boeker, E.R., Dinges, E., He, B. et al. (2008). *Integrated Noise Model (INM) Version 7.0 Technical Manual*. Washington: DC, Federal Aviation Administration, Office of Environment and Energy.

44 Noel, G., Allaire, D., Jacobson, S. et al. (2009). Assessment of the aviation environmental design tool. *Proceedings of the Eighth USA/Europe Air Traffic Management Research and Development Seminar (ATM2009), Napa, CA, (June,* 2009).

45 Khardi, S. and Abdallah, L. (2012). Optimization approaches of aircraft flight path reducing noise: comparison of modeling methods. *Applied Acoustics* 73: 291–301.

46 Camilleri, W., Chircop, K., Zammit-Mangion, D. et al. (2012). Design and validation of a detailed aircraft performance model for trajectory optimization. *AIAA Modeling and Simulation Technologies Conference 2012 (MST 2012)*, Minneapolis, MN, USA,

47 Chircop, K., Zammit-Mangion, D., and Sabatini, R. (2012). Bi-objective pseudospectral optimal control techniques for aircraft trajectory optimisation. *28th Congress of the International Council of the Aeronautical Sciences 2012, ICAS 2012*, Brisbane, Australia, 3546–3555.

48 Cooper, M.A., Lawson, C.P., Quaglia, D. et al. (2012). Towards trajectory prediction and optimisation for energy efficiency of an aircraft with electrical and hydraulic actuation systems. *28th Congress of the International Council of the Aeronautical Sciences 2012, ICAS 2012*, Brisbane, Australia, 3757–3769.

49 Gauci, J., Zammit-Mangion, D., and Sabatini, R. (2012). Correspondence and clustering methods for image-based wing-tip collision avoidance techniques. *28th Congress of the International Council of the Aeronautical Sciences 2012, ICAS 2012*, Brisbane, Australia, 4545–4557.

50 Gu, W., Navaratne, R., Quaglia, D. et al. (2012). Towards the development of a multi-disciplinary flight trajectory optimization tool – GATAC. *ASME Turbo Expo 2012: Turbine Technical Conference and Exposition (GT 2012)*, Copenhagen, Denmark, 415–424.

51 Prats, X., Puig, V., and Quevedo, J. (2011). Equitable aircraft noise-abatement departure procedures. *Journal of Guidance, Control and Dynamics* 34: 192–203.

52 Torres, R., Chaptal, J., Bes, C., and Hiriart-Urruty, J.-B. (2011). Optimal, environmentally friendly departure procedures for civil aircraft. *Journal of Aircraft* 48: 11–22.

53 EU (2002). Directive 2002/30/EC of the European Parliament and of the Council on teh Establishment of Rules and Procedures with Regard to the Introduction of Noise-Related Operating Restrictions at Community Airports. European Commission, Brussels, Belgium.

54 Licitra, G., Gagliardi, P., Fredianelli, L., and Simonetti, D. (2014). Noise mitigation action plan of Pisa civil and military airport and its effects on people exposure. *Applied Acoustics* 84: 25–36.

55 Stevens, N., Baker, D., and Freestone, R. (2010). Airports in their urban settings: towards a conceptual model of interfaces in the Australian context. *Journal of Transport Geography* 18: 276–284.

56 Bent, P., Boeing, R., Snider, R., and Bell, F.F. (2020). Urban Air Mobility Noise: Current Practice, Gaps, and Recommendations.

57 Administration, F.A. (2020). Integration of Civil Unmanned Aircraft Systems (UAS) in the National Airspace System (NAS) Roadmap.

58 Mohamed, A., Watkins, S., Ol, M., and Jones, A. (2020). Flight-relevant gusts: computation-derived guidelines for micro air vehicle ground test unsteady aerodynamics. *Journal of Aircraft*, 58 (3): 693–699.

59 Prudden, S., Fisher, A., Marino, M. et al. (2018). Measuring wind with small unmanned aircraft systems. *Journal of Wind Engineering and Industrial Aerodynamics* 176: 197–210.

60 Kloet, N., Watkins, S., and Clothier, R. (2017). Acoustic signature measurement of small multi-rotor unmanned aircraft systems. *International Journal of Micro Air Vehicles* 9 (1): 3–14.

61 Baars W.J., Bullard, L. and Mohamed, A. (2021). Quantifying modulation in the acoustic field of a small-scale rotor using bispectral analysis. *AIAA Scitech 2021 Forum*, 0713.

62 Job, R. (1988). Community response to noise: a review of factors influencing the relationship between noise exposure and reaction. *The Journal of the Acoustical Society of America* 83 (3): 991–1001.

63 Large, S. and Stigwood, M. (2014). The noise characteristics of 'compliant' wind farms that adversely affect its neighbours. *INTER-NOISE and NOISE-CON Congress and Conference Proceedings*, 6269–6288.

64 Michaud, D.S., Fidell, S., Pearsons, K. et al. (2007). Review of field studies of aircraft noise-induced sleep disturbance. *The Journal of the Acoustical Society of America* 121 (1): 32–41.

65 Namba, S. (1987). On the psychological measurement of loudness, noisiness and annoyance: a review. *Journal of the Acoustical Society of Japan (E)* 8 (6): 211–222.

66 Molino, J.A. (1982). Should helicopter noise be measured differently from other aircraft noise? A review of the psychoacoustic literature.

67 Leverton, J.W. (1975). Helicopter noise: can it be adequately rated? *Journal of Sound and Vibration* 43 (2): 351–361.

68 Gjestland, T. (1994). Assessment of helicopter noise annoyance: a comparison between noise from helicopters and from jet aircraft. *Journal of Sound and Vibration* 171 (4): 453–458.

69 Schomer, P., Hoover, B., and Wagner, L. (1991). Human response to helicopter noise: a test of A-weighting. US Army Construction Engineering Research Laboratory (USACERL), Technical Report N-91/13, Champaign, IL, USA.

70 Cook, G., Jacobson, I., Chang, R., and Melton, R. (1982). Methodology for multiaircraft minimum noise impact landing trajectories. *IEEE Transactions on Aerospace and Electronic Systems AES* 18 (1): 131–146.

71 Zaporozhets, O.I. and Tokarev, V.I. (1998). Predicted flight procedures for minimum noise impact. *Applied Acoustics* 55 (2): 129–143.

72 Hartjes, S., Visser, H.G. and Hebly, S.J. (2009). Optimization of RNAV Noise and Emission Abatement Departure Procedures. *AIAA Aviation Technology, Integration, and Operations conference 2009 (ATIO 2009)*, Hilton Head, SC, USA.

73 Prats i Menendez, X. (2010). Contributions to the optimisation of aircraft noise abatement procedures. PhD, Universitat Politecnica de Catalunya (UPC), Barcelona.

74 Prats, X., Puig, V., and Quevedo, J. (2011b). A multi-objective optimization strategy for designing aircraft noise abatement procedures. Case study at Girona airport. *Transportation Research Part D: Transport and Environment* 16 (1): 31–41.

75 Prats, X., Puig, V., Quevedo, J., and Nejjari, F. (2010). Multi-objective optimisation for aircraft departure trajectories minimising noise annoyance. *Transportation Research Part C: Emerging Technologies* 18 (6): 975–989.

76 Ren, L. (2007). Modeling and managing separation for noise abatement arrival procedures. ScD, Massachusetts Institute of Technology.

77 Visser, H.G. (2005). Generic and site-specific criteria in the optimization of noise abatement trajectories. *Transportation Research Part D: Transport and Environment* 10 (5): 405–419.

78 Visser, H.G. and Wijnen, R. (2001). Optimization of noise abatement arrival trajectories. *AIAA Guidance, Navigation and Control conference 2001 (GNC 2001)*, Montreal, Canada.

79 Gardi, A., Sabatini, R., and Ramasamy, S. (2016). Multi-objective optimisation of aircraft flight trajectories in the ATM and avionics context. *Progress in Aerospace Sciences* 83: 1–36.

80 Chisholm, P. (2021). FIMS (Prototype) System Requirements Specification UTM-REQ-01.

81 Vascik, P.D. and Hansman, R.J. (2018). Scaling constraints for urban air mobility operations: air traffic control, ground infrastructure, and noise. *2018 Aviation Technology, Integration, and Operations Conference*, 3849.

82 ECAC (2016). Report on Standard Method of Computing Noise Contours around Civil Airports Volume 1: Applications Guide, European Civil Aviation Conference (ECAC) SECRETARIAT, Neuilly-sur-Seine, France.

Section II

Systems for Sustainable Aviation

4

Systems Engineering Evolutions

Anthony Zanetti[1], Arun Kumar[1], Alessandro Gardi[1,2], and Roberto Sabatini[1,2]

[1] *School of Engineering, RMIT University, Melbourne, Victoria, Australia*
[2] *Department of Aerospace Engineering, Khalifa University of Science and Technology, Abu Dhabi, UAE*

4.1 Introduction

Civil aviation is set to grow considerably over the coming decades, giving rise to concerns about its impact on the environment. Considering the general need to find an optimal balance between growth and environmental sustainability, this chapter investigates the introduction of green life cycle management (LCM) concepts in aviation. LCM inherently offers a holistic and systematic view of the design, development, operation, and decommissioning phases for both the individual vehicle as well as for the comprehensive transport system, therefore offering a convenient starting point for analysing and mitigating their associated environmental problems. The study presented here takes a systems approach, aimed at extending the knowledge base of systems engineering and sustainable aviation while supplying the reader with insights into the application of principles of the former and the implementation of the latter. In so doing a W methodology is proposed, being the first of its kind that links system requirements and design to a series of tests that simultaneously encompasses validation, verification, and environmental impact assessments (EIAs).

4.1.1 Background

The International Air Transport Association (IATA), the International Civil Aviation Organisation (ICAO), and other government and independent organisations lead the way in directing aviation to a greener future. However, ambiguity still prevails around the enhancements to be pursued, and whether incremental changes will suffice in addressing environmental challenges [1].

Air transportation currently contributes to only 2–3% of global anthropogenic carbon emissions but this could reach 10–15% by the year 2050 [2, 3]. Responding to the challenges of sustainability requires a consideration of the characteristics of a sustainable system, and a fundamental shift in the way products and processes are designed, produced, and operated [4]. Complex systems that have traditionally been conceived and developed as stand-alone entities can no longer be viewed in isolation [5]. LCM plays an essential role

Sustainable Aviation Technology and Operations: Research and Innovation Perspectives, First Edition.
Edited by Roberto Sabatini and Alessandro Gardi.
© 2024 John Wiley & Sons Ltd. Published 2024 by John Wiley & Sons Ltd.
Companion Website: www.wiley.com/go/sustainableaviation

in this enhancement process because it embodies the notion of managing each incremental stage of a system as the approach to achieve the set objectives. The need for a systematic and systemic approach has been intensified through globalisation and the subsequent degree of networking between firms.

This chapter proposes that an appropriate level of mitigation will only be achieved with an in-depth consideration of both direct and indirect contributions, and their associated energy and emission life cycles. Motivation for this research has drawn upon the work of Chester and Horvath [6] who imply that methods to mitigate environmental impacts across a transportation mode must address its entire life cycle. Here, we attempt to implement a more holistic assessment methodology of industry-external factors through greener LCM processes.

4.2 Systems-of-Systems Engineering: Defining the Civil Aviation Boundaries

The interdisciplinary approach of systems engineering has provided useful methodologies to assist in addressing complex real-world problems, [7]. It does, however, come with several shortcomings, particularly in its inability to address problems with a high degree of contextual influence resulting in an inherent incapability to provide complete solutions [5]. Systems conceived and developed on a stand-alone basis to address an individual problem should not be analysed in isolation. Consequently, practitioners are now addressing the challenges that arise from complex system integration, referred to as system-of-systems engineering (SoSE), which provides a more complete, holistic approach to achieving optimality across the independent elements within an enterprise [5, 8–10].

A system-of-systems (SoS) is composed of a collection of elements that are able to function independently. The transition of theory towards a SoSE approach highlights the necessity to analyse the interactions of independent elements as a way of obtaining synergies that lead to higher capabilities and performance. Martin [11] illustrates this shift of thinking: 'Systems engineering is the process that controls system development with the goal of achieving an optimal balance between all system elements'. Shenhar [12] explains SoS as a 'network of systems functioning together to achieve a common goal'. While there is no universally accepted definition of the concept, it is very clear that system-of-systems is becoming ubiquitous throughout the literature. Most authors seem to agree that managing complex systems is a significant challenge, and perhaps becomes an obstacle to achieving certain successes [13]. There is increased interest in finding ways to achieve synergies between the independent elements of a system as a way of optimising the overall system performance. Eisner, et al. [14] highlight that when the objectives of system elements are combined, they form a multifunctional solution to an overall coherent problem. It is, therefore, fundamental to understand the contextual influence of a given system and distinguish between its strategic intent and its unique problem.

While the amount of literature addressing systems engineering is expanding considerably, we still do not have an established SoSE body of knowledge. Literature addressing transportation systems is relatively scarce; however, defining the boundaries and elements of a system is a critical aspect when placed in the context of life cycle management or environmental sustainability. Life cycle management is a core aspect of systems engineering implementing the principle that system design and architectural changes should be traceable to the requirements of the overall system. Most implementations of systems engineering in aviation only involve the design of aircraft and their components [15, 16]. Some have addressed the aviation system as a series of physical (demand and supply) and non-physical components (operating regulations) [17]. Others have adopted a SoS approach to optimise security at airports [18]. All of the above focus on a single element only, and, while it may be practical to do so, the benefits achieved in one particular element may not necessarily benefit the entire system.

Cascetta provides a useful definition of a transportation system: 'a combination of elements and their interactions, which produce the demand for travel and the supply of services to satisfy this demand' [19]. This definition highlights that the supply of the end product, which takes the form of a transportation service from the point of origin to

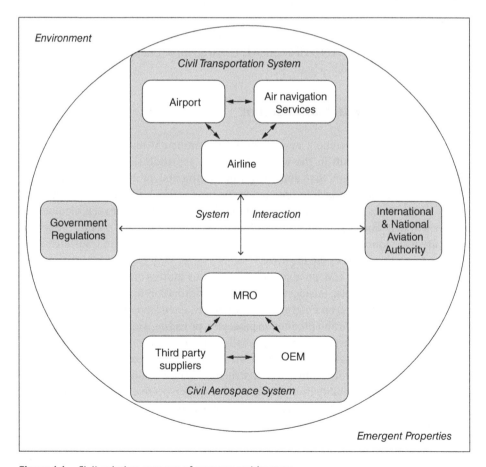

Figure 4.1 Civil aviation systems-of-systems architecture.

destination, requires the provision of complementary services along different stages of the value chain. Franz et al. characterise these complementary services into two elements [20]:

- The civil transportation element: airlines, airports, and navigation service providers.
- The civil aerospace element: manufacturers, suppliers, and approved maintenance organisations.

The list of relevant stakeholders also includes governments, as well as international and national aviation authorities in this depiction as they are the regulators of civil aviation and are the driving force behind environmental optimisation. The identification of these elements endeavours to illustrate and broaden our understanding of the interactions between different aspects of civil aviation. This is a critical factor in mitigating external factors, as each element will be set around unique problems and constraints. This emphasises the requirement to address environmental issues of each element simultaneously. For example, while there has been much research on optimising aircraft technology, the environmental benefits gained from the design phase will be limited if airport and air traffic infrastructures are not equipped to accommodate the growth in air traffic during its operational phase. Thus, establishing a solid foundation for recognising the patterns and structure in SoS problems will help develop improved tools and methodologies with the aim of optimising the overall environmental impact of civil aviation [21]. An architectural depiction of the current aviation system is displayed in Figure 4.1.

4.3 Green Life Cycle Management

Life cycle thinking has become a new norm in environmental management, indicating an important paradigm shift in the way practitioners approach the problem. Life cycle thinking is a broad concept that approaches environmental problems at system level and provides a holistic approach to product development from a cradle-to-grave perspective [22]. This creates a requirement to redefine the relationships between elements of the aviation system and ensures that changes made to the design of the product or system architecture will reduce the total impact on the environment, rather than transferring the externality to another stage of the life cycle.

This draws attention to LCM, an approach that aims to address environmental issues by integrating people, processes, business systems, and information to manage the complete life cycle of a product across an enterprise. Traditionally, environmental management has focused purely on the design and production processes, providing a very limited scope of the environmental impacts of a given product, usually at the expense of other stages in its life cycle [23]. LCM is not a single tool or method but a product management system providing the necessary foundations for organisations to structure activities and product-related information to improve sustainability from an environmental perspective [24]. It can be defined simply as a business initiative that manages a firm's portfolio of products across its entire life cycle; from the conceptualisation of ideas and design, right through to its retirement and disposal [25].

Systems engineering overlaps a significant portion of product LCM. Both disciplines are concerned with managing the multiple view and interrelationships of product information. Sage [7] explains that both concepts enable the accomplishment of desired methodologies

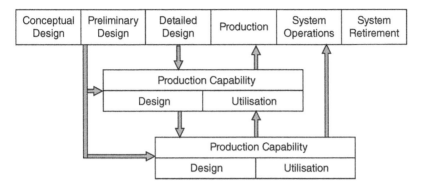

Figure 4.2 Stages of a system life cycle.

that work to fulfil the identified needs of a system. A system life cycle is defined by a number of phases that behave in an interactive and iterative manner. Blanchard [26] affirms this by suggesting that it is essential to consider the overall life cycle in addressing system-level issues, as activities in one phase will interact with those in another.

Figure 4.2 provides a detailed depiction of a life cycle model that is relevant to civil aviation (extracted from Blanchard [26]). Here, the top row denotes the elements of the system that relate directly to the accomplishments of the mission (e.g. an aircraft). However, the diagram also illustrates two life cycle systems closely related to the first. These are specially designed as information loops to address those design and construction capabilities, which may have significant impact on system operations, concurrently [26].

4.4 Supply Chain Architectures

The concept of supply chain management (SCM) is inexorably linked to LCM. Wiley [27] notes that a system will progress through different stages of a predefined life cycle. This highlights the need for product distribution from one phase to the next achieved through a comprehensive structural network of organisations. Here, we can determine the differences between the two management theories, in that LCM is concerned with the product from its design to disposal, and SCM with the processes and systems of distributing the product from the supplier to the end-user.

Sustainable supply-chain architecture differs slightly from that of LCM in that it adopts a cradle-to-cradle approach to product life cycle (opposed to cradle-to-grave). It endeavours to provide a system that is self-regulating and ultimately reduces the requirement for virgin materials. Beamon [28] considers the transition towards reverse logistics, which involves both upstream and downstream flows that provides the necessary infrastructure for recycling and remanufacturing. This opens up the required coordination between two markets: demand from the consumer and the supply of returned products. Reverse logistics encapsulates the process of planning and implementing the flow of raw materials, material waste, and used goods from the point of consumption to the point of origin as a way of recapturing value from recycling [29]. This affectively reduces the systems' reliance on virgin materials and essentially reduces emissions from their synthesis. Figure 4.3 depicts the schematics of

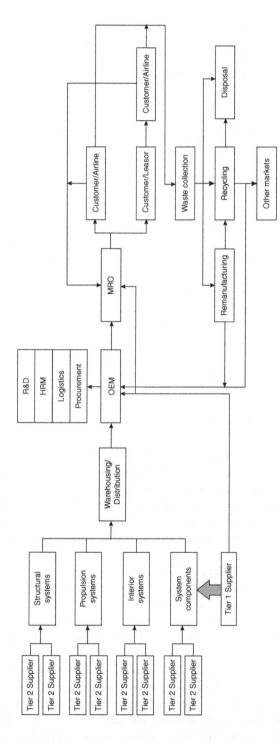

Figure 4.3 Civil aviation reverse logistics architecture.

the aviation supply chain. The black arrows are associated with the forward flow of aircraft components to the manufacturer for central assembly. The finished product (aircraft) is then delivered to its customer (airline). The red arrows highlight the reverse directional flow (reverse logistics) of the used aircraft from the point of consumption to the point of origin.

4.5 Principles for Greener Design

System design will essentially dictate the ecological footprint of a product across its entire life cycle. While a typical life cycle consists of many different phases, systems engineering is primarily performed in the development stage, giving key design considerations to production, operations, and retirement [30]. The traditional development of products has been aimed at achieving improvements in design with respect to economics (cost), functionality, and manufacturing [31]. However, the increasing pressure exerted by governments, stakeholders, and the general public has forced many industries, including aviation, to adopt green engineering practices that require products to be designed and developed with respect to the environment.

Integrating a system engineering approach in design phases determines the overall integrity of the system as whole. It should be used as a systematic review of the total systems performance, rather than of individual elements. This allows us to draw a distinct line between how the product will evolve from its initial design and how its operational use at different life cycle stages will pose a threat to the environment. This presents a number of opportunities to find solutions to environmental issues related to the product through the ability to control the types of inputs required and outputs created [32]. Stapelberg [33] highlights that even if the system is optimised from a design point of view, the performance may be sub-optimal owing to the interaction of other elements of the lifecycle phases. Its integration within the design ensures that relevant technical tasks and analysis are completed in order to pursue system efficiencies.

4.5.1 Design for X (DfX)

The concept of design for X (DfX) is drawn from the collaborated efforts of Gungor and Gupta [34], Ilgin and Gupta [31], Wang and Gupta [35], and Eastman [36]. The term DfX, where X is given as the design parameter, highlights the involvement of different specialities with respect to a specific objective. Wang and Gupta [35], and Ilgin and Gupta [31] focus specifically on environmental issues where product design is related to its ecological impact, its disassembly, and its end-of-life. While the primary focus of this chapter is on green life cycle management, we draw closely on the efforts of the above-mentioned authors in our consideration of optimising designs for the environment.

4.5.2 Design for Environment (DfE)

Design for environment (DfE) is used as a strategic methodology directed at the systematic reduction of environmental impacts inflicted by the whole life cycle of a product [37].

Telenko, et al. [38] indicate that DfE principles are a reflection of the lessons learnt from life cycle analysis. Flaws or potential improvements can then be applied to the product design. DfE is essentially a principle that incorporates environmental considerations into the design and redesign of products and processes [32]. Mitigating environmental impacts can more easily be achieved when environmental issues are identified and resolved during the early stages of product design.

It is important to highlight that the DfE concept alone is not a guarantee to ensure environmental sustainability across a system life cycle. When reviewing green life-cycle and the sustainable supply-chain literature, it becomes apparent that the end-of-life stage is an integral part of environmental sustainability. This phase, which is discussed in more depth in the latter part, provides the general ability for the system to be somewhat self-contained, for example, the recyclability and remanufacturing of an aircraft will have a direct impact on the amount of raw material extracted/energy use from the beginning of the life cycle. There are some key areas to consider. Firstly, it is necessary to be selective with the types of materials used in design and production [35]. Will it minimise waste? Is it easy to separate and recycle? Aircraft architecture is leaning towards the use of carbon fibre-reinforced epoxy resins. The Boeing 787, for example, is constructed of around 50% composites [39]. Its application can reduce the overall aircraft weight by approximately 20% compared to traditional aluminium. Weight has a direct impact on fuel consumption, which is proportional to emissions.

This leads to the concept of designing for recycling, remanufacturing, and disposal, all of which constitute a major part of the wider DfE concept. These focus on the attribute of the design that supports the recycling of the materials in the product. Designing a product for recycling has received considerably less attention from a manufacturer's point of view. This is attributed to the fact that the disposal costs of products are traditionally paid by society as whole [40]. The significance of recycling stems from the reduction of virgin materials – i.e. it forms a basis for a self-regulating system. Recycling can typically provide added value to the aviation value chain, while even affording the opportunity to upgrade aircraft performance through remanufacturing.

4.6 Principles for Greener Manufacturing

The term 'green manufacturing' increasingly appears in the literature, evoking the need to address the manufacturing process with respect to the waste and emissions produced. Melnyk and Smith [41] define 'green manufacturing' as a *system that integrates issues related to product and process design with those of manufacturing planning and control in order to identify, quantify, analyse, and manage the production of environmental waste with an ultimate goal of mitigating environmental impact while also maximising resource efficiency.* Applying these principles to aviation will involve system wide alteration of current industrial processes. There is no unified approach from practitioners as to what needs to change. For example, Sarkar [42] highlights three essential methods to transform aviation manufacturing: (i), the use of green energy, a policy that must be adopted through collaboration across the supply chain, (ii), developing green products, a process heavily reliant on the state-of-art technology and design, and (iii), employing green processes in business operations. Azzone and Noci [43] take a different approach by suggesting that changes are required to enterprise procurement policies, to industrial technologies, and to the transformation of the logistics system. Wang and Gupta [35] share a very similar approach to Azzone and Noci [43] in their framework of a periodic closed-loop production system and green vendor analysis. The Organisation for Economic Co-operation and Development (OECD) [44] highlights the need for a holistic approach to sustainable manufacturing, which extends beyond the boundaries of the enterprise.

Industrial ecology (IE) is a research approach to sustainability that analyses the way in which the manufacturing sector can meet its fundamental requirements by behaving like a natural system [45]. Lunt and Levers [46] highlight that there is no open-ended resource flow in nature, meaning that resources are naturally replenished when they are used. Garetti and Taisch [47] point out that optimising the manufacturing process in such a way as to allow waste materials and used products to be reclaimed will lead to greater environmental efficiency. The optimisation of production systems to accommodate reverse material flow must be considered to epitomise that of a natural ecological system. Wang and Gupta [35] present quite a simplistic idea of what they term a periodic closed-loop production system (Figure 4.4). Similarly, Jayaraman, et al. [49] refer to the recoverability of a product as a process in a system designed to recover materials via methods of separation, recycling, and remanufacturing. This model differs slightly from conventional life cycle management as it adopts a cradle-to-cradle perspective, which minimises the demand for virgin materials in favour of using recycled resources from the product or product components.

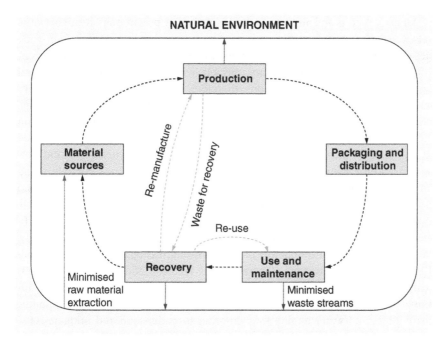

Figure 4.4 Closed-loop production system. Source: [48].

4.7 More Sustainable Operations

There has been a plethora of research on the environmental impacts of air transport operations and associated strategies to reduce these externalities. This is most notably due to the fact that operations contribute to over 75% of carbon emissions across a life cycle. Similar trends are prevalent in all product and service industries. The life cycle assessment (LCA) conducted by Johanning and Scholz [50] highlights the considerable impacts of the operational phase, which consistently accounts for more than 75% of overall impacts. While previous research has been valuable in paving the way for a more sustainable future, the focus is still limited and has not assessed the interactions between operations and other phases of the life cycle. Research has also been narrowed to focus on aircraft and aircraft technologies rather than addressing some other critical aspects, such as predicting the adequacy of current technology and infrastructure in a carbon-neutral air transportation system. It is noted that there are many factors that can reduce emissions from aircraft operations, such as multimodal networks, ground operations, new infrastructure etc. However, this chapter endeavours to provide a brief analysis on one aspect believed to be most significant and promising. The most notable of these is air traffic management (ATM).

4.7.1 Air Traffic Management

ATM is perhaps the most crucial domain of research for enhancing operational efficiency and sustainability, as the routes of multiple aircraft crossing the same airspace regions cannot be planned independently from each other. Therefore any optimisation pursued shall be performed considering a vast surveillance picture in both strategic and tactical timeframes. Airspace sectors and airports are under increasing pressure as a result of traffic growth levels. Current ATM infrastructure is nearing its threshold, evoking system-wide inefficiencies through capacity constraints. Approximately 8% of fuel is wasted on current flight trajectories [51]. Fuel is the primary variable when considering environmental impacts. Less fuel represents fewer emissions; therefore, ATM initiatives are mainly centred on reducing fuel consumption through infrastructure and operational improvements. For example, flight times by one minute of delay could save approximately 4.8 million tonnes of carbon dioxide globally [52].

The most promising strategies that are being implemented to optimise air traffic constraints are focused on a network level with ambitious targets to address environmental sustainability. Frameworks developed by the Advisory Council of Aviation Research and Innovation (ACARE) in Europe are perhaps the most referenced [53, 54]. The Single European Sky Air Traffic Management Research (SESAR) and the Clean Sky Joint Technological Initiative (JTI) are programmes specifically engineered to meet top-tier safety, efficiency, and environmental sustainability objectives by addressing operational and environmental efficiencies [55]. The Next Generation Air Transportation System (NextGen) programme in the U.S.A. is another collaborative research initiative pursing ATM modernisation and operational improvements addressing noise, climate change, local air quality, energy consumption, and water quality [56]. In order to understand and mitigate ATM inefficiencies, we must analyse those individual phases of a flight mission that need to be addressed.

4.7.1.1 Departure Phase

The departure phase includes all ground movements, take-off and climb phases from the gate of departure to the insertion en route. Departure operations vary between airports; however, aircraft may be required to fly less optimal trajectories to comply with congested airspace sectors and noise abatement procedures. This could also be a result of airport-preferred or wind-dictated runways or predefined flight routes. These aspects can often lead to increased fuel consumption due to sub-optimal altitudes and speeds [57].

4.7.1.2 Cruise Phase

A traditional flight path follows pre-assigned waypoints from the departure to the arrival airport. Waypoints can be considered as the 'long way around', covering a greater distance from point A to B. Efforts to better flight paths with great circle arcs (which are the shortest paths) and to exploit tailwind regions while minimising headwinds play a key role in fuel optimisation.

4.7.1.3 Descent Phase

Inefficiencies at the descent phase can be realised by assessing the flight path both vertically and horizontally. Traditional approach practices require aircraft to level off at several points during the phase, translating to longer flight times and increased fuel consumption.

4.7.2 Trajectory Optimisation

There have been several research initiatives to optimise the flight trajectory inefficiencies highlighted above. Continuous ascent and descent profiles are leading the way to green aircraft operations. Criollo, et al. [58] explore the concept of continuous ascent departures, highlighting the fact that these reduce the need to level off during the climb phase, which is common practice to provide separation. Melby and Mayer [59] estimate that total fuel burn during departure can be reduced by 1.67% if continuous ascent procedures are adopted. The SESAR joint undertaking [60] provides results on continuous ascent operations trialled on Air France trans-Atlantic flights in 2009. They state that 135 000 t of carbon dioxide could be saved yearly based on the trials with trans-Atlantic flights alone.

In addition, continuous descent approach is a method use to optimise the vertical profile efficiency of aircraft trajectories. Robinson and Kamgarpour [61] note that this procedure allows aircraft to descend continuously from their cruise altitude, reducing the need to level off. Engines can therefore be at or near idle, bringing about savings in time and fuel consumption. Their analysis finds that continuous descents account for approximately 3% of fuel savings. Furthermore, Clarke, et al. [62] show that there is a 50% reduction in noise pollution compared to a traditional approach.

4.8 Sustainment Practices

Sustainability in aviation crucially entails the provision of all forms of logistics support, including equipment maintenance, that enables capability elements to participate in operations for as long as is required. The life cycle management process has to identify specific changes to improve the effectiveness and efficiency of the sustainment function for aviation. Effective sustainment will have a direct and positive impact on the overall economic, environmental and social sustainability of the sector.

Logistics include activities throughout each phase of the system life cycle. Logistics requirements are applicable in each of the phases of life cycle of systems: conceptual design phase, preliminary system design phase, detail design and development phase, production and/or construction, utilisation and support, and the requirement and disposal phase [63]. Life cycle costing (LCC) and similar methodologies assist decision-makers in making better informed decisions concerning management of assets at any stage throughout the life cycle – from initial planning, through budgeting to source selection, in-service management and, finally, at disposal. Most importantly, the type and amount of resources tend to be determined by early decisions [64].

The life cycle of a system may be broken down into four stages [65]:

i) Design and development.
ii) Production and construction.
iii) Operation and maintenance.
iv) System retirement and phase-out.

According to Blanchard [63], life cycle may be categorised in many different ways, depending on the type of system and the sensitivities desired in cost-effectiveness measurement. Moreover, the elements of the logistics and maintenance support infrastructure can have a major contribution on the total life cycle impacts of a system. The objective is to capture all impacts, including associated future ones, and to create the desired visibility. All applicable costs are estimated for each of the categories in cost breakdown structure (CBS) and for each year in the system life cycle. Furthermore, Blanchard [63] suggests 12 steps of a life cycle cost analysis process as follows:

i) Define systems requirements.
ii) Describe the system life cycle and identify the activities in each phase.
iii) Develop a CBS.
iv) Identify data input requirements.
v) Establish the costs for each category in the CBS.
vi) Select a cost model for the purposes of analysis and evaluation.
vii) Develop a cost profile and summary.
viii) Identify the high cost contributors and establish cause-and-effect relationships.
ix) Conduct a sensitivity analysis.
x) Construct a Pareto diagram and identify priorities for problem resolution.
xi) Identify feasible alternatives for design evaluation.
xii) Evaluate feasible alternatives and select a preferred approach.

The first seven steps lead to the development of a cost profile and a supporting cost summary. In step 11 (evaluation of alternative design configurations) there may be a different profile for each configuration being considered. Once the initial steps of a life cycle cost analysis are completed, the cost contributors are identified, with the objective of introducing appropriate design changes that will, in turn, reduce total system life cycle cost.

The final objective is to design and develop a system that is cost-effective subject to constraints specified by operational and maintenance requirements.

Figure 4.5 illustrates asset management in a general context and can also be extended to aviation. An effective sustainment management methodology would address the identification of several causal factors [64]. It would ensure that:

- Decisions taken during acquisition are based on whole-of-life considerations rather than just the acquisition component.
- Capability planning activities fully consider the costs of maintaining the existing fleet as an input to the decision on what new capability should be acquired.
- Planning is improved to allocate additional sustainment resources as capabilities age.

The informal approaches to sustainment management have the disadvantage of missing the opportunity to move the whole enterprise to a higher level of maturity. The Carnegie Melon University Capability Maturity Model is regarded as a benchmark for process maturity measurement. It defines five levels of maturity as follows:

- *Initial* (chaotic, ad hoc, individual heroics).
- *Repeatable* (process is managed in accordance with agreed metrics).
- *Defined* (process is defined as a standard business process).
- *Managed* (management can control through the process).
- *Optimising* (process includes continual improvement).

Sustainment is an important sub-component of Asset Management and its components are illustrated in light orange in Figure 4.5 [64]. Aircraft operators and design organisations are the main provider of sustainment services to the capability managers, but there is significant diversity in processes across the organisation. This should be addressed through the development of a sustainment management methodology as recommended in the Helmsman Sustainment Complexity Review (2010) [66] with the early and close involvement of the capability managers. This involvement is essential as the capability managers are not only the owners of the assets, but also deliver some essential sustainment services.

Integrated logistics support products (training, spares, documentation, etc.) are sometimes sacrificed by some organisations during acquisition in the complex trade-off between capability and cost. Failure to acquire these products and plan for sustainment seriously increases the whole-of-life costs and the risk of not achieving operational outcomes over an aircraft's planned life. It is critical that the aircraft operators implement the plan to provide a full suite of logistics support products. Blanchard [63] illustrated this issue in his iceberg model, shown in Figure 4.6. It depicts the support aspect of projects under the water and invisible to the unwary, who focus on the part that is visible. There is evidence that in aviation industry, decisions made during acquisition remain largely focussed on the goal of delivery rather than a whole-of-life view. A recommendation to take whole-of-life decisions is included with other recommendations in this chapter.

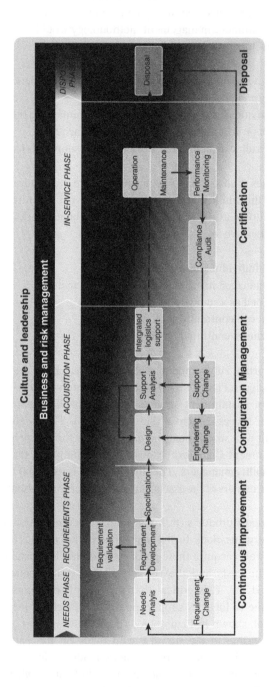

Figure 4.5 Asset management scope. Source: Rizzo [64].

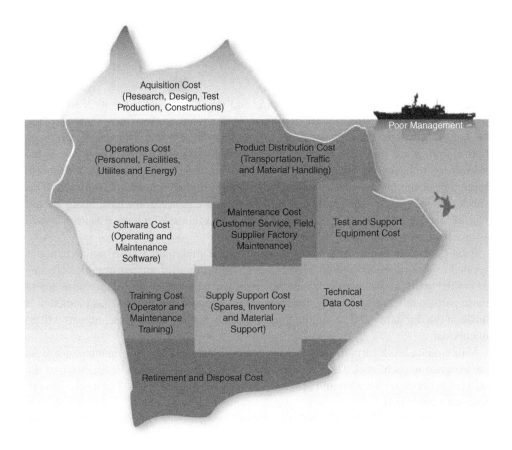

Figure 4.6 Iceberg model. Source: Blanchard [63].

4.9 Logistics Support Concept

A logistics support concept provides the guidance for engineering support, maintenance support, supply support, and training support. The aviation industry's logistics support concept is shaped by a number of factors, including technology; the amount of work that can realistically be performed on aircraft; the relative costs and the strategic necessity of repairing equipment locally versus sending it to overseas repair facilities.

Acquisition sources and technologies employed are quite diverse and therefore separate logistic support concepts may be required for each aircraft model/class. However, logistic support concepts need to be coordinated to ensure that both aviation and industry are allocated a level of work sufficient to maintain and develop the skills of their technical workforces. In the case of aviation, this means a level of work that will be achievable, but challenging, facilitating the retention of tradesmen in the service to assure a viable technical workforce.

4.10 Effective Sustainment

Aircraft will inevitably deteriorate with time and use. Usage will cause wear in moving parts, or structural change in parts that have varying loads applied. Over time oxidation or chemical changes occur. Hence some parts will be checked after a fixed usage, in flying hours, while some parts will be checked after a fixed calendar period. Every part of the aircraft has to be subject to some check over its lifetime, either by testing its function in some way, or the routine replacement of components that are known to deteriorate. This check and/or replacement is known as scheduled maintenance. Some parts on an aircraft will fail, or perform in unexpected ways. Any check on an aircraft for these reasons is known as unscheduled maintenance.

Each aircraft type will have its own maintenance programme. This maintenance is applied to all equipment and systems and is based on the historic performance of same or similar parts in that particular aircraft type and also based on the performance of those parts in other aircraft types. Occasionally on new systems a theoretical prediction is made of the durability and hence the timing of maintenance. In any case, no part will go unchecked.

Maintenance levels form the baseline for determining which specific maintenance tasks are assigned to each level. They are a means to select the scope of maintenance and the skill levels necessary for units and activities at various command levels. The levels of maintenance include [65]:

- Line maintenance.
- Role servicing/change.
- Replenishments.
- Minor maintenance.
- Operational maintenance (OM).
- Deep maintenance.
- Modification production.

Differences in the applicable regulatory frameworks will lead to different requirements on certification of the maintenance work being carried out.

Regulatory frameworks are changing and particularly benefiting from an enhanced integration and uniformity between military and civil aspects.

Indeed, Maintenance Management Plans (MMP) for military and civil are similar [65].

To provide reliable asset capacity, maintenance resources must be balanced with essential workloads and organised in a manner that assures effective performance of all three work types. Deferring critical maintenance inevitably leads to further breakdowns and a perpetuation of reactive maintenance that ultimately leads to failure of the business/company. Most maintenance departments are organised only for reaction to urgent demands. Resources must be provided for the important work that improves future reliability, thereby reducing future urgencies.

Maintenance managers are responsible for supporting maintenance with a variety of services, plus maintenance procurement and inventory management.

Today's maintenance managers face great challenges to increase output, reduce equipment downtime, and lower costs, all without compromising safety and the environment. Smart organization shall manage maintenance effectively to accomplish these objectives. This allows to sustain productivity and fuel growth, while driving down unnecessary overall expenses. Effective maintenance aims to [65]:

- Maximise uptime.
- Maximise accuracy.
- Minimise costs per unit.
- Minimize environmental impacts.
- Prevent safety hazards to employees and the public as much as possible.

In fact, all of the above are strategic necessities to remain in business. The challenge is how best to meet the above demands. In many companies, we have to start at the beginning – put the basics in place – before attempting to achieve excellence and optimise decisions.

Maximising profits while minimizing the environmental impacts depends on keeping production/service assets in good working order. Minimising downtime is essential to maximise the availability of production/service assets. Optimisation helps in finding the right balance. Even though increasing profit, revenues, availability, and reliability while decreasing downtime and cost are all related, we cannot always achieve them simultaneously. The typical trade-off choices in maintenance arise from trying to provide the maximum value to our 'customers'.

In the past, maintenance received little recognition for its substantial impacts in ensuring a cost-effective and environmentally-friendly operation of the system or product. It was viewed only as a necessary and unavoidable standalone process. This perception is progressively changing and efficient maintenance is now believed to be an enhancer of production and service capacity.

4.10.1   Efficient Sustainment

The sustainment of capital assets needs to be cost effective and environmentally sustainable throughout their life cycle. Planning before the acquisition phase of major projects is based on a detailed economic and environmental cost-benefit analysis of continued sustainment versus replacement [64]. Decisions to either purchase new equipment or maintain existing systems should be based on the through-life economic and environmental cost of each option, regardless of whether funding is for the acquisition or sustainment budgets.

Equipment failures are not linear over time but, in fact, rise at an ever increasing rate towards the end of life. This is demonstrated by the bathtub curve in Figure 4.7. This has a direct impact on maintenance requirements, costs and environmental impacts. The time to effect new aircraft capability delivery is typically in excess of 10 years. It is therefore important to determine the wearout failure period well in advance, in order to plan for both the replacement capability and the increase in sustainment required for the ageing capability, in terms of people and funding.

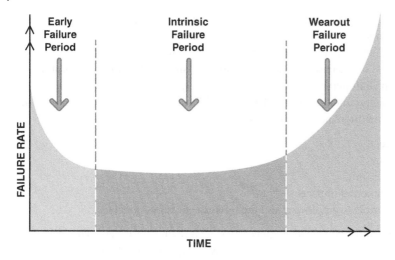

Figure 4.7 The bathtub curve. Source: [67].

4.11 Sustainable End-of-Life Management

Reducing resource consumption on a system-wide level is becoming a necessity in sustainable development practices. The life cycle approach is concerned, to a large degree, with minimising waste across an entire system. It is important to highlight that products have distinct end-of-life strategies based upon their design characteristics. Design objectives that take into account the recyclability and degree of disassembly play a vital role in the amount of material that can be extracted at this particular phase [35]. As Allenby and Graedel [68] state, design has the largest influence on the overall environmental impact of a product.

End-of-life is defined as a particular point in time when a product no longer satisfies user requirements (65). Residual value, quality, and overall economic and environmental performance decrease to an unsatisfactory level over a product's useful life, which can be between 20 to 30 years for aircraft. The significance of the end-of-life phase of the aircraft life cycle stems from the fleet-retirement forecasts presented by Boeing (2014) and Airbus (2014). A total of 27 855 aircraft are expected to be retired by the year 2033, raising the question as to whether traditional methods of aircraft decommissioning satisfy environmental objectives. Only recently has the end-of-life of aircraft been given any considerable amount of attention. This is mostly due to organisations, such as the Aircraft Fleet Recycling Association (AFRA), and such programmes as PAMELA (Process for Advanced Management of End-of-Life Aircraft), which look to optimise value creation from the recycling and remanufacturing of aircraft components. Rose [69] highlights that approximately 50% of aircraft materials have been reused or recovered in the past.

The architectural design of end-of-life systems is specific to the industry and environment in which it operates. A number of strategies and objectives must be taken into account when developing the system architecture that will facilitate end-of-life processes. Rose, Ishii, and

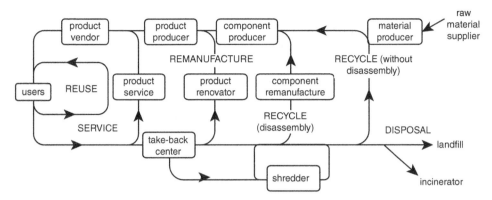

Figure 4.8 General architecture of an end-of-life system. Source: [71].

Stevels [70] present a hierarchical list of strategies based on their overall environmental impact. These are ordered by most favourable:

- *Reuse*: Second-hand use of a component in accordance with its original design intent.
- *Repair*: Applied maintenance to repair a product or component part with the objective of extending its useful life.
- *Remanufacture*: Products are disassembled and inspected. Reusable parts are obtained, cleaned, and repaired to form a remanufactured product.
- *Recycle with disassembly*: Products or component parts are disassembled into material fractions to remove contaminants or hazardous materials.
- *Recycle without disassembly:* Products or component parts undergo shredding to reduce material size to facilitate sorting.
- *Disposal*: Landfill or incinerate a product or component part without energy or value recovery.

Taking into consideration the above strategies, which are relevant to the aviation industry, an environmentally efficient end-of-life system will be designed to accommodate all of the above processes. As addressed in section 2.3, extending the supply chain architecture to allow for reverse product and information flow is an essential requirement of the end-of-life system. Figure 4.8, extracted from Tani [71], illustrates its general architecture. Here, the closed loop makes secondary use of resources and value that already exist within the system. The aim is ultimately to recover as much value as possible and to decrease the system's dependence on virgin materials and energy.

4.12 Life Cycle Models

A number of models have been developed and integrated into system life cycle design forming an abstraction of reality as a way to better the understanding of complexity. Model-based systems engineering (MBSE) is a preferential step towards a holistic view of system dynamics in that it provides the necessary tools and processes that are required to perform collaboration and corroboration earlier in the life cycle. The International Council on Systems Engineering (INCOSE) hasbeen the primary ambassador for model-based methodologies, highlighting its role of supporting system requirements, design, analysis, validation, and operations [72]. The tools and processes act to eliminate ambiguities that arise during complex system design by providing a better ability to define and manage the behaviour and interactions of systems and their subsystems. Given the fact that life cycle phases interact with activities in other phases, modelling addresses system-level requirements through consideration of the phase interactions [73]. There are three commonly cited life cycle models in the systems engineering field; namely the waterfall model [74],; the spiral model [75], and the 'Vee' model [76]. While these models are generic in nature, their application can be integrated in a wide range of fields.

4.12.1   Waterfall Model Evaluation

The waterfall model, depicted in Figure 4.9, is a sequential development methodology introduced by Royce in 1970 [74]. It represents a linear process where system development proceeds through a series of phases beginning with requirements to operation and maintenance. The original model assumed a forward directional progression, and where succession is only viable when the previous phase is complete. Hence, the outputs produced in one phase would become the inputs for the next. The processes are executed in order:

- *Requirement specification*: Identification and detailed assessment of user requirements.
- *Analysis*: Investigation into system requirements and the relevant architectures needed to support them.

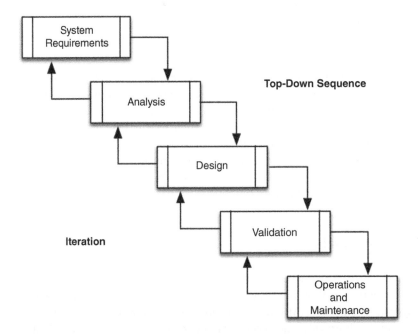

Figure 4.9   Waterfall life cycle model. Source: [74].

– *Design*: Architecting the system design that will satisfy the requirements identified in the first phase.
– *Implementation*: Execution of the system, its specifications, and standards.
– *Validation*: System testing as a tool to confirm requirements and quality are met.
– *Deployment*: Operational system.

While the waterfall model has served as a reference for many other life cycle models, it has met with heavy criticism [77]. The model cannot handle changes to requirements or design that occur downstream. The waterfall is a methodology considered to be inflexible in its nature, with ample evidence showing that sequential steps are often not obeyed, thus rendering the model somewhat unrealistic [78]. Royce later adapted his model to include iterations between each successive step, providing flexibility to changes that may occur downstream [74].

4.12.2 Spiral Life Cycle Model Evaluation

The spiral methodology, depicted in Figure 4.10, was developed by Boehm as a response to address the flaws of the waterfall model [75]. Its spiral characteristic highlights the necessity of performing excessive iterations throughout the system life cycle development, something which the original waterfall model did not allow. Its angular section represents project costs, while the radius represents project maturity. The spiral model was the first to integrate project costs and risk management concepts in its design. The incremental approach of

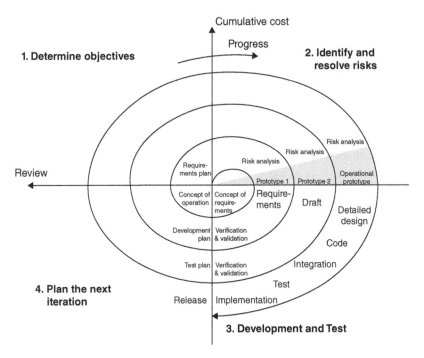

Figure 4.10 The spiral life cycle model. Source: [75].

development and the significant use of risk assessment provide the possibility of detecting and eliminating errors as they occur. Each particular loop depicts a different phase in the system development process, beginning with requirements and design. The number of cycles is specific to the project type; hence the descriptions of the activities are generic to allow for ease in its adoption to organisational specific projects. The quadrants are perhaps the most significant aspect to this model. Each specifies a step or process that is adopted through every cycle and assumes that objectives and requirements are well defined and that risks are eliminated through progression. The basic process is as follows:

Quadrant 1: Determine Objectives. Each program phase begins with requirement and objective elaboration. System requirements are identified in order to assess the alternative means of implementing these requirements into the particular phase [75]. This may mean a cross section analysis of several designs and the constraints imposed on those designs (cost, schedule).

Quadrant 2: Identify and resolve risk. The use of risk evaluation identifies the level of uncertainties that may accompany alternatives acknowledged in the previous quarter [75]. This step involves formulating strategies for resolving sources of risk by means of prototyping, simulation, analytic modelling, or a combination of these.

Quadrant 3: Development and Validation. Development of the designed system at each phase is undertaken. Development processes, such as waterfall or incremental approaches, can be taken at this phase. System testing is performed to validate it has met requirements.

Quadrant 4: Plan for next phase. Reviewing the previous process is an integral part to determine whether it is feasible to go forward or if the end product does not meet requirements and therefore will need to be revised. Issues that need to be resolved are identified and strategies to eliminate these are pursued.

4.12.3 V-Life Cycle Model Evaluation

Perhaps the most widely used methodology by system engineers is the traditional engineering V-model, depicted in Figure 4.11. In 1992, Forsberg and Mooz presented the V-model as a way of modelling the decomposition, integration, and verification of system designs [76]. Many authors have used a number of variants of the model, however, all represent similar activities. It is a top-down development process with a bottom-up implementation approach [73]. The left side of the model represents the activities involved in system development, while the right represents the system integration and verification. While the V-model is a generic representation of system design methodology, it highlights the importance of assessing environmental impacts of a system at its preliminary design stage. The feedback and validation loops (shown as dotted lines) are of key importance to system modification. Franz et al. explain that future ecological impacts are, to a large extent, pre-determined by the design of the system [20]. The main challenge of life cycle engineering in preliminary aircraft design results from the level of complexity of the processes with the aircraft life cycle, as well as the complexity of the aircraft and its subsystems, itself.

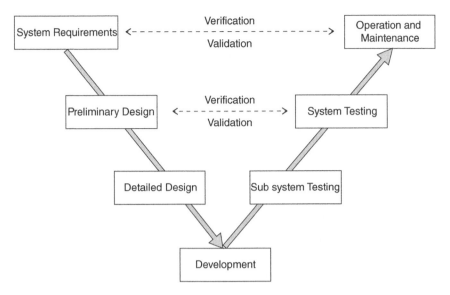

Figure 4.11 The V-model. Source: [76].

4.13 Proposed Life Cycle Methodology

Life cycle models are a valuable tool in the development process of a product or system. Their primary function is to define the processes involved in the evolution of a product. They provide the requirements or criteria for phase transitions – i.e. how to end one stage and begin the next [75]. Some even assist in cost evaluation and others are represented as a function of time. The complexity surrounding system-wide problems necessitates the use of a systematic process, however; these models have been heavily criticised in their simplistic depiction. The spiral, for instance, can be viewed as a sequence of waterfall increments with a specific focus on risk identification. Both the waterfall and V-Model are simple linear processes that can often lead to a false sense of security for project managers. This is a major misapprehension in so far as models should be adapted and extended in practice. Models are not a process themselves, but are, in fact, composed of a series of processes specific to the nature of the actual task. This leads to the important requirement of flexibility and the inherent ability to respond to change. The waterfall, for instance, is limited in its ability to adapt to change and stresses that each incremental step is completed only in succession of the previous one. The V-model places greater emphasis on system validation and provides a series of feedback loops to rectify problems that occur later in the development process. The spiral model, too, focuses on system validation through risk analysis; however, the spiral continues indefinitely and so raises ambiguity and can lead to increases in time and cost.

While it is agreed that the above models have not been developed in an ecological context, there are a series of barriers that need to be overcome. To quote Forrester [79], 'the world is not unidirectional. We live in an on-going circular environment where there is no beginning or end to a process'. With the exception of the spiral model, all previous frameworks adopt a linear architecture, which encompass both a beginning and end. Fiksel [80] advocates that industrial systems require a shift from linear to closed-loop, which resembles the flow of a natural ecosystem. It is agreed in the context of natural systems that energy is transferred in a revolving spiral, however, this does not need to be encompassed within an illustration. Modern technology fails to meet Fiksel's view in so far as it is not yet possible to create an industrial aviation system that does not depend on the use of virgin materials. Furthermore, we argue that there is a beginning and end to industrial development. Processes do not continue indefinitely. The beginning is depicted through requirement analysis and design. This process is succeeded through development and so forth. The overall end of an industrial process is its retirement. However, in view of a circular environment, we argue that extending a model to encompass end-of-life strategies and LCAs does, in fact, resemble the circular flow of a natural system regardless of whether it is depicted as linear or closed-loop.

The W-model has been developed specifically to address the lack of 'sustainable thinking' within current process systems. Project progression is similar in nature to the V-model; however, the 'W' is unique in its divergence, allowing two processes to occur simultaneously. The significance of the W-model stems from its emphasis on LCA. This specifically acts to provide relevant architecture for a self-regulating system (i.e. to close the loop) in addition to its ability to influence the overall environmental design, manufacturing, and operations of a system. The model focuses heavily on validation and verification through specific tests that are directly associated with each development phase and the overall requirement of the system. The 'W' essentially emphasises that LCA should be included within the development stage, so as to ensure that externalities during manufacturing, operations, and end-of-life are minimised at their source.

Utilising the W-model begins with stakeholder collaboration through which operational requirements are identified and later assessed. This is the generic starting point for any systems engineering process and is considered the most crucial in that it will form the basis for subsequent system requirements, design, and development. Operational requirements are based on system functionality, stakeholder needs, and the configuration, role, and environment it may need to satisfy. These requirements establish the expected level of performance and should ultimately address:

– *Where* and *how* will the system be used?
– *What* functions and capabilities are essential to meet desired performance?
– *What* are potential constraints and how can these be rectified?
– *How* will the objectives be accomplished and the requirements be verified?

Extensive analysis and risk assessment are used to develop and assist in defining the top-level characteristics of the system. This essentially dictates the evolutionary strategies to be implemented over a given timeframe. An essential component, often overlooked in systems engineering, is the continuous analytical effort between design and requirements. While the W-model progresses from requirements to preliminary design, a feedback loop between the two components highlights the need for reiteration to help provide further clarification throughout the entire process. This feedback loop acts to verify the preliminary design to system requirements, highlighting the model's responsiveness to change.

This includes performing risk assessments and evaluating alternative system designs through trade-off analysis between different and often conflicting requirements.

The most critical aspect of the W-model is its ability to link system requirements and design to a series of tests that simultaneously encompass generic system validation and verification as well as EIA. Verification and validation of both requirements and design are critical in early development activities. The purpose of these feedback loops is to link early stage activities with testing that would eventually occur later in the process. As suggested, this provides clarification in so far as:

- We can *validate* whether the system specifications will satisfy the requirements of the customer.
- We can *verify* whether the system will behave as expected when simulated against known inputs and if it ultimately upholds the requirements of the customer.
- We can *assess* the environmental externalities (positive or negative) of the processes, designs, or the overall project prior to further development progression.

The overall positive aspect of the W-model is its ability to invoke 'sustainable thinking' throughout system development. LCA is performed on three architectural levels to identify the overall environmental outputs from a product or system through the utilisation of known inputs (inventory analysis). The assessment accounts for all phases of the product life cycle, including such direct and indirect contributions as supply chain transportation, processing raw materials, and water and energy consumption. The depth of the analysis broadly aligns with the notion of 'cradle-to-grave'. EIAs are used to compare current requirements and designs to an alternative that may pose less environmental impact, for example, an aircraft that is designed to be recyclable. These can also be extended to analyse component supply chains and provide logistical solutions (e.g. intermodal logistics networks) that produce less externalities.

LCA is a valuable tool for assessing the cradle-to-grave impacts of a process or system. The inclusion of the LCA process not only differentiates the W-model from existing system-engineering models, but is also the first move towards 'life cycle accountability' – the notion that places responsibility on the manufacturers for their direct production impacts in addition to impacts associated with the product's use, distribution, and disposal. There are four primary components utilised in LCA (Figure 4.12):

1. *Goal definition and scoping*: This step identifies the goals and scope of the LCA and determines boundaries, assumptions, and potential limitations.
2. *Life cycle inventory*: Life cycle inventory analysis involves identifying the flow of materials and activities associated with the completion of the technical system. Here, we can quantify the energy consumed and the environmental externalities produced at each stage of production.
3. *Impact assessment*: Step three involves assessing the impacts identified in the previous phase on either the human health or the environment. These findings are fed into the improvement analysis phase.
4. *Data interpretation and improvement analysis*: An evaluation of the results and findings that were quantified in the previous steps to draw conclusions on opportunities or weaknesses as a way to reduce energy, material inputs, and environmental impacts at each phase of the product life cycle.

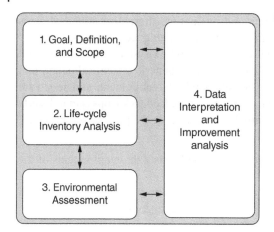

Figure 4.12 Life cycle assessment process.

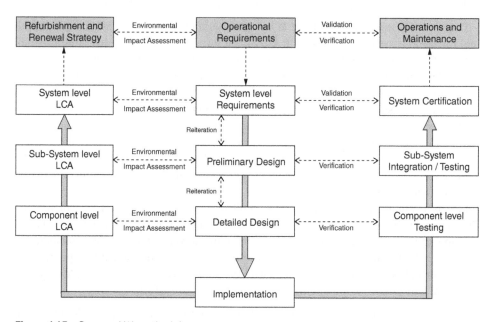

Figure 4.13 Proposed W-methodology.

4.13.1 W-Model Flow

This section provides details on the flow of the W-model illustrated in Figure 4.13.

Phase 1: Operational requirements, operations, maintenance, and end-of-life strategies:
This phase has the primary purpose of defining the top-level, often broad, requirements of the system or product. Acceptance testing and end-of-life strategies are formulated and incorporated in the subsequent processes. The fundamental aspect of this phase is to establish the pre-defined conditions, as per the customer requirements, that will require testing and verification.

Phase 2: System-level requirements, certification testing, and life cycle analysis:

Phase 2 essentially refines the aforementioned requirements and specifically focuses on what the system will require in order to validate these requirements. Requirement analysis is an integral part of this component in so far as alternatives can be identified early in the development process.

Certification testing and system-level LCA are ways to assess the overall integrity of the customer requirements and allows for further clarification and refinement. These tests document deviations between expected results and the actual results.

Phase 3: Preliminary design, sub-system integration/testing, and life cycle analysis:

Further refinement of system and operational requirements established in phases 1 and 2. The primary notion of phase 3 is to develop a conceptual design model based on the above requirements and to plan the testing activities to be performed for system verification. Reiteration may be performed between phase 2 and 3 so as to ensure clarity, focus, and refinement. Life cycle analysis and system testing is performed on a sub-system level as a way of confirming methods of system integration.

Phase 4: Detailed design, component testing, and life cycle analysis:

By phase 4, requirements and alternatives should be well defined and assessed. Detailed design of the system, product, and production and testing processes encompass the primary responsibilities of this phase. Testing and life cycle analysis is completed on a component level as a way of collecting the relevant data for inventory analysis. Component testing also includes supply chain strategies and architectures for component, sub-system, and system distribution.

Phase 5: Implementation:

The project team is responsible for translating requirements, specifications, and design into a physical process. Closed-loop production processes as highlighted in the literature review are an integral component of this phase.

References

1 Charles, M.B., Barnes, P., Ryan, N., and Clayton, J. (2007). Airport futures: towards a critique of the aerotropolis model. *Futures* 39: 1009–1028.

2 Lee, D.S., Fahey, D.W., Forster, P.M. et al. (2009). Aviation and global climate change in the 21st century. *Atmospheric Environment* 43: 3520–3537.

3 Maurice, L. and Lee, D. (2009). Assessing Current Scientific Knowledge, Uncertainties and Gaps in Quantifying Climate Change, Noise and Air Quality Aviation Impacts, final report of the International Civil Aviation Organization (ICAO) Committee on Aviation and Environmental Protection (CAEP) Workshop, Washington DC and Manchester: US Federal Aviation Administration and Manchester Metropolitan University.

4 Bakshi, B.R. and Fiksel, J. (2003). The quest for sustainability: challenges for process systems engineering. *AICHE Journal* 49: 1350–1358.

5 Keating, C., Rogers, R., Unal, R. et al. (2003). System of systems engineering. *Engineering Management Journal* 15: 36–45.

6 Chester, M.V. and Horvath, A. (2009). Environmental assessment of passenger transportation should include infrastructure and supply chains. *Environmental Research Letters* 4.

7 Sage, A.P. (1992). *Systems Engineering*, vol. 6. Wiley.

8 Jamshidi, M. (2008). *Systems of Systems Engineering: Principles and Applications*. CRC press.

9 Sage, A.P. and Cuppan, C.D. (2001). On the systems engineering and management of systems of systems and federations of systems. *Information, Knowledge, Systems Management* 2: 325–345.

10 Keating, C.B. and Katina, P.F. (2011). Systems of systems engineering: prospects and challenges for the emerging field. *International Journal of System of Systems Engineering* 2: 234–256.

11 Martin, J.N. (1996). *Systems Engineering Guidebook: A Process for Developing Systems and Products*, vol. 10. CRC Press.

12 Shenhar, A. (1994). A new systems engineering taxonomy. In: *Proceedings of the 4th International Symposium of the National Council on System Engineering, National Council on System Engineering*, 261–276.

13 Calvano, C.N. and John, P. (2004). Systems engineering in an age of complexity. *Systems Engineering* 7: 25–34.

14 Eisner, H., Marciniak, J., and McMillan, R. (1991). Computer-aided system of systems (S2) engineering. In: *Systems, Man, and Cybernetics, 1991.'Decision Aiding for Complex Systems, Conference Proceedings., 1991 IEEE International Conference on*, 531–537.

15 Jackson, S. (1997). Systems engineering for commercial aircraft. In: *INCOSE International Symposium*, 36–43.

16 Sadraey, M.H. (2012). *Aircraft Design: A Systems Engineering Approach*. Wiley.

17 Janic, M. (2000). *Air Transport System Analysis and Modelling*, vol. 16. CRC Press.

18 Nahavandi, S., Creighton, D., Johnstone, M., and Le, V.T. (2009). Airport operations: a system-of-systems approach. In: *Systems of Systems Engineering: Principles and Applications*, 403–419. CRC Press.

19 Cascetta, E. (2013). *Transportation Systems Engineering: Theory and Methods*, vol. 49. Springer Science & Business Media.

20 Franz, K., Hörnschemeyer, R., Ewert, A. et al. (2012). Life cycle engineering in preliminary aircraft design. In: *19th CIRP Conference on Life Cycle Engineering, LCE 2012*, 473–478.

21 D. DeLaurentis, (2005). "Understanding transportation as a system-of-systems design problem," in *43rd AIAA Aerospace Sciences Meeting and Exhibit*, ed: American Institute of Aeronautics and Astronautics.

22 Mont, O. and Bleischwitz, R. (2007). Sustainable consumption and resource management in the light of life cycle thinking. *European Environment* 17: 59–76.

23 Pesonen, H.-L. (2001). Environmental management of value chains. *Greener Management International* 2001: 45–58.

24 Mclaren, J. and Mclaren, S. (2010). Life cycle management. In: *Hatched: The Capacity for Sustainable Development* (ed. B. Frame, R. Gordon and C. Mortimer). Las Vegas, NV, USA: UNLV.

25 Stark, J. (2011). *Product Life cycle Management*. London, UK: Springer-Verlag.

26 Blanchard, B.S. (2004). *System Engineering Management*. Wiley.

27 Wiley (2015). *INCOSE Systems Engineering Handbook: A Guide for System Life Cycle Processes and Activities*. Wiley.

28 Beamon, B.M. (1999). Designing the green supply chain. *Logistics Information Management* 12: 332–342.

29 Rogers, D.S. and Tibben-Lembke, R. (2001). An examination of reverse logistics practices. *Journal of Business Logistics* 22: 129–148.

30 Stansinoupolos, P., Smith, M.H., Hargroves, K., and Desha, C. (2013). *Whole System Design: An Integrated Approach to Sustainable Engineering*. Routledge.

31 Ilgin, M.A. and Gupta, S.M. (2010). Environmentally conscious manufacturing and product recovery (ECMPRO): a review of the state of the art. *Journal of Environmental Management* 91: 563–591.

32 Sroufe, R., Curkovic, S., Montabon, F., and Melnyk, S.A. (2000). The new product design process and design for environment: "Crossing the chasm". *International Journal of Operations & Production Management* 20: 267–291.

33 Stapelberg, R.F. (2009). *Handbook of Reliability, Availability, Maintainability and Safety in Engineering Design*. Springer Science & Business Media.

34 Gungor, A. and Gupta, S.M. (1999). Issues in environmentally conscious manufacturing and product recovery: a survey. *Computers and Industrial Engineering* 36: 811–853.

35 Wang, H.-F. and Gupta, S.M. (2011). *Green Supply Chain Management: Product Life Cycle Approach*. McGraw Hill Professional.

36 Eastman, C.M. (2012). *Design for X: Concurrent Engineering Imperatives*. Springer Science & Business Media.

37 Giudice, F., La Rosa, G., and Risitano, A. (2006). *Product Design for the Environment: A Life Cycle Approach*. CRC Press.

38 Telenko, C., Seepersad, C.C., and Webber, M.E. (2008). A compilation of design for environment principles and guidelines. In: *ASME 2008 International Design Engineering Technical Conferences and Computers and Information in Engineering Conference*, 289–301. ASME.

39 Beck, A.J., Hodzic, A., Soutis, C., and Wilson, C.W. (2011). "Influence of Implementation of Composite Materials in Civil Aircraft Industry on Reduction of Environmental Pollution and Greenhouse Effect," IOP Conference Series: Materials Science and Engineering, 012015.

40 Alting, L. (1995). Life cycle engineering and design. *CIRP Annals - Manufacturing Technology* 44: 569–580.

41 Melnyk, S. and Smith, R. (1996). Green Manufacturing Society for Manufacturing Engineering, Dearborn, MI.

42 Sarkar, A. (2012). Evolving green aviation transport system: a hoilistic approah to sustainable green market development. *Journal of Climate Change* 1: 164–180.

43 Azzone, G. and Noci, G. (1998). Identifying effective PMSs for the deployment of "green" manufacturing strategies. *International Journal of Operations & Production Management* 18: 308–335.

44 OECD (ed.) (2010). *Eco-Innovation in Industry: Enabling Green Growth*. France: OECD Publishing Paris.

45 Erkman, S. (1997). Industrial ecology: an historical view. *Journal of Cleaner Production* 5: 1–10.

46 Lunt, P. and Levers, A. (2011). Reducing Energy Use in Aircraft Component Manufacture-Applying Best Practice in Sustainable Manufacturing", SAE Technical Paper 0148-7191.

47 Garetti, M. and Taisch, M. (2012). Sustainable manufacturing: trends and research challenges. *Production Planning & Control* 23: 83–104.

48 Vlachos, D., Georgiadis, P., and Iakovou, E. (2007). A system dynamics model for dynamic capacity planning of remanufacturing in closed-loop supply chains. *Computers & Operations Research* 34: 367–394.

49 Jayaraman, V., Jr, V.D.R.G., and Srivastava, R. (1999). A closed-loop logistics model for remanufacturing. *The Journal of the Operational Research Society* 50: 497–508.

50 Johanning, A. and Scholz, D. (2014). *A First Step Towards the Integration of Life Cycle Assessment into Conceptual Aircraft Design*. Deutsche Gesellschaft für Luft-und Raumfahrt-Lilienthal-Oberth eV.

51 ATAG. (2012). Revolutionsing Air Traffic Management. Practical steps to accelerating airspace efficiency in your region, Geneva, Switzerland.

52 Airbus. (cited 6 September, 2012). Environmental Efficiency. Available from: http://www .airbusprosky.com/mission/environmental-efficiency.html

53 ACARE. (cited 5 September 2004). Strategic Research Agenda. Available from: http:// www.acare4europe.com/sites/acare4europe.org/files/document/ASD-Annex-final-211004-out-asd.pdf

54 ACARE. (2012). Strategic Research & Innovation Agenda (SRIA), Advisory Council for Aviation Research and Innovation in Europe (ACARE).

55 Ramasamy, S. and Sabatini, R. (2015). Communication, navigation and surveillance performance criteria for safety-critical avionics and ATM systems. In: *Proceedings of AIAC are available from the Royal Aeronautical Society and Engineers Australia. AIAC 16: Multinatioinal Aerospace Programs-Benefits and Challenges*, 1–12. Melbourne, Australia.

56 Gardi, A., Sabatini, R., and Ramasamy, S. (2016). Multi-objective optimisation of aircraft flight trajectories in the ATM and avionics context. *Progress in Aerospace Sciences* 83: 1–36.

57 CANSO. (2012). ATM Global Environment Efficiency Goals for 2050", Civil Air Navigation Services Organisation (CANSO).

58 Criollo, J., Campos, G.M., Kaloyanov, K., Tefera, J., and Villanueva, R.C. (2013). Optimization and Coordination of the Landing, Departure and Surface Management Operations.

59 Melby, P. and Mayer, R.H. (2008). Benefit potential of continuous climb and descent operations. In: *The 26th Congress of International Council of Aeronautical Sciences (ICAS)*, 14–19. Anchorage, Alaska, USA: ICAS.

60 SESAR. (2011). Modernising the European Sky, SESAR Consortium, Brussels, Belgium.

61 Robinson, J. and Kamgarpour, M. (2010). Benefits of continuous descent operations in high-density terminal airspace under scheduling constraints. In: *10th AIAA Aviation Technology, Integration, and Operations (ATIO) Conference*, 13–15. Fort Worth, TX.

62 Clarke, J.P.B., Ho, N.T., Ren, E. et al. (2004). Continuous descent approach: design and flight test for Louisville international airport. *Journal of Aircraft* 41: 1054–1066.

63 Blanchard, B.S. (2004). *Logistics Engineering and Management*. New Jersey, USA: Pearson Prentice Hall Publishers.

64 Rizzo, P.J. (2011). Plan to reform support ship repair and management practices. Ministerial and Executive Coordination and Communication Division, Defence, Commonwealth of Australia, July 2011.

65 Kumar, A. and Mo, J. (2011). Maintenance and logistics integration. Lecture Notes, RMIT University, Melbourne, Victoria, Australia.

66 Helmsman. (2010). *Helmsman International, The Helmsman Sustainment Complexity Review*. Canberra: Helmsman International Pty Ltd, July 2010.

67 Kapur, K.C. (2014). *Reliability Engineering*. Hoboken, NJ, USA: Wiley.

68 Allenby, B.R. and Graedel, T. (1993). *Industrial Ecology*. Englewood Cliffs, NJ: Prentice-Hall.

69 Rose, C.M. (2000). *Design for Environment: A Method for Formulating Product End-of-Life Strategies*. Stanford University.

70 Rose, C., Ishii, K., and Stevels, A. (2002). Influencing design to improve product end-of-life stage. *Research in Engineering Design* 13: 83–93.

71 Tani, T. (1999). Product development and recycle system for closed substance cycle society. In: *Environmentally Conscious Design and Inverse Manufacturing, 1999. Proceedings. EcoDesign'99: First International Symposium on*, 294–299.

72 INCOSE (2007). *Systems Engineering Handbook: A Guide for System Life Cycle Processes and Activities* (ed. C. Haskins, K. Forsberg and M. Krueger). San Diego, CA: Wiley.

73 Blanchard, B.S. and Fabrycky, W.J. (1990). *Systems Engineering and Analysis*, vol. 4. Englewood Cliffs, NJ: Prentice Hall.

74 Royce, W.W. (1970). Managing the development of large software systems. In: *Proceedings of IEEE WESCON*, 328–388. IEEE.

75 Boehm, B.W. (1988). A spiral model of software development and enhancement. *Computer* 21: 61–72.

76 Forsberg, K. and Mooz, H. (1992). The relationship of systems engineering to the project cycle. *Engineering Management Journal* 4: 36–43.

77 Munassar, N.M.A. and Govardhan, A. (2010). A comparison between five models of software engineering. *IJCSI* 5: 95–101.

78 Vliet, H.V. (2008). *Software Engineering: Principles and Practice*, vol. 3. Wiley.

79 Forrester, J.W. (1998). *Designing the Future," Work presented at Universidad de Sevilla*, vol. 15.

80 Fiksel, J. (2006). Sustainability and resilience: toward a systems approach. *Sustainability: Science Practice and Policy* 2: 14–21.

5

Life Cycle Assessment for Carbon Neutrality

Enda Crossin[1], Alessandro Gardi[2,3], and Roberto Sabatini[2,3]

[1] *University of Canterbury, Christchurch, New Zealand*
[2] *Department of Aerospace Engineering, Khalifa University of Science and Technology, Abu Dhabi, UAE*
[3] *School of Engineering, RMIT University, Melbourne, Victoria, Australia*

5.1 Introduction

The life cycle of a product is comprised of three broad stages:

1. Production, including the extraction of raw material from the ground and conversion of materials into usable form.
2. Use, including the conversion of energy.
3. Disposal, which can include recycling and/or landfill.

This life cycle is often termed 'cradle to grave'. Life cycle assessment (LCA) is a quantitative technique used to assess the environmental performance of products (including service systems) over the full life cycle. The LCA technique is used to identify environmental performance opportunities and to manage environmental risks [1] associated with a product or service system.

This chapter explores the history of LCA, the key principals of undertaking LCA, and methodological aspects of LCA with a focus on applications in the aviation sector. The chapter concludes with a review of LCA trends and an outlook of how LCA can contribute to sustainability in the aviation sector.

5.2 History

The LCA technique has its inception in the United States in 1969, when Coca-Cola commissioned a resource environmental profile analysis (REPA) of different packaging formats [2, 3]. The REPA technique focuses on resource extraction and emission flows, and does not translate these flows into environmental impacts. In the mid-1970s to mid-1980s, the REPA technique was further developed in parallel in Europe and the United States. In the early 1990s, regulatory concerns regarding the misuse of REPA for marketing purposes, and the need to translate resource extraction and emission flows into environmental impacts led to the standardisation of methodologies and the establishment of the

Sustainable Aviation Technology and Operations: Research and Innovation Perspectives, First Edition.
Edited by Roberto Sabatini and Alessandro Gardi.
© 2024 John Wiley & Sons Ltd. Published 2024 by John Wiley & Sons Ltd.
Companion Website: www.wiley.com/go/sustainableaviation

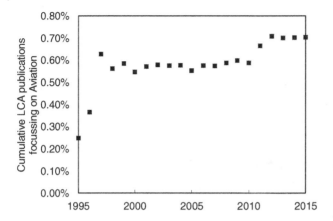

Figure 5.1 Cumulative percentage of LCA publications focussing on aviation and aerospace. Source: Data from Google Scholar.

first LCA standards by the International Standards Organization, under the ISO 14000 environmental management standards (1997 to 2002). These standards were later replaced by ISO 14040:2006 and ISO 14044:2006 [4, 5], which remain as the best practice in guiding the application of LCA. The 1990s and 2000s saw the adoption of LCA into government policy frameworks, particularly for packaging, the development of life cycle inventory (LCI) databases, the evolution of alternative LCA methodologies, and advancements in environmental impact (life cycle impact assessment) modelling techniques [6]. In the past decade (2010s), LCA techniques were applied to include social impact assessment, and began to incorporate geographic information system (GIS) data into inventory models, allowing for more appropriate accounting of regional-specific environmental impacts. The major focal points for the application of LCA are in packaging design, waste management, energy systems, and the built environment. Figure 5.1 depicts the percentage of LCA publications focussing on aviation and aerospace. The graph depicts the beneficial effect that these two important framework evolutions had on the sector. In particular, the first substantial increase occurred between 1995 and 1998, while the second increase from 2010 to 2012 is attributable to LCA publications that focus on the assessment of alternative fuel systems.

5.3 LCA Standards

The ISO 14040:2006 and ISO 14044:2006 standards fit within the ISO 14000 series of environmental standards. ISO 14040:2006 and ISO 14044:2006 outline the framework for undertaking LCA, as illustrated in Figure 5.2. This framework is based on four steps: 1. establishing the goal and scope of the study, 2. inventory analysis, 3. assessing the environmental impacts, and 4. interpretation. The framework allows for the outcomes of these four steps to be reviewed and modified as an LCA study progresses. The details associated with these steps, including how these can be applied to aviation-related LCAs, will be discussed in Section 5.3.

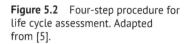

Figure 5.2 Four-step procedure for life cycle assessment. Adapted from [5].

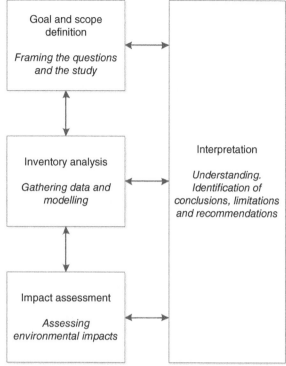

The ISO 14040:2006 and ISO 14044:2006 standards provide an important platform for the application of LCA. However, a common criticism of the ISO standards is that the methods outlined are open to interpretation, allowing for divergent LCA approaches, which potentially undermine the legitimacy of LCA studies [7, 8]. In response to the need for more prescriptive LCA approaches, a number of complimentary LCA standards have been developed, including greenhouse gas ('carbon') foot-printing standards PAS 2050:2011 [9] and ISO/TS 14067:2013 [10]. In addition, a suite of product-specific LCA standards and guidance documents has also been developed, including those for assessing the environmental performance of biofuels [11, 12].

5.4 Overview of LCA Applications

Key motivations for applying LCA include product differentiation (on environmental performance), identification of eco-efficiency cost savings throughout a supply chain, improved decision support (including for procurement), market and/or legislative compliance, and policy support [13]. The main applications of LCA include [14]:

- Product certification.
- Comparing the environmental performance of comparable systems, including the use of Environmental Product Declarations (EPDs).

- Design of products and systems, including the improvement of existing designs.
- Identifying environmental issues associated with a product or system, including potential 'burden shifting'.

As a method that quantifies environmental impacts, LCA is particularly suitable for application in sustainability certification schemes [15]. The Roundtable on Sustainable Biomaterials (RSB; formerly known as the Roundtable on Sustainable Biofuels) manages one such scheme, the RSB Standard. The RSB Standard is used for the certification of materials or fuels produced from biomass, including aviation biofuels. The RSB Standard is based on four major elements, covering social, legal, management, and environmental performance [16]. Within the environmental performance elements, LCA is applied to quantify the potential greenhouse-gas emission reductions [12].

EPDs are independently verified reports used by companies to detail the environmental profile of a product or service over the full life cycle. EPDs allow for a fairer comparison of the environmental performance of systems, based on a common set of rules. EPDs are largely based on LCA outcomes, but can often include other environmental indicators not readily quantified by LCA. For example, an EPD of a laser printer includes LCA-type metrics (e.g. greenhouse gas emissions, acidification potential), as well as metrics on the mass of hazardous waste generated in the life cycle, and the mass of material which will likely be recycled at the end of life [17]. The framework for EPDs are covered under ISO 14025:2006 [18], which then refers to the requirements of the ISO 14040:2006 and ISO14044:2006 LCA standards [4, 18]. Because EPDs are primarily used for comparing the environmental profile of systems, it is important that the methods used to generate the LCA outcomes are based on a common set of guidelines. These guidelines are termed product category rules (PCRs) and different PCRs exist for different products and services. The process for developing PCRs is governed by ISO 14025:2006 and includes an open consultation process with interested parties (e.g. competitors, non-government organisations) to minimise the likelihood of methodological disputes. PCRs are typically managed by independent organisations, often termed programme operators. A number of PCR programme operators exist, including The International EPD® System (www.environdec.com), IERE Earthsure (http://iere.org) and Institut Bauen und Umwelt e.V. (bau-umwelt.de). The applications of EPDs are dominated by the food and agriculture as well as building product industries [19]. The uptake of EPDs in the transport vehicles sector is currently limited, accounting for less than 5% of published EPDs [19]. In the aviation sector, there are currently no published EPDs. However, PCRs have been developed for passenger aircraft [20] and Bombardier published the first EPD of a passenger aircraft with the launch of its C-Series jet in 2016 [21].

5.5 Types of LCA

5.5.1 Process-Based LCA

Most LCA studies are process-based (often termed conventional LCA, bottom-up, or process LCA). In this approach, the life cycle of a system is modelled in discrete, yet interconnected

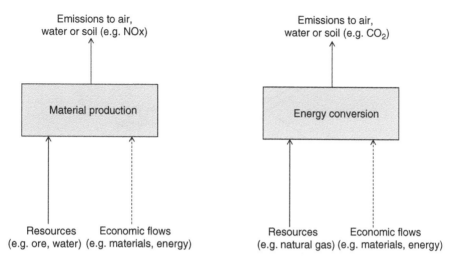

Figure 5.3 System processes for material production and energy conversion in process-based LCA.

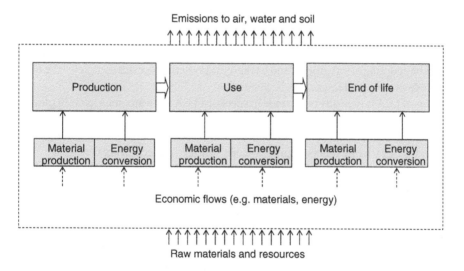

Figure 5.4 Schematic of flows associated with the life cycle of a product system.

system processes. Each system process represents a particular part of the product life cycle, and can include the production of material, energy conversion, transportation of goods etc. These system processes can have associated environmental flow inputs (e.g., resources from nature) or outputs (e.g., emissions to air, water, or soil), as well flows from the economy (e.g., material and energy inputs), Figure 5.3.

The system processes associated with a product life cycle are then interconnected, completing an inventory of environmental flows for the system, Figure 5.4.

Process-based LCA requires the collection of data associated with foreground system processes (e.g., production data from an OEM), together with the coupling of these foreground

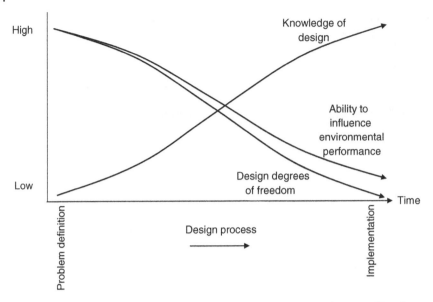

Figure 5.5 Relationship between the progress of a design and the degrees of freedom possible to reduce environmental impacts. Adapted from [22].

system processes with background system processes (e.g. energy), typically available in LCI databases. The coupling of foreground and background system processes completes the LCI, allowing for environmental impacts to be assessed.

Process-based LCA is typically performed in specialist LCA software, such as GaBi (developed by ThinkStep), SimaPro (developed by PRe Consultants), or openLCA (an open source tool). A suite of LCI databases is available for each of these tools. Some of these databases are free (e.g. US LCI), while others require a licence for use (e.g. ecoinvent).

5.5.2 Streamlined LCA

In order for LCA to be most effective in the design of new products, it needs to be implemented early in the design stage [22], Figure 5.5. However, detailed specifications regarding the materials and processes of the design are often not established until further in the design process, meaning that detailed LCAs are not always suitable for the design of new products [23]. Streamlined LCA is a simplified LCA methodology that allows non-LCA experts to explore potential environmental impacts early in the design process. Streamlined LCA can be complemented by life cycle design strategies, based on Design for Environment guidelines. Although the accuracy of a streamlined LCA may be lower than those of a traditional process-based LCA, streamlined LCA plays an important role in facilitating the uptake of life cycle thinking and LCA in design [24].

5.5.3 Attributional vs. Consequential Approaches

Attributional LCA is based on an LCI that models the environmental flows associated with a life cycle and its sub-systems, whilst consequential LCA is based on an LCI that models the changes in environmental flows in response to a decision (i.e. the consequences) [25, 26]. The consequential LCA approach typically involves forecasting changes in systems directly and indirectly related to the product or service being assessed, considering market dynamics (e.g. marginal suppliers). The modelling of indirect land use change (LUC) (refer to Section 5.6.6 in this chapter) is a form of consequential LCA. The application of LCA is often associated with making decisions to reduce environmental impacts in the future. As such, it is often suggested that consequential LCA is more appropriate for policy and decision support than attributional LCA [27]. However, consequential LCA is inherently more uncertain than attributional LCA [28], which can potentially limit its application. Despite this limitation, consequential LCA has the potential to identify environmental issues that may arise as a result of a change in a system, which may have otherwise been overlooked in attributional LCA.

5.5.4   Economic Input-Output LCA

Economic input-output life cycle assessment (EIO-LCA) is based on a system of national accounts, which map monetary flows between different sectors within a given economy. In addition to monetary flows, each economic sector also has environmental flows (e.g. extraction of resources and emissions). When these two aspects are combined, the environmental flows associated with expenditure in a particular economic sector, along with the economic sectors supporting the main economic sector, can be determined. The EIO-LCA approach allows for a rapid determination of environmental impacts, without the need to collect data relating to all processes associated with a product's life cycle. The environmental flows included in an EIO-LCA are not limited to those of specific processes. Rather, EIO-LCA data are limited by an economic boundary. Because of this different boundary approach, EIO-LCA are not considered to 'cut-off' important activities (and resultant environmental impacts) associated with a product/service system [29]. EIO-LCA can be viewed as a form of attributional LCA [25] and allows for a rapid assessment of environmental impacts with a broader system boundary.

Despite these advantages, EIO-LCA has a number of limitations. Firstly, EIO-LCA aggregates economic activities into discrete clusters, which limits the interpretation of EIO-LCA outcomes; EIO-LCA is thus not particularly suited to identifying those downstream activities that drive environmental impacts. Secondly, imported products (i.e. those being produced by a different economy) are not easily modelled in EIO-LCA [30]. Finally, care should be taken when modelling the end-of-life in EIO-LCA, particularly if the consumer does not directly pay for disposal costs during procurement.

In practice, EIO-LCA is most typically applied to LCAs on large infrastructure projects, including buildings and water distribution systems [31]. EIO-LCA databases which integrate with specialist LCA software (e.g. SimaPro, openLCA, GaBi) are widely available. EIO-LCA datasets include those for the US, Danish, Swiss, European, and Australian economies. In addition to these databases, a number of EIO-LCA tools are available online (e.g. http://www.eiolca.net/cgi-bin/dft/use.pl). When undertaking EIO-LCA, care should be given to account for inflation/deflation of the economy and exchange rates (if using EIO-LCA from a different region).

5.5.5 Hybrid LCA

Hybrid LCA combines the advantages of process-based and economic input-output LCA approaches. In so doing, hybrid LCA allows for the rapid development of a broader system boundary [32]. A number of hybrid LCA approaches are available, but the most commonly implemented approach is tiered hybrid analysis, which utilises process-based and EIO LCA data for important processes, with background data supported by process-based LCA databases [33]. In practice, tiered hybrid analysis LCA is easily implemented in specialist LCA software.

5.6 Principles of LCA

LCA is based on a four-step procedure, outlined in the ISO 14040:2006 and ISO 14044:2006 standards, Figure 5.2. Each of these steps are interconnected. Although often reported as sequential, the different steps are frequently revisited throughout an LCA study. The remainder of this section will discuss key principles associated with each procedural step of LCA.

5.6.1 Definition of Goal and Scope

The goal and scope form the basis for undertaking an LCA. The ISO 14044:2006 standard identifies critical elements of both the goal and scope definitions. In defining the goal, the intended application and audience must be clearly stated. The scope definition outlines key methodological issues, including defining the metric upon which environmental performance are assessed (the functional unit), which environmental indicators will be assessed, processes and activities to be included/excluded (the system boundary), how multi-function processes will be handled (allocation procedures), data assumptions and limitations, and the quality of data needed to undertake the study. Critically, the ISO 14044:2006 requires that any LCA which makes comparisons between systems must be subject to a peer review; any such requirement needs to be reported in the scope definition. The goal and scope of an LCA can change as the study progresses.

5.6.2 Inventory Modelling

An LCI is the core component of an LCA and consists of modelling the system(s) under assessment. The main outcome from LCI modelling is a list of environmental flows, detailing the resource extraction and emissions associated with the system being studied (typically quantified against the functional unit). The LCI phase includes collecting and modelling foreground (or primary) data, consisting of direct measurements relating to the product system (e.g. mass of material and energy inputs, emission flows, etc.), and background (or secondary) data, comprised of data from existing LCI databases (e.g. ecoinvent or USLCI) or literature. The foreground and background data are coupled in LCA software, completing the LCI. The completed LCI reflects the processes to be included in the study (the system boundary) and the environmental flows associated with these processes. In undertaking LCI modelling, it is important to capture all environmental flows that are assessed in the life cycle impact assessment (LCIA); if this is overlooked, then environmental impacts may be underestimated.

5.6.3 Allocation Approaches

An important aspect of LCI modelling is allocation; that is, how to account for multi-functional processes and recycling. Allocation relates to the attribution of environmental flows to different products or processes. Examples of multi-functional processes included refining of crude oil into fuels, and an aircraft carrying both freight and passengers. Recycling systems can be viewed as multi-function processes, and as such are often managed using allocation.

The ISO 14044:2006 standard outlines a hierarchical approach to managing allocation. The hierarchical steps are to:

1. Avoid allocation by increasing the level of detail in the LCI; in practice this is typically performed as part of the LCI.
2. Undertake system expansion, whereby co-products are typically given credit for products outside of the system boundary that are functionally equivalent. In practice, there are several ways system expansion can be applied, and ISO 14044:2006 provides limited guidance on how system expansion should be applied.
3. Attribute the impacts of co-products based on causality. For example, certain emissions from an incinerator can be attributed back to the specific combustion of materials.
4. Attribute the impacts of co-products based on some other relationship. These relationships can be based on the co-products' energy, mass, or revenue generation (economic allocation). For example, products from an oil refinery can be allocated based on their relative energy contents.

The application of system expansion requires an understanding of how (or even if) the co-products displace products with equivalent functions in the wider economy. In this respect, system expansion has elements of consequential analysis, even if applied in an attributional LCA study.

Although the ISO 14044:2006 standard mandates a hierarchical approach, applying one approach consistently throughout an LCA study is problematic due to a number of factors, including data availability (including for markets, as required for system expansion), and multiple allocation approaches being applied in the background LCI database(s). As such, LCA studies often have a mixture of allocation procedures applied. The allocation procedures used in a study can have a significant impact on the LCA outcomes. For example, Kaufman et al. [34] assessed the effect of different allocation methods on the greenhouse-gas intensity of corn-based and stover-based ethanol biofuels, relative to a fossil-fuel baseline. The authors report that the allocation choice can change the relative performance of the biofuels from 36% to 79% and − 10% to 41%, respectively. This wide range of outcomes can be problematic, particularly if trying to compare studies, or to assess the environmental performance of a system for regulatory or accreditation purposes. Because of these variations, the ISO 14044:2006 requires that the effect of allocation methods on the outcomes of an LCA be tested using sensitivity studies.

Recycling is a form of coproduction. As for co-product allocation, allocation of recycling processes is not resolved in LCA and several different approaches can be applied. The ISO 14044:2006 requires that recycling allocation follow the same hierarchical steps as outlined above. The application of system expansion to recycling systems can be difficult to apply, because intermediate processing steps (e.g. the collection and sorting of metal scrap) could be attributed to either the system generating the recycled product and/or the system receiving the recycled product. Weidema [35, 36] provides guidelines on how system expansion can be applied to recycling systems. There are several physical allocation approaches for recycling, and detailed reviews and descriptions of these methods are available elsewhere [37–39]. This section will provide a brief overview of thosee recycling approaches commonly applied in LCA:

- Cut-off; the environmental burdens of recycling (see Figure 5.6a) are fully ascribed to the product receiving the recycled material.
- Open-loop; the environmental burdens of recycling are ascribed, based on the amount of recycled material in the product and how much material is recycled at the end of the life cycle (Figure 5.6b).
- Closed-loop; the environmental burdens of recycling are ascribed, based on the amount of recycled material in the product, and the product is credited for avoiding virgin production (Figure 5.6c).

The open-loop method, also known as the 'recycled content', 'cut-off', or '100–0' method is typically applied to materials whose properties are reduced by the recycling process. For example, this approach should be used for the modelling of paper recycling, whereby the processes involved reduce fibre length, and thus strength. The life cycle impacts using the open-loop approach are determined using:

$$L = \left(1 - R_1\right) E_V + R_1 E_R + U + \left(1 - R_2\right) E_D$$

where:

L = Life cycle impacts (per functional unit)
U = Use phase impacts
R_1 = Proportion of recycled material input
R_2 = Proportion of material in the product that is recycled at end-of-life
E_V = Impacts from virgin material input
E_R = Impacts from recycled material input
E_D = Disposal impacts

The closed-loop recycling approach, also known as the 'end-of-life', 'recycling substitution', 'closed-loop with displacement', or the '0–100' method is typically applied to materials whose properties are maintained by the recycling process, most commonly metals. The life cycle impacts using the closed-loop approach are determined using:

$$L = E_V + U + R_2 E_R + \left(1 - R_2\right) E_D - R_2 E_V$$

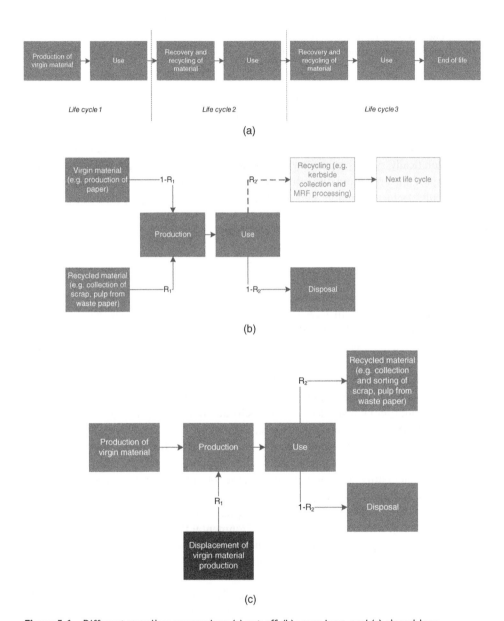

Figure 5.6 Different recycling approaches: (a) cut-off, (b) open-loop, and (c) closed-loop.

where:

L = Life cycle impacts (per functional unit)
U = Use phase impacts
R_2 = Proportion of material in the product that is recycled at end-of-life
E_V = Impacts from virgin material input
E_R = Impacts from recycled material input
E_D = Disposal impacts

The physical recycling approaches inherently assume that the amount of virgin and recycled material in the economy is constant, and that no additional virgin material is required to meet increased demand in the broader economy. As such, it can be argued that the physical allocation approaches do not truly reflect the environmental impacts of recycling systems. Theoretically, system expansion can account for market factors, but these factors are often complex, uncertain and difficult to predict, making the application of system expansion difficult. As for co-product allocation, the choice of recycling approaches can have a significant bearing on the outcomes of an LCA and care should be taken to assess the impacts of alternative approaches on LCA outcomes.

5.6.4   Inventory Modelling in Practice

Collecting, collating, and modelling of the LCI is often considered as the most time-consuming aspect of an LCA study [25, 40]. A range of LCA databases are available for practitioners and include databases developed for specific regions (e.g. USLCI, AusLCI, ELCD), databases developed by industry associations (e.g. PlasticsEurope), and commercially available databases (e.g. ecoinvent). These databases are often underpinned by quality procedures (e.g. for allocation procedures and naming conventions), and allow LCA practitioners to utilise common data, reducing the time needed to model systems which often underpin LCA studies (e.g. electricity and transport systems).

Although these databases are extensive (e.g. ecoinvent contains over 10 000 individual processes), they often do not contain models specifically related to an LCA study. Industry association databases are pivotal in addressing material-specific inventory data gaps. However, confidentiality, and the potential misuse, of production data remain as critical barriers to the supply of inventory data from companies [41]. To address these concerns, data confidentiality can be assured by publishing aggregated inventory data (often termed 'system processes'), which report only the environmental flows associated with a product system, and not company or process-specific data. On the contrary, LCA practitioners can often view industry data as potentially biased and not fully transparent, limiting their applicability [25].

Despite the advancement of LCA databases, key data gaps for specific industries remain. For example, although there is a significant increase in the use of both titanium and carbon fibre materials in aviation, at the time of writing this chapter there are still no life cycle inventories for these materials. This short-coming means that LCA practitioners often need to rely on prior studies to develop inventory models (e.g. for carbon fibre [42–45] or titanium [46]), which can lead to a decrease in the inherent data quality, and an increase in the uncertainty of LCA outcomes.

A key requirement of undertaking LCA is to assess the impact of data uncertainty on the study outcomes. The inclusion of uncertainty data occurs during inventory development, and in the absence of actual uncertainty data (e.g. minimum, maximum, and average values), often takes the form of a pedigree matrix, which translates qualitative descriptions of data quality into quantified standard deviations [47]. These data uncertainty values can be later used in Monte Carlo simulations to investigate the effects of inventory uncertainty on LCA outcomes.

A critical aspect of LCA reporting is to ensure that there is full transparency in the methods used, including all inventory calculations and assumptions [4]. An additional requirement of the ISO standards is that any LCA studies which contain comparative claims that are to be released to the public must be subject to an independent peer review. These two critical requirements have the potential to be misunderstood, with the misunderstanding leading to reluctance to undertake LCA. Although transparency is encouraged, it is not mandatory for reports that are released to the public. The review of sensitive data is typically managed through non-disclosure agreements between parties, allowing for the peer-reviewer to review a fully transparent inventory before the LCA outcomes are finalised and released.

5.6.5 Environmental Impact Assessment Modelling

LCIA aims to translate the LCI (of elementary flows, e.g. resource inputs and emission outputs) of a product system into environmental impact indicators. The LCIA outcomes are often termed 'potential environmental impacts', as LCIA models a potential pathway between a system and impacts, rather than quantifying actual environmental impacts [2].

The selection of environmental indicators occurs at the goal and scope stage of an LCA. A requirement of the ISO 14044:2006 is that the environmental indicators should reflect a set of environmental issues related to the product system(s). In this respect, LCA environmental impacts are typically assessed against multiple impact categories, although it is quite common for assessments to be limited to greenhouse-gas emissions (i.e. carbon footprint studies). These environmental impact categories typically fall under one or more of three areas of protection (or damage categories): human health, ecosystems, and availability of resources. LCIA methods are underpinned by environmental mechanism models. These models account for the fate (and exposure) of environmental flows as they pass through the environment. Environmental impacts can be modelled at the endpoint level, whereby the environmental damage is quantified (e.g. number of species lost), or at the midpoint level, which occur along the impact pathway, Figure 5.7.

ISO 14040:2006 and 14044:2006 outline the framework for LCIA. This framework is represented schematically in Figure 5.8.

The first step in LCIA is to select environmental impact categories that are relevant to the system being assessed. The selection of impact categories and the relevant supporting impact assessment models should be identified during the establishment of the project goal and scope, but can be revisited as the study progresses. The selection of impact categories and models during the goal and scope allows for specific environmental flows to be sought during the inventory data collection stage (refer to Sections 5.6.2 and 5.8.2). In the case of Environmental Product Declaration (EPDs, refer to Section 5.8.2) the impact categories

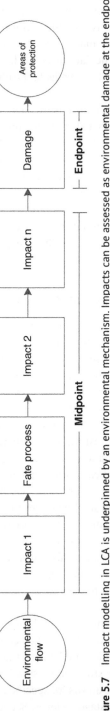

Figure 5.7 Impact modelling in LCA is underpinned by an environmental mechanism. Impacts can be assessed as environmental damage at the endpoint level, or along the impact pathway at the midpoint level. Adapted from [48].

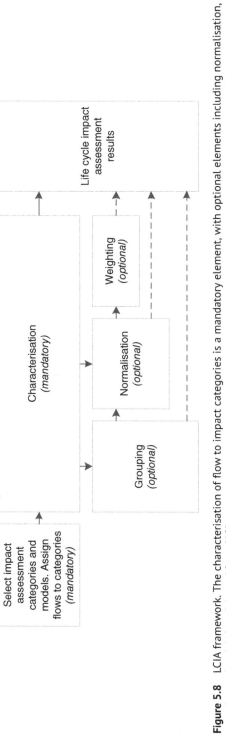

Figure 5.8 LCIA framework. The characterisation of flow to impact categories is a mandatory element, with optional elements including normalisation, grouping and weighting. Adapted from [18].

(and impact assessment methods) to be used are typically pre-defined in the PCRs. Impact categories can be selected at the midpoint or endpoint levels, or both. Most LCAs report impacts at the midpoint level, due to the inherent uncertainty associated with endpoint modelling [49]. Further details of common midpoint category indicators are provided in sections XYZ to XYZ. A schematic showing the relationship between the LCI of environmental flows and common life cycle impact assessment midpoints and endpoints is shown in Figure 5.9.

The second step of LCIA is characterisation. Life cycle impact assessment methods contain tables of characterisation factors (or equivalency factors) for different impact categories. The characterisation factors are derived from the LCIA models and are analogous to the potency of an environmental flow. Characterisation allows for an environmental impact to be quantified in a common unit, often termed the category indicator (e.g. kg CO_2-eq, kg SO_2-eq, etc.). To do this, the quantified individual environmental flows associated with a given impact category (from the LCI) are multiplied by characterisation factors for the given impact category. These characterised flows are then totalled to give a single impact score for the given impact category. This process is repeated for the remaining impact categories. Characterisation allows an LCA practitioner to determine which environmental flows, and subsequently which processes, drive environmental impacts. The following characterised indicators are often included in aviation-related LCA studies:

- Climate change potential/greenhouse gas emissions.
- Depletion of fossil fuel resources.
- Air pollution indicators, including particulate matter (PM), and photochemical oxidation potential.

Further details of these impact categories are provided in sections 5.71.1 to 5.7.3. Other environmental impacts can theoretically be modelled in LCA, including noise [52] and toxicity [51]. However, models for these assessments are still being developed and other techniques may be more appropriate for assessing environmental impacts.

Normalisation is an optional step in LCIA and involves dividing the characterised LCIA results by a characterised baseline. The baseline is usually the characterised impacts of resource extractions and emissions of a geographical region for a given period (e.g. country, global). Normalised LCIA results are dimensionless, but can be expressed on a per-capita basis. Normalisation is useful for error checking of the LCI model, as well as for understanding the relative significance of the indicator results [4]. However, the relative significance of the indicator results should not be interpreted as environmental importance.

The final two optional steps of LCIA are grouping and weighting. Grouping is typically associated with sorting (e.g. into local, regional, or global impacts) or ranking (e.g. by priority of importance) of LCIA results. Weighting accounts for the relative importance of the LCIA results and is commonly used to translate normalised indicator scores into a single indicator result. Weighting relies on a value judgement for the relative importance of indicators and may include monetisation of environmental impacts [14]. Because weightings frequently rely on value judgements, which are not underpinned by science [4], comparative LCAs often avoid using weightings.

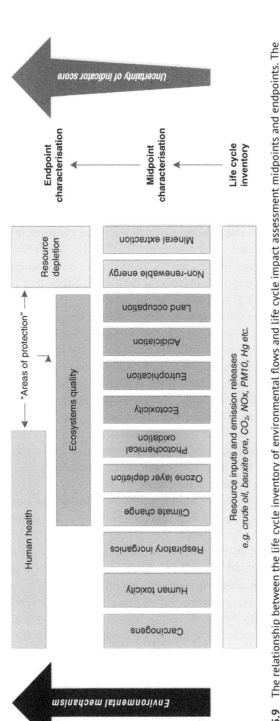

Figure 5.9 The relationship between the life cycle inventory of environmental flows and life cycle impact assessment midpoints and endpoints. The midpoint and endpoint impacts can be sub-categorised into areas of protection. An environmental mechanism underpins the relationship between the life cycle inventory flows, and the midpoint/endpoint impacts. The uncertainty of the life cycle impact assessment increases as the environmental mechanism moves from midpoints to endpoints. Adapted from [7, 50, 51].

5.6.6 Greenhouse Gas Emission Modelling

One of the key impacts normally assessed in LCA studies is climate change, driven by emissions of greenhouse-gas emissions. Greenhouse gases (GHGs) include carbon dioxide, methane, nitrous oxide, hydrofluorocarbons, perfluorocarbons, and sulphur hexafluoride. In LCA, GHGs are typically assessed at the mid-point level using a climate change impact assessment method, which adopts characterisation factors from the global warming potentials (GWPs) of greenhouse gases. The GWP is based on a ratio between the time integral of the radiative forcing potential of a single release of a specific greenhouse gas, and the equivalent time integral of the radiative forcing potential of a single release of carbon dioxide [49]. Thus, climate change impacts are typically reported in kg of carbon dioxide equivalence (kg CO_2-eq). Because GWP is based on a time integral, the time over which the GWP calculation is determined can affect the GWP factors. A time horizon of 100 years is most typically used in LCA and reflects the use of the same horizon by policy makers. However, the choice of the GWP time horizon can affect assessment outcomes [51]. GWPs are periodically updated by the Intergovernmental Panel on Climate Change. Two critical considerations for the GHG modelling of biofuels are the modelling of biogenic carbon, or carbon that is derived from biomass, and potential changes to land use.

There are currently two broad approaches to modelling biogenic carbon. The first approach simply assumes that biogenic uptake (e.g. from growth of plant) and release (e.g. carbon dioxide emissions from combustions) are equal and occur within the time horizon of assessment; they are thus treated as 'carbon neutral' and each have a climate change characterisation factor of zero. Any carbon which is retained in the ground (sequestered) beyond the time horizon is given a credit (characterisation factor of -1 kg CO_2-eq/kg). When using this approach, any biogenic methane emissions must be assessed using an adjusted GWP to account for the biogenic sequestered within the methane molecule. The second approach assumes that biogenic uptake is included as a credit (characterisation factor of -1 kg CO_2-eq/kg), and biogenic carbon dioxide emissions are equivalent to fossil-based carbon dioxide (characterisation factor of $+1$ kg CO_2-eq/kg). In the second approach, any sequestered carbon is already accounted for in the uptake credit and should be assigned a characterisation factor of zero, and the characterisation factor for methane does not need to be adjusted. Theoretically the two approaches should yield the same impact assessment result, but in practice this is often not the case, particularly if the biogenic carbon input and output flows in the background life cycle databases do not balance. The lack of carbon balance in background inventories is typically associated with an allocation of co-products [53] or systems whose inventories do not represent a full life cycle (e.g. when life cycle stages are cut off). Adjustment factors may need to be applied to background inventories to ensure carbon balances are maintained. Despite this difficulty in achieving consistent flows with the second approach, it is the one being adopted by carbon foot-printing standards [9, 10]. These standards typically require that biogenic carbon flows be reported separately from fossil-based carbon flows [54].

LUC relates to the conversion of land from one form to another (e.g. from forest land to cropping land), which typically results in a change in the total carbon stock (i.e. the carbon in the soil, below ground, above ground, or in the air). LUC is typically classified as direct or indirect LUC. Direct LUC are changes in carbon stock that can be directly related to the production of biomass. In contrast, indirect LUC is associated with market dynamics, whereby the production of biomass for the system being assessed results in a change in land use elsewhere in the world (with subsequent GHG impacts), often in a system that is not directly linked to the original production system. Direct and indirect LUC can occur in one system. The application of indirect LUC is however controversial.

An additional aspect that should be considered for assessing greenhouse-gas emissions in the aviation industry is potential variations in GWPs of emission releases at different altitudes. The emissions from aircraft and the interaction of emissions at high altitude are different to ground-level emissions and are not well understood [55, 56]. For example, at altitude, nitrogen oxide emissions (NOx) can accelerate the formation of ozone (O_3), which itself is a greenhouse gas. In addition, this ozone can react with methane, thereby altering the radiative forcing potential (and thus the GWP) of methane at altitude [56]. A commonly adopted approach in LCA standards is to assume that the GWP of greenhouse-gas emissions at altitude is the same as at ground level [9, 10]. However, others have suggested a GWP multiplier factor of approximately two for CO_2 emissions [57, 58]. The exclusion or inclusion of this multiplier factor for CO_2 emissions can have a significant effect on comparing aviation transport with other transportation modes, and consequently the outcomes of LCAs for which aviation contributes to a significant impact, Figure 5.10.

5.6.7 Depletion of Fossil Fuel Resources

A reduction in the dependence on fossil fuels is often a key driver for the development of alternative energy systems. In LCA, this reduction is quantified using a fossil fuel depletion indicator. There are two common midpoint approaches to quantify fossil fuel depletion. The first approach, implemented in the ILCD method [59], is based on fossil fuel scarcity, with the characterisation factors for specific fossil fuels based on known concentrations and depletion rates.

The second approach, used in ecoindicator 99 [60], TRACI [61], and BEES+ [62], characterises fossil-fuel depletion impacts based on the surplus, or marginal, energy required to extract specific fossil fuels in the future, which can also account for the declining quality and quantity of reserves. Fossil-fuel depletion-midpoint impacts using this approach results are typically expressed in MJ. The ReCiPe impact assessment method applies a similar approach, with midpoint impacts expressed in 'oil equivalents' and endpoints expressed in a monetary value associated with increased extraction costs [63].

A third approach assumes that fossil fuels are equally substitutable for a given technology, with the characterisation factors for fossil fuel depletion based on heating values. This approach is implemented in the CML Baseline method [14]. IMPACT 2002+ also adopts this approach, but the indicator is relabelled 'non-renewable energy' [50]. The characterised impact-assessment results units for this approach are also expressed in MJ, but these results should not be compared with results from the surplus energy approach.

The lack of consensus regarding approaches and characterisation factors used for the assessment of fossil-fuel depletions limits the applicability of such methods in LCA [64], and makes comparisons between different studies problematic. Despite the divergent approaches, the relative difference between the characterisation factors across the different methods are limited to less than 5%, Table 5.1. In this respect, the choice of fossil-fuel

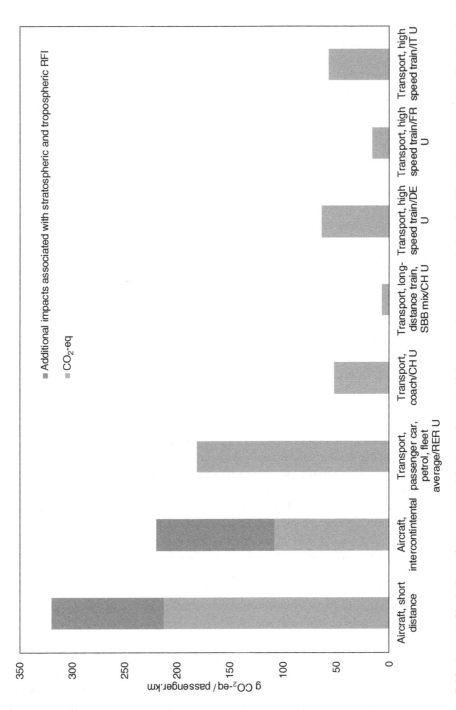

Figure 5.10 Impact of including adjustments to global warming potential on greenhouse gas emissions for different transport modes. The inclusion of an adjustment factor for aviation emissions (orange) results in totals for aircraft being notably higher than for other transport modes. Adapted from [57].

Table 5.1 Normalised characterisation factors for key fossil fuel flows for different impact assessment methods [14, 59, 60, 63, 65].

Environmental flow	Normalised characterisation factor (% of highest characterisation factor)				Maximum variation ($\pm$)
	CML	ReCiPe	ILCD	Ecological Scarcity 2013	
Coal, brown, 8 MJ per kg	17%	16%	17%	17%	0.8%
Coal, brown, 10 MJ per kg	21%	20%	21%	21%	1.0%
Coal, 18 MJ per kg, in ground	38%	37%	38%	38%	1.7%
Coal, 26.4 MJ per kg	56%	54%	56%	56%	2.6%
Coal, 29.3 MJ per kg	63%	60%	62%	63%	2.9%
Gas, natural, 30.3 MJ per kg	65%	65%	65%	65%	0.3%
Oil, crude, 41 MJ per kg	88%	88%	90%	88%	2.8%
Oil, crude, 42 MJ per kg	90%	90%	90%	90%	0.7%
Oil, crude, 42.6 MJ per kg	91%	91%	90%	91%	0.6%
Oil, crude, 42.7 MJ per kg	91%	92%	90%	91%	1.5%
Gas, natural, 46.8 MJ per kg	100%	100%	100%	100%	0.0%

The order of the impact is consistent across the different methods and the maximum variation across the different methods reported is less than 3%, suggesting that the choice of impact assessment method within an LCA study is unlikely to have a significant effect on comparative outcomes for fossil fuel depletion.

depletion-impact assessment method is unlikely to have a significant impact on the direction of comparative outcomes.

5.6.8 Air Pollution Indicators

Models for characterising air pollutants rely on models of fate and exposure. End-point models extend this to account for the health effects of these emissions. Fate models account for the transport of emissions through the environment, transformations (e.g. reactions with other species), and their end deposition (e.g. in air, soil, or water). These fate models rely on specific environmental conditions, such as topology, air flow, and background emissions loading. Exposure models account for the fraction of the emissions within the environment to which humans are exposed.

Particulate matter (PM) emissions are reported as critical air pollutants that adversely affect human health [66, 67]. PM emissions can be classified as either primary PM, resulting from direct emissions, or secondary PM, resulting from atmospheric reactions [68]. The formation of secondary PM is typically associated with organic (e.g. volatile organic compounds, VOCs) or inorganic (e.g. NOx) air emissions [68]. Because of population density differences, exposure factors (and subsequently characterisation factors) for primary PM differ significantly between urban and rural settings [68]. A suite of LCIA methods exists for PM emissions. Given that both the fate and exposure models are region-specific, it is important when applying LCIA to choose an impact assessment model that best reflects local conditions. LCIA-midpoint characterisation factors for PM are included in most impact assessment methods, including eco-indicator 99, ILCD 2011 Midpoint +, TRACI 2.1, and BEES+, while ReCiPe includes both midpoint and endpoint characterisation factors.

Fate and exposure models for air pollutants emitted at cruise altitude are largely lacking, despite their being an acknowledgement that these emissions affect air quality and human health [66, 69]. Because of this modelling limitation, many aviation-related LCA studies report inventory flows (e.g. NOx), rather than with combined inventory indicators (e.g. PM2.5-eq) [70]. This result limits the ability to interpret and understand potential trade-offs occurring across the life cycle. The development of specific fate and exposure models for aircraft-related emissions in LCA would further enhance the applicability of LCA outcomes to the aerospace industry.

5.6.9 Interpretation

The final LCA stage is interpretation, whereby aspects of the goal and scope, inventory analysis, and impact assessment results are used in combination to provide meaningful outcomes from the process. The interpretation stage typically comprises three components: (i) the identification of inventory and impact assessment issues, such as appropriateness of modelling, (ii) an evaluation of the completeness of the study with respect to the goal and scope, and (iii) conclusions and recommendations. The first component of the interpretation stage typically includes a number of sensitivity studies, whereby key variables of the study (most typically associated with the inventory) are changed to assess the effect of these variations on the outcomes. The interpretation stage often triggers a revisit of the other four stages in LCA.

5.7 Aviation LCA Case Studies

5.7.1 Aircraft

Aircraft LCAs are typically performed across the manufacturing, operation, and disposal life cycle stages. Whilst operations can be readily estimated (normally based on fuel-burn projections), the lack of publically available life cycle inventories detailing the types and amounts of materials used in aircraft limits the ability to develop inventories for both the manufacturing and disposal stages. Despite this, there are a number of studies that use qualified estimates as a basis to develop an LCI for aircraft production. For example, Howe et al. [71] and Kolios et al. [72] used process-based LCA to assess the impact of an Airbus A320 across manufacturing, operation, and disposal. Life cycle environmental impacts were dominated by the operations phase, which is consistent with other, similar studies [73]. A question that often arises is whether or not the substitution of aluminium with composites results in environmental savings, given that the production of composites (e.g. carbon-fibre reinforced polymers) can be more energy-intensive than aluminium. Both Kolios et al. [72] and Liu [74] used LCA to demonstrate environmental benefits associated with the use of composite parts in A319 and B737-800 aircraft.

5.7.2 Biofuels

One of the key focal points for LCA is on the environmental assessment of alternative feedstocks and production pathways for aviation biofuels. As reported in Section 1.3, demonstrating environmental impacts using LCA is a critical step in the certification of aviation biofuels. A recent LCA by Crossin [75] is an example of one such approach, whereby the environmental profile of an aviation biofuel, pyrolytically processed from mallee eucalypt biomass, was compared to a Jet A-1 baseline. This study demonstrated that the mallee jet fuel offered a 40% reduction in greenhouse-gas emissions, relative to the baseline, Figure 5.11.

Critically, our previous work [75] highlighted the variation in impacts that can occur with the modelling of food displacement. There remains a lack of consensus on the modelling of food displacement effects.

5.7.3 Airport Operations

There are few LCA studies focussing on airport operations. Lewis [76] utilised hybrid-LCA to assess the environmental profile of operating three different aircraft: an Airbus A320, A330, and A380. The airport construction and operations were assessed using input-output LCA, together with the manufacturing stage of the airliners, while process-based LCA was used for operation of the airliners (fuel production and use). Environmental impacts were assessed across a range of profiles, but the author focussed on the reporting of greenhouse-gas emissions. The airport construction impacts were amortised over an assumed lifetime of 20 years, and allocated based on assumptions regarding passenger movements. Airport operations impacts were modelled on annual spending estimates. Using this approach, airport construction accounted for 0.44% to 2.07% of life-cycle greenhouse-gas impacts, with airport operations accounting for between 0.75% and 3.5% (Figure 5.12).

Although these impacts are limited, the Lewis study highlights the importance of including infrastructure construction and operations impacts in LCA studies; to exclude these impacts would likely underestimate full life cycle impacts.

5.8 Trends and Outlook for LCA

5.8.1 Application in Policy

LCA has long been identified as a technique which can integrate with the development of government policy [14, 18]. Despite this assertion, the evidence of uptake in policy development is limited [77]. The reasons for this lack of uptake are diverse, but include a lack of LCA knowledge on the part of policy-makers, uncertainty, and a lack of trust in LCA methods and outcomes [77, 78], as well as continued debate regarding the applicability of the specific LCA methodology employed (e.g. attributional vs consequential LCA) [79]. Seidel [77] suggests that the lack of integration of LCA in policy development is associated with the processes within which the LCA is to be incorporated, rather than a specific problem

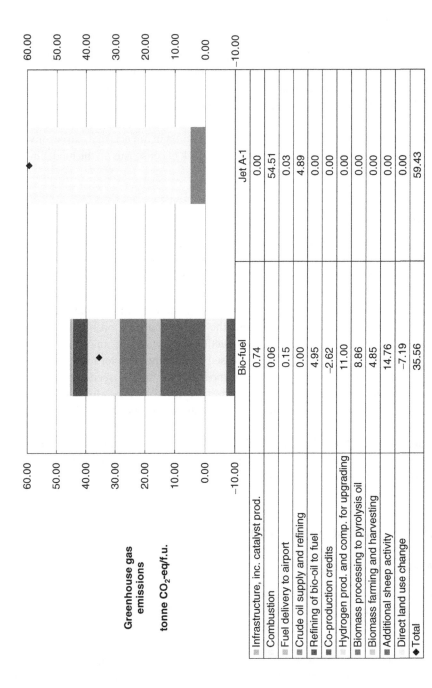

	Bio-fuel	Jet A-1
▨ Infrastructure, inc. catalyst prod.	0.74	0.00
☐ Combustion	0.06	54.51
▨ Fuel delivery to airport	0.15	0.03
▨ Crude oil supply and refining	0.00	4.89
▨ Refining of bio-oil to fuel	4.95	0.00
▨ Co-production credits	−2.62	0.00
▨ Hydrogen prod. and comp. for upgrading	11.00	0.00
▨ Biomass processing to pyrolysis oil	8.86	0.00
▨ Biomass farming and harvesting	4.85	0.00
▨ Additional sheep activity	14.76	0.00
▨ Direct land use change	−7.19	0.00
◆ Total	35.56	59.43

Figure 5.11 Contributors to greenhouse-gas emission profile of mallee eucalypt jet fuel. Source: [75].

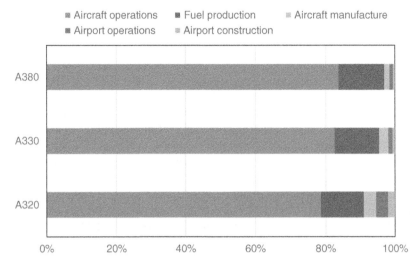

Figure 5.12 Contributors to greenhouse gas emission profile of different aircraft, including airport construction and operations. Source: [76].

with LCA itself. Despite this lack of uptake, the use of LCA by policy-makers can facilitate life cycle thinking, whereby unintended consequences of decisions can be identified [77].

5.8.2 Inventory

The availability of publicly available and transparent life cycle inventories, including for key materials used in aviation (e.g. the production of titanium and carbon-fibre reinforced polymers) has improved markedly in recent years, and it is expected that more inventories will be made available as the uptake and application of LCA increases.

Inventory development issues associated with allocation and recycling procedures remain largely unresolved in LCA [53, 64, 80], despite these issues being identified as significant in the early development of LCA. The development of PCRs which underpin the LCA methodology used in EPDs has improved the variability of allocation and recycling approaches, and it is anticipated that further harmonisation of approaches will occur in the future.

A critical issue associated with biofuel studies are the methods used to account for rebound effects [81], such as indirect LUC, and food displacement. These methods are still being developed and the development of consistent approaches across different industry sectors is expected to lead to more certain LCA outcomes [82], allowing for better uptake in decision-making and policy development.

5.8.3 Life Cycle Impact Assessment

Despite there being recent advances in developing regionalised life cycle inventories, region-specific life cycle impact assessment techniques are currently not readily available for most environmental indicators [81]. In addition to these models, there are no aviation-specific LCIA methods to account for variations in emission releases through the different atmospheric layers. The development of aviation-specific LCIA methods would reduce the uncertainty of LCA outcomes from aviation studies.

5.8.4 Extension of LCA into Other Sustainability Metrics

There is continued interest in applying LCA techniques [81, 82] to the assessment of other aspects of sustainability, including cost and social impacts. Techniques for life cycle costing (LCC) are well-established and are typically based on net-present value techniques, which account for direct costs associated with a system, as well as the change in the value of money over time. The integration of LCC with LCA is often limited to restrictions in LCA software (whose functionalities are often designed around LCA rather than LCC) and availability of cost data. Nevertheless, coupling LCA with LCC allows for a systematic approach to assessing environmental impacts and financial costs. An emerging approach, social LCA (SLCA), uses principles of economic input-output LCA and risk-management techniques to identify both direct and indirect social hotspots, such as labour rights issues, throughout global supply chains [81, 82].

References

1 United Nations Environment Programme (2004). *Why Take a Life Cycle Approach?* Paris: United Nations Environment Programme.

2 Curran, M.A. (2006). *Life Cycle Assessment: Principles and Practice*. United States Environmental Protection Agency, Cincinnati, OH, USA.

3 Horne, R. (2009). Life cycle assessment: origins, principles and context. In: *Life Cycle Assessment: Principles, Practice and Prospects* (eds. R. Horne, T. Grant and K. Verghese), 1–8. Melbourne: CSIRO Publishing.

4 International Organization for Standardization. (2006) ISO 14044:2006(E). *Environmental management - Life cycle assessment - Requirements and guidelines*. Geneva: International Organization for Standardization.

5 International Organization for Standardization. (2006) ISO 14040:2006(E). *Environmental management - Life cycle assessment - Principles and framework*. Geneva: International Organization for Standardization.

6 Guinée, J.B., Heijungs, R., Huppes, G. et al. (2011). Life cycle assessment: past, present, and future. *Environmental Science & Technology* 45: 90–96.

7 European Commission (2010). *ILCD Handbook: General Guide for Life Cycle Assessment - Detailed Guidance*. Luxembourg: European Commission, Joint Research Centre, Institute for Environment and Sustainability.

8 Pryshlakivsky, J. and Searcy, C. (2013). Fifteen years of ISO 14040: a review. *Journal of Cleaner Production* 57: 115–123.

9 British Standards Institute. (2011). PAS 2050:2011. *Specification for the assessment of the life cycle greenhosue gas emissions of goods and services*. British Standards Institute.

10 International Organization for Standardization. (2013) ISO/TS 14067:2013. *Greenhouse gases — Carbon footprint of products — Requirements and guidelines for quantification and communication.* Geneva: International Organization for Standardization,.

11 Allen, D.T., Allport, C., Atkins, K. et al. (2009). Propulsion and power rapid response research and development (R&D) support. Delivery Order 0011: Advanced Propulsion Fuels Research and Development-Subtask: Framework and Guidance for Estimating Greenhouse Gas Footprints of Aviation Fuels. United States Airforce.

12 Roundtable on Sustainable Biomaterials. (2012). *RSB GHG Calculation Methodology.* RSB reference code: RSB-STD-01-003-01 (Version 2.1). Geneva: Roundtable on Sustainable Biomaterials.

13 Madsen, J., Nicholson, A., and Martin, L. (2012). *Enhancing the Value of Life Cycle Assessment.* Deloitte Development LLC.

14 Guinée, J.B. (2002). *Handbook on Life Cycle Assessment. Operational Guide to the ISO Standards.* Dordrecht: Kluwer Academic Publishers.

15 Rubik, F. and Frankl, P. (2000). *Life Cycle Assessment in Industry and Business.* Berlin, Heidelberg: Springer.

16 Roundtable on Sustainable Biomaterials (2015). *A Guide to the RSB Standard.* Geneva: Roundtable on Sustainable Biomaterials.

17 The International EPD System. (2015) *Product category rules according to ISO 14025.* Product Group Classification: UN CPC 45264. Laser Printers Used With Data Processing Machines. Stockholm: The International EPD System.

18 International Organization for Standardization. (2006) ISO 14025:2006(E). *Environmental labels and declarations — Type III environmental declarations — Principles and procedures.* Geneva: International Organization for Standardization.

19 Ibáñez-Forés, V., Pacheco-Blanco, B., Capuz-Rizo, S.F., and Bovea, M.D. (2016). Environmental product declarations: exploring their evolution and the factors affecting their demand in Europe. *Journal of Cleaner Production* 116: 157–169.

20 The International EPD System. (2015) *Product category rules according to ISO 14025. Product Group Classification: UN CPC 49623. Passenger Commercial Aeroplanes.* Stockholm: The International EPD System.

21 Bombardier. (2015) C Series product brochure.

22 Bhander, G.S., Hauschild, M., and McAloone, T. (2003). Implementing life cycle assessment in product development. *Environmental Progress* 22: 255–267.

23 Ramani, K., Ramanujan, D., Bernstein, W.Z. et al. (2010). Integrated sustainable life cycle design: a review. *Journal of Mechanical Design* 132: 091004.

24 Verghese, K.L., Horne, R., and Carre, A. (2010). PIQET: the design and development of an online 'streamlined' LCA tool for sustainable packaging design decision support. *The International Journal of Life Cycle Assessment* 15: 608–620.

25 Finnveden, G., Hauschild, M.Z., Ekvall, T. et al. (2009). Recent developments in Life Cycle Assessment. *Journal of Environmental Management* 91: 1–21.

26 Ekvall, T. and Weidema, B.P. (2004). System boundaries and input data in consequential life cycle inventory analysis. *The International Journal of Life Cycle Assessment* 9: 161–171.

27 Zamagni, A., Guinée, J., Heijungs, R. et al. (2012). Lights and shadows in consequential LCA. *The International Journal of Life Cycle Assessment* 17: 904–918.

28 Brander, M., Tipper, R., Hutchison, C., and Davis, G. (2008). Technical Paper: *Consequential and Attributional Approaches to LCA: A Guide to Policy Makers with Specific Reference to Greenhouse Gas LCA of Biofuels.* Econometrica Press.

29 Matthews, H.S. and Small, M.J. (2000). Extending the boundaries of life-cycle assessment through environmental economic input-output models. *Journal of Industrial Ecology* 4: 7–10.

30 Hedayati, M., Iyer-Raniga, U., and Crossin, E. (2014). A greenhouse gas assessment of a stadium in Australia. *Building Research & Information* 42: 602–615.

31 Herstein, L.M., Filion, Y.R., and Hall, K.R. (2011). Evaluating the environmental impacts of water distribution systems by using EIO-LCA-based multiobjective optimization. *Journal of Water Resources Planning and Management* 137: 162–172.

32 Suh, S. and Huppes, G. (2002). Missing inventory estimation tool using extended input-output analysis. *The International Journal of Life Cycle Assessment* 7: 134–140.

33 Rowley, H.V., Lundie, S., and Peters, G.M. (2009). A hybrid life cycle assessment model for comparison with conventional methodologies in Australia. *The International Journal of Life Cycle Assessment* 14: 508–516.

34 Kaufman, A.S., Meier, P.J., Sinistore, J.C., and Reinemann, D.J. (2010). Applying life-cycle assessment to low carbon fuel standards—how allocation choices influence carbon intensity for renewable transportation fuels. *Energy Policy* 38: 5229–5441.

35 Weidema, B.P. (2003). Market Information in Life Cycle Assessment.

36 Weidema, B.P. (2001). Avoiding co-product allocation in life-cycle assessment. *Journal of Industrial Ecology* 4: 11–33.

37 Ekvall, T. (1999). Key methodological issues for life cycle inventory analysis of paper recycling. *Journal of Cleaner Production* 7: 281–294.

38 Ekvall, T. and Tillman, A.-M. (1997). Open-loop recycling: criteria for allocation procedures. *The International Journal of Life Cycle Assessment* 2: 155–162.

39 Kim, S., Hwang, T., and Lee, K.M. (1997). Allocation for cascade recycling system. *The International Journal of Life Cycle Assessment* 2: 217.

40 Curran, M.A. (2013). Life cycle assessment: a review of the methodology and its application to sustainability. *Current Opinion in Chemical Engineering* 2: 273–277.

41 Frischknecht, R. (2004). Transparency in LCA-a heretical request? *The International Journal of Life Cycle Assessment* 9: 211–213.

42 Schmidt, J.H. and Watson, J. (2013). Eco Island Ferry – Comparative LCA of island ferry with carbon-fibre composite based and steel based structures. Aalbord, Denmark: 2.-0 LCA consultants.

43 Griffing, E. and Overcash, M. (2010). Carbon fiber HS from PAN. Chemical Life Cycle Database, Environmental Clarity.

44 Griffing, E., Eric, V., and Overcash, M. (2014). Life cycle inventory data for carbon fiber and epoxy systems and use in environmentally optimized designs. San Francisco: LCA XIV.

45 Das, S. (2011). Life cycle assessment of carbon fiber-reinforced polymer composites. *The International Journal of Life Cycle Assessment* 16: 268–282.

46 Norgate, T.E., Jahanshahi, S., and Rankin, W.J. (2007). Assessing the environmental impact of metal production processes. *Journal of Cleaner Production* 15: 838–848.

47 Ciroth, A., Muller, S., Weidema, B., and Lesage, P. (2016). Empirically based uncertainty factors for the pedigree matrix in ecoinvent. *The International Journal of Life Cycle Assessment* 21: 1338–1348.

48 Hauschild, M.Z. (2005). Assessing environmental impacts in a life-cycle perspective. *Environmental Science & Technology* 39: 81A–88A.

49 Shine, K.P., Fuglestvedt, J.S., Hailemariam, K., and Stuber, N. (2005). Alternatives to the global warming potential for comparing climate impacts of emissions of greenhouse gases. *Climatic Change* 68: 281–302.

50 Jolliet, O., Margni, M., Charles, R. et al. (2003). IMPACT 2002+: a new life cycle impact assessment methodology. *The International Journal of Life Cycle Assessment* 8: 324–330.

51 Hauschild, M.Z., Goedkoop, M., Guinée, J. et al. (2013). Identifying best existing practice for characterization modeling in life cycle impact assessment. *The International Journal of Life Cycle Assessment* 18: 683–697.

52 Cucurachi, S. and Heijungs, R. (2014). Characterisation factors for life cycle impact assessment of sound emissions. *Science of the Total Environment* 468–469: 280–291.

53 Finkbeiner, M., Ackermann, R., Bach, V. et al. (2014). Challenges in life cycle assessment: an overview of current gaps and research needs. In: *Background and Future Prospects in Life Cycle Assessment* (ed. W. Klöpffer), 207–258. Springer-Verlag.

54 Renouf, M.A., Sevenster, M., Logie, K. et al. (2015). *Best Practice Guide for Life Cycle Impact Assessment (LCIA) in Australia*. Australian Life Cycle Assessment Society.

55 Intergovernmental Panel on Climate Change (1999). *Aviation and the Global Atmosphere. Summary for Policymakers*. Cambridge: Intergovernmental Panel on Climate Change.

56 Jardine, C.N. (2005). *Calculating the Environmental Impact of Aviation Emissions*. Environmental Change Institute, Oxford University Centre for the Environment.

57 Jungbluth, N. (2013). Aviation and Climate Change: Best practice for calculation of the global warming potential.

58 Sausen, R., Isaksen, I., Grewe, V. et al. (2005). Aviation radiative forcing in 2000: an update on IPCC (1999). *Meteorologische Zeitschrift* 14: 555–561.

59 European Commission. (2011). ILCD Handbook: Recommendations for Life Cycle Impact Assessment in the European context, European Commission, Joint Research Centre, Institute for Environment and Sustainability, Luxembourg.

60 Ministry of Housing, Spatial Planning and the Environment. (n.d.). *The Eco-indicator 99. Manual for Designers*. The Hague: Ministry of Housing, Spatial Planning and the Environment.

61 Bare, J.C. (2012). *Tool for the Reduction and Assessment of Chemical and Other Environmental Impacts (TRACI), Version 2.1 - User's Manual*. Cincinnati: Environmental Protection Agency.

62 NIST. (n.d.). BEES Online Tutorial. Fossil Fuel Depletion. Gaithersburg: National Institute of Standards and Technology.

63 Goedkoop, M., Heijungs, R., Huijbregts, M. et al. (2013). ReCiPe 2008. A life cycle impact assessment method which comprises harmonised category indicators at the midpoint and the endpoint level. Ministry of Housing, Spatial Planning and Environment (VROM).

64 Klinglmair, M., Sala, S., and Brandão, M. (2014). Assessing resource depletion in LCA: a review of methods and methodological issues. *The International Journal of Life Cycle Assessment* 19: 580–592.

65 Frischknecht, R. and Büsser Knöpfel, S. (2013). Swiss eco-factors 2013 according to the ecological scarcity method, Methodological fundamentals and their application in Switzerland. *Environmental studies*.

66 FAA. (2015). Aviation Emissions, Impacts & Mitigation. A Primer. Federal Aviation Administration, Office of Environment and Energy.

67 EASA, EEA, and EUROCONTROL. (2016). European Aviation Environmental Report. European Aviation Safety Agency, European Environment Agency, and EUROCONTROL.

68 Renouf, M.A., Grant, T., Sevenster, M. et al. (2016). Best Practice Guide for Mid-Point Life Cycle Impact Assessment in Australia. Australian Life Cycle Assessment Society.

69 Barrett, S.R.H., Britter, R.E., and Waitz, I.A. (2010). Global mortality attributable to aircraft cruise emissions. *Environmental Science & Technology* 44: 7736–7742.

70 Mikhail, V.C. and Arpad, H. (2009). Environmental assessment of passenger transportation should include infrastructure and supply chains. *Environmental Research Letters* 4: 024008.

71 Howe, S., Kolios, A.J., and Brennan, F.P. (2013). Environmental life cycle assessment of commercial passenger jet airliners. *Transportation Research Part D: Transport and Environment* 19: 34–41.

72 Kolios, A.J., Howe, S., Asproulis, N., and Salonitis, K. (2013). Environmental impact assessment of the manufacturing of a commercial aircraft. In: *Proceedings of the 11th International Conference on Manufacturing Research: Advances in Manufacturing Technology*, 19–20, Cranfield, UK.

73 de Oliveira Fernandes Lopes, J.V. (2010). Life Cycle Assessment of the Airbus A330–200 Aircraft. Universidade Tecnica de Lisboa.

74 Liu, Z. (2013). Life cycle assessment of composites and aluminium use in aircraft systems, MSc by Research, School of Engineering, Cranfield University.

75 Crossin, E. (2017). Life cycle assessment of a mallee eucalypt jet fuel. *Biomass and Bioenergy* 96: 162–171.

76 Lewis, T. (2013). A Life Cycle Assessment of the Passenger Air Transport System Using Three Flight Scenarios. Master in Industrial Ecology, Department of Energy and Process Engineering, Norwegian University of Science and Technology.

77 Seidel, C. (2016). The application of life cycle assessment to public policy development. *The International Journal of Life Cycle Assessment* 21: 337–348.

78 Herrmann, I.T., Hauschild, M.Z., Sohn, M.D., and McKone, T.E. (2014). Confronting uncertainty in life cycle assessment used for decision support. *Journal of Industrial Ecology* 18: 366–379.

79 Plevin, R.J., Delucchi, M.A., and Creutzig, F. (2014). Using attributional life cycle assessment to estimate climate-change mitigation benefits misleads policy makers. *Journal of Industrial Ecology* 18: 73–83.

80 Wiloso, E.I., Heijungs, R., and de Snoo, G.R. (2012). LCA of second generation bioethanol: a review and some issues to be resolved for good LCA practice. *Renewable and Sustainable Energy Reviews* 16: 5295–5308.

81 Hellweg, S. and Milà i Canals, L. (2014). Emerging approaches, challenges and opportunities in life cycle assessment. *Science* 344: 1109–1113.

82 McManus, M.C. and Taylor, C.M. (2015). The changing nature of life cycle assessment. *Biomass and Bioenergy* 82: 13–26.

6

Air Traffic Management and Avionics Systems Evolutions

Alessandro Gardi[1,2], Yixiang Lim[3], Nichakorn Pongsakornsathien[4], Roberto Sabatini[1,2], and Trevor Kistan[5]

[1]*Department of Aerospace Engineering, Khalifa University of Science and Technology, Abu Dhabi, UAE*
[2]*School of Engineering, RMIT University, Melbourne, Victoria, Australia*
[3]*Agency for Science, Technology and Research (ASTAR), Singapore*
[4]*School of Engineering, RMIT University, Bundoora, Victoria, Australia*
[5]*Thales Australia, Melbourne, Victoria, Australia*

6.1 Introduction

Various contemporary trends are stimulating a generational change in the aviation sector. On the one hand, conventional civil air transport has seen a rapid recovery from the pandemic groundings, compounding the pre-existing growth and renovation trends which led to several new airports and air routes being opened around the world. On the other hand, the recent proliferation of innovative vehicle and service concepts targeting both urban air mobility (UAM) and high-altitude airspace operations are prompting a comprehensive overhaul of conventional regulations and procedures. In no area is such change as evident and impactful as in air traffic management (ATM), which for many decades has evolved in line with its original post-WW2 practices, thereby now posing significant challenges to aviation regulators, service providers, and operators in ensuring continuous improvements to the levels of safety, efficiency, and environmental sustainability.

ATM services rely on a set of operational measures to fulfil their statutory mission of preventing collisions and promoting an ordered and expedite flow of air traffic [1, 2]. These measures are based on communication, navigation and surveillance (CNS) systems to identify the conflicts, negotiate a resolution, and follow an alternative flight trajectory, thereby supporting the safe deconfliction, sequencing, and spacing of traffic by means of lateral, vertical, and longitudinal route amendments (Figure 6.1). In the conventional paradigms, human operators are still heavily involved in the majority of ATM duties, which are more tactical than strategic in nature.

Air traffic flow management (ATFM) is the service with the greatest strategic nature overall and involves a number of measures to accomplish the overarching ATM mission. ATFM specifically aims at continuously matching the air traffic demand with the available capacity of airspace and airport resources. Strategic and tactical ATFM problems are all formulated as mathematical optimisation problems and global solution approaches are developed to promote optimised flows across a large number of airports and wider airspace

Sustainable Aviation Technology and Operations: Research and Innovation Perspectives, First Edition.
Edited by Roberto Sabatini and Alessandro Gardi.
© 2024 John Wiley & Sons Ltd. Published 2024 by John Wiley & Sons Ltd.
Companion Website: www.wiley.com/go/sustainableaviation

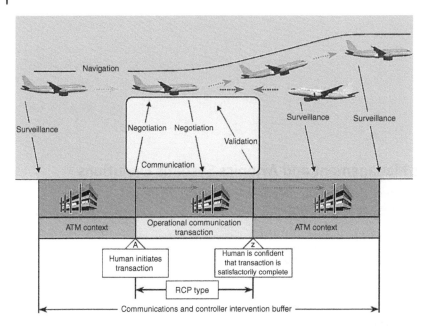

Figure 6.1 CNS/ATM concept of operation. Source: Adapted from [3].

regions. Both long-term (strategic) and short-term (pre-tactical) measures can be adopted to resolve perturbations arising due to unpredicted weather and capacity disruptions. The effectiveness of these measures largely depends on the amount, accuracy, and timeliness of the information exchanged. As a result, human operators crucially depend on accurate data, forecast models, and technology enablers for making better-informed and more effective decisions.

6.2 Current Progress in the Modernisation Efforts

At the time of writing, a comprehensive ATM renovation process is well underway, with such large-scale initiatives as NextGen in the United States and SESAR in Europe completing several of the initially targeted upgrades [4, 5]. These programmes focussed on the most promising concepts and technologies in the ATM and avionics domain, in line with the evolutions originally envisioned within the Advisory Group for Aerospace Research and Development (AGARD) of the North Atlantic Treaty Organization (NATO), the Group for Aeronautical Research and Technology in Europe (GARTEUR), and the Future Air Navigation Systems (FANS) special committee of the International Civil Aviation Organization (ICAO) in the 1980s [6]. Among other operational enhancements, these programmes focused on various technological advances [7]:

- A substantial growth in the adoption of contemporary information and communication technologies (ICT) and networked resources in both ground-based and airborne systems;

- Higher levels of collaborative decision making (CDM) to allow all involved parties to participate in the enhancement of system performance by sharing and accessing more accurate and updated information;
- Four-dimensional (4D) trajectory-based operations (TBO) and performance-based operations (PBO);
- Dynamic airspace management (DAM) for an optimised exploitation of airspace capacity;
- Improvements in the capacity and efficiency of airports.

In order to implement these innovative concepts and progress along the planned evolutionary pathways, a number of new CNS/ATM and avionics (CNS + A) improvements were deemed essential:

- Avionics and ATM decision support systems (DSS), featuring automation-assisted 4D trajectory (4DT) planning and negotiation/validation functionalities;
- Enhanced ground and satellite datalink-based aeronautical communications systems, such as the L-band digital aeronautical communication system (LDACS);
- Global navigation satellite systems (GNSS) as a primary means of navigation, supported by suitable ground-, avionics- and satellite-based augmentation systems (GBAS/ABAS/SBAS);
- Enhanced ground-based and satellite-based surveillance, including automated dependent surveillance broadcast (ADS-B) and self-separation;
- A system wide information management (SWIM) network;
- Evolutionary human machine interface and interaction (HMI^2) design, interoperability and higher levels of automation.

The implementation of these novel CNS + A technologies and operational concepts was planned in a phased manner. For instance, ever since the original implementation stages (time-, trajectory-, and performance-based operations), SESAR's strategic vision evolved towards tackling the transition to the 'Digital European Sky' (DES) in the following four phases [4]:

- Phase A: addressing immediate network criticalities with a focus on more efficient and effective capacity and demand balancing;
- Phase B: optimization of the overall network efficiency by introducing free-route operations, airport performance management and initial U-Space services;
- Phase C: defragmentation through virtualisation, increasing automation, and dynamic adaptation of ATM services;
- Phase D: implementation of DES through the delivery of a fully scalable digital ecosystem supporting distributed services, high levels of automation and truly unsegregated manned/unmanned operations.

6.3 Role of ATM and Operational Innovations in Increasing Aviation Sustainability

As already covered in other sections, the aviation sector is directly responsible for a number of adverse effects on the environment and on living beings. While some emissions such as carbon dioxide (CO_2), unburnt hydrocarbons (HC), soot, and sulphur oxides (SO_X) adversely affect the environment along the entire flight, others either do not always eventuate, as in the case of condensation trails (contrails), or their adverse effects are deemed significant mostly in proximity of the ground, as in the case of aircraft noise and carbon monoxide (CO). Nitrogen oxides (NO_X) in addition to their toxicity to living beings when emitted at higher concentrations in proximity of the ground (tropospheric NO_X) are also known to trigger a family of chemical processes ultimately causing the depletion of the ozone layer, and are thereby associated with a positive radiative forcing (stratospheric NO_X).

While major ATM renovation programmes focus on supporting a cohesive and widespread implementation of capacity and efficiency improvements, several of the targeted upgrades also had recognised benefits in mitigating environmental and health impacts of air transport. Some of these improvements were also deemed potentially more effective than aircraft design enhancements, such as in the case of perceived noise and contrails. Additionally, these programmes introduce a force-multiplier effect by simultaneous implementation of the upgrades across continents and macroregions, as opposed to standalone local initiatives. When addressing environmental sustainability, ATM renovation programmes adhered to the paradigm '*less fuel consumed equates to less emissions*' as this was aligned with the efficiency and economic interests of aircraft operators. However, it became increasingly evident that a number of additional ATM innovations and operational measures could be very effective in mitigating all environmental and health impacts, potentially more than design enhancements, such as in the case of perceived noise and contrails. ATFM enhancements result in both flight time and fuel savings on a large-scale and these figures are continuously monitored, though it is difficult to ascertain the effectiveness of individual enhancements from real data.

6.4 ATFM and Demand-Capacity Balancing Evolutions

Air traffic demand is highly variable in space and time, with significant fluctuations occurring over the day, weeks, and seasons. This highly variable demand faces a mostly inflexible capacity supply due to ground infrastructures and airspace configuration, and this underlying mismatch contributes to major disruptions and delays if not properly addressed. Perturbations or disruptions occurring locally due to a variety of reasons can quickly affect a growing number of airports, airspace sectors, and flights, unless effective mitigation measures are put in place. Demand-capacity balancing (DCB) is the key mission of ATFM and aims to regulate the flow of traffic and the airspace configuration to address these issues and achieve an optimal balance. DCB is most useful when imbalances are predicted early enough to avoid tactical ATC intervention. A number of progressive improvements were proposed for the fundamental ATFM approaches and are currently being developed as part of all major ATM modernisation initiatives.

At present, state-of-the-art ATFM DSS are based on either analytic estimation or numerical simulation [8, 9], with fast-time simulation (FTS) and the associated critical evaluation of the resulting information being a very popular approach.

The implementation of 4D TBO and several other CNS + A technologies is the critical technological step in the realisation of a resilient, fully interoperable ATM system supporting strategic DCB initiatives that have the highest effectiveness overall. In line with the implementation of 4D-TBO, ATM modernisation will tackle the concurrent evolution of the airspace into a dynamically-optimised resource to overcome the rigidity of the conventional airspace structure, so as to actively manage the denominator of the DCB and consequently release further capacity. The long-term evolution in this perspective is represented by the DAM paradigm, which can be achieved by dynamically modifying the geographic extent of the airspace sectors, morphing their boundaries to accommodate shifting traffic patterns, weather, or other dynamic factors [9]. The capacity of the neighbouring ATC sectors may be temporarily decreased as a result, but this is a far more flexible concept thanconsolidating or de-consolidating sectors onto ATC positions. Regulatory considerations (ATCo ratings, traffic mix, etc.) and international letters of agreement (LOA) still constrain the extent to which the sector boundaries could change, but relatively moderate shifts could still yield significant operational benefits for both ANSPs (maximise ATCo performance and manage staffing levels) and airline operators (minimise DCB measures such as reroutes and holding). More data sharing and interoperability amongst ATM systems are, in any case, crucial to enable a progressive relaxation of cross-border arrangements and the airspace changes will be increasingly based on the available CNS performance level rather than human workload. In this perspective, DAM is considered an essential component of PBOs.

6.4.1 Dynamic Airspace Morphing

DAM concepts can be classified based on the kind of airspace modulation, which can address time-only, space-only (2D/3D) or space-time (4D). While discrete-time strategic and tactical DAM measures for airspace capacity modulation are already implemented in ATFM, such as the flexible use of airspace (FUA) [10], tactical space and space-time modulations for DAM have been the subjects of active research for many years [11–17]. The most far-fetched space- and space-time modulation concepts are envisioned to implement some kind of morphing, where the boundaries of individual sectors shift in time as a function of the relative traffic densities [9]. Limited research has studied the factors which should drive such modulation, which will necessarily be unsteady and non-isotropic in nature so as to be able to realise an effective space-time adaptation [9, 18, 19]. A morphing-based (4D) DAM solution would be capable of both accommodating evolving traffic trends in the future and to synergistically mitigate the effects that unforeseen perturbations have on the nominal capacities, which can easily lead to greater disruptions on wider regions of airspace.

Eulerian flow theories such as the Lighthill-Witham-Richards (LWR) model efficiently represent the traffic flow as a continuum, which is an accurate approximation at the macroscopic scale. The continuity equation at the core of these methods is a linear partial differential equation that can be applied in discretised form to various airway segments. This approach was successfully demonstrated by Work and Bayen [20], who used an Eulerian model for flow prediction and to determine speed control measures. We present here a promising extension of such an approach to implement a fully 4D dynamic morphing of airspace sectors in the enroute context, which was originally outlined in [21]. The generalised continuity equation is applied to model the flow of air traffic within a given airspace, assumed to be a finite control volume which is fixed in space. The flow is assumed to be inviscid, incompressible, and irrotational, yielding a very simple yet accurate approximation, as demonstrated in various road transport studies [22]. The flux of aircraft into and out of the control volume is given as a function of time, with inflows considered to be negative and outflows considered to be positive. The generalised continuity equation can be expressed in differential form as:

$$\frac{\partial \rho}{\partial t} + \nabla \cdot (\rho \boldsymbol{v}) = 0 \tag{6.1}$$

where ρ is the air traffic density (aircraft count per NM^3), $\boldsymbol{v}$ is mean air traffic flow velocity (kts) such that $\rho \boldsymbol{v}$ is the flux (aircraft count per NM^2 hour). Assuming incompressibility ($\nabla \cdot \boldsymbol{v} = 0$), and defining a flow potential φ such that $\boldsymbol{v} = \nabla \varphi$, Eq. (6.1) can be re-written as:

$$\frac{\partial \rho}{\partial t} + \nabla \varphi \cdot \nabla \rho = 0 \tag{6.2}$$

Additionally, for an arbitrary element i within the control volume, the rate of change of the element's aircraft count is given according to the following equation:

$$\frac{\partial n_i}{\partial t} = \frac{\partial (\rho_i V_i)}{\partial t} = V_i \frac{\partial \rho_i}{\partial t} + \rho_i \frac{\partial V_i}{\partial t} \tag{6.3}$$

where n_i is the aircraft count, and ρ_i and V_i are respectively the air traffic density and volume (NM^3) of element i. Rearranging Eq. (6.3) and substituting into Eq. (6.2), one obtains the continuity equation as applied to element i:

$$\frac{1}{V_i} \left(\frac{\partial n_i}{\partial t} - \rho_i \frac{\partial V_i}{\partial t} \right) + \nabla \varphi_i \cdot \nabla \rho_i = 0 \tag{6.4}$$

where the subscript i is used to denote properties belonging to element i. Finally, multiplying throughout by the element volume, one obtains:

$$\frac{\partial n_i}{\partial t} - \rho_i \frac{\partial V_i}{\partial t} + V_i (\nabla \varphi_i \cdot \nabla \rho_i) = 0 \tag{6.5}$$

Equation (6.5) can be formulated as an optimal control problem by considering a control volume with 4-D fields $\rho(x, y, z, t)$ and $\varphi(x, y, z, t)$ and containing I sectors. The system dynamics are given by:

$$\boldsymbol{x} \triangleq \begin{cases} x_i = n_i, i \in [1, \dots, I] \\ x_j = V_i, j \in [I+1, \dots, 2I] \end{cases} \tag{6.6}$$

$$\boldsymbol{u} \triangleq \{\dot{V}_i\}, i \in [1, \dots, I] \tag{6.7}$$

$$\dot{\boldsymbol{x}} \triangleq \begin{cases} \dot{x}_i = \rho_i u_i - x_j (\nabla \varphi_i \cdot \rho_i), \\ \dot{x}_j = u_i \end{cases} \tag{6.8}$$

Minimising the quadratic cost function:

$$J = \sum_{i=1}^{I} (n_{Max} - \overline{n}_i)^2 \tag{6.9}$$

where $\overline{n}_i$ is the time-averaged aircraft count in element i, n_{Max} is the maximum number of aircraft in an individual sector (presently limited by human workload considerations, in the future by CNS + A computational complexity). The optimisation should be subject to the following constraints:

$$x_i \geq \frac{n_{Max}}{2}, \forall i \in [1, \dots, I] \tag{6.10}$$

$$x_j \geq V_{Min}, \forall j \in [I+1, \dots, 2I] \tag{6.11}$$

$$-\dot{V}_{Max} \leq u_i \leq \dot{V}_{Max} \forall i \in [1, \dots, I] \tag{6.12}$$

Equation (6.10) is specified to ensure that the traffic load for each sector remains between 50% and 100%. Equation (6.11) is the minimum volume V_{Min}, which can be a function of the sector characteristic dimension multiplied by a minimum altitude range. The minimum characteristic dimension is the one corresponding to a transit time of approximately 11 minutes, which marks the transition from the maximum number of aircraft constraints to the maximum throughput constraint in the Federal Aviation Administration (FAA) Monitor Alert Parameter (MAP) model [18]. Equation (6.12) is a heuristic constraint introduced to prevent excessively abrupt changes in the sector and represents the maximum volume variation per hour $\dot{V}_{Max}$. The control parameter $\dot{V} = [\dot{V}_1, \dots, \dot{V}_I]^T$ is used to control the sector morphing by updating the position of airspace nodes, which are the intersections between different sector edges as well as the airspace boundary. Given an airspace configuration containing I sectors and J nodes, a system of equations can be established as:

$$\boldsymbol{A} \cdot \boldsymbol{x} = \boldsymbol{b} \tag{6.13}$$

where $\boldsymbol{x}_{2J \times 1} = [x_1, y_1, x_2, y_2, \dots, x_J, y_J]^T$ is the node translation matrix containing the x and y coordinates describing the translation of all nodes; $\boldsymbol{A} = \begin{bmatrix} A1 \\ A2 \end{bmatrix}$ contains the inflation matrix $\boldsymbol{A1}_{I \times 2J}$ which describes the change in volume of sector i when node j is perturbed by a unit vector in the x- and y-directions, and the boundary constraint matrix $\boldsymbol{A2}_{J_b \times 2J}$ which contains constraints on the movement of boundary nodes. Figure 6.2 illustrates the method used to determine the coefficients of $\boldsymbol{A1}$.

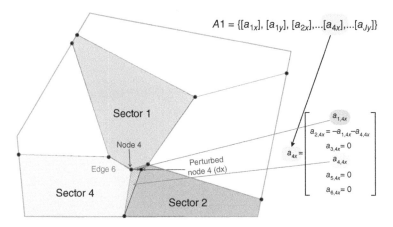

Figure 6.2 Method for determining the coefficients of inflation matrix A1 [21].

Node *j* is perturbed by unit vectors in the x and y directions to obtain a vector containing the volume changes of each sector due to this perturbation. This process is iterated through all nodes. The vector $\boldsymbol{b} = \begin{bmatrix} \boldsymbol{b1} \\ \boldsymbol{b2} \end{bmatrix}$ contains $\boldsymbol{b1}_{I\times1} = [\Delta\tilde{V}_1, \Delta\tilde{V}_2, \dots, \Delta\tilde{V}_I]^T$, the vector of normalised inflation in each sector and $\boldsymbol{b2}_{J_b\times1} = [\emptyset]$, the boundary constraint vector. $\boldsymbol{b1}$ is obtained by integrating the normalised control parameter $\dot{\tilde{V}}$ over time – $\dot{\tilde{V}}_i$, in turn, can obtained from normalising $\dot{V}$ with respect to each sector's relative volume such that the inflation in all sectors is equal to zero:

$$\sum_I \dot{\tilde{V}}_i = \sum_I (\dot{V}_i - \Lambda_i) = 0 \; where \; \Lambda_i = \frac{V_i \cdot \sum_I \dot{V}_i}{\sum_I V} \tag{6.14}$$

The morphing algorithm is illustrated in Figure 6.3. At a specified time period, the control parameter $\dot{V}$ is given by the solution the optimal control problem and is normalised and integrated to obtain $\boldsymbol{b}$. At the same time, the sector and node parameters are extracted from the existing sector configuration and used in generating the $\boldsymbol{A}$ matrix. As the actual inflation of the sectors do not scale linearly with translations in x and y, (i.e. $\boldsymbol{A1}$ is an approximation of the actual inflation and is accurate only for small perturbations about the nodes), the actual node translations are computed in a stepwise iterative manner. This is achieved by subdividing $\boldsymbol{b1}$ into intervals such that the sector inflation at each interval does not exceed a given threshold. The solution of $\boldsymbol{x}$ obtained at each interval is used to update the sector configuration and regenerate $\boldsymbol{A1}$ and $\boldsymbol{b1}$ for the subsequent iteration. The morphing results

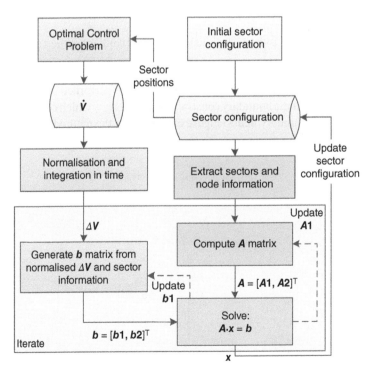

Figure 6.3 Sector morphing algorithm [21].

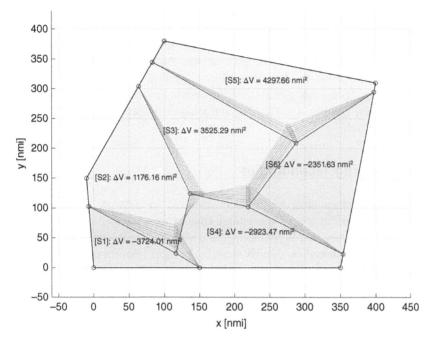

Figure 6.4 Results of the proposed morphing algorithm (negative volume variations are depicted in blue) [21].

are shown in Figure 6.4. Fairly large values of $\dot{V}$ were generated to exaggerate the change in sector sizes. As shown in the figure, the algorithm successfully computes the required node translation to achieve the specified normalised volume changes. Although the boundary nodes were constrained by *A2* and *b2* such that they could move along the boundary of the airspace, the algorithm did not exploit these degrees of freedom and instead focussed on interior nodes to achieve the required volume changes.

6.5 4D Trajectory Optimisation Strategies

The increase in flexibility resulting from introducing TBO functionalities such as real-time 4DT planning and negotiation processes will enable the planning of more efficient and environmentally-friendly trajectories even in the presence of unforeseen perturbations [23]. In particular, avionics and ATM DSS for TBO can support user-preferred optimal flight paths (intents) and limit human interventions to strategic high-level management and emergency assistance, so as to decrease the overall workload without compromising the situational awareness. In order to do so, these DSS are expected to dependably produce optimal and feasible 4DT solutions that resolve the impending conflicts, while complying with the timeframe allocated for the complete ATM transaction and minimise applicable environmental impacts. Increased efficiencies and higher throughputs are obtained in a CNS + A context by actively managing 4DT. In this respect, trajectory optimisation techniques implemented in an online process (as opposed to conventional offline flight planning) demonstrated great potential in terms of environmental and economic benefits [24]. This aspect is discussed in much greater detail in several of the other sections in this book.

As the dynamics involved in 4D-TBO and DAM are mutually interdependent and both concepts are fundamentally based on real-time optimisation algorithms, significant research is required to address the loop dependence, stability, and optimality metrics, promoting rapid convergence to global economic/environmental optimality. In the medium-long term, new combined traffic flow and airspace models integrating 4D-TBO and DAM will be required.

6.6 Other Emerging Technologies

Notwithstanding the various objectives already achieved by SESAR, NextGen, and analogous programmes so far, a very substantial amount of work is still necessary to relieve aviation from its legacy post-WW2 heritage and align itself with contemporary ICT standards. The following subsections discuss some of the most relevant trends currently being observed.

6.6.1 Artificial Intelligence and Task Redistribution

The effectiveness of several operational measures is compromised by excessive uncertainties in trajectory prediction and weather forecasts over the longer ranges involved. The most effective solutions being developed in this respect revolve around big data analytics and artificial intelligence (AI) algorithms of the machine learning (ML) kind, as these are capable of capturing several factors not included in the theoretical models [25–27]. As in several other sectors, the amounts of information collected, analysed, and shared among all relevant ground-based and airborne entities will have to grow much further to support sophisticated optimisation processes and predictive analytics and thus optimally deal with unpredicted events and mitigate disruptions.

A parallel shift in automation support and a move away from the centralised command and control-oriented ATM paradigm towards more distributed/collaborative management are also critically necessary. ATM will undertake a sensible progression in this respect, which will involve a redistribution of current functions and services from the ATM system to other entities including airborne systems, airline operation centres (AOC), and flight crews, and from humans to machines, to improve the efficiency of the system as a whole. The most promising concepts formulated in this respect emerged from recent UAS traffic management (UTM), U-Space and advanced air mobility (AAM) programmes [28–34]. For instance, as depicted in Figure 6.5, several new operational paradigms promote the reallocation of tactical deconfliction, separation assurance and collision avoidance (SA&CA) to vehicles, which will be equipped with both longer-range vehicle-to-infrastructure (V2I) and shorter-range vehicle-to-vehicle (V2V) data communications in line with the most advanced automotive concepts. This allows the ATM/UTM system to focus on demand/capacity monitoring, traffic clustering, and flow management initiatives. In turn, this relieves the human ATM/UTM operator from tactical deconfliction and communication/handshake duties so that they can focus on system integrity monitoring and strategic planning over wide regions.

Figure 6.5 Proposed redistribution of ATM services and duties among the various entities [35].

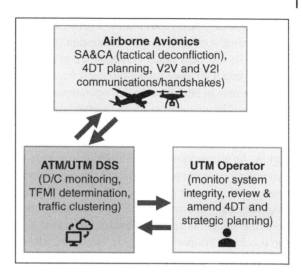

6.6.1.1 Airspace Restructuring

Airspace structure design is an important aspect of UAM/UAS integration and to date remains an open gap. Contemporary research in ATM/UTM is therefore looking to redefine airspace structures and classes to match urban and suburban traffic requirements and with sufficient flexibility to accommodate unsegregated operations of UAM, conventional aircraft, high-altitude platforms, and space launch/re-entry vehicles in high-density operational contexts [36, 37]. Although lessons learned in conventional ATM are useful, many of its solutions cannot be scaled down and applied directly to the local management of UAM and UAS [34]: the differences in operational complexity, traffic volumes, fleet mixes, and supporting infrastructure make airspace structure and sectorisation strategies for conventional air traffic inapplicable to the UAS/UAM context. While various sectorisation concepts and models have been proposed in the literature, to date, none has been standardised or even agreed upon to cover the whole spectrum of UTM operations. Among the many concepts that have been proposed in the literature, four are readily discernible: full mix, layers, zones, and tubes [38, 39]. The division can also be performed on the basis of the services provided, and on the level of overall system performance required to support a given category of operation. Access to a particular zone is then contingent on the UAS meeting the level of system performance stipulated for that zone. This mirrors the implementation of Performance-Based Navigation (PBN) for manned aircraft, wherein the employed navigation systems are required to meet a certain level of performance depending on the region and phase of flight. In fact, PBN has also been endorsed by NASA for UTM as part of the AAM initiative [40, 41]. Navigation performance for UAS can vary to a much greater extent than manned aircraft and a successful adaptation of PBN to the UTM context should accommodate this widely varying performance. Further, the PBN concept can be extended more generally to include not only navigation but also (as a minimum) communication and surveillance systems in line with the broader concept of PBO, which proves suitable in safely accommodating the highly dynamic nature of operations as envisaged under the UTM framework.

The following paragraphs outline a promising concept proposed by the authors [36, 42]. The doctrine applied is that the separation of UTM traffic and the airspace management strategy should be a direct consequence of the infrastructure supporting the operation [9]. The urban airspace model proposed here is specifically conceived to accommodate a high diversity in performance and operational characteristics among different UAM/UAS platforms. The airspace design and sectorisation strategies are centred on the discretisation of the airspace into elementary three-dimensional (3D) cells. The dimension of each cell is a function of the performance of the infrastructure supporting a given airspace region. As a consequence, the airspace structure and sector volumes are driven by the dynamically varying performance of the separation services provided. Two representative operational cases are considered: flight over skyline and flight below skyline. The model is conceptually illustrated in Figure 6.6 and consists of two stages. During the offline planning stage, the reference grid is generated as a set of *elementary cells* according to the baseline CNS performance. In our approach, the elementary cells are parameterised as cuboids with a square base, but the methodology could easily be extended to other 3D shapes. The length (r_x), width (r_y), and height (r_z) of the cuboidal cells are formulated as a function of the expected baseline CNS performance for the region. The model is applied in two timeframes: offline planning and online airspace management.

An occupancy grid can successively be derived from the set of elementary cells and from the active CNS protection volumes. Cells that are contained within and on the boundary of an aircraft's protection volume correspond to occupied space and the remaining cells correspond to unoccupied space. This is illustrated in Figure 6.7. The occupancy grid supports the DCB by not just considering the number of aircraft simultaneously active in a region, which was the traditional approach for human-centric ATM, but also their specific

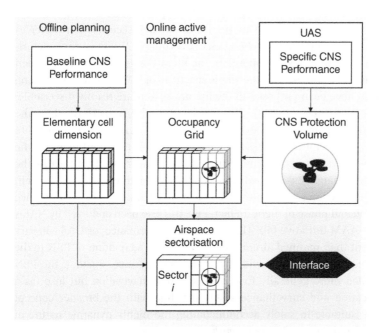

Figure 6.6 High-level overview of the performance-based airspace model [36].

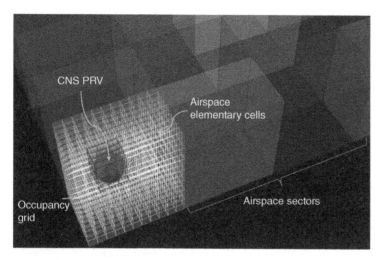

Figure 6.7 Overall urban airspace concept [36].

CNS performance, which supports more automated and autonomous SA/CA concepts. *Airspace sectors* are then generated as clusters of elementary cells to support the efficient management of traffic across the urban region. These sectors need to consist of an optimal number of elementary cells to balance the traffic complexity and the communication overhead due to sector handovers. All the elements of this performance-based airspace model, i.e. elementary cells, CNS protection volume, occupancy grid, and airspace sectors, are visualised in Figure 6.7. The following sections introduce the underpinning theoretical framework and mathematical models for all these entities. However, the CNS protection volumes will be addressed up front as they form the theoretical basis upon which the overall framework and the elementary cells were defined.

This airspace model provides the flexibility for implementation in the planning stage of three different operational timeframes: offline, pre-tactical online, and tactical online. The proposed airspace model can be applied to achieve different goals. In particular, the airspace capacity can be determined based on the performance of the CNS systems supporting the operations or, alternatively, the model can be applied to evaluate the CNS performance requirements to be enforced given a desired airspace capacity target. This two-way approach is conceptually illustrated in Figure 6.8.

The model is intended to enhance not only safety and efficiency but also ease the UTM operator's interpretability by using CNS performance as the main driver in airspace structure design, determining spacing, sector dimensions, and capacity altogether, while also supporting an intuitive visualisation. Therefore, as previously indicated, the elementary cell dimensions shall be parameterised as functions of the CNS performance for a given airspace region.

To facilitate ATM/UTM system processing and enhanced capability to balance airspace demand and capacity dynamically, the occupancy grid concept is proposed to approximate the volumetric demand. A grid of elementary cells is constructed for the entire urban region. The CNS protection volumes around each active aircraft are then superimposed onto the grid as illustrated in Figure 6.9. An elementary cell is designated as occupied when it contains at least one point of the CNS protection volume bounding each aircraft.

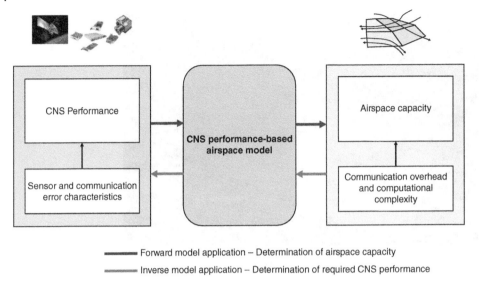

Forward model application – Determination of airspace capacity
Inverse model application – Determination of required CNS performance

Figure 6.8 Dual approach of performance-based airspace model [36].

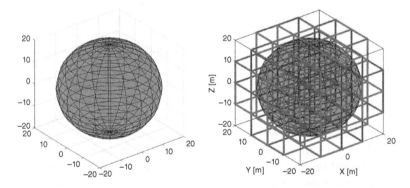

Figure 6.9 UAS protection volume (**left**) and relative occupancy grid (**right**) [36].

Each occupied cell, therefore, is a result of the demand placed on the airspace by virtue of the CNS performance. For visualisation purposes, occupied and free elementary cells are designated different colours (red and grey respectively).

The occupancy grid approach allows for a rapid assessment of the occupied space against the overall free space. A count performed over the cells supports the calculation of the theoretical remaining capacity to service the airspace with the given CNS infrastructure. A sectorisation scheme based on the previously defined elementary cells can optimally balance computational complexity and communication overheads. A UAS operation spanning multiple sectors requires a handover to be performed at each sector boundary. The number of cells allocated to a given sector is determined so as to optimise two key factors: the communication overhead required to perform all the required handovers for all involved traffic (which increase proportionally with the density of sectors) and the

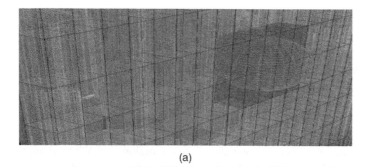

(a)

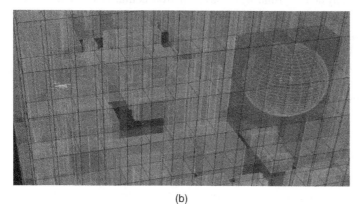

(b)

Figure 6.10 UAS protection volume with occupancy grid in a below-skyline scenario.

computational complexity associated with computing an avoidance volume and avoidance trajectory (which increases proportionally with sector dimensions). The protection volume and occupancy grid are illustrated in Figure 6.10.

6.6.1.2 Human-Autonomy Interactions

The development of more effective visualisation approaches for the complex underlying 4D phenomena remains a major aspect for ATFM DSS evolution, with considerable research tackling the active monitoring of human operator cognitive states [43–48]. Additionally, the adoption of AI 'black-box' models which can impede the human supervisory efforts is stimulating research in explainable AI (X-AI) [35, 49–52]. Consequently, major areas of contemporary research in ATFM DSS include HMI[2] to visualise complexity, ML/AI predictions, and dynamic density concepts, also in preparation for accommodating UAM/AAM and high-altitude airspace operations.

The safety certification of an AI-based system is still a challenge and substantial standardisation efforts are envisioned in this respect [53, 54]. To this end, the implementation of suitable integrity monitoring and augmentation techniques for both CNS systems, DSS, and human operators is expected to support the design, development, test and evaluation (DDT&E) of emerging cyber-physical-human architectures [53, 55], also supporting the certification case for AI-based CNS + A systems.

6.7 Conclusions

ATM is well underway on a generational journey away from its historical heritage thanks to major large-scale initiatives, such as in Europe and in North America. However, significant action is still required in several areas, ranging from the full digitisation of CNS infrastructures to the integration of AI-based automation and associated certification and human interfaces. This chapter overviewed some of the most important evolutionary trends in ATM/UTM, prioritising those which are expected to introduce the greatest changes or yield the most significant environmental benefits overall. While the majority of ATM enhancements being pursued globally do not primarily focus on environmental sustainability, their impacts are nonetheless envisioned to yield significant gains in this area, by removing inefficiencies, promoting resilience, and enhancing flexibility. The long-term goal of the ATM community is to achieve a scalable, distributed, and fully-interoperable system, which can accommodate any current and likely future air/space platforms and therefore facilitate the uptake of other important sustainable aviation technologies (e.g., aerostructures, propulsion and airport evolutions).

References

1 ICAO (2001). Annex 11 to the Convention on International Civil Aviation - air traffic services. In: *International Civil Aviation Organization (ICAO)*. Montreal, QC, Canada.

2 ICAO Procedures for air navigation services. In: *Air Traffic Management* (ed. Q.C. Montreal). Canada: The International Civil Aviation Organization (ICAO) Doc. 4444.

3 ICAO. (2006). Doc. 9869 - Manual on Required Communication Performance (RCP). The International Civil Aviation Organization (ICAO), Montreal, Canada.

4 SESAR (2020). European ATM Master Plan (Edition 2020): Digitalising Europe's Aviation Infrastructure. SESAR Joint Undertaking, Luxemburg.

5 FAA (2020). NextGen Annual Report (Fiscal Year 2020): A Report on the History, Current Status, and Future of National Airspace System Modernization. US Federal Aviation Administration (FAA), Washington, DC, USA.

6 Bradbury, J.N. (1991). ICAO and future air navigation systems. In: *Automation and Systems Issues in Air Traffic Control* (eds. J.A. Wise, V.D. Hopkin, and M.L. Smith), pp. 79–99. Springer.

7 Sabatini, R., Gardi, A., Ramasamy, S. et al. (2015). Modern avionics and ATM Systems for Green Operations. In: *Encyclopedia of Aerospace Engineering* (ed. R. Blockley and W. Shyy). Wiley.

8 Majumdar, A., Ochieng, W.Y., Bentham, J., and Richards, M. (2005). En-route sector capacity estimation methodologies: an international survey. *Journal of Air Transport Management* 11: 375–387.

9 Kistan, T., Gardi, A., Sabatini, R. et al. (2017). An evolutionary outlook of air traffic flow management techniques. *Progress in Aerospace Sciences* 88: 15–42.

10 EUROCONTROL (2017). European Route Network Improvement Plan - Part 3: Airspace Management Handbook - Guidelines for Airspace Management. EUROCONTROL. Brussels, Belgium.

11 Martinez, S.A., Chatterji, G.B., Sun, D., and Bayen, A.M. (2007). A weighted-graph approach for dynamic airspace configuration. *AIAA Guidance, Navigation, and Control Conference 2007*, Hilton Head, SC, 1476–1491, American Institute of Aeronautics and Astronautics (AIAA), Reston, VA, USA.

12 P. Cheng and R. Geng, "Dynamic Airspace Management - Models and Algorithms," in *Air Traffic Control*, M. Mulder, Ed., ed: InTech, 2010.

13 Sherali, H.D. and Hill, J.M. (2011). Configuration of airspace sectors for balancing air traffic controller workload. *Annals of Operations Research* 203: 3–31.

14 Tang, J., Alam, S., Lokan, C., and Abbass, H.A. (2012). A multi-objective approach for dynamic airspace sectorization using agent based and geometric models. *Transportation Research Part C: Emerging Technologies* 21: 89–121.

15 Chen, Y., Bi, H., Zhang, D., and Song, Z. (2013). Dynamic airspace sectorization via improved genetic algorithm. *Journal of Modern Transportation* 21: 117–124.

16 Bloem, M., Drew, M., Lai, C.F., and Bilimoria, K.D. (2014). Advisory algorithm for scheduling open sectors, operating positions, and workstations. *Journal of Guidance, Control, and Dynamics* 37: 1158–1169.

17 Chen, Y. and Zhang, D. (2014). Dynamic airspace configuration method based on a weighted graph model. *Chinese Journal of Aeronautics* 27: 903–912.

18 Welch, J.D., Cho, J.Y.N., Underhill, N.K., and DeLaura, R.A. (2013). Sector workload model for benefits analysis and convective weather capacity prediction. *10th USA/Europe ATM R&D Seminar, ATM* 2013.

19 Welch, J.D., Andrews, J.W., Martin, B.D., and Sridhar, B. (2007). Macroscopic workload model for estimating en route sector capacity. *7th USA/Europe ATM Research and Development Seminar, ATM2007*, Barcelona, Spain.

20 Work, D.B. and Bayen, A.M. (2008). Convex formulations of air traffic flow optimization problems. *Proceedings of the IEEE* 96: 2096–2112.

21 Lim, Y., Premlal, N., Gardi, A., and Sabatini, R. (2018). Eulerian Optimal Control Formulation for Dynamic Morphing of Airspace Sectors. *31st Congress of the International Council of the Aeronautical Sciences, ICAS 2018*, Belo Horizonte, Brazil.

22 Treiber, M. and Kesting, A. (ed.) (2013). *Traffic Flow Dynamics*. Berlin, Germany: Springer-Verlag.

23 Gardi, A., Sabatini, R., and Ramasamy, S. (2016). Multi-objective optimisation of aircraft flight trajectories in the ATM and avionics context. *Progress in Aerospace Sciences* 83: 1–36.

24 Gardi, A., Sabatini, R., and Kistan, T. (2019). Multiobjective 4D trajectory optimization for integrated avionics and air traffic management systems. *IEEE Transactions on Aerospace and Electronic Systems* 55: 170–181.

25 EUROCONTROL (2020). FLY AI Report: Demystifying and Accelerating AI in Aviation/ATM. European Aviation Artificial Intelligence High Level Group.

26 SESAR (2020). SESAR JU Webinar: Artificial Intelligence in ATM (part 1). Youtube: SESAR,

27 EASA (2020). Artificial intelligence roadmap: a human-centric approach to AI in aviation. *European Aviation Safety Agency*.

28 SESAR (2021). Drone DCB Concept and Process. DACUS, SESAR-ER4-31-2019--U-space.

29 EUROCONTROL (2019). Concept of operations for European UTM systems. EUROCONTROL.

30 FAA (2020). Federal Aviation Administration - UTM Concept of Operations v2.0, Department of Transportation, Washington D.C.. (D. o. transportation, ed.) Washington D.C.

31 FAA (2020). Unmanned Aircraft System Traffic Management (UTM). 800 Independence Avenue, SW Washington, DC 20591: United States Federal Aviation Administration.

32 Ippolito, C.A. (2021). Unmanned Aircraft Systems (UAS) Traffic Management (UTM) Project: Advanced Airborne Autonomy for Urban Flight Operations. Presented at the UTM Technical Interchange Meeting, NASA AAM Working Group

33 Johnson, M. (2021). UTM Conflict Management. NASA Advanced Air Mobility, UTM Technical Interchange Meeting.

34 Hackenberg, D. (2020). NASA advanced air mobility (AAM) Mission [PowerPoint presentation]. In: *Presented at the AAM Ecosystem Working Groups (AEWG), National Aeronautical and Space Administration (NASA)*. AAM Mission Office.

35 Pongsakornsathien, N., Gardi, A., Bijjahalli, S. et al. A Multi-Criteria Clustering Method for UAS Traffic Management and Urban Air Mobility. *IEEE/AIAA 40th Digital Avionics Systems Conference, DASC 2021*, San Antonio, TX, USA, Institute of Electrical and Electronics Engineers (IEEE), Piscataway, NJ, USA.

36 Pongsakornsathien, N., Bijjahalli, S., Gardi, A. et al. (2020). A performance-based airspace model for unmanned aircraft systems traffic management. *Aerospace* 7: 1–25.

37 Bijjahalli, S., Gardi, A., Pongsakornsathien, N. et al. (2022). A Unified Airspace Risk Management Framework for UAS Operations. *Drones*, 6.

38 Sunil, E., Hoekstra, J., Ellerbroek, J. et al. (2015). Metropolis: Relating airspace structure and capacity for extreme traffic densities. *ATM seminar 2015, 11th USA/EUROPE Air Traffic Management R&D Seminar*. United States Federal Aviation Administration (FAA) and European Organization for the Safety of Air Navigation (Eurocontrol).

39 Schneider, O., Kern, S., Knabe, F. et al. (2014). *METROPOLIS Concept Design Report*. Kluyverweg, The Netherlands: Delft University of Technology.

40 Kopaderkar, P.H. (2016). Safely enabling UAS operations in low-altitude airspace. *IEEE/AIAA 35th Digital Avionics Systems Conference, DASC* 2016, Sacramento, CA, USA, 33.

41 Mendonca, N., Metcalfe, M., and Wiggins, S. (2020). AAM Ecosystem Working Groups (AEWG): Urban Air Mobility (UAM) Concept of Operations (ConOps) Airspace Breakout [PowerPoint presentation]. Presented at the AAM Airspace Working Group Kickoff, National Aeronautical and Space Administration (NASA) AAM Mission and Deloitte.

42 Pongsakornsathien, N., Bijjahalli, S., and Gardi, A. (2020). A Novel Navigation Performance-based Airspace Model for Urban Air Mobility. *IEEE/AIAA 39th Digital Avionics Systems Conference, DASC 2020*, San Antonio, TX, USA, Institute of Electrical and Electronics Engineers (IEEE), Piscataway, NJ, USA.

43 Lim, Y., Gardi, A., Sabatini, R. et al. (2018). Avionics human-machine interfaces and interactions for manned and unmanned aircraft. *Progress in Aerospace Sciences* 102: 1–46.

44 Pongsakornsathien, N., Lim, Y., Gardi, A. et al. (2019). Sensor networks for aerospace human-machine systems. *Sensors* 22 (13): 4673.

45 Pongsakornsathien, N., Gardi, A., Sabatini, R., et al. (2020). Human-Machine Interactions in Very-Low-Level UAS Operations and Traffic Management. *IEEE/AIAA 39th Digital Avionics Systems Conference, DASC 2020*, San Antonio, TX, USA, Institute of Electrical and Electronics Engineers (IEEE), Piscataway, NJ, USA.

46 Planke, L.J., Gardi, A., Sabatini, R. et al. (2021). Online multimodal inference of mental workload for cognitive human machine systems. *Computers* 10 (6): 81.

47 Pongsakornsathien, N., Gardi, A., Sabatini, R., and Kistan, T. (2021). Evolutionary Human-Machine Interactions for UAS Traffic Management. *AIAA Aviation Forum 2021, Virtual Conference*, American Institute of Aeronautics and Astronautics (AIAA), Reston, VA, USA.

48 Pongsakornsathien, N., Gardi, A., Lim, Y. et al. (2022). Wearable cardiorespiratory sensors for aerospace applications. *Sensors* 22 (13): 4673.

49 Xie, Y. and Pongsakornsathien, N.. (2020). Artificial Intelligence Explanation for Decision Support System in Low Level Air Traffic Management. *AIAA 2020 Region VII Student Paper Conference*, American Institute of Aeronautics and Astronautics (AIAA), Reston, VA, USA.

50 Mathews, S.M. (2019). *Explainable Artificial Intelligence Applications in NLP, Biomedical, and Malware Classification: A Literature Review*, 1269–1292. Cham: Intelligent Computing.

51 Lertworawanich, P., Pongsakornsathien, N., and Xie, Y. (2021). Artificial Intelligence and Human-Machine Interactions for Stream-Based Air Traffic Flow Management. *32nd Congress of the International Council of the Aeronautical Sciences, ICAS 2021*, Shanghai, China.

52 Pongsakornsathien, N., Gardi, A., Sabatini, R., and Kistan, T. (2021). Interpretable Human-Machine Interactions for UAS Traffic Management. *AIAA Aviation and Aeronautics Forum and Exposition, AVIATION 2021*, Virtual Conference, American Institute of Aeronautics and Astronautics (AIAA), Reston, VA, USA.

53 Sabatini, R., Roy, A., Blasch, E. et al. (2020). Avionics systems panel research and innovation perspectives. *IEEE Aerospace and Electronic Systems Magazine* 35: 58–72.

54 Sabatini, R., Kramer, K., Blasch, E. et al. (2021). From the editors of the special issue on avionics systems: future challenges. *IEEE Aerospace and Electronic Systems Magazine* 36: 5–6.

55 Ranasinghe, K., Sabatini, R., Gardi, A. et al. (2022). Advances in integrated system health management for mission-essential and safety-critical aerospace applications. *Progress in Aerospace Sciences* 128.

7

Optimisation of Flight Trajectories and Airspace

Alessandro Gardi[1,2], Yixiang Lim[3], and Roberto Sabatini[1,2]

[1]*Department of Aerospace Engineering, Khalifa University of Science and Technology, Abu Dhabi, UAE*
[2]*School of Engineering, RMIT University, Melbourne, Victoria, Australia*
[3]*Agency for Science, Technology and Research (ASTAR), Singapore*

7.1 Introduction

The origins of the trajectory optimization problem can be traced to the Brachistochrone problem posed by Johann Bernoulli in 1696. The first noteworthy application of the concept to aviation was in relation to the problems posed and first solved by Zermelo as early as the 1930s [1, 2]. These studies have effectively underpinned the establishment of wind-optimal flight planning techniques that have been used for decades. Another phase of significant historical development is the one related to the determination of the optimal climb and descent performance (vertical profile optimization) in the 1970s [3–8], which underpinned the implementation of onboard Flight Management Systems (FMS) that have assisted pilots in their duties for decades.

With the deployment of technological communication, navigation, and surveillance (CNS) advancements such as satellite navigation and data-links, and the concurrent introduction of enhanced operational paradigms, the significance of trajectory optimisation in aviation has evolved and received renewed interest in conjunction with the concept of 4-Dimensional Trajectory-based Operations (4D-TBO). This chapter, originally published in [9, 10], introduces the most relevant literature, algorithms and models that are commonly adopted for trajectory optimization research around the world. Visser [11] described in detail the application of trajectory optimisation in the context of forthcoming 4D time-based operations. Betts performed considerable work on the implementation of direct transcription and collocation for aerospace trajectory optimisation problems [12–17], for which the implementation of efficient sparse nonlinear programming was an essential aspect. Hagelauer and Mora-Camino investigated the application of dynamic programming and identified that by applying expert knowledge to limit the search space and state transitions, and by introducing neural networks, it was possible to considerably restrict the computation times and enable future implementation for next generation flight management systems (FMS) [18]. Yokoyama and Suzuki studied the application of genetic algorithms to identify a good initial guess before introducing a direct collocation solved by means of a Block Diagonal Hessian (BDH) method [19]. Jardin identified

Sustainable Aviation Technology and Operations: Research and Innovation Perspectives, First Edition.
Edited by Roberto Sabatini and Alessandro Gardi.
© 2024 John Wiley & Sons Ltd. Published 2024 by John Wiley & Sons Ltd.
Companion Website: www.wiley.com/go/sustainableaviation

that some key issues in the development of real-time optimal deconflicted trajectory generation algorithms are associated with wind-optimal routing, conflict detection, and optimal conflict resolution, and consequently proposed the adoption of a discrete dynamic programming approach featuring neighbouring optimal wind routing and a deterministic conflict grid [20, 21]. The application of genetic algorithms for multi-objective trajectory optimisation (MOTO) was discussed in [22]. Jacobsen and Ringertz studied the modelling of airspace constraints for horizontal trajectory optimisation, specifically addressing polygon shaped constraints [23]. Eele and Richards proposed the adoption of a branch-and-bound step to efficiently solve optimal path planning problems involving obstacle avoidance or airspace constraints [24]. This allows splitting the horizontal path planning problem into a number of simpler subproblems, and potentially permits the identification of the global optimality region. Nine different branching strategies were considered. Soler et al. studied various promising approaches throughout the years, where the practical application to aircraft 4D trajectory planning was a core aspect [25–31]. Clarke applied trajectory optimisation techniques for the development of continuous climb and descent profiles addressing multiple aircraft types in representative traffic flow scenarios, which are being implemented as standard departure and arrival procedures at a number of US airports [32–34].

A recent review of trajectory design models was performed by Delahaye et al. [35], which outlined that evolutionary algorithms are recommended for the strategic planning, in conjunction with the very large dimensions and complexity of the problem. An approach based on penalty volumes and light propagation algorithms is recommended for the pre-tactical and tactical stages, whereas optimal control is preferred for the emergency trajectory design.

7.1.1 Theoretical Framework

In general, an optimisation process involves the adoption of suitable algorithms, decision logics, and heuristics to identify the best element from a finite or infinite set of available alternatives. Trajectory optimisation studies the methods to determine the best possible trajectory of a dynamical system in a finite-dimensional manifold, in terms of specific objectives and adhering to given constraints and boundary conditions. By this very definition, a substantial similarity between the statements of the trajectory optimisation problem (TOP) and of the optimal control problem (OCP) can be noted. Consequently, the two designations are frequently interchanged in the literature. One distinction noted concerns the mathematical nature of the unknown solutions; in OCP the interest surrounds the identification of optimal input functions, whereas in TOP the search is typically restricted to a finite set of static input parameters [36]. This consideration partly justifies the rapid success of TOP solution methods based on parameterisations. Furthermore, the formal distinction of *control, state,* and *output variables* is not always implemented in TOP, possibly due to the investigation of solution strategies beyond the framework of the optimal control theory, which will be briefly mentioned here. Notwithstanding, the continued familiarity with the OCP formulation has proved very fruitful, as the theoretical and computational advances in optimal control were directly transposed to enhanced TOP solution strategies and more generally in the aerospace vehicle guidance, navigation, and control (GNC) domain. Following the early works on the solution of the Brachistochrone problem, further studies by Euler and Lagrange led to the establishment of the *calculus of variations*. Since, as previously mentioned, the optimal solutions sought are in the form of functions, the calculus of variations effectively refers to *functional optimisation*. Therefore, an early-developed approach to the solution of OCP that is still commonly considered at present is based on the calculus of variations. Conversely, the transcription of OCP in the finite-dimensional non-linear programming (NLP) problem is increasingly popular as a computationally fast OCP solution strategy. As expanded later, most OCP solution methods can be ascribed to either of these approaches, and therefore are conventionally categorised as either *direct methods*, if based on the transcription to a finite NLP problem, or *indirect methods*, if theoretical derivations based on the calculus of variation are implemented to formulate a boundary value problem (BVP), and both approaches have led to very successful numerical implementations [11, 36–38]. The distinction between direct and indirect solution methods is less clear-cut than originally postulated, as some methods either fall in between or are intentionally hybridised. Additionally, more recent studies address the theoretical analogies between the two philosophies, and the findings are progressively overruling the distinction [36]. A third class of OCP solution strategies is represented by heuristic methods, such as simulated annealing, evolutionary algorithms, tree/graph/pattern search, and particle-swarm.

Theoretical solution strategies for OCP – such as the ones developed exploiting parameterisations and the calculus of variation – were highly instrumental for the development of numerical solution techniques for TOP. Notwithstanding, in principle, TOP can also encompass alternative non OCP-based formulations, including linear/nonlinear parametric optimisation problems on both continuous and discrete search spaces. This is particularly noteworthy in the aviation context, due to the widespread reliance on a geographical organisation of airspace and air routes, upon which all flight trajectory descriptors have been based, as well as the rules of the air, piloting techniques, traffic separation criteria, and air traffic management (ATM) lexicon. In particular, the current air navigation procedures, as well as the air route network, rely on discrete geographical descriptors for the horizontal flight path – known as Lateral NAVigation (LNAV) – and on altitude and airspeed constraints for the vertical flight profile – known as Vertical NAVigation (VNAV). Consequently, some studies in the operational aviation domain have approached the optimisation of flight trajectories within the current air navigation framework. This category of studies has been informally called *procedural optimisation*, as most frequently it involves the optimization of standard instrument departure (SID) procedures and standard terminal arrival routes (STAR).

7.1.2 Optimal Control Problem

The most general and convenient way to formulate the TOP is based on optimal control. Consequently, in line with the control theory, we introduce the vector of time-dependent state variables $x(t) \in \mathbb{R}^n$, the vector of time-dependent control variables $u(t) \in \mathbb{R}^m$, the vector of system parameters $p \in \mathbb{R}^q$ and the time $t \in [t_0; t_f]$. In the following subsections we introduce the dynamic constraints, the path constraints, the boundary conditions, and the cost functions. A consistent definition of all these components is fundamental to formulating a well-posed OCP and to perform an appropriate selection of the numerical solution method and of the multi-objective technique.

7.1.3 Dynamic Constraints

The specificity of the trajectory optimisation and optimal control with respect to other mathematical optimisation branches is the application to dynamical systems, i.e. in motion or transition over time. Therefore, a key component in the TOP formulation is the set of dynamic constraints, which are intended to reproduce the feasible motion of the system (i.e. the aircraft, in our case) within the TOP. A system of differential algebraic equations (DAEs), consisting of the time derivatives of the state variables, is usually adopted to introduce the system dynamics, and the dynamic constraints are therefore written as:

$$\dot{x}(t) = f[x(t), u(t), t, p] \tag{7.1}$$

Nonlinear dynamics are natively encompassed, while other cases, such as discrete-time dynamics, may also be accounted for by adopting adequately relaxed formulations. More details on optimal control of discrete-time systems can be found in [39–41].

7.1.3.1 Path Constraints

In the generalised TOP formulation, all non-differential constraints insisting on the system between the initial and final conditions are classified as *path constraints*, as they restrict the path, i.e. the space of states and controls, of the dynamical system. In order to represent all possible non-differential restrictions on the vehicle motion, two types of algebraic path constraints are considered: *inequality constraints* and *equality constraints*. A generalised expression of an inequality constraint is:

$$g_i(x(t), u(t), t; p) \le 0 \tag{7.2}$$

whereas an equality constraint can be written as:

$$h_i(x(t), u(t), t; p) = 0 \tag{7.3}$$

Equality constraints (Eq. (7.3)) can be considered a subset of inequality constraints, as they can be the result of two opposite inequalities such as in:

$$\begin{cases} g_{i,a}(x(t), u(t), t; p) \le 0 \\ g_{i,b}(x(t), u(t), t; p) \le 0 \end{cases} \tag{7.4}$$

where $g_{i,a} = -g_{i,b}$, hence Eq. (7.2) can account for both types. A compact representation of growing popularity is [42]:

$$C_{min} \le C[x(t), u(t), t; p] \le C_{max} \tag{7.5}$$

for which equality constraints are simply encompassed by imposing $(C_{min})_i = (C_{max})_i$. Each of the i-th compound inequalities of Eq. (7.5) can be split in two opposite inequality constraints similarly to Eq. (7.4). Despite the generality of the unified formulations (Eqs. (7.2) and (7.5)), it is typically preferable to treat equality constraints separately, as they introduce an additional relationship between some state and/or control variables, hence they usually allow the reduction of the number of states or controls in the TOP. Inequality constraints have a more ambiguous impact on the TOP, as sometimes the path of the dynamical system lies far from the constraint, and hence the latter could be safely ignored since it does not have any influence on the solution. Therefore, at each instant in the TOP time domain, an inequality constraint can either be *active*, if the system path intersects the constraint such that $g_i(\ ) = 0$ has to be enforced, or *inactive*, when $g_i(\ ) < 0$ is naturally verified, so that the constraint can be safely ignored. Consequently, for each inequality constraint the path of the dynamical system can be subdivided into *constrained subarcs* and *unconstrained subarcs*. Unfortunately, this is computationally nontrivial: usually, the occurrence and the exact number of constrained subarcs along the path is not known a priori, as are also very frequently the locations of junction points between unconstrained and constrained subarcs, or between two diversely constrained subarcs.

7.1.3.2 Boundary Conditions

Boundary conditions specify the values that state and control variables will have at the initial and final times. Since boundary conditions are not always necessarily restricted to definite values, it is useful to adopt a generalised expression including relaxed conditions. Similar to the already introduced expression of Eq. (7.5) for the path constraints, we then write the boundary conditions as:

$$\mathcal{B}_{min} \leq \mathcal{B}[\mathbf{x}(t_0), \mathbf{x}(t_f), \mathbf{u}(t_0), \mathbf{u}(t_f); \mathbf{p}] \leq \mathcal{B}_{max} \qquad (7.6)$$

where equality conditions are still encompassed by imposing $(\mathcal{B}_{min})_i = (\mathcal{B}_{max})_i$.

7.1.3.3 Cost Functions and Performance Indexes

In order to optimise a given performance, it is necessary to introduce a scalar value, the *performance index*, which by means of a suitably defined *cost function* quantifies the achievement of that particular objective. The optimisation process can then be translated into the mathematical minimisation (or maximisation) of one such performance index. Generally speaking, a performance index may depend on a function of terminal state values and parameters, $\Phi[\mathbf{x}(t_f), \mathbf{u}(t_f), \mathbf{p}]$, or on running costs along the system time, expressed as the integral function of one or multiple state variables, control variables, and parameters as $\int_{t_0}^{t_f} \Psi[\mathbf{x}(t), \mathbf{u}(t), \mathbf{p}]dt$. The generic TOP formulation involving a performance index J_i that takes into account both components was introduced by Bolza [43–45], and is expressed as:

$$J_i = \Phi[\mathbf{x}(t_f), \mathbf{u}(t_f), \mathbf{p}] + \int_{t_0}^{t_f} \Psi[\mathbf{x}(t), \mathbf{u}(t), \mathbf{p}]dt \qquad (7.7)$$

The optimisation is classified as *single-objective* when an individual performance index J is introduced and *multi-objective* when two or multiple performance indexes J_i are defined. Different objectives typically conflict, that is the attainment of a better J_k would lead to a worse J_h, $\{h, k \in [1; n_J], h \neq k\}$. Hence, the optimisation in terms of two or more objectives generates a number of possible compromises. Therefore, a trade-off decision logic must be introduced to identify an individual solution, and this this is the subject of multi-objective optimisation theory, which will be discussed in Section 7.1.3.4.

7.1.3.4 Resulting Mathematical Formulation

In summary, having introduced the dynamics, the path constraints, and the boundary conditions as well as the cost functions, the trajectory optimisation problem can be analytically stated as [42]:

Determine the states $\mathbf{x}(t) \in \mathbb{R}^n$, the controls $\mathbf{u}(t) \in \mathbb{R}^m$, the parameters $\mathbf{p} \in \mathbb{R}^q$, the initial time $t_0 \in \mathbb{R}$ and the final time $t_f \in \mathbb{R} \mid t_f > t_0$, that optimise the performance indexes

$$J = \Phi[\mathbf{x}(t_f), \mathbf{u}(t_f), \mathbf{p}] + \int_{t_0}^{t_f} \Psi[\mathbf{x}(t), \mathbf{u}(t), \mathbf{p}]dt$$

subject to the dynamic constraints

$$\dot{x}(t) = f[x(t), u(t), t, p],$$

to the path constraints

$$C_{min} \leq C[x(t), u(t), t; p] \leq C_{max},$$

and to the boundary constraints

$$B_{min} \leq B[x(t_0), x(t_f), u(t_0), u(t_f); p] \leq B_{max}.$$

7.1.4 Numerical Solution Techniques

Figure 7.1 outlines a comprehensive tree of techniques that were proposed for the solution of TOP, of which the most commonly adopted will be described in this section. Some of the techniques will not be discussed in detail due to their limited diffusion. Details on temporal finite element methods based on weak Hamiltonian formulation for the solution of OCP are available in [11, 46]. As briefly mentioned in Section 7.1.3, two mainstream strategies have been extensively adopted for the solution of the TOP, namely *direct methods* and *indirect methods* [11, 36–38]. In the first class, also defined by the motto '*discretise then optimise*' the determination of the unknown control function is attempted directly, and this involves the discretisation of the infinite-dimensional TOP into a finite-dimensional NLP problem. In the indirect methods, which emerged before and are defined by the motto '*optimise then discretise*', analytical manipulations based on the calculus of variations are exploited to transform the OCP into a nonlinear BVP.

7.1.4.1 Lagrangian Relaxation and First Order Optimality Conditions

The *Lagrangian relaxation* consists of approximating a constrained optimisation problem with an unconstrained one. The process involves constructing a new function $\mathcal{L}$, called *Lagrangian*, which incorporates the constraints c multiplied by a vector of additional unknowns, the Lagrange multipliers, λ_i:

$$\mathcal{L}(x, \lambda) = F(x) - \lambda^T c(x) \tag{7.8}$$

The optimisation process is then applied to the Lagrangian. Some additional conditions are introduced to ensure that the solution of the relaxed optimisation problem converges on the solution of the constrained optimisation. The augmented functional in the case of the problem of Bolza can be written as:

$$J = \int_{t_0}^{t_f} [\psi + \lambda^T(\dot{x} - f) + \mu^T(x - g_x)] \, dt + [\phi + v^T(x - b_x)]_{t_f} \tag{7.9}$$

where λ, μ, and v are the vectors of Lagrange multipliers for dynamic constraints, path constraints, and boundary conditions, respectively, with dimensions consistent with the related constraints. In the BVP solution process, some of these Lagrange multipliers are promoted to the rank of *adjoints* or '*co-states*' and treated separately. Convergence to the optimality

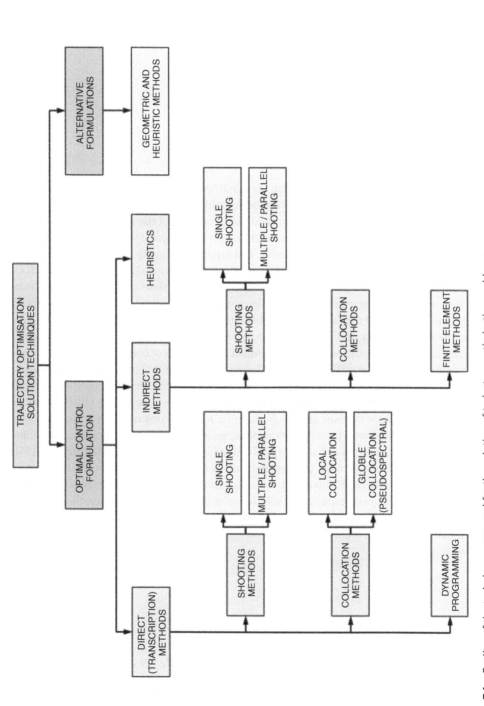

Figure 7.1 Outline of the techniques proposed for the solution of trajectory optimisation problems.

of the augmented functional is ensured by *first-order optimality conditions*. As an example, the set of conditions for a path and boundary constrained problem is:

$$\boldsymbol{f_u} - \boldsymbol{\psi_u}\lambda = 0 \tag{7.10}$$

$$\int_{t_0}^{t_f} (\boldsymbol{f_p} - \boldsymbol{\psi_p}\lambda)\, dt + (\boldsymbol{b_p} + \boldsymbol{\phi_p}v)_{t_f} = 0 \tag{7.11}$$

$$\dot{\lambda} - \boldsymbol{f_x} + \boldsymbol{\psi_x}\lambda = 0 \tag{7.12}$$

$$(\lambda + \boldsymbol{b_x} + \boldsymbol{\phi_x}\lambda)_{t_f} = 0 \tag{7.13}$$

7.1.4.2 Boundary-Value Problems

BVP are differential problems for which the conditions on the value of the unknown solution are prescribed at the boundary(ies) of the domain. In TOP applications, BVP are usually multi-dimensional and involve *Dirichlet* boundary conditions (also called *first type* or *fixed*). As BVP in the aerospace domain are most commonly encountered when dealing with problems of fluid dynamics, thermodynamics, and structural analysis, a large portion of the solution methods were developed for these applications. A general introduction to the BVP theory can be found in [47–50], whereas more details on state-of-the-art solution techniques, with particular emphasis on problems arising in the fluid-dynamics domain, can be found in [51–53]. If the Lagrange and Mayer terms are quadratic in $\{x, u, p\}$ and the constraints are linear in $\{x, u, p\}$ the BVP is linear and some originally proposed techniques to attain a numerical solution of the TOP included the methods of complementary functions, the adjoint variables, and the particular solutions [54]. Given that considerable nonlinearities are present in most of the models involved in aerospace TOP, the reduction of nontrivial cases into linear quadratic BVP is typically precluded. Systems involving either Mayer or Lagrange functionals that are nonquadratic in $\{x, u, p\}$ or constraints that are nonlinear in $\{x, u, p\}$ can be solved iteratively or exploiting some kind of heuristics. Iterative solution techniques can be either based on gradient methods or on higher order methods [54]. Gradient methods use at most the first derivatives of the function under analysis, while *quasilinearisation* methods involve at most the second derivatives of the function under analysis. Both categories require the solution of a linear BVP at each iteration. A scalar numerical convergence performance has to be defined to assess the efficiency of a specific iterative BVP solution algorithm. The following definitions of convergence performances were proposed [54]:

$$P = \int_{t_0}^{t_f} N(\dot{x} - \boldsymbol{\phi})\, dt + N(\boldsymbol{\psi}) \tag{7.14}$$

$$Q = \int_{t_0}^{t_f} N(\dot{\lambda} - \boldsymbol{f_x} + \boldsymbol{\phi_x}\lambda)\, dt + \int_{t_0}^{t_f} N(\boldsymbol{f_u} + \boldsymbol{\phi_u}\lambda)\, dt$$
$$+ N\left[\int_{t_0}^{t_f} (\boldsymbol{f_p} - \boldsymbol{\phi_p}\lambda)dt + (\boldsymbol{g_p} + \boldsymbol{\psi_p}\mu)_{t_f}\right] + N(\lambda + \boldsymbol{g_x} + \boldsymbol{\psi_x}\mu) \tag{7.15}$$

where N is the norm operator. The algorithm should stop when either P or Q attain a certain desired tolerance, i.e.:

$$P \le \varepsilon_1 \lor Q \le \varepsilon_2 \tag{7.16}$$

with $\varepsilon_1 \ll \varepsilon_2$, since the compliance to the constraints (metric P), shall prevail over optimality (metric Q) in most practical implementations.

7.1.4.3 Iterative Solution of Unconstrained Nonlinear Programming Problems

Similar to the nonlinear BVP arising in indirect solution approaches, the NLP problems involved in the direct solution of TOP are typically solved by iterative algorithms, for which the basic theory is briefly mentioned here, or by some kind of heuristics. Extensive details on NLP theory and on the development of computationally efficient NLP solution algorithms can be found in [49, 55, 56]. Adopting the n-dimensional Taylor series expansion of $F(\mathbf{x})$ to the third term we may write:

$$F(\mathbf{x}^{(k+1)}) \cong F(\mathbf{x}^{(k)}) + \nabla_x^T F(\mathbf{x}^{(k)}) \cdot \mathbf{s}^{(k)} + \frac{1}{2} \mathbf{s}^{(k)^T} H(\mathbf{x}^{(k)}) \mathbf{s}^{(k)} \tag{7.17}$$

Where

$$\mathbf{s}^{(k)} \triangleq \mathbf{x}^{(k+1)} - \mathbf{x}^{(k)} \tag{7.18}$$

$$\nabla_x F(\mathbf{x}) = \left[\frac{\partial F(x)}{\partial x_1}; \dots \frac{\partial F(x)}{\partial x_n} \right]^T \tag{7.19}$$

$$H(F(\mathbf{x})) = \begin{bmatrix} \dfrac{\partial^2 F(\mathbf{x})}{\partial x_1^2} & \cdots & \dfrac{\partial^2 F(\mathbf{x})}{\partial x_1 \partial x_n} \\ \vdots & \ddots & \vdots \\ \dfrac{\partial^2 F(\mathbf{x})}{\partial x_n \partial x_1} & \cdots & \dfrac{\partial^2 F(\mathbf{x})}{\partial x_n^2} \end{bmatrix} \tag{7.20}$$

The necessary conditions for a local minimiser, $\check{\mathbf{x}}$, are:

$$\nabla_x F(\check{\mathbf{x}}) = \mathbf{0} \tag{7.21}$$

$$\mathbf{s}^T H|_{\check{\mathbf{x}}} \mathbf{s} \geq 0 \tag{7.22}$$

Whilst the sufficient conditions for a strong local minimiser are:

$$\nabla_x F(\check{\mathbf{x}}) = \mathbf{0} \tag{7.23}$$

$$\mathbf{s}^T H|_{\check{\mathbf{x}}} \mathbf{s} > 0 \tag{7.24}$$

An iterative NLP solution can thus be formulated so that the search direction at step k based on the n-dimensional Newton method is written:

$$\mathbf{s}^{(k)} = -[H^{-1} \nabla_x F(\mathbf{x})]_{\mathbf{x}^{(k)}} \tag{7.25}$$

In a number of aerospace applications, the evaluation of derivatives of the objective function or of the constraints at every step may be computationally cumbersome; therefore various heuristic strategies were studied and implemented in NLP solvers. A number of strategies are based on the recursive update of the Jacobian or Hessian matrixes, as [56]:

$$D^{(k+1)} = D^{(k)} + \mathcal{U}(\Delta F, \Delta \mathbf{x}) \tag{7.26}$$

For instance, an n-dimensional generalisation of the secant method providing a recursive update of the Jacobian that minimises the Frobenius norm, proposed by Broyden, is:

$$D^{(k+1)} = D^{(k)} + \frac{(\Delta F - D^{(k)} \Delta \mathbf{x})}{\Delta \mathbf{x}^T \Delta \mathbf{x}} \tag{7.27}$$

For the Hessian matrix, the Broyden-Fletcher-Goldfarb-Shanno (BFGS) update can be implemented, which is a rank two positive definite secant approximation:

$$D^{(k+1)} = D^{(k)} + \frac{\Delta g \, \Delta g^T}{\Delta g^T \, \Delta x} - \frac{D^{(k)} \, \Delta x \, \Delta x^T \, D^{(k)}}{\Delta x^T \, D^{(k)} \, \Delta x} \tag{7.28}$$

Heuristic strategies formulated to recursively determine the inverse of the matrixes were less successful. It is also important to observe in this context that the quasi-Newton methods based on recursive updates manifest a superlinear rather than a quadratic convergence rate.

In addition to recursive updates of matrixes, other notable factors for the development of computationally efficient NLP solvers include the treatment of unfeasible, redundant, or rank-deficient constraints and of discontinuities, scaling, sparsity, and mesh refinements, above others. Considerable factors also involve the integration schemes and the matrix calculus algorithms adopted. More details on fast and efficient implementations of iterative NLP solution algorithms are given in [56].

7.1.5 Indirect Methods

As previously mentioned, indirect methods are a family of TOP solution methods in which the solution is attempted by applying the theory of the calculus of variations. Therefore, by means of suitable analytical derivations, the problem of Bolza is transformed into an augmented BVP. The *Hamiltonian* function, augmented with the dynamic and path constraints by means of the *Lagrangian relaxation*, hence including the co-state λ is:

$$\mathcal{H}(x, \lambda, \mu, u, t) = \mathcal{L} + \lambda^T f[x, u, t, p] - \mu^T C[x, u, t; p] \tag{7.29}$$

where μ are the Lagrangian multipliers associated with the path constraints. Boundary conditions are also augmented with the co-state:

$$\omega_{min} \le \omega[x(t_0), x(t_f), \lambda(t_0), \lambda(t_f), t_0, t_f, p] \le \omega_{max} \tag{7.30}$$

The dynamic equations of the resulting Hamiltonian Boundary-Value Problem (HBVP) are:

$$\begin{cases} \dot{x} = \dfrac{\partial \mathcal{H}}{\partial \lambda} \\[2mm] \dot{\lambda} = \dfrac{-\partial \mathcal{H}}{\partial x} \end{cases} \tag{7.31}$$

The TOP is therefore brought back to a two-point or multi-point BVP. The adoption of variational calculus for the approximate solution of OCP is described in detail in [57]. An example of the variational approach applied for the transcription of a TOP with no path constraints on the states, and a representative application to the minimization of the direct operating costs (DOCs), i.e. fuel costs and time costs, in the two-dimensional (2D) case, is described in [58]. The first step taken is to adjoin the dynamic constraints to the cost function as follows:

$$J = \phi(x(t_f), t_f) + \int_{t_0}^{t_f} \{\psi(x, u, t) + \lambda^T(t) \, (f(x, u, t) - \dot{x})\} \, dt \tag{7.32}$$

This leads to the definition of the following Hamiltonian function:

$$\mathcal{H}(x, u, t) = \psi(x, u, t) + \lambda^T f(x, u, t) \tag{7.33}$$

so that:

$$J = \phi(\mathbf{x}(t_f), t_f) + \int_{t_0}^{t_f} \{\mathcal{H}(\mathbf{x}, \mathbf{u}, t) - \lambda^T(t)\dot{\mathbf{x}}\} dt \tag{7.34}$$

Integrating by parts yields:

$$J = \phi(\mathbf{x}(t_f), t_f) - \lambda^T(t_f)\mathbf{x}(t_f) + \lambda^T(t_0)\mathbf{x}(t_0) + \int_{t_0}^{t_f} \{\mathcal{H}(\mathbf{x}, \mathbf{u}, t) + \dot{\lambda}^T(t)\mathbf{x}(t)\} dt \tag{7.35}$$

The variation of the performance index with respect to the states and controls is:

$$\delta J = \left[(\nabla_{\mathbf{x}}^T \phi - \lambda^T)\,\delta \mathbf{x}\right]_{t_f} + [\lambda^T \delta \mathbf{x}]_{t_0} + \int_{t_0}^{t_f} \left\{ \left(\nabla_{\mathbf{x}}^T \mathcal{H} + \dot{\lambda}^T\right)\delta \mathbf{x} + \nabla_{\mathbf{u}}^T \mathcal{H}\,\delta \mathbf{u} \right\} dt \tag{7.36}$$

The co-state equations are therefore written as:

$$\nabla_{\mathbf{x}}^T \mathcal{H} + \dot{\lambda}^T = 0 \rightarrow \dot{\lambda}^T = -\nabla_{\mathbf{x}}^T \mathcal{L} - \lambda^T \cdot \nabla_{\mathbf{x}} \mathbf{f} \tag{7.37}$$

$$\lambda^T(t_f) = \nabla_{\mathbf{x}}^T \phi \Big|_{t_f} \tag{7.38}$$

hence Eq. (7.36) can be rewritten as:

$$\delta J = [\lambda^T \delta \mathbf{x}]_{t_0} + \int_{t_0}^{t_f} \nabla_{\mathbf{u}}^T \mathcal{H}\,\delta \mathbf{u}\,dt \tag{7.39}$$

the necessary condition for J to be minimum is that:

$$\nabla_{\mathbf{u}}^T \mathcal{H} = 0\,t \in [t_0, t_f] \tag{7.40}$$

Whenever the controls are constrained as

$$\mathbf{C}(\mathbf{u}, t) \leq \mathbf{0} \tag{7.41}$$

then for all admissible $\delta \mathbf{u}$, i.e. simultaneously fulfilling all constraints $\mathbf{C}$, we shall have:

$$\nabla_{\mathbf{u}}^T \mathcal{H} \triangleq \delta \mathcal{H} \geq \mathbf{0} \tag{7.42}$$

which represents Pontryagin's Maximum Principle [59]. Although indirect methods for the solution of TOP were overlooked with the rise of efficient direct solution methods and of the computational capability to quickly solve very large NLP problems, recent research has identified some advantages that a combined approach exploiting both the calculus of variation and NLP may provide. Indirect collocation methods mentioned hereafter are an example of one such combined approach, which is actively being researched at present.

7.1.5.1 Indirect Shooting

Indirect shooting, also known as simple shooting, is one of the most basic iterative methods to attempt the solution of the BVP. The system of Hamiltonian dynamics is integrated numerically together with an initial guess from t_0 to t_f. If the terminal conditions thus obtained differ from the desired boundary conditions by more than a set tolerance ε, an updated initial guess set is generated and another integration is performed. In practice, indirect shooting is hardly ever viable for the solution of aerospace TOP, as the Hamiltonian dynamics have a very bad conditioning, which is further exacerbated as the integration intervals $t \in [t_0; t_f]$ increase. Furthermore, the integration on long time intervals may require a considerable computation time.

7.1.5.2 Indirect Multiple Shooting

As an attempt to resolve the limitations of indirect shooting, with indirect multiple shooting the time interval is divided into $n_i + 1$ subintervals. The indirect shooting method is then applied individually to each subinterval, that is, the HBVP are integrated together with the initial guesses on the state and co-state values. After each successful integration, the interface conditions

$$x\left(t_i^-\right) = x\left(t_i^+\right)$$
$$\lambda\left(t_i^-\right) = \lambda\left(t_i^+\right)$$

(7.43)

are imposed and a root-finding iteration is introduced to minimise the objective function associated with the discrepancies $\left(x\left(t_i^-\right), \lambda\left(t_i^-\right)\right) - \left(x\left(t_i^+\right), \lambda\left(t_i^+\right)\right)$ until a satisfactory threshold is attained. The size of the HBVP is effectively increased due to the inclusion of the interface values of the state and co-state at the boundaries of each interval. By partitioning the integration interval in a sufficiently high number of subintervals, the issues associated with the hypersensitivity of the Hamiltonian dynamics can be mitigated and a rapid convergence may eventually be attained.

7.1.5.3 Indirect Collocation

Collocation methods, such as those introduced in the following section in the context of direct methods, can be applied to parametrise the state and co-state of the HBVP associated with the OCP. Methods following this approach are termed indirect collocation methods. Indirect collocation methods were recently considered for the solution of aerospace TOP [36, 60]. These approaches have promising qualities when the application to TOP is considered, although the limitations in terms of flexibility intrinsic to indirect methods are still compounding.

7.1.5.4 Limitations

The most severe limitations encountered when adopting indirect solution methods are related to the initial guess and the overall flexibility of the approach. In particular, since the co-state variables are not representative of real physical entities, their initial guess is challenging. Additionally, the analytical derivation performed prior to establishing the HBVP is significantly dependent on the original TOP formulation. This notably restricts the applicability and flexibility of the approach, effectively limiting the application of indirect methods to specific problems where their adoption is advantageous, such as in the case of trajectory optimisation in the presence of wind, as discussed in Section 8.

7.1.6 Direct Methods

As previously mentioned, direct methods involve the transcription of the infinite-dimensional problem in a finite-dimensional NLP problem, hence following the approach summarised as '*discretise then optimise*'. The transcription into a finite-dimensional NLP problem can be either performed by introducing a control parameterisation based on arbitrarily chosen analytical functions, as in *transcription* methods, or by adopting a generalised piecewise approximation of both control and state variables based on a polynomial sequence to an arbitrary degree, as in *collocation* methods. In both cases the transcribed dynamical system is integrated along the time interval $[t_0; t_f]$. The search for the optimal set of discretisation parameters is formulated as a NLP problem, which is solved computationally by exploiting the most efficient numerical optimisation algorithms available. In direct transcription methods, a basis of known linearly independent functions $q_k(t)$ with unknown coefficients a_k is adopted as the parameterisation in the general form:

$$z(t) = \sum_{k=1}^{N} a_k q_k(t) \tag{7.44}$$

Considerable research on computationally efficient TOP solution algorithms based on direct transcription was performed by Betts et al. [12–17, 61–63]. Analogously, remarkable work on computationally efficient TOP solution algorithms based on direct collocation methods was performed by Rao et al. [64–71].

7.1.6.1 Direct Shooting

In direct shooting and multiple direct shooting, the parameterisation is performed on the controls $u(t)$ only; the dynamic constraints are integrated with traditional numerical methods, such as the Runge-Kutta family, and the Lagrange term in the cost function is approximated by a quadrature approximation. A generalised parameterisation of control variables can be written as:

$$u(t) = \sum_{k=1}^{N} c_k q_k(t) \tag{7.45}$$

7.1.6.2 Multiple Direct Shooting

With multiple shooting, the analysed time interval is partitioned into $n_i + 1$ subintervals. A direct shooting method is then applied to each subinterval. Continuity of the state is enforced at the interfaces similarly to multiple indirect shooting, as in the following expression:

$$x\left(t_i^-\right) = x\left(t_i^+\right) \tag{7.46}$$

7.1.6.3 Local Collocation Methods

By introducing distinct parameterisations on both the control variables and the state variables, it is possible to implicitly integrate the dynamics, that is approximate the integral of dynamics with a quadrature as:

$$\int_{t_i}^{t_{i+1}} f[x(s), u(s), s, p] ds \approx \sum_{j=1}^{Q} \beta_{ij} f[x(\tau_j), u(\tau_j), \tau_j, p] \tag{7.47}$$

where τ_j are the nodes of the quadrature approximation. The Lagrange term of the cost function ψ is also approximated with a numerical quadrature. This methodology is at the basis of a family of direct OCP solution methods, called collocation methods.

7.1.6.4 Global Collocation Methods

In global collocation methods, the direct solution of the OCP is attempted by enforcing the evaluation of the state and control vectors in discrete collocation points across the entire problem domain. Pseudospectral methods (PSMs) are considered one of the most computationally effective families of global collocation techniques available for the direct solution of large nonlinear OCP. They are based on a global collocation of orthogonal (spectral) interpolating functions. Due to their considerable success, we will review their working principle in more detail. Further information is given in [36, 67]. As an initial step, their application involves the introduction of the non-dimensional scaled time $\tau \in [-1, 1]$ so that:

$$t = \frac{t_f - t_0}{2} \tau + \frac{t_f + t_0}{2} \tag{7.48}$$

Such a change of variable requires that the differential operator is transformed as follows:

$$\frac{d}{dt} = \frac{2}{t_f - t_0} \frac{d}{d\tau} \tag{7.49}$$

If the final time is not known (i.e. it is either unconstrained or inequality-constrained in the boundary conditions), t_f will be an additional unknown variable that will have to be determined by the NLP solver. The states and control variables of the OCP are approximated by a set of polynomials of order N, and the problem is thereby discretised in $N+1$ nodes. These interpolation polynomials must be an orthogonal basis in the discretised space. Hence, they have to satisfy the null scalar product property:

$$P_i(x_j) * P_k(x_l) = 0 \ \forall \ i \neq j, \ \forall \ k, l \in \{1, \dots N+1\} \tag{7.50}$$

Although various families of interpolating polynomials can be successfully adopted and comprehensive dissertations may be found in [72, 73], the best implementations of PSM in terms of computational efficiency adopt simple interpolation polynomials in conjunction with a careful selection of the collocation nodes distribution. For these reasons, the basic Lagrange polynomials are frequently adopted for the interpolation of states and controls. A Gaussian quadrature rule guarantees the order of accuracy of the discretisation associated with the quadrature. Adopting the interpolation polynomials $P_k(\tau)$ on the $N+1$ nodes τ_k, the states are approximated as:

$$\tilde{x}_i(\tau) = \sum_{k=1}^{N} \tilde{x}_i(\tau_k) \cdot P_{i,k}(\tau) \tag{7.51}$$

and the controls are approximated as:

$$\tilde{u}_j(\tau) = \sum_{k=1}^{N} \tilde{u}_j(\tau_k) \cdot P_{j,k}(\tau) \tag{7.52}$$

The evaluation of the dynamic constraints (i.e. the state equations) is then performed in the nodes only, leading to a problem of finite dimensions. The dimension of the discrete problem is not the same in all cases, however. Lagrange polynomials of order N are expressed as:

$$P_k(\tau) = \prod_{j \neq k} \frac{\tau - \tau_j}{\tau_k - \tau_j}, \forall j \in [0, N] \tag{7.53}$$

As an example, a representation of Lagrange polynomials of the fourth order for equally spaced nodes (basic case) is given in Figure 7.2, whereas Figure 7.3 depicts the same polynomials applied to Chebyshev nodes.

Figure 7.2 Fourth order Lagrange interpolation polynomials for equally spaced nodes.

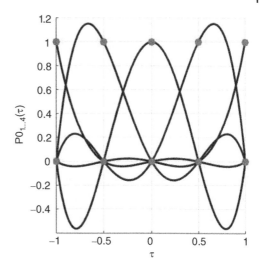

Figure 7.3 Fourth order Lagrange polynomials for Chebyshev nodes.

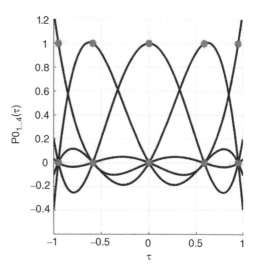

Chebyshev pseudospectral methods (CPMs) adopt Chebyshev polynomials of order N. An application of the Chebychev pseudospectral method to aircraft dynamics is found in [74] and involves the evaluation of the Chebyshev trigonometric polynomials:

$$P_N(\tau) = \cos(N \cos^{-1} \tau) \tag{7.54}$$

in the $N+1$ nodes:

$$\tau_k = \cos \frac{k \pi}{N}, k \in [0, N]. \tag{7.55}$$

Two recently adopted PSM variants are the Gauss PSM and the Legendre-Gauss-Lobatto (LGL) PSM [75]. Gauss PSM are based on the Gauss-Legendre quadrature, whereas the LGL PSM are based on the LGL quadrature, also simply known as the Lobatto quadrature. Gauss PSM are specifically conceived to ensure that the Karush-Kuhn-Tucker (KKT) conditions are identical to the discretised first-order optimality conditions. Legendre polynomials may be calculated by using the Rodrigues formula:

$$P_N(\tau) = \frac{1}{2^N k!} \frac{d^{(N)}}{d\tau^{(N)}} [(\tau^2 - 1)^N]$$ (7.56)

The LGL nodes are the $N + 1$ zeros of the polynomial:

$$L_N(\tau) = (1 - \tau^2)\dot{P}_N(\tau)$$ (7.57)

where $\dot{P}_N(\tau)$ is the first derivative of the Legendre polynomial of degree N [76]. In Gauss PSM the dynamic constraints are not collocated at the boundary nodes, whereas in the LGL PSM the evaluation of states and controls is also performed at the boundary nodes, thereby the dimension of the NLP problem is increased by two additional nodes.

7.1.7 Heuristic Methods

Although heuristics are involved to a certain extent in most numerical solution algorithms at various levels, in this section we briefly mention the solution strategies for aerospace TOP that involve a substantial exploitation of heuristics and metaheuristics for the optimisation of aircraft flight trajectories. A number of researchers proposed TOP solution strategies exploiting evolutionary algorithms, as they can natively feature nonlocal search scopes, hence detecting global optimality regions. Evolutionary algorithms (EAs) refer to a class of algorithms emulating the natural evolution processes, which are considered computationally intelligent since their pruning of unsuitable solutions emulates a machine-learning process. For mathematical optimisation purposes, a particularly useful evolution process is natural selection, for which a population of individuals (*genotypes*) evolves towards the fittest. *Crossovers* and *mutations* are the genetic processes most commonly implemented to generate new *off-springs* to be added to the population, which may or may not prove fitter. Mutations and crossovers are of a nondeterministic nature and allow EA to overcome the nonglobal convergence manifested by most iterative solution methods, causing convergence to local minimisers if the initial guess is not within the region of global optimality. The most desired quality of EA is, therefore, that they are natively suitable for the research of the global optima, since the method produces an arbitrary number of non-deterministic initial guesses and the algorithm itself is not restricted to the concept of the search scope. Among other research activities, EA were exploited as part of the research within the SGO-ITD of Clean Sky by the Green Systems for Aircraft Foundation (GSAF) academic cluster [77–88].

7.1.8 Trajectory Optimisation in the Presence of Wind

A trajectory optimisation problem of significant interest in the aviation domain consists of determining the optimal routing in the presence of winds, which was originally tackled by Zermelo as early as the 1930s [1, 2], and triggered several decades of active research. The potential environmental and economic benefits offered by an optimal routing in any given wind field are very substantial in most cases and all the more so in the case of long-range air transport routes. That notwithstanding, the limited accuracy of medium-to-long term wind forecasts and increasing air traffic densities, which progressively restrict the freedom of lateral and longitudinal routing, may compromise, and sometimes revert, these gains. The simplest case consists of determining the optimal routing of an aircraft travelling horizontally at a constant cruise airspeed between two known points in a known and constant wind field, ignoring the fuel consumption as well as variations in all other meteorological variables. In addition to loosening the constant altitude constraint, other generalisations involve the adoption of constant or variable non-uniform wind fields (i.e. 2D/3D/4D), more sophisticate modelling of the aircraft dynamics and of the atmosphere, multiply-connected search domains, and more complex cost functions addressing fuel consumption or other objectives in addition to flight time. As this research was largely performed before the recent widespread adoption of direct TOP solution methods, a considerable portion of the proposed numerical solution strategies are based on the calculus of variation. A very comprehensive treatment addressing the theory and solution of the trajectory optimisation problem under varying space-time meteorological conditions was produced by De Jong and the Royal Netherlands Meteorological Institute in 1974 [89]. The study led to the definition of a unified theoretical approach to the problem formulation, taking into account both space non-uniformities and time variations of the meteorological conditions and flight performances. Theoretical manipulations led to viable alternative formulations involving gradient equations for the time of transfer and a phase velocity equation for the airspeed. Considerations of the real operational conditions restricting the available choices in flight routing highlighted the suitability of graph optimisation methods for global optimality. Some iterative solution algorithms were proposed, including a special graph algorithm, and their viability in flight planning was numerically demonstrated. Bijlsma discussed the case of optimal aircraft routing in general wind fields and the most promising strategies for the development of a computational solution algorithm [90]. The convergence issues and non-global convergence of iterative solution methods are raised and support the adoption of algorithms based on graph theory. One such method capitalising on the advantages of the calculus of variation and of the graph theory is outlined. The general case with variable cruise speed v and wind speed u is introduced by the following dynamic constraints, where the sole control variable, θ, is the ground track azimuth:

$$\dot{x}_1 = v(t, x_1, x_2, \theta) \cos \theta + u_1(t, x_1, x_2, \theta) \tag{7.58}$$

$$\dot{x}_2 = v(t, x_1, x_2, \theta) \sin \theta + u_2(t, x_1, x_2, \theta) \tag{7.59}$$

Leading to the following Euler-Lagrange equations:

$$\dot{\lambda}_1 = -\lambda_1 \left(\frac{\partial v}{\partial x_1} \cos \theta + \frac{\partial u_1}{\partial x_1} \right) - \lambda_2 \left(\frac{\partial v}{\partial x_1} \sin \theta + \frac{\partial u_2}{\partial x_1} \right) \tag{7.60}$$

$$\dot{\lambda}_2 = -\lambda_1 \left(\frac{\partial v}{\partial x_2} \cos \theta + \frac{\partial u_1}{\partial x_2} \right) - \lambda_2 \left(\frac{\partial v}{\partial x_2} \sin \theta + \frac{\partial u_2}{\partial x_2} \right) \tag{7.61}$$

$$-\lambda_1 v \sin \theta + \lambda_2 v \cos \theta = 0 \tag{7.62}$$

Due to the impracticality of developing trajectory optimisers strictly based on the solution of BVP with the required flexibility to be implemented in avionics and ATM systems, researchers have investigated alternative approaches and heuristics. A method based on mesh discretisation that does not require an iterative solution is presented in [91]. Another approach based on wind networking was also recently proposed [92, 93]. This approach is especially valuable in increasing the trajectory predictability in oceanic airspace and other extents outside primary radar surveillance.

7.1.9 Multi-Objective Optimality

As mentioned in Section 7.1.3, optimisation in terms of multiple conflicting objectives $J_k = Q_k(\boldsymbol{p})$, $k \in [1, n_J]$ leads to large set of solutions that can be considered optimal in the sense that will be expanded on in the following section. Therefore, a trade-off selection strategy must be introduced in order to identify a single optimal solution from the large set of compromise solutions, and this is the purpose of multi-objective optimisation theory, discussed in this section. Conflicting objectives arise when introducing multiple environmental, economic, and operational criteria. Furthermore, as previously mentioned in Section 7.1.4, the implementation of constraints that are either unfeasible or opposing the attainment of better optimality can hinder the optimisation process or introduce numerical ambiguity, which also has to be addressed by adopting suitable multi-objective optimality techniques. This is particularly noteworthy if the operation in a real traffic environment is considered, as the arbitrary number of constraints introduced in the online tactical timeframe to resolve traffic conflicts can hamper the optimisation process. In all these instances it is necessary to complement the optimisation algorithm with predefined heuristics, decision logics and human machine interface and interaction (HMI^2) formats that increase the versatility and effectiveness of the trajectory planning algorithm.

A comprehensive and detailed review covering a large number of multi-objective optimisation strategies for engineering applications is given in [94]. This section resumes the techniques that are particularly suited to the optimisation of the aircraft trajectory with respect to multiple operational, economic, and environmental criteria, while complying with an arbitrary number of constraints. As we will describe in more detail, this decision-making process is usually performed either by means of an a priori articulation of preferences (i.e. beforehand), or of an a posteriori articulation of preferences (i.e. afterwards). Some other strategies were proposed and adopted, including progressive articulation of preferences and no articulation of preferences [94]. In the aerospace domain, bi-objective trajectory optimisation, i.e. in terms of fuel costs and flight time costs, has been widely studied and has led to results that have already been exploited for some time. An important advantage is that by limiting the study to two objectives it is possible to introduce a single scalar value, such as the cost index (*CI*) that is implemented in most of the current generation FMS, to account for the trade-off between the two conflicting objectives. In the emerging environmentally sustainable aviation research, it is nonetheless valuable to account for several different objectives in a flexible multi-model/multi-objective optimisation framework. Pollutant emissions, fuel consumption, perceived noise, convective weather and turbulence avoidance, operative costs, and contrail formation are all examples of the most common objectives currently studied.

7.1.9.1 Pareto Optimality

A point in the design space $\boldsymbol{p}^* \in \boldsymbol{P}$ is defined *Pareto optimal* if and only if there is no other point $\boldsymbol{p} \in \boldsymbol{P}$ such that $Q(\boldsymbol{p}) \leq Q(\boldsymbol{p}^*)$ and $Q_i(\boldsymbol{p}) < Q_i(\boldsymbol{p}^*)$ for at least one i. The *Pareto front*, also called the *Pareto frontier*, is the set of all the Pareto optimal points $\boldsymbol{p}^*$. Pareto optimal points are *non-dominated*, that is, there is no other solution that strictly dominates the Pareto optimal solution in terms of any objective. The Pareto front is the multi-objective and multi-dimensional equivalent of the single optimal solution of single objective optimisation. Due to the fact that in many applications a single solution is ultimately pursued even in large complex problems, multi-objective optimisation techniques lead to the identification of a single optimal solution that must be Pareto optimal, eventually at least in the weak sense, and thus must belong to the Pareto front.

7.1.9.2 A Priori Articulation of Preferences

In the a priori *articulation of preferences* approach, the user adopts a formulation of the multi-objective optimality that involves either a quantitative or a qualitative combination or prioritisation of the various objectives J_k, $k \in [1, n_k]$ and hence leads to the definition of a single combined objective $\widetilde{J}$. A single-objective TOP solution method is employed to optimise the combined objective, ultimately leading to an individual optimal solution. This approach is schematically represented in Figure 7.4. Methods falling in the a priori category analyse ways to define the combined objective starting from the various possible user-defined preference articulations.

Weighted Global Criterion Method

A very common choice for the condensation of the various objectives into a single cost function is to assign a weight to each objective function $Q_i(p)$ or to a functional of the objective and sum them together. By defining Q_i^0 the utopia point, that is $Q_i^0 = \min_{\boldsymbol{p}}\{Q_i(\boldsymbol{p}) \,|\, \boldsymbol{p} \in P\}$, we can introduce the following combined weighting:

$$\widetilde{J} = \left\{ \sum_{i=1}^{n_J} w_i^r \left[Q_i(\boldsymbol{p}) - Q_i^0 \right]^s \right\}^{1/s} \tag{7.63}$$

where $w_i \neq 0 \,\forall i$ is the weight assigned by the user to each single objective. Generally $\sum_i^{n_J} w_i = 1$, but this is not strictly necessarily. An important subcase is when $w_i = 1 \,\forall i$, $r = 1$ and $s = 2$, so that the weighted global criterion is actually the *geometric distance* (modulus)

Figure 7.4 Schematic representation of the a priori articulation of preferences.

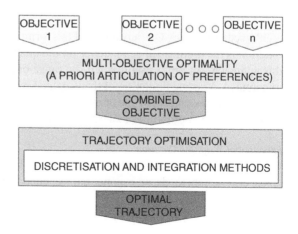

of the optimal solutions from the utopia point. Another subcase, which is the simplest and perhaps the most common method to combine the various objectives in a single one, is the *weighted sum method*, for which $Q_i^0 = 0 \ \forall \ i$, and $r = s = 1$, so that

$$\tilde{J} = \sum_{i=1}^{n_J} w_i Q_i(\boldsymbol{p}) \tag{7.64}$$

For all the weighted global criterion methods, an adequate normalisation of the objective functions $\boldsymbol{Q}$, even if not explicitly mandated, is fundamental to attaining the intended balance of importance. This aspect is nontrivial, as the results will significantly depend on the choice of reasonable reference values for normalisation. The presence of a nonzero utopia point in the generalised formulation partially alleviates this criticality, by replacing absolute magnitudes with relative ones.

Weighted Min-Max

The weighted min-max, also called weighted Chebychev method, endeavours to optimise the worst performance among the various objectives at any given time, as in:

$$\tilde{J} = \max_k \left(w_k \left[Q_k - Q_k^0 \right] \right) \tag{7.65}$$

The weighted min-max may frequently require more iterations than the weighted global criterion to reach the optimal solution, but at every step the structure of the evaluated $\tilde{J} = Q(\boldsymbol{p})$ will be simpler than the weighted global criterion, potentially improving the computational speed when complex nonlinear objective functions are considered.

Weighted Product

A different kind of weighting was proposed for which the combined performance index is defined as a product of all the objective functions, to the power of the assigned weight w_i, as in:

$$\tilde{J} = \prod_{i=1}^{n_J} [Q_i(\boldsymbol{p})]^{w_i} \tag{7.66}$$

It is important to note that in order to attain a non-trivial solution, it is strictly necessary that the condition

$$Q_i(\boldsymbol{p}) \neq 0, \forall i \tag{7.67}$$

is enforced at all times. One advantage of this formulation compared to the weighted global criterion is the diminished dependence on the quality of the normalisation performed, so this approach can prove useful when the range of the objective function is unknown or unbounded. A strong limitation of the weighted product, on the other hand, is that when nonlinearities are present in one or more of the objective functions $\boldsymbol{Q}(\boldsymbol{p})$, the computational complexity of the optimisation may increase considerably.

Exponential Weighted Criterion

To overcome the limitation to convex portions of the search domain imposed by weighted global criterion methods, an exponential weighted formulation can be adopted, as in:

$$\tilde{J} = \sum_{i=1}^{n_J} \left(e^{pw_i} - 1 \right) e^{pQ_i(\boldsymbol{p})} \tag{7.68}$$

As in the weighted product, a significant advantage is the diminished dependence on the quality of the normalisation performed, while nonlinearities in the objective functions still significantly affect the computational performances.

Lexicographic and Sequential Goal Programming Methods

Another strategy for a priori articulation of preferences is the definition of an order of importance for the various objectives. A single objective optimisation step is then performed for each objective J_k in the defined order, ensuring that the performance of the new step $Q(p^{(k)})$ is equal or better than the previous step $Q(p^{(k-1)})$ for all the objectives between 1 and (k), as in:

$$Q_j(p^{(k)}) \leq Q_j(p^{(k-1)}), \forall j \in [1; k] \tag{7.69}$$

Since the original version of the method appears heavily unbalanced towards the first objective, a relaxed version, for which in each new step there is a limited freedom to worsen the performance of the previous step, is introduced in order to attain a more balanced optimal solution

$$Q_j(p^{(k)}) \leq \left(1 + \frac{\delta_j}{100}\right) Q_j(p^{(k-1)}), \forall j \in [1; k] \tag{7.70}$$

Physical Programming

A notable physical programming approach was proposed, for which the user can introduce intuitive considerations and unstructured information that are used as *design metrics* to construct the combined objective function. In particular, the user may define quantitative ranges to particular parameters, based on qualitative considerations, such as desirable, tolerable, undesirable, and unacceptable. The user-defined classifications and ranges are then translated in a number of structured numerical objective functions as:

$$Q_a(p) = \log \left\{ \frac{1}{dm} \sum_i^{dm} \overline{Q}_i[Q_i(p)] \right\} \tag{7.71}$$

The most interesting advantage of physical programming is that the user can directly intervene in the solution region and restrict the part of the Pareto front studied.

7.1.9.3 A Posteriori Articulation of Preferences

In the a posteriori *articulation of preferences* approach, a single optimal solution, belonging to the Pareto front, is chosen after the whole set or a portion of it has already been determined. This approach is schematically represented in Figure 7.5. Methods in this category aim essentially at *populating* the Pareto front with an even distribution of points, in order to reduce the computational requirements and increase the effectiveness of the selection such that the user can perform the final choice from a set of sufficiently diversified solutions.

Physical Programming

Similarly to the a priori implementation, physical programming can be applied to translate unstructured information supplied by the user into decision criteria. As already discussed, the physical programming approach involves the definition of desirable, tolerable, undesirable, and unacceptable ranges of objective functions and constraint values (ranges for the

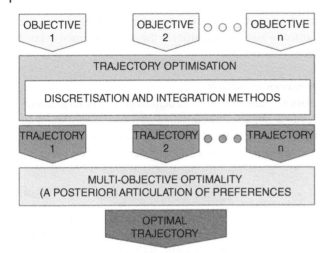

Figure 7.5 Schematic representation of the a posteriori articulation of preferences.

state and control variables are defined by inequality constraints of TOP). One advantage of the a posteriori physical programming approach in a closed-loop implementation is that the method restricts the search domain in addition to the solution region, by discarding the combination of performance indexes that produced unacceptable solutions. This can prevent computational resources from being wasted in calculating Pareto optimal points lying outside the acceptable portions of the Pareto front.

Normal Boundary Intersection

In the normal boundary intersection (NBI) strategy, the user-supplied weightings $\boldsymbol{w}$ are modulated in order to obtain an even distribution of Pareto optimal points. The NBI strategy can be formulated as:

$$\text{Minimise } \lambda, \text{ subject to: } \boldsymbol{\Phi}\boldsymbol{w} + \lambda\mathbf{n} = \mathbf{Q}(\mathbf{p}) - \boldsymbol{Q}^0 \tag{7.72}$$

where $\boldsymbol{\Phi}$ is the pay-off matrix made by the column vectors of the objective functions $\mathbf{Q}\left(\boldsymbol{p}_i^*\right) - \boldsymbol{Q}^0$ at the minimum of the i-th objective function. The vector $\boldsymbol{w} \geq \mathbf{0}$, $\sum_i^{n_J} w_i = 1$ are parameters to be systematically modified to obtain the complete Pareto front. It must be pointed out that this method does not provide sufficient conditions for Pareto optimality, so that some generated solutions are actually weakly optimal or suboptimal points.

Normal Constraint Method

The normal constraint (NC) method is a modification of the NBI method encompassing a tactic to filter suboptimal solutions. For this method, an arbitrary number of evenly distributed sample points are determined in the utopia hyperplane as a linear combination of the vertices with consistently varied weights. Each sample is then correlated to a Pareto optimal solution through a single objective optimisation process.

7.1.10   Conclusions

The theoretical results achieved and the very efficient solution methods proposed for the optimal control of nonlinear dynamical systems offer unique opportunities for exploitation in aircraft path planning. The greatest advantage lies in the fact that the optimality of the solution is mathematically supported. Pseudospectral transcription methods are emerging as the most efficient solution techniques and yield considerable flexibility in their implementation. Suitable multi-objective formulations are, however, essential to capture multiple conflicting objectives and unfeasible constraints. This chapter reviewed the fundamental solution techniques and multi-objective optimality strategies that can be adopted for aircraft trajectory optimisation applications. The knowledge gathered from this review is exploited when designing the multi-objective 4D trajectory optimisation (MOTO-4D) algorithms presented in chapter 5.

7.2   Emission Models and Environmental Optimality Criteria for Trajectory Optimisation

7.2.1   Introduction

Optimality criteria and constraints are introduced in the optimisation problem by means of suitable models. Figure 7.6 outlines the interdependencies introduced in principle by models typically adopted in multi-objective trajectory optimisation (MOTO) for environmental sustainability studies. These models can either consist of one or more mathematical functions, which are either analytically derived from physical/geometrical principles or empirically formulated, or consist of raw numerical data, usually in tabular form. From a theoretical point of view, it is desirable to identify representative functions in analytical form, i.e. being accurate, simple, versatile, and universal, though these qualities are frequently conflicting in nature. The identification of functions in analytical form has the advantage of leading to results of higher scientific significance and/or of more general applicability, which are the most relevant from the research perspective. Another advantage of adopting functions in analytical form is that the effectiveness of numerical solution strategies can be accurately estimated. In practice, in a number of applications, it is necessary to deal with numerical data in tabular or other forms. Some detail is therefore given to techniques for the exploitation of raw data, and to the necessary pre-arrangements to ensure numerical stability and convergence.

7.2.2   Flight Dynamics

As the geometric design of flight trajectories is deeply related to the dynamics of the body in aerial motion, flight dynamics are a core component, and are discussed beforehand. The focus of this work is on fixed-wing transport aircraft; hence the models introduced are specifically tailored to this category of flying platforms. A detailed discussion on the derivation of flight dynamics equations from first principles is described in [95–98].

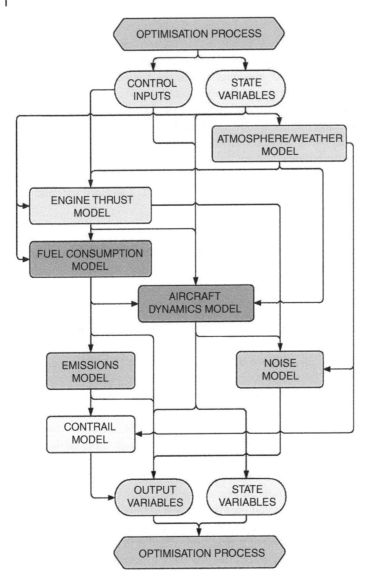

Figure 7.6 Layout of the typical model interdependencies in multi-objective trajectory optimisation studies.

7.2.2.1 Rigid Body Models

Assuming the aircraft to be a rigid body with a static mass distribution, an accurate model of its flight dynamics can be introduced, which complements the equilibrium of forces along the coordinate axes of a suitable Cartesian reference frame located in the centre of mass of the aircraft, named *body frame*, with the equilibrium of their momentums. This model involves a high number of parameters to define the properties of inertia and of aerodynamic stability and control forces. Adequate experimental and numerical investigations are typically required in order to define the parameters with a high degree of precision. For the implementation in MOTO problems, and for such applications as flight simulation and trajectory estimation, flight dynamics are typically transcribed in a set of differential algebraic equations (DAE). An introductive description of such derivation may be found in [95]. The set of DAE and complementary kinematic relations defining the six degrees of freedom (6DOF) rigid body dynamics of a fixed-wing aircraft are [99]:

$$
\begin{bmatrix} \dot{u} \\ \dot{v} \\ \dot{w} \end{bmatrix} = \frac{g}{W} \begin{bmatrix} F_{X_b} \\ F_{Y_b} \\ F_{Z_b} \end{bmatrix} + g \begin{bmatrix} -s\theta \\ s\phi\, c\theta \\ c\phi\, c\theta \end{bmatrix} + \begin{bmatrix} rv - qw \\ pw - ru \\ qu - pv \end{bmatrix}
\tag{7.73}
$$

$$
\begin{bmatrix} \dot{p} \\ \dot{q} \\ \dot{r} \end{bmatrix} = \begin{bmatrix} I_{XX_b} & 0 & -I_{XZ_b} \\ 0 & I_{YY_b} & 0 \\ -I_{ZX_b} & 0 & I_{ZZ_b} \end{bmatrix}^{-1} \cdot
$$
$$
\begin{bmatrix} M_{X_b} + (I_{YY_b} - I_{ZZ_b})qr + I_{XZ_b}pq \\ M_{Y_b} + (I_{ZZ_b} - I_{XX_b})pr + I_{XZ_b}(r^2 - p^2) \\ M_{Z_b} + (I_{XX_b} - I_{YY_b})pq - I_{XZ_b}qr \end{bmatrix}
\tag{7.74}
$$

$$
\begin{bmatrix} \dot{x}_f \\ \dot{y}_f \\ \dot{z}_f \end{bmatrix} = \begin{bmatrix} c\theta c\psi & s\phi s\theta c\psi - c\phi s\psi & c\phi s\theta c\psi + s\phi s\psi \\ c\theta s\psi & s\phi s\theta s\psi + c\phi c\psi & c\phi s\theta s\psi - s\phi c\psi \\ -s\theta & s\phi c\theta & c\phi c\theta \end{bmatrix} \begin{bmatrix} u \\ v \\ w \end{bmatrix} + \begin{bmatrix} v_{WX_f} \\ v_{WY_f} \\ v_{WZ_f} \end{bmatrix}
\tag{7.75}
$$

$$
\begin{bmatrix} \dot{\phi} \\ \dot{\theta} \\ \dot{\psi} \end{bmatrix} = \begin{bmatrix} 1 & s\phi s\theta/c\theta & c\phi s\theta/c\theta \\ 0 & c\phi & -s\phi \\ 0 & s\phi/c\theta & c\phi/c\theta \end{bmatrix} \cdot \begin{bmatrix} p \\ q \\ r \end{bmatrix}
\tag{7.76}
$$

where:

u, v, w : translational velocity components along the three axes of the body reference frame (m s^{-1});

p, q, r : rotational velocity components around the three axes of the body reference frame; representing rolling, pitching, and yawing rates (rad s^{-1}), respectively;

ϕ, θ, ψ : Euler angles, respectively representing bank, pitch, and yaw Euler rotations (rad);

x_f, y_f, z_f : components of the relative position vector between the Earth-fixed reference frame and the body centre of mass (m);

$F_{X_b}, F_{Y_b}, F_{Z_b}$: resultants of the aerodynamic and propulsive forces acting along the three axes of the body reference frame (N);

$M_{X_b}, M_{Y_b}, M_{Z_b}$: resultants of the aerodynamic and propulsive moments acting around the three axes of the body reference frame (N m);

$I_{XX_b}, I_{YY_b}, I_{ZZ_b}, I_{XZ_b}$: non-null components of the inertia tensor (kg m^2);

$v_{WX_f}, v_{WY_f}, v_{WZ_f}$: components of the wind vector along the three axes of the Earth-fixed reference frame (m s^{-1});

W : aircraft weight (N), which may be either constant or subject to fuel consumption.

In particular, Eq. (7.73) represent the translational dynamics, Eq. (7.74) the rotational dynamics, Eq. (7.75) the translational kinematics and Eq. (7.76) the rotational kinematics. Rigid-body models are considered unsuitable for the calculation of trajectories on medium-long timeframes. This is due both to the large dimensions and complexity of the resulting MOTO/estimation problem, as well as to the presence of short period modes that are keen to introduce numerical instabilities [100]. Rigid body models are nevertheless fundamental for the study of transition manoeuvres and more in general for the dynamic stability and control analysis and design of aircraft, and are successfully adopted in a number of trajectory optimisation studies, in combination with a careful selection of path constraints.

7.2.2.2 Point-Mass Models

A commonly adopted approach to derive a simplified set of equations of motion for atmospheric flight is based on the approximation of the aircraft as a point-mass object, thereby neglecting all aspects associated with its rotational dynamics. The resulting dynamics are characterised by only three degrees of freedom (3DOF) – i.e. the three spatial coordinates – which is the name adopted to define this family of models. These models are based on Newton's second law expressed along the coordinate axes of the *body frame*, and on the expression of the motion of such frame with respect to an inertial reference frame of convenience. All aspects associated with the rotational state of the aircraft are neglected. The model can involve either a constant mass or a variable mass. Models belonging to the first category are adopted when the analysed timeframe is relatively limited, such that the fuel consumption may be neglected, or when no fuel is consumed, such as in the case of sailplanes or total engine failures. As an example, the following set of DAE is associated with a variable mass 3DOF model [95]:

$$
\begin{cases}
\dot{v} = \dfrac{g}{W}\,(T\cos\varepsilon - D - W\sin\gamma) \\[2mm]
\dot{\gamma} = \dfrac{g}{v\,W}\cdot[(T\sin\varepsilon + L)\cos\mu - W\cos\gamma] \\[2mm]
\dot{\chi} = \dfrac{g}{v\,W}\cdot\dfrac{(T\sin\varepsilon + L)\sin\mu}{\cos\gamma} \\[2mm]
\dot{\phi} = \dfrac{v\cos\gamma\sin\chi + v_{w_\phi}}{R_E + z} \\[2mm]
\dot{\lambda} = \dfrac{v\cos\gamma\cos\chi + v_{w_\lambda}}{(R_E + z)\cos\phi} \\[2mm]
\dot{z} = v\sin\gamma + v_{w_z} \\[2mm]
\dot{m} = -FF
\end{cases}
\tag{7.77}
$$

where the state vector consists of the following variables:

v : longitudinal velocity (scalar) (m s^{-1});
γ : flight path angle (scalar) (rad);
χ : track angle (scalar) (rad);
ϕ : geographic latitude (rad);
λ : geographic longitude (rad);
z : flight altitude (m);
ε : thrust angle of attack (rad);
m : aircraft mass (kg);

and the variables forming the control vector are:

T : thrust force (N);
N : load factor [ ];
μ : bank angle (rad);

Other variables and parameters include:

D : aerodynamic drag (N);
v_w : wind velocity, in its three scalar components (m s^{-1});
g : gravitational acceleration (m s^{-2})
R_E : Earth radius (m);
FF : fuel flow (kg s^{-1}).

Frequently the modelling is restricted to the flight profile in the vertical plane or in the horizontal plane only.

7.2.3 Turbofan and Turboprop Engine Models

This section will briefly outline useful models for fuel consumption and thrust with respect to current generation turbofan/turbojet and turboprop engines. Different propulsive technologies are potentially associated with vastly dissimilar energy sources and consumption rates, and this is increasingly evident with the appearance of novel alternative propulsion systems. That notwithstanding, most of the aircraft in current operation rely on air-breathing, internal combustion engines of the turbofan or turboprop type, employing hydrocarbon fuels, which are an established technology with efficient and cost-effective production and supply chains. Within the MOTO context, steady state empirical models are commonly employed to reproduce the dependencies of thrust and fuel flow on altitude (z), true airspeed (v_{TAS}), temperature, and throttle (τ), which are the only ones directly associated with the aircraft trajectory. The following empirical expressions were adopted in the development of Eurocontrol's Base of Aircraft Data (BADA), to determine the climb thrust and the fuel flow FF of a turbofan propelled aircraft, which operationally equates to the maximum thrust T_{MAX} in all flight phases excluding take-off [101]:

$$T_{MAX} = C_{T1}\left(1 - \frac{H_P}{C_{T2}} + C_{T3}H_P^2\right)[1 - C_{T5}(\Delta T - C_{T4})] \tag{7.78}$$

$$FF = \max\left[\tau\, T_{MAX}\, C_{f1}\left(1 + \frac{v_{TAS}}{C_{f2}}\right), C_{f3}\left(1 - \frac{H_P}{C_{f4}}\right)\right] \tag{7.79}$$

where τ is the throttle control, H_P is the geopotential pressure altitude in feet, ΔT is the deviation from the standard atmosphere temperature in kelvin, v_{TAS} is the true airspeed. $C_{T1}...C_{T5}$, $C_{f1}...C_{f4}$ are the empirical thrust and fuel flow coefficients, which are also supplied as part of BADA for a considerable number of current aircraft [101]. Similarly, the following empirical expressions were adopted in BADA to determine the maximum thrust and fuel flow of a turboprop propelled aircraft:

$$T_{MAX} = \frac{C_{T1}}{v_{TAS}}\left(1 - \frac{H_P}{C_{T2}}\right) + C_{T3} \tag{7.80}$$

$$FF = \max\left[\begin{array}{c} \tau\, T_{MAX}\, C_{f1}\left(1 - \frac{v_{TAS}}{C_{f2}}\right)\cdot\left(\frac{v_{TAS}}{1000}\right), \\[2ex] C_{f3}\left(1 - \frac{H_P}{C_{f4}}\right) \end{array}\right] \tag{7.81}$$

where again $C_{T1}...C_{T3}$, $C_{f1}...C_{f4}$ are also supplied as part of BADA for a considerable number of aircraft [101]. The accuracy of the empirical models and of the coefficients supplied as part of Eurocontrol's BADA were analysed in [102].

Due to the very competitive specific energies and energy densities, which are additionally important in the weight/volume sensitive aerospace domain, no significant change in the market-share of hydrocarbon fuels is expected in the near future. These aspects, combined with the lengthy and costly development and certification processes to be undertaken by any innovation in the aviation domain, introduce a considerable technological inertia. Given the generational lifespan of aircraft models, it is expected that petroleum-based fuels will remain the largest source of chemical energy for aeronautical propulsion for at least the next two decades. For all these reasons, current generation jet fuels are widely targeted in trajectory optimisation studies for the assessment of aviation environmental impacts in the future.

7.2.4 Pollutant Emissions

Emission indexes (*EI*) specific to each atmospheric pollutant (*AP*) species were introduced in order to distinguish the dependencies of pollutant emissions from the fuel flow (*FF*), and are very frequently adopted in trajectory optimisation studies. The general expression to calculate the total emission of the *AP* from its associated emission index (EI_{AP}) expressed in $[Kg_{AP}/Kg_{Jet A-1}]$ is:

$$AP = \int_{t_0}^{t_f} EI_{AP}(\tau) \cdot FF(\tau, v, z)\, dt \quad [Kg] \tag{7.82}$$

where *FF* is the fuel flow ($Kg\ s^{-1}$).

The International Civil Aviation Organization (ICAO) has established an extensive and constantly updated databank for engine emissions based on data collected independently by a number of entities, which proves to be a valuable starting point for simplified empirical engine emission models [103]. The fuel-specific *EI* are measured at the standard throttle settings defined in ICAO Annex 16 volume 2 [104]. In particular, for an exclusively subsonic engine, the reference throttle settings are take-off (100% of rated engine thrust), climb (85%), approach (30%), and idle (7%).

Eurocontrol's BADA employs an empirical calculation method for the *FF* as a function of the engine thrust (τ), of the ambient pressure and of the true airspeed conditions [101]. For the implementation of Eq. (7.82) in a trajectory optimiser, it is convenient to refer to the differential formulation:

$$\frac{d\,AP}{dt} = EI_{AP}(\tau) \cdot FF(\tau, v, z) \quad \left[\frac{Kg}{s}\right] \tag{7.83}$$

The various aviation-related AP taken into consideration have different dependencies, which are discussed separately. The most relevant carbon-related exhaust products of fossil fuels are carbon dioxide (CO_2), carbon monoxide (CO), and unburned hydrocarbons (HC). CO and HC are significantly noxious for both the environment and for living creatures, and are therefore the primary target. Significant amounts of CO and HC are generated during the incomplete combustion incurred at low throttle settings [105]. An empirical model for

CO and HC emissions ($EI_{CO/HC}$) at mean sea level, based on a nonlinear fit of turbofan engines experimental data available in the ICAO emissions databank, is:

$$EI_{CO/HC}(\tau) = c_1 + \exp(-c_2\tau + c_3) \quad \left[\text{g}/\text{kg}\right] \tag{7.84}$$

where, as a first estimate, the fitting parameters $c_{1,2,3}$ accounting for the CO emissions of 165 current civil turbofan engines from the ICAO emissions database are $c = \{0.556, 10.208, 4.068\}$ for CO and $c = \{0.083, 13.202, 1.967\}$ for HC [106]. Figures 7.7 and 7.8 represent the experimental data and the empirical models.

All the carbon contents of the fuel that are not transformed into CO or HC are transformed into CO_2, which has a positive radiative forcing impact and therefore is a major contributor to the greenhouse effect. The reference value is $3.16 \, \text{ton}_{CO_2}/\text{ton}_{\text{Jet-A1}}$ [108]. As combustion temperatures rise the atmospheric nitrogen increasingly reacts with oxygen, generating a family of nitrogen-based combustion products, nitrogen oxides (NO_X), which are associated with important impacts and shall therefore be mitigated. Based on the ICAO emissions databank, an empirical curve fit model can be introduced for the NO_X emission index at mean sea level based on the throttle setting. The following expression, plotted in Figure 7.9, is an example of such a curve fitting comprehensively accounting for 177 current civil aircraft engines in operation [106]:

$$EI_{NO_X}(\tau) = 7.32 \, \tau^2 + 17.07 \, \tau + 3.53 \quad \left[\text{g}/\text{kg}\right] \tag{7.85}$$

In order to obtain an accurate estimate of pollutant emissions at height, a methodology commonly referred to as '*Method 2*' was developed by Boeing in 1995 [109, 110]. This

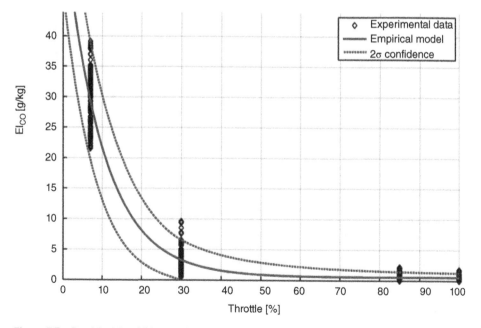

Figure 7.7 Empirical fit of CO emissions as a function of the throttle for a number of current turbofan engines [107].

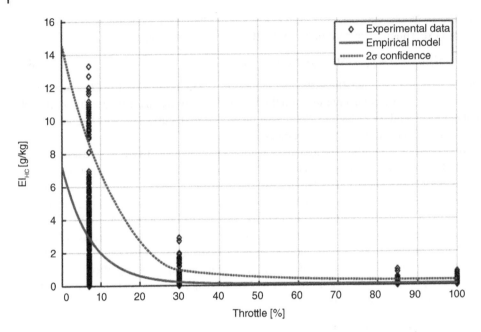

Figure 7.8 Empirical fit of HC emissions as a function of the throttle for a number of current turbofan engines [107].

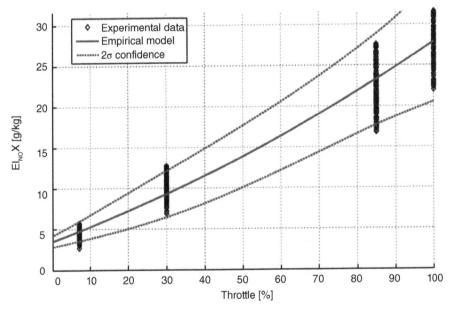

Figure 7.9 Empirical fit of NO_x emissions as a function of the throttle for a number of current turbofan engines [107].

method proposes an empirical correction to account for installation effects, and subsequently introduces *EI* corrections based on ambient temperature, pressure, and relative humidity. The turbofan and turboprop engine models adopted in BADA and presented in Section 6.3 implement the altitude dependency as part of the empirical fuel flow models following this philosophy. Further information on the modelling of aircraft emissions is available in [86, 111].

7.2.5 Operational Costs

The operational costs of aircraft are a fundamental aspect of trajectory optimisation. It is necessary, in particular, to capture the economic dependencies on flight time, which is typically a conflicting criterion with respect to fuel consumption and gaseous emissions. Moreover, some taxation schemes have dependencies with the flown trajectory, and should also be modelled. By taking into consideration maintenance costs (MC), flight crew costs (CC), schedule costs (SC), cabin services (CS) and fees/taxes (FT), the total costs (TC) of a commercial transport mission can be expressed as:

$$TC = MC(t) + CC(t) + SC(t) + CS(t) + FT(t, \boldsymbol{x}) \tag{7.86}$$

Maintenance, crew, and ownership costs are generally characterised by a linear dependence on flight time, but crew rotations, and shifts introduce stepped increases when the section is long enough. Schedule costs are a highly nonlinear component, capturing all implications of a suboptimal timetable and fleet exploitation, including unnecessary stopovers in the rotation. Cabin services are also typically nonlinear, for example involving steps when additional on-board meals are introduced. Fees and taxes encompass a wide range of different dependencies such as landing/parking fees, carbon taxes etc.

7.2.6 Atmosphere and Weather

Diversions around unpredicted hazardous weather cells and extensive flight periods in regions of headwind or crosswind conditions have substantial negative effects on all the environmental and economic performances of a flight, in addition to potentially causing delays that disturb operations and negatively impact passenger satisfaction. Furthermore, many of the models introduced in the trajectory optimization framework require local atmospheric data as an input. For all these reasons, accurate and updated weather data are essential for advanced Air Traffic Management (ATM) Decision Support Systems (DSS). Meteorological data handled by ground-based ATM systems shall correspond to data handled by airborne avionics systems as far as practically possible. This ensures consistency in the 4DT planning and negotiation/validation processes and supports full interoperability, which is an essential aspect for the functional air/ground integration that is implemented as part of the CNS+A roadmap. It is therefore convenient to briefly review the most recent standards and planned evolutions of weather data services for aviation and to propose their implementation in the development of ground-based ATM DSS [112].

The Radio Technical Commission for Aeronautics (RTCA) standard DO-340 introduces an advanced concept of use for meteorological (MET) data link services [113]. These are specified in terms of service category, method of delivery, and of the weather information involved. The MET services are classified into two categories. Category 1 services are the primary means of delivering MET information and may be exclusively relied on to support decisions without questioning their validity, while Category 2 services are useful for making non-critical decisions but should not be relied upon as the sole source of information. Category 1 services comprise both MET and Aeronautical Information Service (AIS) data links. MET data links are used on board aircraft to provide weather information for supporting flight crew decisions. There are three types of pilot decision support services, which reflect the different planning and execution needs of the flight crew: a weather planning decision service (WPDS), a weather near-term decision service (WNDS) and a weather immediate decision service (WIDS). WPDS provides weather information for strategic planning, such as in the case of changes in routing or cruise altitude due to ATFM initiatives or destination airport closures. In such scenarios, it is assumed that the flight crew has an advance time of 20 minutes or more to comprehensively evaluate the situation and to plan/validate diversions or route amendments. WNDS provides weather information for tactical decision-making, such as avoiding hazardous weather cells (including cumulonimbus, icing, turbulence, etc.) especially in terminal arrival/departure operations. In these cases, the flight crew has limited time for replanning and co-ordination with ATM operators, typically between 3 and 20 minutes. WIDS is conceived to provide weather information for freshly detected or quickly evolving weather hazards in order to allow the flight crew to initiate an emergency avoidance or abort take-off/landing. These decisions are assumed to require immediate action from within a few seconds to three minutes.

All these services are supported by three delivery modes: broadcast, demand, and contract. Broadcast data link delivery service issues continuous regular transmissions of MET information to all aircraft within range. Demand and contract data links delivery services require two-way communication between the aircraft and ground station, where the flight crew initiates a request for specific MET information that the ground station then responds to. For the demand service, the ground station only needs to respond to an initiated request, while for the contract service, the information request is usually pre-coordinated and this requires the ground station to monitor the aircraft and provide the MET information at predefined time intervals or positions.

RTCA DO-308 specifies four different categories for MET data formats [114]. These are: point data, area data, vector graphics, and gridded data. Point and area data are given in alpha-numeric strings and, as their name suggests, provide weather information on either a single geospatially located point or over an area delimited by a polygonal line. Examples of point data include the conventional meteorological terminal air report (METAR) and terminal aerodrome forecast (TAF), whereas examples of area data include significant meteorological information (SIGMET). Vector graphic data are images, represented by vectors, points, lines, or other geometric entities and can be used to mark out volumes of interest. Gridded data, typically in the form of general regularly-distributed information in binary format (GRIB), consist of a four-dimensional structured grid (latitude, longitude, altitude, time) of weather data with a forecast time dimension. The amount of MET information exchanged is also dependent on the delivery method. Broadcast services always transmit the full set of weather information, while demand and contract will usually transmit a subset of information based on the initiated request and thus have shorter transmission times. The nature of the decision service also affects the type of information provided. For example, Table 7.1 shows the information provided by the three pilot decision support services.

Table 7.1 MET information classified according to decision service [115].

MET service	Weather planning decision service (WPDS)	Weather near-term decision service (WNDS)	Weather immediate decision service (WIDS)
Time horizon	Greater than 20 min	3 min to 20 min	Less than 3 min
Profile grid-point data	Offline and strategic online operations	Tactical online operations	Emergency operations
Airport equivalent	METAR (incl. RVR), TAF		RVR, gusts, and wind shear

As WPDS is used for route planning and optimisation, the information transmitted has a significantly longer time horizon than WNDS and WIDS. The geographic extent of WPDS MET data is also significantly larger as WPDS data is conceived for strategic large-scale replanning and diversions, whereas WNDS information is used for local tactical rerouting or altitude changes and might involve higher fidelity and resolution, while these tasks are not supported by WIDS. The profile grid-point data is mainly used for arrival and departure planning and involves MET information along the planned flight profile only. The information provided in WIDS is used for more tactical operations, such as airborne separation assurance systems (ASASs). Information regarding hazardous weather is similar across all the three services, while for airport weather, WPDS and WNDS provide more comprehensive information in the form of METAR, TAF, and Runway Visual Range (RVR), while WIDS provides only visibility and wind shear weather data. RTCA DO-324 specifies the required communication performance (RCP) for the service delivery [115]. These are determined in terms of transaction time (TT) and can be either the RCP transaction time (TT_{RCP}) or the normal transaction time (TT_{95}), both expressed in seconds. A transaction is defined as the basic unit of an interaction between peer parties that includes one or more operational messages transmitted from one party to the other. TT_{RCP} is the maximum time for completion of a transaction, while TT_{95} is the time before which 95% of all transactions should be completed. Different TT requirements are defined for airport, terminal, and en-route domains. These are tabulated in Tables D7 to D15 of RTCA DO-324 [115].

The recommended quality of MET information is defined in ICAO Annex 3 [116], which lists the desirable accuracies of measurements of forecasted data for a range of weather data. A partial list is provided in Table 7.2, which shows the desired quality of measured data and also the TAF data. The rest of the document also provides recommendations regarding forecast data for general trends, take-off, and en-route cases. However, these are not operational requirements but rather desirable accuracy figures to fulfil typical operational needs. RTCA DO-308 also identifies a list of candidate MET products [114]. In particular, the Global Forecast System (GFS), which is used for planning and offline/strategic online optimisation purposes, falls under the World Area Forecast Centre group shown in Table 7.3.

The WPDS data described above have been sufficient for flight planning and offline/strategic online trajectory optimisation applications within airborne avionics and ground-based ATM DSS, as the main factors driving the lateral/vertical planning are the 4D wind and temperature fields. Nevertheless, a significant gap is represented by the lack of information regarding weather cells and other adverse phenomena. Consequently, in terms of future CNS+A evolutions, ICAO's ASBU roadmap acknowledges that further

Table 7.2 Recommended quality of MET information [116].

Weather data	Accuracy of measurement	Accuracy of forecast	Minimum percentage within range
Wind direction	**± 10°**	**± 20°**	
Mean surface wind	± 1 knot up to 10 knots ± 10% above 10 knots	± 10 5 knots	80% of cases
Visibility	± 50 m up to 600 m ± 10% between 600 m and 1500 m ± 20% above 1500 m	± 200 m up to 800 m ± 30% between 800 m and 10 km	80% of cases
Cloud amount	±1 okta	One okta below 1500 ft Occurrence of BKN or OVC between 1500 ft and 10 000 ft	80% of cases
Cloud height	±33 ft up to 330 ft ±10% above 330 ft	± 100 ft up to 1000 ft ± 30% between 1000 ft and 10 000 ft	70% of cases
Air temperature	±1 °C	±1 °C	70% of cases

Table 7.3 Reference global forecast MET data for WPDS [114].

World Area Forecast Centre Information	Data format	Refresh rate (h)	Validity (h)
Wind (latitudinal and longitudinal)	Gridded	6	36
Temperature	Gridded	6	36
Humidity	Gridded	6	36
Tropopause – height, temperature, direction	Gridded	6	36
Maximum wind – speed, direction, height	Gridded	6	36
Significant weather charts	Vector	6	13

improved meteorological services are required to implement such advanced functionalities as the ones involved in 4D-TBO. Long-range weather forecasts (such as GRIB and METAR) are already being supplemented by nowcasting techniques for the provision of ATM and ATFM services, particularly in the terminal area. Some of these advanced weather services are already acknowledged by RTCA in DO-308, as shown in Table 7.4. In particular, the US National Centre for Atmospheric Research's Thunderstorm Identification, Tracking, Analysis and Nowcasting (TITAN) system is an example of a nowcasting system belonging to the National Convective Weather Forecast (NCWF) class and used by ANSPs in the US, Australia, and South Africa.

Nowcasting provides detailed current weather information extrapolated up to six hours ahead, typically with sub-kilometre spatial resolution and a temporal resolution in

Table 7.4 Advanced METLINK products for flight planning in the USA and Europe [114].

	Data format	Refresh rate	Validity (h)
National Weather Service (NOAA)			
National Convective Weather Forecast (NCWF)	Gridded/ Vector	5 min	1
Graphical Turbulence Guidance (GTG)	Gridded	15 min	0.25
Current icing product (CIP)	Gridded	1 h	N/A
Forecast icing potential (FIP)	Gridded	1 h	3
WIMS (FLYSAFE)			
WIMS thunderstorm	Gridded/Vector	5 min to 6 h	0.2–1
WIMS turbulence	Gridded/Vector	6 h	36
WIMS icing	Gridded/Vector	15 min to 12 h	0.24–24
WIMS wake vortex	Gridded/Vector	1 to 6 h	2–12

the order of minutes. Nowcasting systems use sophisticated algorithms to track and extrapolate individual storm cells from weather radar information. Aviation weather service providers are increasingly supplementing this information with satellite imagery and ground-based sensors such as lightning detectors. The ability to downlink weather data turns datalink-equipped aircraft into mobile weather sensors able to report on the weather in their vicinity. Current-generation ATM systems are capable of receiving this data and exporting it to aviation weather service providers. Such data can contribute to a '4D weather cube', supplementing weather data along major air routes and allowing those sections of the cube that are of the most interest to aviation users to be updated more frequently and more accurately. However, the availability of a near real-time, high-resolution full 4D weather cube is not far off. In 2015 SESAR successfully demonstrated a web-services-enabled SWIM implementation of the FAA NextGen 4D weather cube concept in their 'Optimising trajectories over the 4DWeatherCube' SWIM Masterclass. Further research and development is currently being carried out as part of SESAR's TOPMET project.

The weather models most commonly adopted in MOTO studies exploit the global weather data available from the National Climatic Data Center (NCDC) of the National Oceanic and Atmospheric Administration (NOAA) and selectively extrapolate the required information on structured geospatial/spatiotemporal (e.g. 4D) grids. Data extracted from the GFS are typically collected on a 0.25° latitude and longitude resolution grid, updated every 6 hours (4 times daily) and including projection of up to 180-hours in 3-hour intervals. Figure 7.10 depicts the 3-dimensional wind and temperature fields at some of the typical cruise altitudes of a jetliner for a 15-hour advance forecast sampled at 0600 UTC on April 3rd 2015. Figure 7.11 depicts the three-dimensional relative humidity field under the same conditions.

Whenever an interpolation is performed to translate data tabulated at pressure altitudes into Cartesian altitudes and to extrapolate data at higher resolution grids, the following

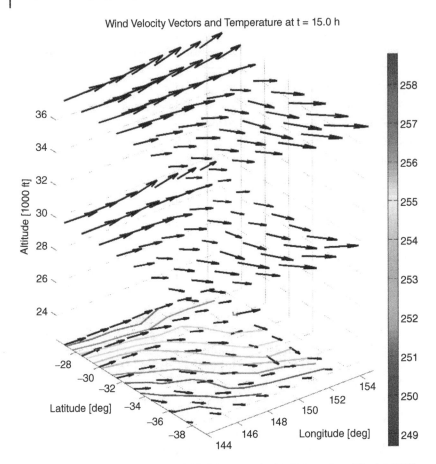

Figure 7.10 Wind and temperature 3D fields at typical jetliner cruise altitudes [106].

expression (known as barometric formula) should be adopted to determine intermediate values instead of linear interpolation:

$$\frac{P}{P_0} = \left[\frac{T_0}{T_0 + L(h - h_0)}\right]^{\frac{g}{L \cdot R^*}} \tag{7.87}$$

where P is the extrapolated pressure, P_0 and T_0 are reference pressure and temperature, h and h_0 are the altitude of extrapolation and reference points, L is the locally evaluated lapse rate and R^* is the specific gas constant for air.

The calculation and exploitation of weather conditions and forecasts are actively researched topics, as their influence on the operational efficiency and the accuracy of trajectory optimisation results is recognised. Further details are given in [117].

7.2.7 Noise

The noise generated by various mechanical, thermochemical, and fluid dynamic processes within the engine as well as around the aircraft structure is propagated through the atmosphere, which acts as a low pass filter due to thermo-fluid dynamics and molecular

Relative Humidity Fields at t = 15.0 h

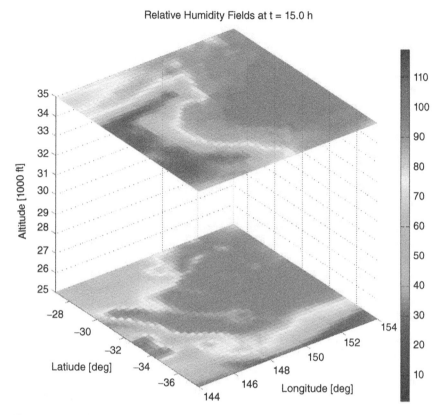

Figure 7.11   Relative humidity 3D field at typical jetliner cruise altitudes [106].

processes [118]. When the receiver is at a sufficient distance, most contributors can be merged into a single point source with non-uniform distribution [118]. By means of a suitable propagation model, and thanks to the adoption of apposite demographic distribution databases (D3) and digital terrain elevation databases (DTEDs), noise footprints on the ground can be determined. However, the high physical complexity associated with the emission and propagation of aircraft noise limits the adoption of accurate models to offline optimisation studies, typically performed to optimise the departure and approach paths of aircraft, which are those portions of the flight in close proximity to the ground. The specific noise problem formulation also depends on the governing regulations, which mandate different output metrics [119–121]. More specifically, due to the interactions between the sound waves emitted by various locations at different frequencies and the aircraft structure at various attitudes and airspeeds, noise emissions are characterised by highly uneven and nonlinear 4D profiles. The vagaries of the atmosphere in terms of unsteady and non-uniform composition and thermodynamic state will further affect the propagation of sound waves and thus alter the noise footprint on the ground. That notwithstanding, some assumptions may enable the integration of simplified noise models in real-time avionics and ATM system implementations for the evaluation of an estimated noise footprint of generated trajectories. For instance, the propagation of noise generated by a point source

Figure 7.12 Noise footprint of a typical flight departure from London Heathrow, highlighting the 70 dB sound exposure level (SEL) area [130].

with non-uniform distribution can be formulated adopting a noise-power-distance (NPD) model. The integration of this model with a simplified worst-case emission and with limited resolution and computationally optimised DTED and D3 can enable a quasi-real-time noise model implementation. A number of optimisation case studies specifically targeted the minimisation of perceived noise emissions in the design and redesign of departure and arrival procedures [122–128]. More details on the modelling of noise emission and propagation are given in [118, 120, 121, 128, 129] (Figures 7.12 and 7.13).

7.2.8 Condensation Trails

Contrails are increasingly being addressed in aviation sustainability research due to their recognised radiative forcing potential [131]. Contrails are formed when the combination of local atmospheric temperature and relative humidity are favourable, such that the hot water vapour in the engine exhausts leads to locally attained or exceeded liquid saturation conditions, generating droplets that subsequently freeze. The discriminant for such phenomenon is referred to as the Schmidt-Appleman criterion, and is based on the evaluation of the slope of the exhaust mixing curve (MS):

$$MS = \frac{p \; c_P \; EI_{H_2O}}{\varepsilon \left(Q - \frac{v}{\kappa} \right)} \quad \left[{}^{Pa}\!/_{K} \right] \tag{7.88}$$

where p is the local atmospheric pressure (Pa), c_P is the specific heat capacity of air (J Kg^{-1} K^{-1}), EI_{H_2O} is the water vapour emission index, ε is the ratio of molar masses of water vapour and air, Q is the specific fuel combustion heat (J Kg^{-1}), v is the true

Figure 7.13 Noise footprint of an optimised departure trajectory, reducing the 70 dB SEL area by 47% [130].

airspeed (m s^{-1}) and κ is the thrust specific fuel consumption (Kg N^{-1} s^{-1}). The contrails so formed are susceptible to convective and thermodynamic processes, which may lead to a rapid dissipation of the formed contrail or to a belated formation. In particular, the theory distinguishes three phases: jet, vortex, and dispersion phases, which are governed by different physical processes and are hence modelled individually. More details are

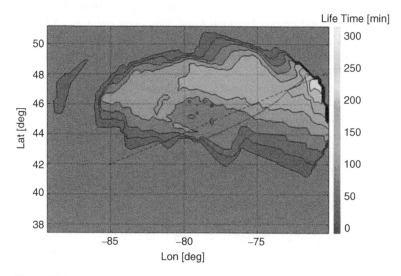

Figure 7.14 2D + T MOTO with respect to contrail lifetime and fuel consumption [142].

given in [132–136]. In relation to the growing environmental and social concerns, over the past decade research in aircraft trajectory optimisation has increasingly addressed contrail avoidance, and a number of approaches were proposed [137–143]. Among other solutions, a 4D mapping of the lifetime and radiative forcing associated with contrails was proposed by the authors for real-time implementations of contrail models [142, 143]. An altitude-constrained two-dimensional plus time (2D + T) trajectory optimisation case study addressing the conflicting objectives of long-lasting contrails and of fuel consumption is depicted in Figures 7.14.

7.2.9 Conclusions

This chapter introduced and briefly described the various models that are typically adopted in trajectory optimisation studies. While flight dynamics, engine thrust, and, to a lesser extent, weather models are applicable across the entire trajectory optimisation spectrum, the models for pollutants, operational costs, noise, condensation trails, and airspace/ATM constraints are less common but are crucially important for environmental sustainability assessments. The trajectory optimisation algorithms implemented in several case studies of this book integrate the models presented in this chapter in suitable arrangements.

References

1 Zermelo, E. (1930). On navigation in the air as a problem in the calculus of variations (orig. Über die Navigation in der Luft als Problem der Variationsrechnung). *Jahresbericht der Deutschen Mathematiker-Vereinigung* 39: 44–48.

2 Zermelo, E. (1931). On the navigation problem for a calm or variable wind distribution (orig. Über das Navigationsproblem bei ruhender oder veränderlicher Windverteilung). *Zeitschrift für Angewandte Mathematik und Mechanik* 11 (2): 114–124.

3 Erzberger, H., McLean, J.D., and Barman, J.F. (1975). *Fixed-Range Optimum Trajectories for Short-Haul Aircraft, National Aeronautics and Space Administration (NASA) TND-8115.* Moffett Field, CA, USA: Ames Research Center.

4 Barman, J.F. and Erzberger, H. (1976). Fixed-range optimum trajectories for short-haul aircraft. *Journal of Aircraft* 13: 748–754.

5 Erzberger, H. and Lee, H. (1978). *Characteristics of Constrained Optimum Trajectories with Specified Range, National Aeronautics and Space Administration (NASA) Technical Memorandum TM-78519.* Moffett Field, CA, USA: Ames Research Center.

6 Sorensen, J.A., Morello, S.A., Erzberger, H. (1979). Application of trajectory optimization principles to minimize aircraft operating costs, *Proceedings of the 18th IEEE Conference on Decision and Control*, 415–421.

7 Lee, H.Q. and Erzberger, H. (1980). *Algorithm for Fixed-Range Optimal Trajectories, National Aeronautics and Space Administration (NASA), Technical Paper TP-1565.* Moffett Field, CA, USA: Ames Research Center.

8 Erzberger, H. and Lee, H. (1980). Constrained optimum trajectories with specified range. *Journal of Guidance, Control and Dynamics* 3: 78–85.

9 Gardi, A. (2017). A Novel Air Traffic Management Decision Support System - Multi-Objective 4-Dimensional Trajectory Optimisation for Intent-Based Operations in Dynamic Airspace. PhD Thesis, School of Engineering, RMIT University, Melbourne, Australia.

10 Gardi, A., Sabatini, R., and Ramasamy, S. (2016). Multi-objective optimisation of aircraft flight trajectories in the ATM and avionics context. *Progress in Aerospace Sciences* 83: 1–36.

11 Visser, H.G. (1994). *A 4D Trajectory Optimization and Guidance Technique for Terminal Area Traffic Management*. Delft University of Technology.

12 Betts, J.T. and Cramer, E.J. (1995). Application of direct transcription to commercial aircraft trajectory optimization. *Journal of Guidance, Control, and Dynamics* 18: 151–159.

13 Betts, J.T. and Huffman, W.P. (1998). Mesh refinement in direct transcription methods for optimal control. *Optimal Control Applications and Methods* 19: 1–21.

14 Betts, J.T. and Huffman, W.P. (1999). Exploiting sparsity in the direct transcription method for optimal control. *Computational Optimization and Applications* 14: 179–201.

15 Betts, J.T., Biehn, N., Campbell, S.L., and Huffman, W.P. (2000). Compensating for order variation in mesh refinement for direct transcription methods. *Journal of Computational and Applied Mathematics* 125: 147–158.

16 Betts, J.T., Biehn, N., Campbell, S.L., and Huffman, W.P. (2002). Compensating for order variation in mesh refinement for direct transcription methods II: computational experience. *Journal of Computational and Applied Mathematics* 143: 237–261.

17 Betts, J.T. and Huffman, W.P. (2004). Large scale parameter estimation using sparse nonlinear programming methods. *SIAM Journal on Optimization* 14: 223–244.

18 Hagelauer, P. and Mora-Camino, F. (1998). A soft dynamic programming approach for on-line aircraft 4D-trajectory optimization. *European Journal of Operational Research* 107: 87–95.

19 Yokoyama, N. and Suzuki, S. (2001). Flight Trajectory Optimization using Genetic Algorithm Combined with Gradient Method. *Information Technology for Economics and Management* 1.

20 Jardin, M.R. and Bryson, A.E.J. (2003). Neighboring optimal aircraft guidance in winds. *Journal of Guidance, Control, and Dynamics* 24: 710–715, 2001.

21 Jardin, M.R. (2003). Real-Time Conflict-Free Trajectory Optimization. *Proceedings of 5th USA/Europe Air Traffic Management Research and Development Seminar (ATM 2003)*, Budapest, Hungary.

22 Bower, G.C. and Kroo, I.M. (2008). Multi-Objective Aircraft Optimization for Minimum Cost and Emissions over Specific Route Networks. *Proceedings of 26th International Congress of the Aeronautical Sciences (ICAS 2008)*, Anchorage, AK, USA.

23 Jacobsen, M. and Ringertz, U.T. (2010). Airspace Constraints in Aircraft Emission Trajectory Optimization. *Journal of Aircraft* 47: 1256–1265.

24 Eele, A.J. and Richards, A. (2009). Path-Planning with Avoidance Using Nonlinear Branch-and-Bound Optimization. *Journal of Guidance, Control, and Dynamics* 32: 384–394.

25 Soler, M., Olivares, A., and Staffetti, E. (2010). Hybrid Optimal Control Approach to Commercial Aircraft Trajectory Planning. *Journal of Guidance, Control, and Dynamics* 33 (3): 985–991.

26 Soler, M., Olivares, A., Staffetti, E., and Zapata, D. (2012). Framework for aircraft trajectory planning toward an efficient air traffic management. *Journal of Aircraft* 49 (1): 341–348.

27 Bonami, P., Olivares, A., Soler, M., and Staffetti, E. (2013). Multiphase mixed-integer optimal control approach to aircraft trajectory optimization. *Journal of Guidance, Control, and Dynamics* 36 (5): 1267–1277.

28 Soler, M., Olivares, A., and Staffetti, E. (2014). Multiphase Optimal Control Framework for Commercial Aircraft Four-Dimensional Flight-Planning Problems. *Journal of Aircraft* 52 (1): 274–286

29 González-Arribas, D., Soler, M., and Sanjurjo-Rivo, M. (2018). Robust aircraft trajectory planning under wind uncertainty using optimal control. *Journal of Guidance, Control, and Dynamics* 41 (3): 673–688.

30 García-Heras, J., Soler, M., González-Arribas, D. et al. (2021). Robust flight planning impact assessment considering convective phenomena. *Transportation Research Part C: Emerging Technologies* 123: 102968.

31 Simorgh, A., Soler, M., González-Arribas, D. et al. (2022). A Comprehensive Survey on Climate Optimal Aircraft Trajectory Planning. *Aerospace* 9 (3): 146.

32 Clarke, J.P.B., Ren, L., McClain, E. et al. (2012). Evaluating concepts for operations in metroplex terminal area airspace. *Journal of Aircraft* 49: 758–773.

33 Clarke, J.P., Brooks, J., Nagle, G. et al. (2013). Optimized profile descent arrivals at Los Angeles international airport. *Journal of Aircraft* 50: 360–369.

34 Park, S.G. and Clarke, J.P. (2015). Optimal control based vertical trajectory determination for continuous descent arrival procedures. *Journal of Aircraft* 52: 1469–1480.

35 Delahaye, D., Puechmorel, S., Tsiotras, P. and Feron, E. (2014). Mathematical models for aircraft trajectory design: A survey. In *3rd ENRI International Workshop on ATM/CNS, EIWAC 2013*, vol. 290 LNEE, 205–247. Tokyo: Springer Verlag.

36 Rao, A.V. (2010). Survey of numerical methods for optimal control. *Advances in the Astronautical Sciences* 135: 497–528.

37 von Stryk, O. and Bulirsch, R. (1992). Direct and indirect methods for trajectory optimization. *Annals of Operations Research* 37: 357–373.

38 Betts, J.T. (1998). Survey of numerical methods for trajectory optimization. *Journal of Guidance, Control, and Dynamics* 21: 193–207.

39 Zhou, K., Doyle, J.C., and Glover, K. (1996). *Robust and Optimal Control*. Upper Saddle River, NJ, USA: Prentice Hall.

40 Lewis, F.L., Vrabie, D., and Syrmos, V.L. (2012). *Optimal Control*, 3e. New York City, NY, USA: Wiley.

41 Friesz, T.L. (2010). *Dynamic Optimization and Differential Games*. Heidelberg/Berlin, Germany: Springer Science.

42 Rao, A.V. (2010). Trajectory optimization. In: *Encyclopedia of Aerospace Engineering* (ed. R. Blockley and W. Shyy) Wiley.

43 Ben-Asher, J.Z. (2010). *Optimal Control Theory with Aerospace Applications*. Reston, VA, USA: American Institute of Aeronautics and Astronautics (AIAA).

44 Bolza, O. (1909). *Lectures on the Calculus of Variations (orig.: Vorlesungen über Variationsrechnung)*. Leipzig, Germany: B. G. Teubner.

45 Berkovitz, L.D. (1961). Variational methods in problems of control and programming. *Journal of Mathematical Analysis and Applications* 3: 145–169.

46 Hodges, D.H. and Bless, R.R. (1991). Weak Hamiltonian finite element method for optimal control problems. *Journal of Guidance, Control, and Dynamics* 14: 148–156.

47 Haug, E.J. and Choi, K. (1993). *Methods of Engineering Mathematics*. Englewood Cliffs, NJ, USA: Prentice-Hall.

48 Kelley, W.G. and Peterson, A.C. (2001). *Difference Equations - An Introduction with Applications*, 2e. San Diego, CA, USA: Harcourt/Academic Press.

49 Quarteroni, A., Sacco, R., and Saleri, F. (2007). *Numerical Mathematics*, 2e. Heidelberg, Germany: Springer Berlin Heidelberg.

50 Bronshtein, I.N., Semendyayev, K.A., Musiol, G., and Muehlig, H. (2007). *Handbook of Mathematics*, 5e. Heidelberg, Germany: Springer-Verlag Berlin Heidelberg New-York.

51 Quarteroni, A. (2014). *Numerical Models for Differential Problems*, 2e. Milan, Italy: Springer-Verlag.

52 Canuto, C., Hussaini, M.Y., Quarteroni, A., and Zang, T.A. (2007). *Spectral Methods - Fundamentals in Single Domains*. Berlin, Heidelberg, Germany: Springer-Verlag.

53 Canuto, C., Hussaini, M.Y., Quarteroni, A., and Zang, T.A. (2007). *Spectral Methods - Evolution to Complex Geometries and Applications to Fluid Dynamics*. Heidelberg, Germany: Springer-Verlag.

54 Miele, A. (1975). Recent advances in gradient algorithms for optimal control problems. *Journal of Optimization Theory and Applications* 17: 361–430.

55 Bertsekas, D.P. (1999). *Nonlinear Programming*, 2e. Belmont, MA, USA: Athena Scientific.

56 Betts, J.T. (2010). *Practical Methods for Optimal Control and Estimation Using Nonlinear Programming*, 2e, vol. 19. SIAM.

57 Hull, D.G. (2001). Variational calculus and approximate solution of optimal control problems. *Journal of Optimization Theory and Applications* 108: 483–497.

58 Sorensen, J.A., Morello, S.A., and Erzberger, H. (1979). Application of trajectory optimization principles to minimize aircraft operating costs. In: *18th IEEE Conference on Decision and Control*, vol. 1, 415–421.

59 Pontryagin, L.S., Boltyanskii, V.G., Gamkrelidze, R.V. et al. (1986). *The Mathematical Theory of Optimal Processes*, 3e, vol. 4. Montreux, Switzerland: Gordon and Breach Science Pub. S.A.

60 Fahroo, F. and Ross, I. M. (2000). Trajectory optimization by indirect spectral collocation methods. *Astrodynamics specialist conference*, Denver, CO, USA.

61 Gherman, I., Schulz, V., and Betts, J.T. (2006). Optimal flight trajectories for the validation of aerodynamic models. *Optimization Methods and Software* 21: 889–900.

62 Engelsone, A., Campbell, S.L., and Betts, J.T. (2007). Direct transcription solution of higher-index optimal control problems and the virtual index. *Applied Numerical Mathematics* 57: 281–296.

63 Betts, J.T., Campbell, S.L., and Engelsone, A. (2007). Direct transcription solution of optimal control problems with higher order state constraints: theory vs practice. *Optimization and Engineering* 8: 1–19.

64 Benson, D.A., Huntington, G.T., Thorvaldsen, T.P., and Rao, A.V. (2006). Direct trajectory optimization and costate estimation via an orthogonal collocation method. *Journal of Guidance, Control, and Dynamics* 29: 1435–1440.

65 Huntington, G.T. and Rao, A.V. (2008). Comparison of global and local collocation methods for optimal control. *Journal of Guidance, Control, and Dynamics* 31: 432–436.

66 Garg, D., Patterson, M., Hager, W.W. et al. (2010). A unified framework for the numerical solution of optimal control problems using pseudospectral methods. *Automatica* 46: 1843–1851.

67 Rao, A.V., Benson, D.A., Darby, C. et al. (2010). Algorithm 902: GPOPS, a MATLAB software for solving multiple-phase optimal control problems using the gauss pseudospectral method. *ACM Transactions on Mathematical Software* 37.

68 Darby, C.L., Hager, W.W., and Rao, A.V. (2011). An hp-adaptive pseudospectral method for solving optimal control problems. *Optimal Control Applications and Methods* 32: 476–502.

69 Darby, C.L., Hager, W.W., and Rao, A.V. (2011). Direct trajectory optimization using a variable low-order adaptive pseudospectral method. *Journal of Spacecraft and Rockets* 48: 433–445.

70 Garg, D., Hager, W.W., and Rao, A.V. (2011). Pseudospectral methods for solving infinite-horizon optimal control problems. *Automatica* 47: 829–837.

71 Patterson, M.A. and Rao, A.V. (2012). Exploiting sparsity in direct collocation pseudospectral methods for solving optimal control problems. *Journal of Spacecraft and Rockets* 49: 364–377.

72 Boyd, J.P. (2000). *Chebyshev and Fourier Spectral Methods*, 2e. Mineola, NY, USA: Dover publications.

73 Funaro, D. (1992). *Polynomial Approximation of Differential Equations*. Berlin, Germany: Springer-Verlag.

74 Bousson, K. and Machado, P. (2010). 4D Flight trajectory optimization based on pseudospectral methods. *World Academy of Science, Engineering and Technology* 70: 551–557.

75 Basset, G., Xu, Y., and Yakimenko, O.A. (2010). Computing short-time aircraft maneuvers using direct methods. *Journal of Computer and Systems Sciences International* 49: 481–513.

76 Brix, K., Canuto, C., and Dahmen, W. (2013). Legendre-Gauss-Lobatto grids and associated nested dyadic grids. *Aachen Institute for Advanced Study in Computational Engineering Science* arXiv:1311.0028.

77 Chircop, K., Xuereb, M., Zammit-Mangion, D., and Cachia, E. (2010). A generic framework for multi-parameter optimization of flight trajectories. In: *27th International Congress of the Aeronautical Sciences (ICAS 2010)*, 5118–5127. Nice, France.

78 Pervier, H., Nalianda, D., Espi, R. et al. (2011). Application of genetic algorithm for preliminary trajectory optimization. *SAE International Journal of Aerospace* 4: 973–987.

79 Sammut, M., Zammit-Mangion, D., and Sabatini, R. (2012). Optimization of fuel consumption in climb trajectories using genetic algorithm techniques. *AIAA Guidance, Navigation, and Control Conference (GNC 2012)*, Minneapolis, MN, USA.

80 Gu, W., Navaratne, R., Quaglia, D. et al. (2012). Towards the development of a multi-disciplinary flight trajectory optimization tool - GATAC. In: *ASME Turbo Expo 2012: Turbine Technical Conference and Exposition (GT 2012)*, Copenhagen, Denmark, 415–424.

81 Camilleri, W., Chircop, K., Zammit-Mangion, D., Sabatini, R., and Sethi, V. (2012). Design and validation of a detailed aircraft performance model for trajectory optimization. *AIAA Modeling and Simulation Technologies Conference 2012 (MST 2012)*, Minneapolis, MN, USA.

82 Navaratne, R., Tessaro, M., Gu, W. et al. (2012). Generic framework for multi-disciplinary trajectory optimization of aircraft and power plant integrated systems. *Journal of Aeronautics & Aerospace Engineering* 2: 1–14.

83 Xuereb, M., Chircop, K., and Zammit-Mangion, D. (2012). GATAC - A generic framework for multi-parameter optimization of flight trajectories. *AIAA Modeling and Simulation Technologies Conference 2012 (MST 2012)*, Minneapolis, MN, USA.

84 Pisani, D., Zammit-Mangion, D., and Sabatini, R. (2013). City-pair trajectory optimization in the presence of winds using the GATAC framework. *AIAA Guidance, Navigation, and Control Conference (GNC 2013)*, Boston, MA, USA.

85 Xuereb, M., Zammit Mangion, D., Sammut, M., and Chircop, K. (2013). GATAC - Version 3 of the generic framework for multi-parameter optimization of flight trajectories. *AIAA Modeling and Simulation Technologies conference 2013 (MST 2013)*, Boston, MA, USA.

86 Celis, C., Sethi, V., Zammit-Mangion, D. et al. (2014). Theoretical optimal trajectories for reducing the environmental impact of commercial aircraft operations. *Journal of Aerospace Technology and Management* 6: 29–42.

87 Tsotskas, C., Kipouros, T., and Savill, A. M. (2014). The design and implementation of a GPUenabled multi-objective Tabu-search intended for real world and high-dimensional applications. *14th Annual International Conference on Computational Science, ICCS 2014*, Cairns, QLD: 2152–2161.

88 Tsotskas, C., Kipouros, T., and Savill, M. (2013). Biobjective optimisation of preliminary aircraft trajectories. In: *7th International Conference on Evolutionary Multi-Criterion Optimization, EMO 2013* vol. 7811 LNCS (ed. Sheffield), 741–755.

89 De Jong, H. M. (1974). Optimal track selection and 3-dimensional flight planning - Theory and practice of the optimization problem in air navigation under space-time varying meteorological conditions. Royal Netherlands Meteorological Institute (K.N.M.I.), De Bilt, Netherlands 93.

90 Bijlsma, S.J. (2009). Optimal aircraft routing in general wind fields. *Journal of Guidance, Control, and Dynamics* 32: 1025–1028.

91 Girardet, B., Lapasset, L., Delahaye, D., and Rabut, C. (2014). Wind-optimal path planning: application to aircraft trajectories. In: *13th International Conference on Control Automation Robotics and Vision, ICARCV 2014, 2014*, 1403–1408.

92 Rodionova, O., Sbihi, M., Delahaye, D., and Mongeau, M. (2014). North Atlantic aircraft trajectory optimization. *IEEE Transactions on Intelligent Transportation Systems* 15: 2202–2212.

93 Rodionova, O., Delahaye, D., Sbihi, M., and Mongeau, M. (2014). Trajectory prediction in North Atlantic oceanic airspace by wind networking. In: *33rd Digital Avionics Systems Conference, DASC 2014*, 7A31–7A315.

94 Marler, R.T. and Arora, J.S. (2004). Survey of multi-objective optimization methods for engineering. *Structural and Multidisciplinary Optimization* 26: 369–395.

95 Hull, D.G. (2010). *Fundamentals of Airplane Flight Mechanics*. Berlin/Heidelberg, Germany: Springer-Verlag.

96 Zipfel, P.H. (2007). *Modeling and Simulation of Aerospace Vehicle Dynamics*, 2e. Reston, VA, USA: American Institute of Aeronautics and Astronautics Inc. (AIAA).

97 Pamadi, B.N. (2004). *Performance, Stability, Dynamics and Control of Airplanes*, 2e. Blacksburg, VA, USA: American Institute of Aeronautics and Astronautics Inc. (AIAA).

98 Stevens, B.L. and Lewis, F.L. (2003). *Aircraft Control and Simulation*, 2e. Hoboken, NJ, USA: Wiley.

99 Phillips, W.F. (2010). *Mechanics of Flight*, 2e. Hoboken, NJ, USA: Wiley.

100 Imado, F., Heike, Y., and Kinoshita, T. (2011). Research on a new aircraft point-mass model. *Journal of Aircraft* 48: 1121–1130.

101 Eurocontrol Experimental Centre (EEC). (2013) "User Manual for the Base of Aircraft Data (BADA) Revision 3.11", Brétigny-sur-Orge, France Technical/Scientific Report No. 13/04/16-01.

102 Battipede, M., Sirigu, G., Cassaro, M., and Gili, P. (2015). *"Analysis of the impact of performance model accuracy on 4D trajectory optimization", AIAA Modeling and Simulation Technoogies conference 2015 (MST 2015)*. FL, USA: Orlando.

103 ICAO. ICAO AIRCRAFT ENGINE EMISSIONS DATABANK [Online]. (Access date: 23/02/2023). Available: http://easa.europa.eu/node/15672

104 ICAO (2008). *"Annex 16 to the Convention on International Civil Aviation - Environmental Protection - Volume II: Aircraft Engine Emissions", The International Civil Aviation Organization (ICAO)*. QC, Canada: Montreal.

105 Wulff, A. and Hourmouziadis, J. (1997). Technology review of aeroengine pollutant emissions. *Aerospace Science and Technology* 1: 557–572.

106 Ramasamy, S., Sabatini, R., and Gardi, A. (2015) "Novel flight management systems for improved safety and sustainability in the CNS+A context", *Integrated Communication, Navigation and Surveillance Conference (ICNS 2015)*, Herndon, VA, USA.

107 Sabatini, R., Gardi, A., Ramasamy, S. et al. (2015). Modern avionics and ATM systems for green operations. In: *Encyclopedia of Aerospace Engineering* (ed. R. Blockley and W. Shyy). Wiley.

108 Hupe, J., Ferrier, B., Thrasher, T. et al. (2013). *ICAO Environmental Report 2013: Destination Green - Aviation and Climate Change*. Montreal, Canada: ICAO Environmental Branch.

109 Martin, R.L. (1996). Appendix D. Boeing method 2 fuel flow methodology description - presentation to CAEP working group III certification subgroup. In: *Scheduled Civil Aircraft Emission Inventories for 1992: Database Development and Analysis - NASA Contractor Report 4700* (ed. S.L. Baughcum, T.G. Tritz, S.C. Henderson and D.C. Pickett). Hampton, VA, USA: NASA Langley Research Center.

110 Dubois, D. and Paynter, G. C. (2006). "Fuel flow method 2″ for estimating aircraft emissions", *SAE Technical Paper 2006-01-1987*.

111 Ruijgrok, G.J.J. and van Paassen, D.M. (2012). *Elements of Aircraft Pollution, 3 Delft*. Netherlands: VSSD.

112 Gardi, A., Marino, M., Ramasamy, S., Sabatini, R., and Kistan, T. (2016). "4-Dimensional trajectory optimisation algorithms for air traffic management systems",

35th IEEE/AIAA Digital Avionics Systems Conference (DASC 2016), Sacramento, CA, USA.

113 RTCA. (2012). "RTCA DO-340: Concept of Use (CONUSE) for Aeronautical Information Services (AIS) and Meteorological (MET) Data Link Services", ed: SC-206.

114 RTCA. (2007). "RTCA DO-308: Operational Services and Environment Definition (OSED) for Aeronautical Information Services (AIS) and Meteorological (MET) Data Link Services", ed: SC-206.

115 RTCA. (2010). "RTCA DO-324: Safety and Performance Requirements (SPR) for Aeronautical Information Services (AIS) and Meteorological (MET) Data Link Services", ed: SC-206.

116 ICAO. (2010). "Annex 3 to the Convention on International Civil Aviation - Meteorological Service for International Air Navigation", The International Civil Aviation Organization (ICAO), Montreal, Canada.

117 Troxel, S., and Reynolds, T. (2015). "Use of numerical weather prediction models for NextGen ATC Wind Impact Studies", *7th AIAA Atmospheric and Space Environment conference*, Dallas, TX, USA.

118 Zaporozhets, O., Tokarev, V., and Attenborough, K. (2011). *Aircraft Noise: Assessment, Prediction and Control*. CRC Press.

119 ICAO. (2012) "Doc. 9501 - Environmental Technical Manual - Volume I: Procedures for the Noise Certification of Aircraft", The International Civil Aviation Organization (ICAO), Montreal, Canada.

120 ECAC. (2005). CEAC Doc 29 – Report on Standard Method of Computing Noise Contours around Civil Airports, E. C. A. C. (ECAC).

121 Boeker, E. R., Dinges, E., He, B., Fleming, G., Roof, C. J., Gerbi, P. J. et al., (2008). "Integrated Noise Model Technical Manual v. 7.0".

122 Visser, H. G. and Wijnen, R. (2001). "Optimization of noise abatement arrival trajectories", *AIAA Guidance, Navigation and Control conference 2001 (GNC 2001)*, Montreal, Canada.

123 Visser, H.G. (2005). Generic and site-specific criteria in the optimization of noise abatement trajectories. *Transportation Research Part D: Transport and Environment* 10: 405–419.

124 Ren, L. (2007). "Modeling and Managing Separation for Noise Abatement Arrival Procedures", ScD, Department of Aeronautics and Astronautics, Massachusetts Institute of Technology.

125 Hartjes, S., Visser, H. G., and Hebly, S. J. (2009) "Optimization of RNAV noise and emission abatement departure procedures", AIAA Aviation Technology, Integration, and Operations conference 2009 (ATIO 2009), Hilton Head, SC, USA.

126 Prats, X., Puig, V., Quevedo, J., and Nejjari, F. (2010). Multi-objective optimisation for aircraft departure trajectories minimising noise annoyance. *Transportation Research Part C: Emerging Technologies* 18: 975–989.

127 Prats, X., Puig, V., and Quevedo, J. (2011). A multi-objective optimization strategy for designing aircraft noise abatement procedures. Case study at Girona airport. *Transportation Research Part D: Transport and Environment* 16: 31–41.

128 Khardi, S. and Abdallah, L. (2012). Optimization approaches of aircraft flight path reducing noise: comparison of modeling methods. *Applied Acoustics* 73: 291–301.

129 Ruijgrok, G.J.J. (2007). *Elements of Aviation Acoustics*, 2e. Delft, Netherlands: VSSD.

130 Sabatini, R. (2013). "Cranfield University in Clean Sky: Avionics and CNS/ATM Research Focus", presented at the Clean Sky 2 Academia and Clusters, Brussels, Belgium.

131 Lee, D.S., Pitari, G., Grewe, V. et al. (2010). Transport impacts on atmosphere and climate: aviation. *Atmospheric Environment* 44: 4678–4734.

132 Schumann, U. (2005). Formation, properties and climatic effects of contrails. *Comptes Rendus Physique* 6: 549–565.

133 Penin, G. (2012). "Formation of Contrails", TU Delft, Delft, Netherlands, 3 Oct 2012.

134 Ponater, M., Marquart, S., and Sausen, R. (2002). Contrails in a comprehensive global climate model: parameterization and radiative forcing results. *Journal of Geophysical Research* 107: ACL 2-1–ACL 2-15.

135 Detwiler, A.G. and Jackson, A. (2002). Contrail formation and propulsion efficiency. *Journal of Aircraft* 39: 638–644.

136 Schrader, M.L. (1997). Calculations of aircraft contrail formation critical temperatures. *Journal of Applied Meteorology* 36: 1725–1728.

137 Fichter, C., Marquart, S., Sausen, R., and Lee, D.S. (2005). The impact of cruise altitude on contrails and related radiative forcing. *Meteorologische Zeitschrift* 14: 563–572.

138 Zolata, H., Celis, C., Sethi, V. et al. A multi-criteria simulation framework for civil aircraft trajectory optimisation. In: *ASME International Mechanical Engineering Congress and Exposition 2010 (IMECE 2010)*, Vancouver, BC, Canada, vol. 2010, 95–105.

139 Sridhar, B., Ng, H., and Chen, N. (2011). Aircraft trajectory optimization and contrails avoidance in the presence of winds. *Journal of Guidance, Control, and Dynamics* 34: 1577–1584.

140 Chen, N. Y., Sridhar, B., and Ng, H. K. (2011). "Contrail reduction strategies using different weather resources", *AIAA Guidance, Navigation, and Control Conference 2011 (GNC 2011)*, Portland, OR, USA.

141 Deuber, O., Sigrun, M., Robert, S. et al. (2013). A physical metric-based framework for evaluating the climate trade-off between CO_2 and contrails—the case of lowering aircraft flight trajectories. *Environmental Science and Policy* 25: 176–185.

142 Lim, Y., Gardi, A., and Sabatini, R. (2015). Modelling and evaluation of aircraft contrails for 4-dimensional trajectory optimisation. *SAE International Journal of Aerospace* 8, pp. Technical Paper 2015-01-2538.

143 Lim, Y., Gardi, A., Marino, M., and Sabatini, R. (2015). "Modelling and evaluation of persistent contrail formation regions for offline and online strategic flight trajectory planning", *International Symposium on Sustainable Aviation (ISSA 2015)*, Istanbul, Turkey.

Section III
Aerostructures and Propulsive Technologies

8

Advanced Aerodynamic Configurations

Matthew Marino[1], Alessandro Gardi[2,3], Roberto Sabatini[2,3], and Yixiang Lim[4]

[1] *School of Engineering, RMIT University, Bundoora, Victoria, Australia*
[2] *Department of Aerospace Engineering, Khalifa University of Science and Technology, Abu Dhabi, UAE*
[3] *School of Engineering, RMIT University, Melbourne, Victoria, Australia*
[4] *Agency for Science, Technology and Research (ASTAR), Singapore*

8.1 Introduction

More sustainable aviation operations are a primary focus for many local and international bodies. The increasing price of fossil fuel, various carbon taxes/trading schemes, and contributions to global warming are concerns that need to be addressed for a sustainable aviation future [1]. International aviation operations contribute 2–3% to global warming [2]. If aviation remains on its current path with no significant operational changes, its contribution is predicted to increase to 10–15% by 2050 due to the forecasted demand [1]. The contributions of aircraft and airport emissions to global warming have been well acknowledged in the past two decades, with significant investments in research and development to reduce aviation's carbon footprint and also to provide a foundation for a sustainable aviation industry [2]. The International Civil Aviation Organization (ICAO) has identified various requirements to develop aviation operations into a more efficient model and yield significantly greater efficiencies to reduce aviation's emissions by 50% by 2050 [1]. One of these recommendations related to new aircraft design configurations in the near future.

Even though aerodynamic advancements in conventional aircraft can only offer small efficiency improvements, new aircraft configurations have shown potential for significant emission reductions by exploiting designs that substantially increase lift/drag ratios and allow for lighter aircraft structures [3]. Although novel aircraft configurations still require further research and development, the realisation of these concepts is predicted to significantly reduce emissions and grow confidence in achieving a 50% reduction in carbon-based emissions [4].

In this chapter, some of the fundamental technical aspects pertaining to aerodynamics are introduced to support a meaningful discussion and an assessment of the progress achieved to date. The lift and drag of an aircraft can be determined using the following equations:

$$L = \frac{1}{2}\rho V^2 S C_L \tag{8.1}$$

$$D = \frac{1}{2}\rho V^2 S C_D \tag{8.2}$$

Sustainable Aviation Technology and Operations: Research and Innovation Perspectives, First Edition.
Edited by Roberto Sabatini and Alessandro Gardi.
© 2024 John Wiley & Sons Ltd. Published 2024 by John Wiley & Sons Ltd.
Companion Website: www.wiley.com/go/sustainableaviation

where:

L : Lift force (N)
D : Drag force (N)
ρ : Air density (kg m^{-3})
V : Velocity (m s^{-1})
S : Wing surface area (m^2)
C_L : Lift coefficient (dimensionless)
C_D : Drag coefficient (dimensionless)

By analysing the drag force as a function of the airspeed for a trimmed aircraft in steady, level flight, it can be easily verified that the drag coefficient C_D is the combination of two distinct components – the **induced drag** (C_{Di}) and the **skin friction** (C_{D0}) **drag** (also called **parasite drag**). By its very nature, the induced drag coefficient depends on the lift coefficient and this led to the establishment of the widely adopted **parabolic drag** approximation, for which the drag coefficient can be expressed as:

$$C_D = C_{Di} + C_{D0}$$

$$C_D = \frac{C_L^2}{\pi\, e\, AR} + C_{D0} \tag{8.3}$$

where:

e : Oswald efficiency factor
AR : the aspect ratio of the wing

The induced drag is associated with the turbulent trail wake of the aircraft, which is due to the '*finiteness*' of the wing. In particular, the AR defines the infiniteness of the wing, that is, how close it is to the ideal infinite wing, which would be characterised by a 2D flow (with components in the vertical plane only). The finite wing (i.e. limited AR), on the other hand, does not prevent the air from moving in the span-wise direction (i.e. perpendicular to the vertical plane), allowing it to 'spill' from the high-pressure regions (below) and enter the low-pressure regions (above) thanks to the pressure difference. This parasite effect gives birth to tip vortices. These tip vortices combine to form the turbulent trail wake of the aircraft. All the kinetic energy stored in the trail wake is energy lost from the aircraft motion, causing drag. The tip vortex phenomenon and the resulting wake is depicted in Figure 8.1.

Moreover, since high C_L are produced by a larger pressure difference over the wing, these yield stronger pressure spillage and hence stronger vortices. Finally, e is determined by the shape of the wing. Some wing shape designs are better than others in preventing spillage or reducing its effects. For instance, high AR wings are long and slender – the effects of the vortices are less significant when compared to low AR wings.

Figure 8.2 summarises the relative effect that the frontal shape of the wing system has on the baseline Oswald coefficient, whereas Figure 8.3 illustrates the effect of Hoerner wing tips and winglets.

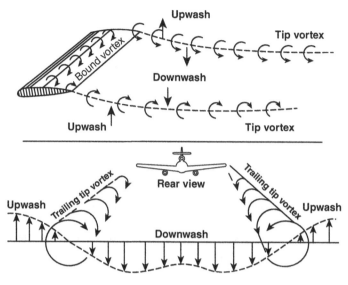

Figure 8.1 Key aspects of the tip vortex phenomenon.

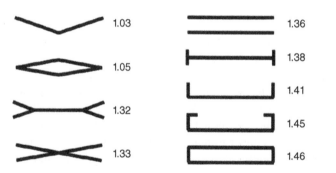

Figure 8.2 Relative enhancement of the Oswald coefficient due to the frontal configuration of the wing system.

8.2 Wing Tip Design

The design of wing tips aims to reduce the effect of the wing tip vortexes, hence minimise induced drag. The best wing tip designs afford the following advantages simultaneously:

1. Increase the effective wingspan and thus increases the aspect ratio of the wing. Increased aspect ratio will reduce the effect of the vortex on the wing lifting surface.
2. Winglets or tip fences reduce air movement at the wing tips, consequently reducing the strength of the tip vortices

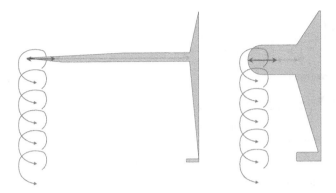

Figure 8.3 Tip vortex mitigation by advanced wing tip designs: Hoemer (left) and winglets (right).

New concepts include tip sails and minix wings, which generate counter-rotating vortices to neutralise the tip vortex. Figure 8.4 conceptually illustrates the effect of wing tip devices on the tip vortex.

8.3 Blended Wing-Body

The development of the blended wing-body (BWB) commenced in 1994 as a joint venture between NASA and McDonnel Douglas to evaluate if the configuration offered a more efficient means of passenger and cargo transportation in a high-capacity, long-haul transport aircraft [5]. The BWB aircraft is significantly different from common transport aircraft types. The design approach is taken from a '*Flying Wing*' concept where the majority of the aircraft's surface is used to produce lift as opposed to current transport aircraft where the pressurised cabin only exists as a chamber to hold passengers and/or cargo. This is conceptually illustrated in Figure 8.5. The BWB design maximises internal space inside the main cabin. This is due to the utilisation of available internal space allowed by the aircraft's design. All cargo, passengers and flight systems are positioned within the BWB structure. The BWB design also allows for the integration of alternative propulsive engines, such as the high-bypass ratio turbo fan engines, open rotor, electric motor propulsion blended body propulsion and other engines using alternative fuels. The Boeing X48C scale model (Figure 8.6) was developed to experimentally investigate a number of aspects specific to the BWB concept.

The BWB is unique in that the fuselage is an integration platform for wing, control surfaces, and engines inlets. The design is conceptualised to carry approximately 800 passengers within an operational radius of 7000 nautical miles (nm) at a cruise velocity of Mach 0.85 [5]. These operation requirements are similar to current aircraft used on intercontinental flights, such as the Airbus A350 and Boeing 777. It is designed to be a high-capacity passenger or cargo aircraft that is more fuel efficient and environmentally sustainable.

Due to the shape and configuration of the BWB, the aircraft was envisioned to have an advantageous operational efficiency. Preliminary findings approximated 27% less fuel burn, 15% weight saving, 12% lower empty weight, 27% less total thrust required and 20% higher

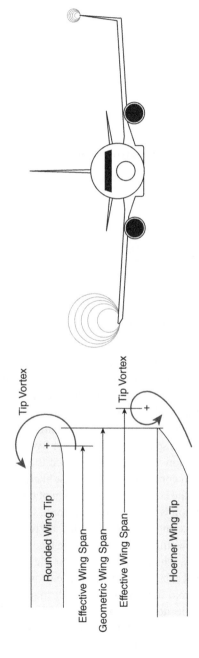

Figure 8.4 Tip vortex mitigation by advanced wing tip designs: Hoerner (left) and winglets (right).

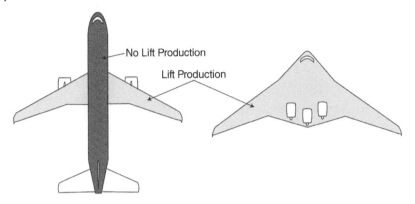

Figure 8.5 Lift production comparison between conventional aircraft and blended wing body.

Figure 8.6 The Boeing X48C BWB prototype [6].

lift to drag ratio relative to current commercial aircraft of similar size [5]. The savings are a direct result of the aerodynamic design of the BWB body. As most of the aircraft is used to generate lift, the BWB is more aerodynamically efficient that conventional commercial aircraft and has the potential to significantly reduce aviation's contribution to climate change. The internal usable space of the BWB is unmistakably larger relative to aircraft of similar size. The centre structure allows for multiple cabin seating arrangements and can be distributed laterally across the BWB span (Figure 8.7). The BWB configuration also allows for a multi-level cabin configuration which can be utilised to expand either passenger or cargo capacity. The internal cabin configuration will most likely be governed by an operational business model suited for the commercial utilisation of the BWB.

The aircraft seating configuration is likely to be of an arena style arrangement. Columns or walls, separating groups of passengers longitudinally, are required to form an internal support structure and assist in resisting the cabin pressure loads. The design also limits the amount and location of windows, which can only be near the aircraft's leading edge. The unorthodox nature of the internal cabin layout will affect customer flight satisfaction, since windows and arena style seating may not appeal to the wide majority (Figure 8.8). There are also social issues associated with its design such as the significant reduction in the amount of passenger windows and arena style seating [7]. To deal with customer satisfaction, the internal design of the aircraft must be considered and treated as a priority to encourage and support potential paying passengers. To appeal to the passenger's sense of comfort, organic light emitting diode (OLED) high-definition screens can be applied to the internal structure, displaying the outside views from the aircraft. It is anticipated that the real-life images will

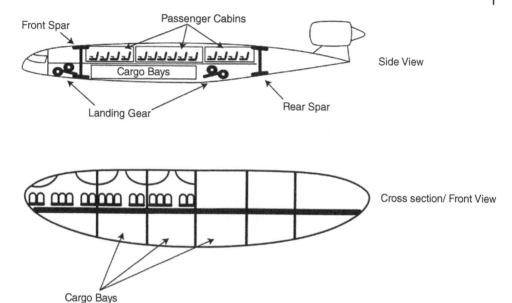

Figure 8.7 An example of the internal structure of the BWB.

provide passengers with a full 360° view outside the aircraft and alleviate their requirements for windows throughout the passenger cabin. This design feature could deal with passenger claustrophobia by creating the illusion of a larger internal space. This idea has recently been studied and is currently under development by Airbus and other aircraft manufacturers. The seating arrangement also brings issues of safety as standard emergency procedures would not apply. New emergency procedures must be formulated for the evacuation and response to a variety of emergency situations in accordance with US Federal Aviation Regulations (FAR) requirements. BWB emergency and evacuation procedures have not been extensively studied and thus this is a significant area for future research efforts.

Due to the design of the BWB, the shape of the aircraft places more material away from the aircraft centroid, increasing the strength of the overall structure. The majority of the material is located on the top and bottom surfaces of the BWB structure, which maintains bending stiffness in an efficient manner while the internal structure acts as a shear web to resist shear loads. The BWB will also be reinforced with a forward and aft spar to form the traditional '*Wing Box*' structure and will offer bending and torsional stiffness needed for gusts, manoeuvering, or high-lift device loads. This structural design results in a structure that requires less material while maintaining the required structural performance characteristics to satisfy aviation certification requirements. As the weight of the aircraft is reduced (mg), the amount of lift needed for steady, level flight also decreases. Similarly, as the wing area increases, the lift coefficient required to maintain steady, level flight is also reduced. This has a direct effect on the amount of aircraft drag as induced drag has a squared relationship to lift coefficient:

$$C_L = \frac{mg}{\frac{1}{2}\rho V^2 S} \qquad\qquad C_D = C_{D_o} + \frac{C_L^2}{\pi e AR} \qquad\qquad (8.4)$$

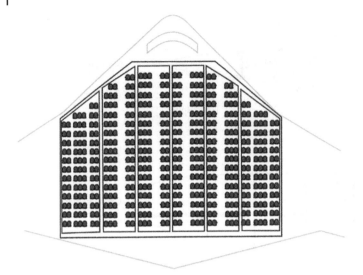

Figure 8.8 An example of the internal seating arrangement of the BWB.

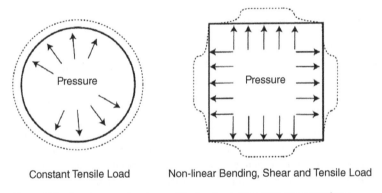

Constant Tensile Load   Non-linear Bending, Shear and Tensile Load

Figure 8.9 Loads on a structure with circular and square cross sections.

The BWB design does not come without design challenges. Pressurisation remains an important design consideration for high-altitude flight. With conventional aircraft designs, the tubular structure naturally distributes the load in an even fashion and guarantees that fuselage panels are constantly under a tension load. The design itself is efficient and offers a low weight and stable structure with predictable failure modes. Other non-circular structures are stressed in a non-uniform fashion when internally pressurised. This type of structure requires additional material to strengthen areas of high bending, shear, and tension stresses (Figure 8.9).

The BWB design departs from the tubular cabin design and results in a non-uniform load distribution due to the internal pressure exerting on the passenger/cargo cabin structure (Figure 8.10). One of the more challenging aspects is its non-circular fuselage structure, which requires extensive structural reinforcement to maintain pressurization [8]. This is a non-issue with conventional aircraft due to the structural efficiency inherent in the tubular cabin design.

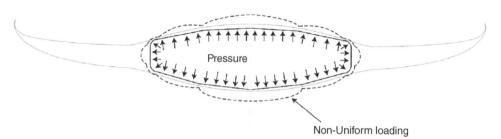

Figure 8.10 Pressure induced load on the BWB structure.

In order to adequately maintain the structural rigidity of the cabin area, the load associated with cabin pressure must be accounted for. This would most likely cause localised stresses, which will require more material to maintain the required structural strength of the cabin structure. There is also a degree of load failure uncertainty due to the complex load distribution prompting greater safety factors and additional material. In a normal cylindrical aircraft cabin, the circular structure maintains hoop stress and resolved load to tensile. Recent research has taken this concept and applied it to a BWB internal cabin structure. Multiple cylindrical cabins are constructed and merged together with small passageways between them, as illustrated in Figure 8.11. This allows the individual cabins to naturally form hoop stress in the critical areas of the cabin, thus allowing for a lighter design.

The integration between the body and wing requires a substantial rethinking of the classical design methodologies used for the design of commercial aircraft. One important structure that carried the majority of the aerodynamic loads is the wing box structure. On a conventional aircraft the wings connect to the fuselage through a main load bearing structure. In a BWB configuration the body and the wing are one and, as such, a load transfer structure is not needed. The general structural design consensus is a wing box structure that exists through the entire length of the span. The forward and rear spar would function similarly to that of a normal wing in providing structural rigidity in shear and bending. Bucking is avoided in the normal manner by the introduction of ribs throughout the outer sections of the wing while the overall structure maintains torsional strength in line with traditional and proven structural design methods. Larger ribs are placed at greater relative pitch to the centre body where passengers and cargo are held. Although their primary function is to resist buckling, they are also needed to maintain structural strength due to pressure forces from the pressurised cabin. These considerations are conceptually illustrated in Figure 8.12

8.3.1 BWB Noise Benefits

The design of the BWB also offers significant reductions in aircraft ground noise as the engines are envisioned to be located above the rear section of the aircraft, which allows the fuselage to act as a reflective sound barrier to propagate the majority of noise towards the sky [10]. This appeals to residents around the terminal area and will significantly reduce perceived noise. Reduced noise levels will improve resident satisfaction. Stability and control of the aircraft remains an important technical and design challenge due to the absence of a vertical and horizontal stabiliser as found on conventional aircraft. The design of the BWB

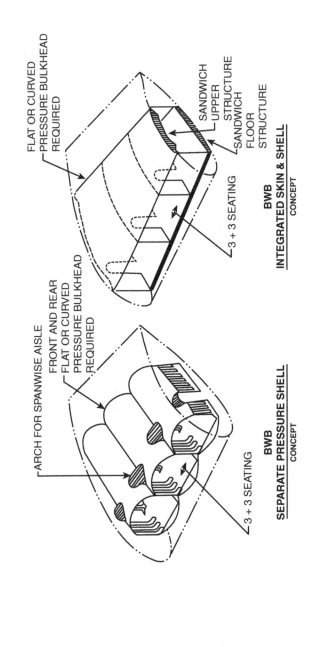

Figure 8.11 Pressure shell and integrated skin conceptual designs for BWB main cabin [5].

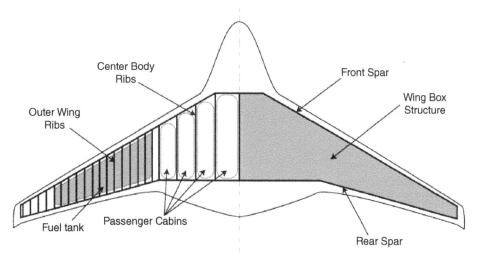

Figure 8.12 The general internal structure of the BWB [5, 9].

wing must be optimised to produce a neutral static stability with trim to be provided by the control surfaces along the BWB trailing edge. The BWB is also less laterally stable relative to conventional configurations due to the absence of a vertical stabiliser. The use of large wing tips has shown promise in providing degrees of passive lateral stability, however stability augmentation systems may be needed to provide active lateral stabilisation thought all flight phases. The BWB utilises an elevon control system layout where the lateral and longitudinal control is supplied by the trailing edge control surfaces. As previously stated, the wing tips may offer passive lateral stability and provide a location for rudders for active yaw control.

8.3.2 Propulsion Configurations

The design and layout of the BWB allows for various propulsion configurations. However, the traditional engine-pylon system on the top rear surface of the BWB was identified as the ideal solution [5]. Its large wing platform area not only allows for multiple possibilities to situate aircraft engines, it also allows for distributed propulsion, since a greater number of engines and or electric motors can be used. Multi-engine designs are desirable because of their benefits in terms of redundancy and the mitigation of noise. In particular, as the number of engines increases, the perceived noise of the aircraft decreases. This is a result of using smaller engines that produce a higher frequency noise relative to larger engines. As such, the noise dissipates at a higher rate with sound waves propagating in a more focused and narrow manner. There have also been studies that show efficiency benefits due to the advantageous aerodynamic effects of distributed propulsion [9, 10]. In terms of safety, the multi-engine design is beneficial as a failing engine will reduce total thrust marginally compared to twin-engine designs. Other engines will most likely be able to adequately substitute for the loss in propulsive power and will do so in a redundant manner. For a 7-engine design a single engine failure will result in a 14% reduction in maximum available power. The opposite case is a 2-engine aircraft configuration where an engine failure

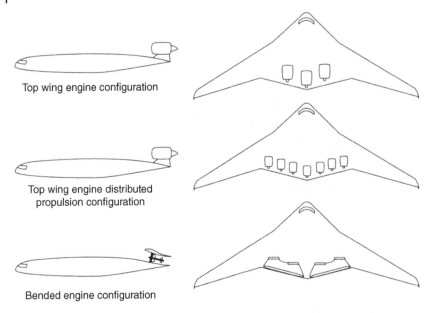

Top wing engine configuration

Top wing engine distributed
propulsion configuration

Bended engine configuration

Figure 8.13 Propulsive configurations of the BWB.

results in a 50% loss in available power. The multi-engine design also allows the application of various complementary propulsion technologies, such as hybrid, electric, biofuel, or hydrogen engines. Because of the lower safety-criticality of individual engines/motors, the BWB with a multi-engine configuration provides a safer and more viable approach in testing new propulsive technologies. This is because highly mature, conventional, gas-turbine engines can be used in conjunction with new engine types. The large number of smaller engines permits high levels of redundancy and offers an effective means of transitioning new propulsive technologies into the current air travel operation. Figure 8.13 illustrates the versatility of the BWB design in terms of propulsive configurations, with higher levels of distributed propulsion eventually leading to a 'blended engine' configuration.

In particular, the design attributes of the BWB allow for imbedded propulsion systems. This type of propulsive configuration is directly inspired by fighter jet designs. The inlet of the engines utilises the boundary layer created by the body of the aircraft. The relative flow velocity of the BWB boundary layer is general slower than the free stream velocity, which in turn can be accelerated at greater rates to increase propulsive efficiency. The position of the exhaust provides high flow and pressure, which mixes with the aircrafts wake turbulence and reduces the induced drag created from the BWB. Safety is a concern with this configuration as various engines failures can impinge on the surrounding structure, which is in close proximity to the main cabin. Other issues include complexity of maintenance, upgrade compatibility, cabin/wing-to-engine redundancy, and the initial compatibility of available engines.

8.3.3 Emergency Exits

There have been a few proposals regarding emergency exit passages, one such is by Zhou (Figure 8.14). The design shown is for the 250-seat example, with sufficient emergency exits

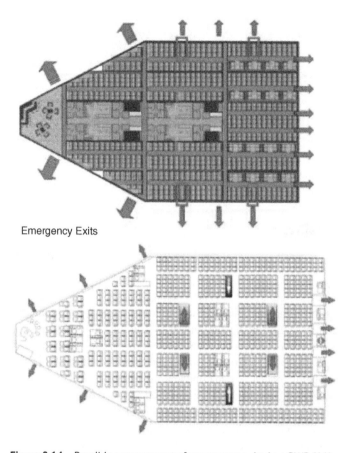

Emergency Exits

Figure 8.14 Possible arrangement of emergency exits in a BWB [11].

set in terms of the size, position, access in all possible locations, and the configuration of the aircraft. One of the problems with the BWB is that it is usually proposed to be much wider than conventional aircraft and this results in an increased distance between the passenger and the exits. In Zhou's concept, emergency exits are provided all around the cabin area with wing and aft exits. This proposal is according to the FAR 25.807 to type B dimensions and two type A doors at the leading edge of the wing. [11].

This kind of a proposed seating arrangement is called multi-bubble architecture, in which six seats abreast are used for a single aisle compartment. There are two passenger and service doors arranged on each side of the leading edge and two ventral exits in each side of the rear body. The proposed evacuation exits are as per the airworthiness regulation [11].

8.3.4 Cargo Considerations

The BWB aircraft configuration is not restricted to a passenger carrier. The BWB cargo aircraft has also been explored and offers flexibility due to the non-circular shape of the fuselage. Initial studies compared the BWB airlifter concept to the C-17 and, as such, the configurations were conceptualised to carry 19 463 L pallets. The internal structure of the BWB aircraft was found to be advantageous for cargo transport, due to a greater

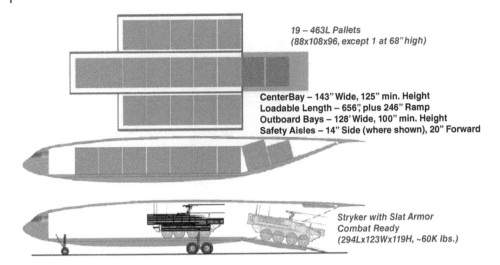

Figure 8.15 Military cargo BWB design [12].

availability of lateral space. The cargo configuration was split into three individual cargo compartments, with the centre section being the largest in volume (Figure 8.15). The aircraft was also designed with a rear opening door which would give access to the central loading bay. The side loading bays would require the cargo to be manoeuvred into place and filled prior to the central bay being loaded. Pre-loading analysis is required as the aircraft must be balanced both laterally and longitudinally around its CG to avoid stability and control issues when in flight. This would also require pre-loading logistics to strategically place cargo in the side cargo bays before loading the central bay. Research into cargo loading time and loading logistics is needed to evaluate the time and monetary cost of loading the BWB cargo aircraft relative to common airlifters, such as the C-17 and Globe Master aircraft.

The BWB would be able to load combat vehicles as shown in the figure above and would have a maximum load of 115 000 pounds (Figure 8.15). Skorpura suggested that for loading and unloading a rear ramp would be the best solution but front loading is also possible.

Furthermore, to keep the size of the BWB similar to conventional airlifters, the wings would be able to fold. Due to the shorter aircraft length and the addition of a folding wing, this would reduce the airfield infrastructure required to park such an aircraft.

8.4 Box Wing

The box-wing configuration consists of two horizontal wings joined at the tips by vertical wings, and has its roots in the aerodynamic work of Ludwig Prandtl and Max Munk from the 1920s. Prandtl-Munk's best wing theory noted that this kind of wing system would drastically reduce induced drag but the configuration fell by the wayside as the conventional wing-and-tube planform became popular. It has recently been revived due to research studies from the likes of Lockheed Martin. This configuration is also known by some designers as the Prandtlplane, in honour of Ludwing Prandtl [13].

The primary advantage the box-wing offers over conventional aircraft is that it inhibits the formation of wingtip vortices, which are the primary cause of induced drag on conventional aircraft. The closed wing system is designed to keep the airflow around the wing attached all the way around the wing system. However, such a radical design also comes with a number of challenges in terms of structural design and analysis of the wing system and ensuring the aircraft is stable and controllable. It is likely that the box-wing configuration will be able to take advantage of a number of new technologies in the development of new aircraft, such as laminar flow and new propulsion technologies including high-bypass engines. For Unmanned Aerial Vehicles (UAVs) and special missions, it must be noted that the box wing has an advantage in terms of offering the same lifting area as a conventional design but with a shorter wingspan, meaning it can operate in environments where that is a limiting factor, such as urban surroundings.

The aerodynamics and structures of the box wing are highly integrated and interdependent, meaning that, unlike a conventional configuration, the planform cannot be analysed with regard to one discipline independently and then the other, but must consider both at the same time. This kind of multidisciplinary analysis, even at the conceptual design stage, is necessary because of the highly interlinked nature of the configuration. For example, the aerodynamic performance of the box wing improves as the vertical separation between the two horizontal wings improves. However, having a long, thin vertical wing connecting the two horizontal wings can lead to problems with flutter and buckling, meaning that components must be designed to be heavier and stiffer as the vertical distance increases. An optimum amount of vertical separation will hence exist for the design, which will change depending on the size of the aircraft and the mission requirements for that particular design.

One of the first serious studies on the concept of the box-wing airliner was conducted at Lockheed in the mid-1970s, for NASA [14]. The study looked at a transonic biplane concept, joined at the tips, to fly at Mach 0.95 for a range of 5000 nm and carrying a payload of 400 passengers. It is clear that this mission requirement was for the dimensioning and design of a very large box-wing configuration. A parametric design space exploration was carried out, based on the Prandtl-Munk biplane theorem in the most part for the aerodynamics, while structural analysis was conducted using a primitive structural code which mimicked conventional configuration analysis for the box-wing layout. The findings were somewhat questionable, but did establish the validity of the box wing as an unconventional wing configuration in need of further study. The basic outcome was that an equal ramp weight was predicted for the box wing, as compared to a conventional aircraft that would meet the same mission requirements, and that flutter would prove to be a large problem at the chosen cruise Mach number [14].

The box-wing configuration is currently under investigation for a variety of different missions, from very large passenger airliners, to smaller, single-aisle airliners to regional business jets and even UAVs. While the possible reduction in induced drag is very high, up to 40%, more holistic design studies such as [15] have found that a reduction in fuel burn of around 5% or 6% is possible over short-range missions. This takes into account the disadvantages of the box wing, such as the higher parasitic drag and the complex structure of the wing planform which can mean heavier weight, over the course of the mission when compared to a conventional aircraft. The current design analysis and optimisation studies with regard to the box-wing aircraft indicate that it is more likely to be an efficient and effective alternative to the conventional configuration for short-range missions with lower requirements in terms of cruise altitude and Mach number.

There are a number of different configuration integration and design challenges that will need to be overcome with regard to the box wing while it is still in the proposed design stage. The landing-gear integration and the placement of the front and rear wings is a large challenge. The Prandtl-Munk theorem, on which the box-wing is based, states that the most efficient aerodynamic composition of the aircraft is one where the lift is equally divided between the front and rear horizontal wings. This leads to the maximum reduction in induced drag. However, once stability and handling considerations are taken into consideration, this condition can rarely be met as the aircraft is generally unstable and will not rotate with half the lift at the rear of the fuselage. This also limits the placement of the front wing as these are usually placed more towards the front of the fuselage, meaning that that landing gear cannot be integrated underneath the wing and hence must be placed in the belly of the fuselage. This drives up the weight of the fuselage and landing gear, which means some of the overall efficiency gains are lost.

There is another challenge with regard to the wings of the box-wing configuration, which results from splitting up the lifting area over two wings and hence leading to wings with smaller chords than conventional configurations. This can mean a number of different issues for the wing design, including the fact that there are significant changes in the Reynolds number effects and the structural impacts of having less space and volume for control surfaces, fuel systems, de-icing systems and so forth. These can be even more problematic if the rear wing is placed high on the tailplane, meaning the vertical tailplane needs to be heavier and stronger to support these systems and the structural loads from the rear wing itself. Some box-wing designs propose a V-tail empennage to alleviate some of these issues and improve controllability of the aircraft, too, although the superiority of the V-tail is not completely proven for the box-wing.

In contrast to the blended-wing-body configuration, the fuselage design of the box wing does not differ greatly from the design of the fuselage for conventional configurations, but the engine placement for the box wing is another design challenge that needs to be overcome. The conventional option of below the lower wing still exists, but limits how low the lower wing can be placed for ground clearance reasons. If the engines are placed on the upper wing, there is significantly more space and high-bypass engines can be used, but access and maintenance may become a significant issue. Some designs have engine placement at the rear of the fuselage, but that can often lead to concerns with noise in the cabin. One unconventional option is the use of the over-the-wing nacelle configuration, as seen in the HondaJet, which alleviates the issues of access and clearance, but presents its own challenges in terms of wing aerodynamics and structures.

Generally, the box wing continues to present a more efficient and hence attractive alternative to the conventional configuration, especially for short-range missions, but given the significant design and technical challenges that still need to be overcome before it is fully understood and reliable as a design, it is probably still some decades away from flying.

8.5 Wing Morphing Technology

Morphing wing technology is a concept that allows the wings to actively or passively change shape during flight. This can be done through mechanical or organic means. For instance, bird's wings can actively change shape due to its organic nature or muscles and skeletal structure, while the F-111 actively changes its wing shape by altering the sweep angle. By modifying the shape of the wing during the mission, morphing wing technology allows greater aerodynamic efficiency to be achieved in different phases of flight. This technology has been in use since the time of the **Wright Flyer** in 1903, where cables were used to provide twist to the wings. The concepts that are currently being investigated include:

1. Variable wing sweep – sweeping the wings back during supersonic flight can prevent the generation of shockwaves. This is typically achieved by a pivoting mechanism. The **Messerschmitt** was a prototype aircraft developed in WW2 and the **Bell X-5** was another experimental aircraft developed to evaluate aerodynamic effects.
2. Variable-span – wings with a low span can fly at faster speeds and are highly manoeuvrable, while wings with a large span have good range and fuel efficiency.
3. Variable chord – current mechanisms, such as flaps and slats, can be extended or retracted to modify the aircraft chord, but are aerodynamically inefficient and use heavy and complex actuators. Macro-fibre composite (MFC) or shape memory alloys (SMA) allow for the actuators to be embedded in the wing structure.
4. Variable aerofoil shape – the camber, thickness is adjusted to modify the aerodynamic properties of the wing at different flight phases. An example of this concept is provided Fig. 8.16.

The current standing organisation review of morphing wings can be visualised as in Figure 8.17.

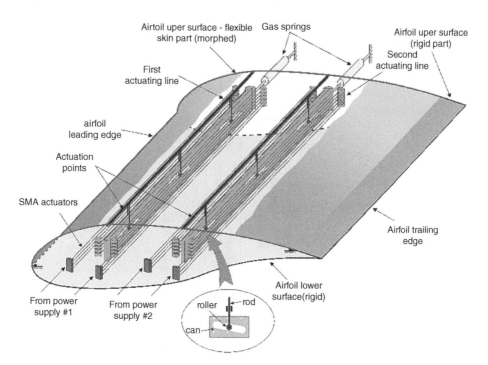

Figure 8.16 Example of morphing wing smart actuation system [25].

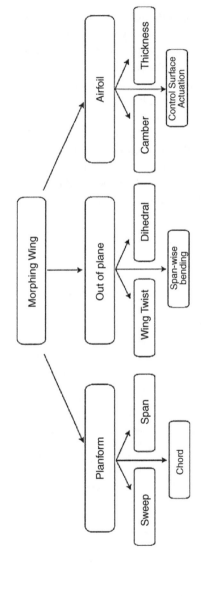

Figure 8.17 Morphing wing technologies.

The morphing wing aircraft concept is a relatively old design with its first inception in the Wright Flyer of 1909. The wings were allowed to twist under the load of various cables running through a complex pulley system throughout the aircraft. This, of course, was only allowed due to the flexible nature of the wing itself. As time passed aircraft wings became solid structures which allowed for greater strength and flight performance. The morphing concept was re-introduced on the Bell X-5 by adding variable sweeping wings that could be actuated in flight to change the aircraft's configuration. This was later applied to such military aircraft as the B-1 bomber, F111 Aardvark, and F-14 Tomcat. The wing morphing capability was considered advantageous, as it allowed a full forward sweep configuration for improved slow speed flight dynamics while the sweep back configuration allowed for improved high-speed flight dynamics. The concept of variable configurations is of interest to the commercial aviation industry, as it allows greater operational efficiencies due to the idea that the wing structure can change to suit the flight phase itself.

Current technology exhibits passive forms of wing morphing. Span-wise bending of the aircraft's wing has offered various performance benefits, such as lighter wing structures and increased passenger comfort due to enhanced gust damping. As the wing of the aircraft becomes more flexible, the relative weight of the aircraft is reduced too. Current airliner aircraft utilise various composite alloy materials in the manufacture of light wings that are able to bend in a passive manner when in flight and are approximately 20% lighter than past aircraft with stiff wing designs. Further weight savings are possible when other morphing wing designs, such as variable camber wing, morphing control surfaces and blended high life devices, are implemented. Increasing wing flexibility may also cause issues with aerodynamic flutter as the frequency of vibration modes decreases due to reduced weight and stiffness. Active flutter suppression may be needed to combat wing vibrations.

8.5.1 Camber Morphing

Active camber changes are an effective and efficient means of changing the wing's lifting capability with a highly cambered wing for take-off and landing phases. It enhances the wing's lifting performance actively varying the camber of the airfoil while reducing the parasite, form, and induced drag penalties as attached flow favours gradual surfaces. This technology also has the capability of reducing the formation, size, and persistence of aircraft wake turbulence. High-lift devices on current airliners, such as flaps and slats, are inherently inefficient as significant amounts of drag is produced due to the increased amount of wing tip vortices brought upon through individual and segmented lifting surfaces.

Current research is focused on replacing conventional high-lift devices with variable camber morphing wing technology for both reduced weight and noise [16]. This morphing technology facilitates the active and controlled change of airfoil shape. This is particularly important as it allows for a thicker and more highly cambered airfoil for slow flight phases with the ability to reconfigure into a thin straight airfoil for high-velocity/supersonic flight operations [17].

There are also concepts of wing twist morphing which allow for greater aerodynamic efficiency over the conventional aileron design and also allow variable airfoil geometry over the span of the wing [17]. This type of morphing is enabled due to the advancements in material science. The use of organic materials and smart material alloys provides a controllable means of contracting or retracting a material [18]. In other words, the material's surface area can be increased or decreased in a controlled manner by electronic means. This is supported by novel internal structures which allow bending in the required direction while maintaining wing strength and rigidity.

8.5.2 Morphing Control Surfaces

Much like the variable camber morphing wing, the virtual camber of the wing can also be altered by the mechanical actuation of the control surfaces [18]. Morphing the trailing edge of an airfoil and elongating this actuation across the span is known as shape adaptive control surfaces and would replace the conventional mechanical control surfaces on an aircraft. Current control surfaces utilise a segmented control surface design which conforms to the overall airfoil shape. The control surface itself serves as a trailing edge and control surface of either the wing or stabiliser. The control surface itself is stiff and rotates around a hinge as represented in Figure 8.18. This type of mechanical actuation requires either hydraulic, pneumatic, or electrical systems that rotate the control surface around the hinge point. These systems are complex and require substantial power for their operation. Although this design has proven to be successful in control surface function for the aileron, elevator, rudder and other surfaces, flow is accelerated violently around the hinge where a sharp change in airfoil surface direction occurs. In high-deflection situations, the flow can separate and significantly increase the amount of drag. A control surface design that continuously conforms to the airfoil shape while providing a flexible and controllable means of deflecting the trailing edge would be more efficient. Shape adaptive control surfaces may provide an evolutionary design feature with initial studies showing designs that are lighter and more aerodynamically efficient than conventional designs [18–22]. One of the challenges of introducing shape adaptive control surfaces is integrating the active and non-active surfaces along the span. Flexible material joining the active and non-active tailing edge is currently under research with flight tests carried out by the NASA research facility. The material properties of the joining material must be flexible enough to allow morphing actuation while maintaining enough strength to maintain lift.

The BWB provides an adequate platform for the morphing structure, which is only limited by the engineer's imagination. Although current research is mainly focus on wing morphing technology, there is scope to expand the concept to various parts of the airframe and provide lighter and more efficient aerodynamic structures that will contribute to reducing aircraft weight, fuel consumption, and emissions.

8.5.3 Span-Wise Morphing

Span-wise morphing is related to the concept of actively changing the span of an aircraft in flight. Within this research field this has mainly been done mechanically using telescopic mechanical devices. Devices like these are inspired by telescopes and allow sections of the telescope to slide into each other to reduce the overall length. Recent studies have used telescope wing spars in which ribs and other mechanical devices are attached, too. These spars allow for span-wise changes while maintaining the structural rigidity. The constitution of such a wing primarily consists of several overlapping wing segments that extend and retract with minimum sliding clearance.

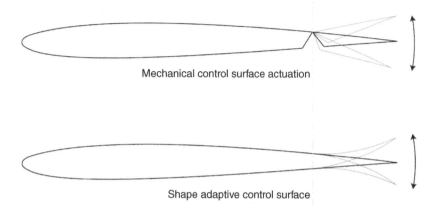

Figure 8.18 Mechanical and shape adaptive control surface deflection.

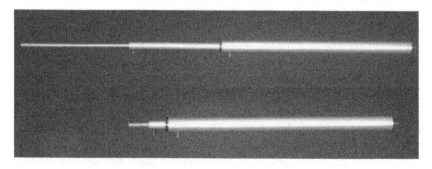

Figure 8.19 Telescopic spar in extended/retracted configuration [17].

The telescopic wing concept consists of multiple segments. These are constructed in a manner that allows the length to increase while maintaining the required structural rigidity [17]. An example is featured in (Figure 8.19). The wing in this design consists of: pneumatic telescopic spar, rigid airfoil skins, and rib elements. During the testing phase the concept was able to demonstrate a 114% increase in aspect ratio. The design was proven to be feasible for a small-scale UAV design. Moreover, wind tunnel testing demonstrates the parasitic drag increase near the joints between panels.

Another approach to modify the wingspan utilises a scissors-like mechanism for the wing box. [18] This mechanism is used to alter the aircraft span and sweep, which could undergo 55% span change by employing a DC motor to actuate the spooling screw.

8.5.4 Wing Twisting

The Defense Advanced Research Projects Agency (DARPA) Smart wing design program utilised a gradual airfoil camber alteration to generate wing twisting. The wing was split

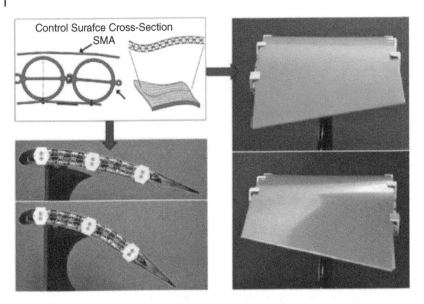

Figure 8.20 Bio-Inspired actuation system using memory alloys and cellular core structure [24].

into 10 segments, with each segment undergoing out-of-plane shape alteration utilising an eccentuator concept [23]. The eccentuator primarily operates by transferring rotary input motion at one end of the bent beam into vertical and lateral displacement at the other end. According to [16] lateral motion, however, is highly undesirable, due to the addition of a sliding joint to the system.

Elzey et al. [24] have described a bio-inspired metal vertebrate actuation system used to achieve a fully reversed twist shape deformation (Fig. 8.20). The actuator is constructed of SMA face sheets with an internal cellular flexible core structure [9]. The core design utilised a vertebrate actuation mechanism through cylindrical or pin joints that allowed rotation relative to each other. The actuator rib's position remains fixed by means of tubular spacers that are connected to vertebrae at both ends. Having a fixed joint at both ends prevents the actuator from out of plane rotation (roll) [9].

8.6 Boundary Layer Control

Fundamentally, boundary layer control involves the active control of the boundary layer to improve the aerodynamic performance of the aircraft. The main strategies being pursued are:

1. Laminar flow – minimising the turbulence on the surface of the wing allows a reduction of skin friction drag. The laminar flow wing has been practically demonstrated by the Airbus Breakthrough Laminar Aircraft Demonstrator in Europe (BLADE) project and further work is now being pursued to bring the technology to market.
2. Prevention of flow separation – separation occurs at high angles of attack and leads to loss of lift. Delaying the flow separation may yield some drag penalty (more skin friction) but substantially enhances the flexibility in the design of the wing, which also becomes

more stable in a broader range of velocities. Overall drag reduction can be achieved if these two advantages are carefully harnessed in a design optimisation.

The technologies proposed, conceptually illustrated in Figure 8.21, are:

1. Vacuum: the boundary layer is kept on the surface of the wing through suction. Slow moving air is sped up through suction to prevent separation and reducing drag.
2. Pressure: the boundary layer is re-energised through blowing. A boundary layer with low energy tends to separate from the surface.
3. Rough surface: some friction is introduced to make the boundary layer turbulent, delaying separation.
4. Mirror: a smooth surface minimises skin friction but causes separation easily, as the boundary layer does not 'stick' to the surface.

These concepts have been widely investigated for different aircraft/engine configurations. While laminar and rough surfaces are now mature technologies, various studies have shown that embedding vacuum and/or pressure chambers in the wing might not allow to achieve the desired environmental benefits, mainly due to a high energy consumption.

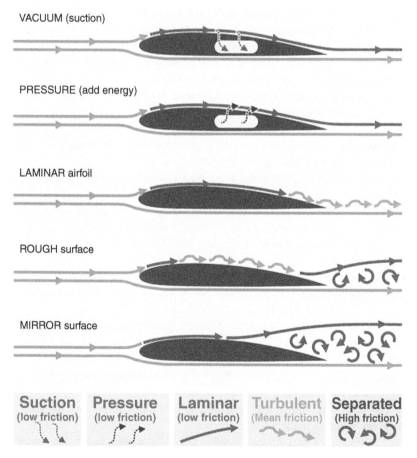

Figure 8.21 Outline of the most commonly investigated techniques for boundary layer control (active and passive).

8.7 Conclusions

The customer demand for air travel is constantly increasing and with it aviation's contributions to carbon-based emissions. Current aircraft technologies are proving effective in reducing aircraft emissions. However, new materials, efficient propulsion, and green operations can only reduce emission up to a certain degree. To meet the 2050 goal of 50% less carbon emission, new and innovative aircraft designs need to be introduced in air travel operations. Advanced aircraft configurations have shown a clear potential to substantially reduce carbon emissions and change the way we fly. However, in most cases, the uptake of these technologies in future civil air transport would require significant advancements of the associated airworthiness and operational management frameworks.

References

1 Maurice, L. and Lee, D. (2009). *Assessing Current Scientific Knowledge, Uncertainties and Gaps in Quantifying Climate Change, Noise and Air Quality Aviation Impacts, final report of the International Civil Aviation Organization (ICAO) Committee on Aviation and Environmental Protection (CAEP) Workshop.* Washington, DC and Manchester: US Federal Aviation Administration and Manchester Metropolitan University.

2 Lee, D.S., Fahey, D.W., Forster, P.M. et al. (2009). Aviation and global climate change in the 21st century. *Atmospheric Environment* 43: 3520–3537.

3 Ordoukhanian, E. and Madni, A.M. (2014). Blended wing body architecting and design: current status and future prospects. In: *12th Annual Conference on SystemsEngineering Research, CSER 2014*, Redondo Beach, CA, 619–625.

4 Hileman, J., Spakovszky, Z., Drela, M. et al. (2010). Airframe design for silent fuel-efficient aircraft. *Journal of Aircraft* 47: 956–969.

5 Liebeck, R.H. (2004). Design of the Blended Wing Body Subsonic Transport. *Journal of Aircraft* 41: 10–25.

6 Vicroy, D. (2011). X-48B Blended Wing Body Ground to Flight Correlation Update.

7 Hall, A., Mayer, T., Wuggetzer, I., and Childs, P. (2013). Future aircraft cabins and design thinking: optimisation vs. win-win scenarios. *Propulsion and Power Research* 2: 85–95.

8 Roman, D., Allen, J. B., and Liebeck, R. H. (2000). Aerodynamic design challenges of the blended-wing-body subsonic transport. *18th Applied Aerodynamics Conference 2000*, Denver, CO.

9 Leifsson, L., Ko, A., Mason, W.H. et al. (2013). Multidisciplinary design optimization of blended-wing-body transport aircraft with distributed propulsion. *Aerospace Science and Technology* 25: 16–28.

10 Ko, A., Leifsson, L.T., Schetz, J. et al. (2003). MDO of a blended-wing-body transport aircraft with distributed propulsion. *AIAA Paper* 6732: 2003.

11 Zhou, W. (2011). Cabin environment and air quality in civil transport aircraft. Master thesis. Cranfield University, Cranfield, UK.

12 Buch, G.M. Jr. (2010). AMC's next strategic airlifter: the blended wing body?. Thesis. US Air Force Institute of Technology, Wright-Patterson AFB, OH.

13 Frediani, A., Cipolla, V., Salem, K.A. et al. (2020). Conceptual design of PrandtlPlane civil transport aircraft. *Proceedings of the Institution of Mechanical Engineers, Part G: Journal of Aerospace Engineering* 234 (10): 1675–1687.

14 Lange, R.H., Cahill, J.F., Bradley, E.S. et al. (1974). Feasibility study of the transonic biplane concept for transport aircraft application.

15 Andrews, S.A. and Perez, R.E. (2018). Comparison of box-wing and conventional aircraft mission performance using multidisciplinary analysis and optimization. *Aerospace Science and Technology* 79: 336–351.

16 Van Dam, C. (2002). The aerodynamic design of multi-element high-lift systems for transport airplanes. *Progress in Aerospace Science* 38: 101–144.

17 Joshi, S.P., Tidwell, Z., Crossley, W.A., and Ramakrishnan, S. (2004). Comparison of morphing wing strategies based upon aircraft performance impacts. *Sea* 2: 32.

18 Sofla, A.Y.N., Meguid, S.A., Tan, K.T., and Yeo, W.K. (2010). Shape morphing of aircraft wing: status and challenges. *Materials and Design* 31: 1284–1292.

19 Barbarino, S., Bilgen, O., Ajaj, R.M. et al. (2011). A review of morphing aircraft. *Journal of Intelligent Material Systems and Structures* 22: 823–877.

20 Gilbert, W.W. (1981). Mission adaptive wing system for tactical aircraft. *Journal of Aircraft* 18: 597–602.

21 Braga, D.F.O., Tavares, S.M.O., da Silva, L.F.M. et al. (2014). Advanced design for lightweight structures: review and prospects. *Progress in Aerospace Science* 69: 29–39.

22 Bartley-Cho, J.D., Wang, D.P., Martin, C.A. et al. (2004). Development of high-rate, adaptive trailing edge control surface for the smart wing phase 2 wind tunnel model. *Journal of Intelligent Material Systems and Structures* 15: 279–291.

23 Kudva, J.N. (2004). Overview of the DARPA smart wing project. *Journal of Intelligent Material Systems and Structures* 15 (4): 261–267.

24 Elzey, D.M., Sofla, A.Y.N., and Wadley, H.N.G. (2003). A bio-inspired high-authority actuator for shape morphing structures. *Proceedings of SPIE 5053, Smart Structures and Materials 2003: Active Materials: Behavior and Mechanics.* San Diego, California, USA.

25 Grigorie, T.L., Botez, R.M., and Popov, A.V. (2012). Fuzzy Logic Control of a Smart Actuation System in a Morphing Wing. In: *Fuzzy Controllers – Recent Advances in Theory and Applications.* (ed. S. Iqbal). IntechOpen.

9

Lightweight Structures and Advanced Materials

Raj Das[1] and Joel Galos[2]

[1]*School of Engineering, RMIT University, Melbourne, Victoria, Australia*
[2]*Department of Materials Engineering, California Polytechnic State University, San Luis Obispo, California, USA*

9.1 Sustainability in Aerospace Materials and Structures

The aerospace sector is a prime industry where advanced materials and manufacturing play a major role. Sustainability, eco-efficiency, and green chemistry are controlling the development of the next generation of materials. Consumer demands, as well as environmental legislation in many countries, are pressuring manufacturing processes to develop environmental-friendly products. The ecological impact of an aerospace component is not only determined by the materials chosen, but also by in-service usage and disposal or recyclability of the product itself. There are several approaches and strategies that can be adopted to ensure sustainability in aerospace materials and structures, with the crucial ones being optimised structural designs of lightweight and multifunctional materials, use of biodegradable and recyclable materials, and damage-tolerant designs of highly durable and long-life structures.

Lightweight structures lead to fuel savings and hence reduce both costs and carbon emissions, contributing to environmentally friendly systems. Consequently, this has enabled the creation of transport vehicles possessing improved payload capability, effective range, and energy efficiency; all of which are desirable attributes for aircraft and spacecraft. Light alloys and composite materials have become a vital part of modern aircraft material systems to achieve lighter structural parts. The design of lightweight material systems is a key factor in developing fuel-efficient, light structures that provide the required mechanical properties and structural integrity. This is naturally accompanied by the need for more energy-efficient materials and structure-fabrication processes as a complementary aspect for sustainable-materials technology in the aerospace domain. To develop lightweight designs, structural optimisation is regularly used as an integrated design tool that involves the application of optimisation methods to the design synthesis of structures with a view to identifying the 'best' structural configuration. This enables the design of a lightweight structure with the requisite strength and stiffness constraints, which will simultaneously deliver the desired performance and satisfy the design constraints in a logical and systematic manner.

Sustainable Aviation Technology and Operations: Research and Innovation Perspectives, First Edition.
Edited by Roberto Sabatini and Alessandro Gardi.
© 2024 John Wiley & Sons Ltd. Published 2024 by John Wiley & Sons Ltd.
Companion Website: www.wiley.com/go/sustainableaviation

Material selection and manufacturing process are particularly important for aerospace materials. Moreover, biodegradability and recyclability have emerged as important features in many engineering applications [1]. Green economies and green policies have emerged over the past two decades and are supported by the vision of developing eco-friendly products. This has led to extensive utilisation of biodegradable polymer-based composites in various large-scale industries and in targeted products used in the aerospace and aviation sector. Hence there is a growing need to develop novel biodegradable and recyclable products, which in turn can contribute to sustainable aerospace structures.

Lastly, damage-tolerant and failure-resistant materials and structures lead to safety, durability, and long operational life, thus contributing to sustainable aerospace systems. Fatigue failure can pose a threat to the integrity, safety, and reliability of many aerospace structures subjected to high stress levels. The consequences can be severe in that, apart from economic loss, they may include serious injuries, loss of life, and environmental damage. The potential cost savings, as a result of incorporating fracture/fatigue analysis into the design and maintenance of structures can be significant. The sources for these huge costs include over-design with higher safety factors to account for design uncertainties, additional repair, rework and maintenance costs, premature replacement of useable components, and lastly productivity or performance loss due to failures.

The notable thing about failure by fracture is its unpredictability due to the multiple potential failure sites and the consequences in the event of an actual failure. The Aloha Airlines failure in 1988 highlighted the problem of fatigue failure and its threat to the aircraft in flight. An Aloha Airline Boeing 737 aircraft (flight 243 bound for Honolulu) suffered a fatigue failure at an altitude of 24 000 ft. Fatigue failure of a fuselage lap joint led to the separation of an 18-ft section of the fuselage upper lobe. The high stress concentration at the knife edge of the countersunk rivet holes, coupled with localised corrosion, resulted in the initiation of fatigue cracks. This was one of the numerous cases of fatigue failures that prompted a more comprehensive implementation of fracture-based design principles. As a result of these and other recent failures, there has been an increasing emphasis on the use of damage-tolerance design concepts, which include fracture strength and fatigue life as the design objectives.

This chapter aims to provide a description of advanced materials and structural design approaches to promote green technology and establish sustainable systems in the aerospace and aviation industry.

9.2 Structural Design Methodology

The aim of light weight structural design is to produce a structural part that has an ideal combination of strength and stiffness, usually using relatively low-density materials. The primary design methodology employed is new lightweight materials combined with design optimisation techniques.

9.2.1 Lightweight Structures and Materials

Lightweight aircraft structures require less energy in motion and thus have a reduced environmental impact. The aim of lightweight construction is to maintain or increase structural

Figure 9.1 Aloha Airlines flight 243 failure. Source: NTSB, Public domain, via Wikimedia Commons.

Figure 9.2 Examples of lightweight structures include (a) stiffened panels, (b) sandwich panels, and (c) cellular materials like balsa wood as shown.

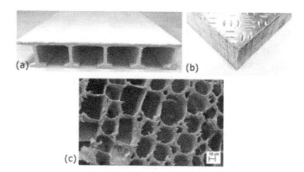

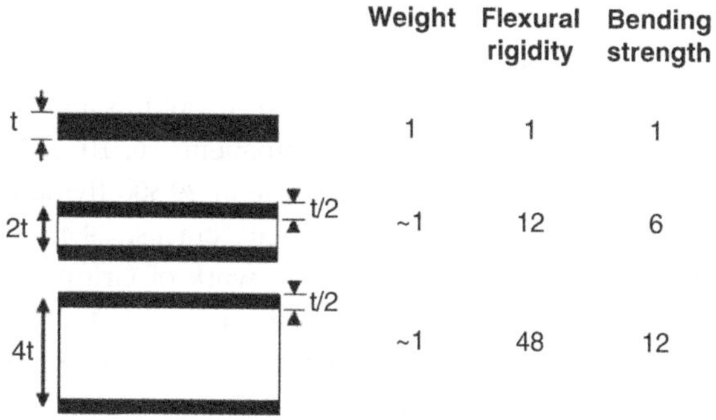

Figure 9.3 Examples of improvements to the flexural rigidity and bending strength of sandwich composites for negligible increases in weight. Adapted from [2].

	Weight	Flexural rigidity	Bending strength
t	1	1	1
2t, t/2	~1	12	6
4t, t/2	~1	48	12

functionality while decreasing overall mass. Conventional approaches for reducing structural mass in aircraft include the use of stiffened panels (Figure 9.1a), sandwich panels (Figure 9.1b), cellular structures, and hybrid structures. Less dense materials, such as cellular materials (Figure 9.1c), honeycombs and foams, are also used to achieve structural light-weighting. These materials are commonly used in sandwich panels as lightweight core materials, bonded between two thin, stiff, and strong face sheets (often made from an aluminium alloy or a composite material). Sandwich panels are a prime example of lightweight structures used in aircraft that can improve structural performance (flexural rigidity and bending strength) without significantly increasing overall mass (Figures 9.2 and 9.3). Traditional materials and new materials used to achieve aircraft light weighting are discussed in more detail in Sections 9.4 and 9.5, respectively.

9.2.2 Lightweight Structural Design

Globalisation has generated an urgent need to create and develop new products and processes, and to improve existing products. This has had a significant impact on the aerospace engineering design process and resulted in a substantial review of traditional design methods to meet the requirements of next generation aircraft. Traditional design methodologies are inherently iterative, trial-and-error processes that essentially use an engineer's creativity, ingenuity, and experience. The growing demands and complexity of the aerospace technology have initiated the development of more systematic and robust design processes, based on cutting-edge science and technology. Examples of these are the inclusion of optimisation techniques and damage-tolerance criteria into the design process. To aid lightweight structural designs, a range of high-performance computational tools coupled with optimisation methodologies of structures (topology and shape) are routinely employed [3–9].

The need for the structures to be stronger, lighter, safer, and more durable was first acknowledged in aerospace industries, although it is now recognised that other capital-intensive industries, like automotive, rail, marine, and mining, also have similar requirements in order to be commercially competitive in a global environment. To achieve such a design goal, it is necessary to implement an automated design cycle with provisions for continuous product improvement. Structural optimisation techniques provide a rational and systematic approach to accomplish design automations. Several commercial computer aided engineering (CAE) packages (e.g. GENESIS) with fully integrated CAD modelling, structural analysis, and optimisation capability have been developed to realise this.

Most optimisation techniques are limited to performance criteria based on weight, stress, displacement, frequency, etc. However, in many cases structures fail by fracture or fatigue. Hence damage-tolerance optimisation has been applied for the design of critical aerospace structural components [6, 10, 11]. A study on the repair of 71 Boeing 747 fuselages, with an average life of 29 500 flying hours concluded that 57.6% of these required repairs because of fatigue cracks, 29.4% for corrosion, and 13.0% for impact damage [12]. Based on the early work of Griffith [13] and subsequently of Irwin [14], it was shown that conventional strength-based design methods were inadequate for lightweight engineering structures, such as those used in aerospace applications, with small to large cracks. After World War II the increased use of high strength materials, to realise weight savings, also led to failure-prone structures. Structures designed using high strength materials frequently experienced service stress-induced cracking. Out of 2500 Liberty ships built during World War II, 145 broke in two, and almost 700 experienced serious failures [15]. It was also found that high strength materials often had a low fracture strength (low residual strength in the presence of cracks) and an increased susceptibility to cracking.

It was found that failures in aerospace structures could result from cracks initiated by pre-existing material flaws, or manufacturing defects. This led to the emergence of a new design concept based on engineering fracture mechanics. This design philosophy known as *damage-tolerance or durability-based design* is built around an understanding of structural behaviour in the presence of defects or cracks, thus compensating for the shortcomings inherent in the conventional design [16].

9.3   Damage Tolerant Structural Design

Aerospace structures are the most safety-critical amongst all types of engineering structures. In the damage-tolerance design philosophy, the structure is assumed to contain defects or flaws. Each potential 'hot spot' is assumed to contain initial flaws that can result from material or manufacturing defects [6, 11, 17]. These flaws may be present in any arbitrary location of the structure. This is the most important and practical assumption, since any structure, no matter how advanced its manufacturing process is, will inevitably contain flaws, inclusions or macro/micro-voids, or cracks may also develop during the structure's operational life. These flaws are allowed to grow to an acceptable size. The stress intensity factors corresponding to each crack configuration during crack growth are evaluated and are used to determine the crack growth rate and the residual strength of the structure. For satisfactory operation of a component, its residual strength should always be greater than the peak operational load (limit load). It is thought that if a component has small initial flaws and a large critical crack length, then the stress optimised shapes will be similar to the fracture strength optimised shapes. This is often the case for many aerospace applications. However, for some aerospace components and other structures (e.g., in marine, rail, and mining industries), initial flaws can be 5–15 mm deep and 10–40 mm long. In such cases, optimised shapes or topologies based on stress and residual strength can differ considerably [6, 18]. In order to develop lighter designs with increased damage tolerance, it is imperative that the designs be optimised for their fracture strength and fatigue life. The damage-tolerance design philosophy aims to achieve the following goals:

- An aircraft structure should perform safely under operational loads and constraints in the presence of flaws/defects.
- Materials with high strength and high crack propagation-resistance should be used.
- The residual strength of the structure must not fall below a given safe limit.
- Cracks must be detected before they reach the critical size.

In order to apply the concepts of fracture mechanics in the design of damage tolerant aerospace structures, the following information must be known [15]:

- Fracture toughness data for a wide range of materials, specimen geometries, environmental, and manufacturing conditions.
- The operational load spectra.
- Efficient methods for evaluating the stress intensity factors.
- A relationship between the crack growth rate (da/dN) and the variation in the effective stress intensity factor (ΔK) for a wide variety of materials, specimen thicknesses, and environmental conditions. This can either be obtained through testing or by using an appropriate fatigue crack growth model.

The USAF Damage Tolerance Design Handbook [16] outlines the detailed damage tolerance design requirements, and these are also presented in [19, 20]. In the damage-tolerance design process, a relationship is needed between the crack length and the service life. This relationship is shown schematically in Figure 9.4a. Based on the geometry, fracture toughness, and loading, a relation between crack size (a) and residual strength should also be

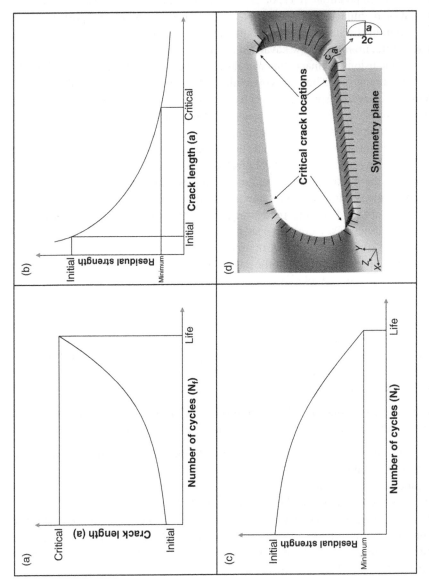

Figure 9.4 (a) Variation of crack length with service life (number of cycles), (b) relationship between residual strength and crack length, (c) relationship between residual strength and service life (number of cycles), and (d) inclusion of numerous cracks in the damage tolerance design optimisation [21].

established (Figure 9.4b). The combination of these two plots produces Figure 9.4c, showing the variation of the residual strength of the structure with service life. This indicates that as the operational life increases, the cracks grow, and the residual strength drops.

The critical crack size can be determined from the knowledge of the fracture toughness, and the relationship between the crack size and the residual strength (Figure 9.4b). However, a central point of the damage-tolerance design philosophy is that the residual strength should always exceed the limit load. Damage tolerance requires an inspection schedule such that the cracks are detected before they reach their critical size. As has been stated, cracks are assumed to exist throughout the structure and the initial flaw size is assumed to be equal to the minimum detectable crack size using the available non-destructive inspection methods. The time required for a crack to grow from the minimum detectable size to the critical length determines the maximum available inspection period. Within this interval, at least one inspection must be carried out at all fracture-critical locations, and cracks, if found, should be repaired. All these measures are required to ensure that the crack propagation does not cause fracture failure within the expected service life.

There are two major factors to be considered in implementing the damage tolerance criteria, inspection technique, and material properties. While establishing the inspection intervals, it should be noted that the capability of the non-destructive evaluation technique to detect smaller flaws is more effective in gaining inspection time compared to an increase in critical crack length. It has been shown that decreasing the minimum detectable crack length by half provides almost twice the detection period [15]. A longer inspection period can lead to a substantial reduction in the maintenance cost for an aircraft fleet.

Damage-tolerance optimisation is considered a growing and important field of research. The advances in optimisation methods, fracture mechanics, and high-performance computing have made this work realisable in practice and will continue to augment research in this area. When designing or optimising an aerospace structure with the aim of extracting the best performance, we need to remember that flaws will inevitably be present in some structures or may develop during a structure's operational life. It is necessary to take this fact into account and to explicitly include fracture strength and fatigue life as the design objectives. Unless complemented by durability-based design optimisation, the basic purpose of a conventional strength-based design may be defeated.

9.4 Traditional Materials for Light Weighting

Aluminium and its alloys typically have a relatively low density ($\sim$2700 kg m^{-3}), high ductility, as well as high electrical and thermal conductivities. The mechanical properties of aluminium can be improved by cold work and alloying. The most common alloying elements include copper, manganese, zinc, magnesium, and silicon. There are over 500 different commercial aluminium alloys which are separated into eight categories called series (1000, 2000, 3000, 4000, 5000, 6000, 7000, 8000). The 7000 and 2000 series alloys are the most common aluminium alloys used in aircraft. The 2000 series alloys are used in damage-tolerant

Table 9.1 Compositions, mechanical properties, and typical applications of the common aluminium alloys used in aircraft.

Aluminium Alloy	Composition[a]	Condition (Temper)	Yield strength (MPa)	Elongation in 50 mm (%)	Typical applications in aircraft
2024	4.4 Cu, 1.5 Mg, 0.6 Mn	Heat-treated (T4)	325	20	Fuselage structure, wing skins
7075	5.6 Zn, 2.5 Mg, 1.6 Cu, 0.23 Cr	Heat-treated (T6)	505	11	Spars, stringers, framework, pressure bulkheads

a) Remainder of the composition is aluminium [22].

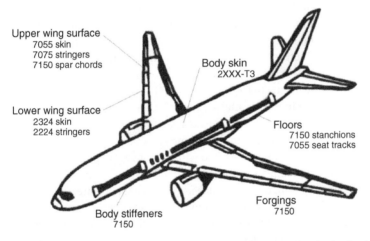

Upper wing surface
7055 skin
7075 stringers
7150 spar chords

Body skin
2XXX-T3

Lower wing surface
2324 skin
2224 stringers

Floors
7150 stanchions
7055 seat tracks

Body stiffeners
7150

Forgings
7150

Figure 9.5 Examples of aluminium alloy structural components in the Boeing 777 aircraft. Source: adapted from [23].

applications due to their high strength, fatigue resistance, and toughness. 7000 series alloys generally have higher strength than the 2000 series alloys and are used in highly stressed aircraft structures. The properties of the two most common aluminium alloys used in aircraft are described in Table 9.1.

Aluminium alloys have been important materials in the construction of lightweight airframes for aircraft since the 1930s. Aluminium alloys used in airframe components like the fuselage and wings have helped enable modern flight. Compared to other structural materials, aluminium is still the most widely used material in aircraft and is likely to remain an important structural material, even with the growing use of composites in new aircraft. Many new aircraft continue to be constructed primarily of aluminium, including large long-range passenger aircraft, such as the Boeing 777 (Figure 9.5). Aluminium alloys will likely remain critical to lightweight airframe construction for the foreseeable future.

9.5 New Materials for Light Weighting

Composite materials are man-made or naturally occurring materials prepared from two or more materials with considerably different physical or chemical properties that remain distinct and separate within the resultant material. Composite materials have been widely used in different fields for many decades and have become primary components for various aerospace applications. From World War II, composites have gained in industrial exposure for aircraft [24] and since then have formed an integral part of the aerospace industry [25–30], because of their high strength-to-weight ratio [25, 31], stiffness-to-weight ratio [32–34], high fatigue resistance [35–38], and corrosion resistance [39–41]. However, their performance under impact loading is often poorer compared to metal alloys [25]. This led to the development of fibre-reinforced composites, in the early 1950s, in order to increase the fracture toughness of the aircraft structures [42]. Composite materials are designed to provide the ideal properties for many aerospace applications and are replacing metallic components. Two kinds of composites, namely polymer composites and hybrid composites, are commonly used in aerospace applications.

9.5.1 Polymer Composites

High performance polymer composites are one of the major modern material developments and these materials are now reaching maturity. Polymer composite materials consist of a dispersed reinforcement phase (e.g. glass fibres or carbon fibres) set in a continuous polymer matrix phase (e.g. epoxy resin or polyester resin). The dispersed reinforcement phase provides mainly stiffness and strength, but also can render other mechanical and functional properties, such as creep, wear, fracture toughness, fatigue, etc. The matrix phase protects the reinforcement phase and transfer applied loads. The reinforcement phase can be discontinuous (e.g. particles or whiskers) or continuous fibres, as shown in Figure 9.6. Two types of fibre reinforcements are commonly used, synthetic fibres and natural fibres, for fibre-reinforced polymer (FRP) composites. The synthetic reinforcement fibres are usually carbon, glass, or aramid, and the natural fibres are flax, jute, hemp, bamboo, and wood among others. Conventional aerospace composites usually have glass or carbon fibres based on a thermosetting or thermoplastic polymer matrix [43]. There are supplementary classes of hybrid composites in which the matrix is a metal or a ceramic, such as fibre metal laminates (FMLs).

Polymer composites reinforced with long, continuous fibres typically exhibit the highest mechanical properties. A sheet of long fibre-reinforced material is anisotropic; i.e. its mechanical properties depend on the orientation of the fibres. Therefore, in structures two or more plies (laminae) are stacked together to form a composite laminate (Figure 9.7), which enables the alignment of fibres to the direction of loading. Carbon-fibre reinforced

Figure 9.6 Polymer composites can be classified by the architecture of their reinforcement phase, which can be (a) particulate, (b) whisker (or short fibre), or (c) long, continuous fibre.

(a)   (b)   (c)

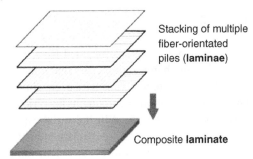

Stacking of multiple fiber-orientated piles (**laminae**)

Figure 9.7 Schematic of a composite laminate formed from stacked plies (laminae).

Composite **laminate**

Table 9.2 Properties of typical matrix materials used in polymer composites compared to aluminium alloy (2024-T4) [22].

Property (units)	Quasi-isotropic carbon fibre-epoxy laminate	Quasi-isotropic glass fibre-epoxy laminate	Aluminium alloy (2024-T4)
Density (kg m^{-3})	1600	1900	2800
Price estimate (USD kg^{-1})	40	25	3
Young's modulus (GPa)	55	20	75
Tensile strength (MPa)	670	480	400
Elongation (% strain)	0.34	2.0	11
Fracture toughness (MPa.m$^{0.5}$)	16	2.3	39
Fatigue endurance limit (% of ultimate strength)	70–80	50–70	20–35
Maximum service temperature (°C)	175	175	184
Thermal conductivity (W/m °C)	1.8	0.5	121
Thermal expansion coefficient (μstrain/°C)	1.2	9.0	23

epoxy laminates are the most common type of polymer composite used in aircraft structural components.

The properties of common aerospace polymer composites are compared with a common aerospace aluminium alloy in Table 9.2. FRP composites are often selected over aluminium in aircraft structural applications due to their high specific mechanical properties (mechanical properties divided by density).

FRP composites, especially carbon fibre laminates have several advantages compared with aluminium alloys including:

- High stiffness and strength (for a given weight).
- Properties can be tailored by choice of matrix, reinforcement, and its orientation, and stacking sequence.
- Good corrosion resistance and fatigue properties.
- Toughness (due to the creation of multiple crack paths).
- Heat insulation and low coefficient of thermal expansion.
- Radar absorption properties.

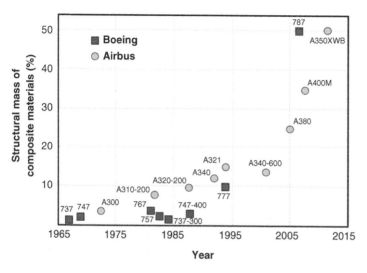

Figure 9.8 Percentage of structural mass of composite materials in selected Boeing and Airbus aircraft over the past 50 years.

On the other hand, FRP composites have several notable disadvantages compared to aluminium alloys that can limit their application. These broadly include:

- High raw material costs.
- Slow and costly manufacturing processes.
- Low strength, damage tolerance, and toughness in certain directions.
- High temperature instability.
- Directionality requires carefully detailed design. Low electrical conductivity.

Polymer composites have been used in passenger aircraft since the 1960s (Figure 9.8), with the first applications being in components such as undercarriage doors and fairings. Prior to passenger aircraft, FRP composites were used as lightweight load-bearing structures in military aircraft. The horizontal stabiliser of the Boeing 737-300 was the first primary load-bearing composite structure in a passenger aircraft and was made from a carbon fibre-reinforced epoxy laminate. With improvements to materials and processing methods, as well as improved practices in engineering design, the use of polymer composites in airframe structures developed gradually during the 1990s and accelerated at the turn of the century, as shown in Figure 9.8. A significant increase in the use of polymer composites occurred with the Airbus A380, which is built from around 25% composite materials. More recently, polymer composites have been used extensively in the construction of the Boeing 787 and the Airbus A350XWB, which are both built from around 50% composite materials. Carbon FRP composites and sandwich panels utilising these materials are now used in many major structural components of these and other new passenger aircraft.

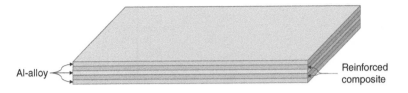

Al-alloy — Reinforced composite

Figure 9.9 A typical fibre metal laminate (FML).

9.5.2 Hybrid Composites – Fibre Metal Laminates (FMLs)

Hybrid composites belong to a class of composite materials that consist of two or more different categories of reinforcements or matrices. FMLs are a new kind of hybrid composite, combining suitable properties of both fibre reinforced composites and metal alloys, developed over the past three decades [44]. FMLs have been effectively used for various applications, particularly in aeronautical and space structures. The concept of this category of composite is to add layers of metals within layers of FRPs or pure polymers to improve their mechanical properties [27, 28], while maintaining their light weight. Thus, FMLs are composed of alternative layers of metal and fibre reinforced composites [45] and as a result belong to the family of hybrid composites [46–48] (Figure 9.9).

FMLs have several advantages over conventional fibre reinforced composites, such as:

- Sustainability and durability of a structure under fatigue loading.
- Superior corrosion and impact resistance.
- Higher damage tolerance and lower density of composites [24, 49].
- The excellent fatigue resistance is attributed to the fibre bridging property [50], and the exceptional resistance to corrosion is due to the action of the composite prepregs as a barrier to moisture in-between the layers of aluminium [51].

Several categories of FML composites have been designed and manufactured to tailor specific properties and meet desired performance criteria. The most widely used FMLs are GLARE, ARALL, and CARALL, which use glass, aramid, and carbon fibres, respectively, reinforced in a polymer matrix along with aluminium (most commonly) as the metal layer.

FMLs can be manufactured with metal layers made of aluminium, magnesium, or titanium and FRP layers consisting of glass, carbon, or aramid fibres. As a result, FMLs exhibit mechanical, machining, and forming properties of the metals along with superior fatigue performances of FRPs [52]. Different types of FMLs along with their classification are shown in Figure 9.10. The first ever type of FML were called ARALLs (aramid reinforced aluminium laminates) developed in the early 1980s and used for aircraft. This was followed by the development of GLARE (glass laminate aluminium reinforced epoxy) [53, 54]. The other commonly used FML is called CARALL (carbon reinforced aluminium laminate), which is based on aramid, glass, and carbon fibres [24, 55]. Different types of FMLs, such as those based on different fibres (glass, aramid, or carbon fibres), polymers (thermoset, thermoplastic, or elastomer matrices), and metals (aluminium, titanium, or magnesium) are referred to in Figure 9.10 [53].

ARALL was first developed as the aircraft wing material and primarily consisted of 0.2–0.4 mm thick Al 7075-T6/2024-T3 plate and bi/uni-directional prepregs of aramid

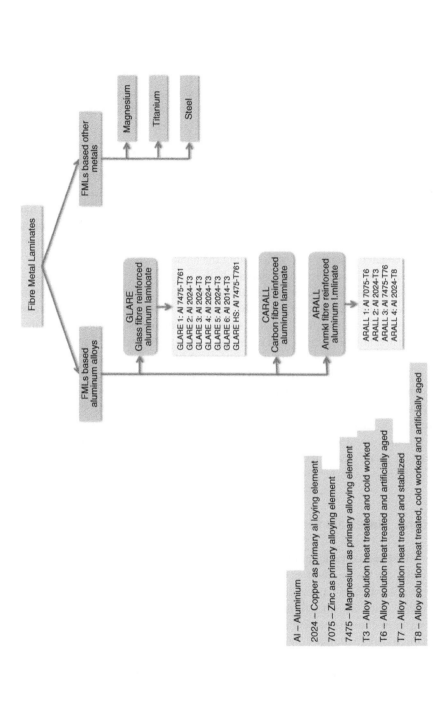

Figure 9.10 Typical classes of FMLs where 'Al xxxx' defines the major aluminium alloying constituent and 'Txxx' demonstrates the type of heat treatment and tempering performed to the alloy solution [53].

fibres [53]. These FMLs were found to provide around a 20% weight saving when compared to monolithic aluminium and were first applied as a material for cargo doors [52].

In spite of having many advantages, the aramid fibres have inadequate compression properties and this led to the development of the second-generation FMLs, called the GLARE, in the late 1980s [56], used in the fuselage of aircraft [57]. It consisted primarily of layers of S-2 glass/epoxy composites and Al 7475-T761/2024-T3. Further research also led to the production of GLARE with different standard grades for various aerospace related applications [58], one of the most recent ones being the Airbus A380's upper fuselage, leading to a weight saving of about 794 kg [49]. Another type of FML, namely CARALL, which is based on carbon fibre-reinforced polymers combined with an aluminium alloy was also developed [59]. The applications of CARALL are, however, quite limited because of the galvanic corrosion experienced between the layers of aluminium alloy and carbon, which may result in a reduction of strength. Further, the dissimilarity in the coefficients of thermal expansion of the carbon FRP and the alloy results in the residual thermal stresses [59]. The GLARE has better impact tolerance than ARALL, because glass-epoxy prepregs are stronger and more ductile than the aramid epoxy ones. The GLARE also possesses smaller damage sizes (the smallest circle diameter that can be drawn around the damaged area), when compared to both ARALL and CARALL under impact loading [60].

FMLs show significantly superior properties under impact when compared to monolithic metals of the same areal densities. One of the most vital reasons behind the failure of aircraft structures is the impact phenomenon, among other damages, such as damage due to inefficient repairs, fatigue damage, and corrosion damage [60]. Indeed, impact damage was observed in about 13% of repairs to the fuselages of 71 Boeing 747 aircraft [50], and most commonly impact damage was observed in and around aircraft doors, on their noses, at their tails and in cargo compartments of aircrafts [61]. Although different sources, such as debris in the runway, maintenance damage, hail, collisions between the structure and the cargo or the service cars, bird strikes, engine debris, and others [61], are responsible for impact damages, there are still specific impact regimes for aircraft and these need to be considered for a safe design [60]. Consequently, FMLs have been increasingly used in structural applications over the past two decades, especially in the aeronautical industry.

9.6 Natural Materials for Aerospace Applications

With a strong emphasis on environmental awareness, the development of biodegradable composite materials has received considerable attention in the past two decades. The growing awareness of sustainability and environmental issues have led to an increasing interest in green materials, specifically in relation to composites. The eco-friendly way to degrade waste material is biodegradation. Biodegradability traditionally describes the process of natural degradation of compounds present in a material by micro-organisms in the environment. Recyclability refers to the ability to re-use the product or raw material after the lifecycle of an original product [62]. There is growing interest in bio-composites particularly in the context of sustainability. They can be defined as composites that have at least one phase that is derived from biological or natural sources. Shells (e.g. abalone) are good examples of bio-composites and are formed from living organisms that synthesise layered composite materials, exhibiting high mechanical properties. The main advantages of bio-composites include biodegradability, renewability, and often relatively low cost.

Significant research and development efforts are underway to incorporate biodegradable materials into novel composites. These blend traditional advantages of composites (e.g. lightweight, high specific mechanical, and functional properties) with biodegradability and sustainability. When fibre reinforced composites are used in aircraft as a replacement for metals, the ecological impact in the use phase will be reduced due to less energy consumption resulting from lower structural weight. The impact on the environment in the disposal phase of the composites, however, has highlighted the hazardous nature of some of the composite constituents, leading to an adverse impact on the environment. To address this, biodegradable polymers have a vital role in realising either partially or fully biodegradable composite materials [1]. Cost analysis and of the additional manufacturing cost against gains in performance of biodegradable polyesters [63] concluded that the cost of biodegradable polymers (hence the composites) is higher than non-biodegradable ones; however, the difference in price has significantly narrowed over the past 10 years [64]. Biodegradable polyesters are thus commonly used biopolymers, and their composites have been progressively researched over the past few years as a potential aerospace material.

For these materials, the manufacturing phases include diminution of raw materials (mining) and manufacture (machining) of products. The consumption phase comprises use of energy for the maintenance of the structure. The discarding phase contains incineration, degradation, and recycling. To this end, the usage of natural fibre-reinforced thermoplastics [65] reduces the problem associated with the discarding reinforcement phase because of their environmentally-friendly properties.

Bio-composites, based on these attributes, are environmentally friendly. However, like natural fibre composites (NFCs), their commercial success depends on their ability to meet service requirements, which can be particularly demanding in aircraft applications. To this end, there are several significant issues preventing their widespread adoption in aircraft, including high moisture absorption, high viscosity of bio-resins that makes component processing difficult, and poor wettability of fibres leading to poor adhesion with many matrix materials. The two main potential biodegradable constituents in a composite are fibres and matrix, with either of them potentially biodegradable.

9.6.1 Natural Fibre Composites

Natural fibres have been used by humans for thousands of years and are now receiving increased attention for use as a reinforcement in polymer composites due to their potential to create sustainable materials. Changes in regulations and attitudes have pushed the development of NFCs and have drawn industry attention. Some examples of natural fibres and their properties are shown in Table 9.3. The properties of natural fibre composites vary with fibre type, fibre source, fibre micro-structure, and the level of adhesion between the fibre and matrix phase.

Natural fibre composites combine biodegradable natural fibre reinforcements from a renewable source with a polymer matrix. Replacing synthetic fibres such as carbon or glass by plant fibres implicates the inherent advantages of renewability, biodegradability, and CO_2 neutrality combined with good specific mechanical performance. Besides

Table 9.3 Properties of typical natural fibres used as reinforcements in composites [22].

Fibre	Density (kg m^{-3})	Young's modulus (GPa)	Tensile strength (MPa)	Elongation (% strain)
Flax	1500	47	840	1.9
Hemp	1500	62	700	1.8
Jute	1500	31	555	1.8
Sisal	1400	15	660	2.4
Cotton	1500	9.1	490	7.4

technical criteria, economic reasons are of critical importance for today's industry. The fact that natural fibres come from a low-cost biomass resource provides a key benefit for bio-composites. Research has already shown that NFCs are environmentally superior to glass fibre reinforced composites [66].

Despite their potential, there are several significant disadvantages of natural fibre composites that limit their widespread adoption in aircraft. These disadvantages include high water absorption, flammability, and relatively low and highly variable mechanical properties. However, chemical treatment of the natural fibres can be used to improve adhesion between the fibres and the polymer matrix, which can enhance the mechanical and thermal properties of the composite.

9.6.2 Bio-polymer Composites

The biopolymer industry has a promising future, driven mainly by the ecological benefits of using renewable resources. A wide range of biodegradable polyester-based composites are available in the new era of novel materials. The polyester-based composites are mainly manufactured by reinforcing with glass or carbon or natural fibres using several techniques, such as compression moulding and injection moulding methods. The properties, such as strength and stiffness, are improved. Although these composites have been initially used in biomedical and civil engineering applications, they are rapidly finding their place in aerospace components as an attractive group of bio-composites.

The use of natural fibres as a reinforcement has increased in the last two decades, due to their easy availability and degradable property. Polyester composites reinforced with natural fibres are manufactured by the ply stacking method followed by compression moulding or injection moulding techniques. Polymer-polymer composites based on polyesters are commonly manufactured by pultrusion, compression moulding, and injection moulding methods. The mechanical properties of polyester composites reinforced with natural fibres, such as tensile strength, impact strength, and stiffness, are usually satisfactory and suitable for a range of applications in aerospace, including several non-critical or low load-bearing structural components used in aircraft.

The aliphatic polyesters have improved biodegradability properties when compared to other polymers. PLA, PGA, PCL, and PBS are extensively used polyesters to fabricate composite materials. Polyester-based biodegradable composites offer ease of handling, dimensional stability, and chemical resistance, as well as good mechanical and functional properties. The rate of biodegradation, mechanical properties, and piezoelectric properties are the key properties required for aerospace applications.

The focus of future research and development in biodegradable composite materials will be to develop materials with optimum mechanical and other functional properties, in conjunction with a high level of biodegradability. This is highlighted by the current extensive use of biodegradable composite materials for a wide range of applications, including medical, automotive, industrial, construction, and food packaging applications. From the eco-friendly perspective, the significance and utility of biodegradable composite materials cannot be overstated. In this context, polyester-based biodegradable composites will have an important role in future aerospace material developments.

9.7 Summary and Outlook

Lightweight structures and advanced materials have played an important role in the realisation of modern aircraft and will be critical in the future of sustainable aviation. Structural optimisation is used extensively to design lightweight materials and structures.

The application of damage-tolerance optimisation in the design and maintenance of structures can dramatically reduce the loss of structural integrity and the likelihood of fracture or fatigue failures. This will allow the operation of these structures more effectively and efficiently with a greater durability and an increased operational life.

Competition between lightweight aluminium alloys and FRP composites for use in primary aircraft structures has been increasing in recent years, which is likely to continue over the coming decades. There are a number of issues preventing more widespread adoption of polymer composites, including cost and recyclability [67]. However, such emerging technologies as thin-ply composite laminates [68], and multifunctional composite structures for electrical energy storage [69] are likely to play an important role in future sustainable aircraft. Multifunctional composites can cater for new properties, such as fire resistance and blast/impact resistance of structural parts [70–73]. New processing techniques like microwave processing [74] are also being investigated to address slow and costly manufacturing processes associated with polymer composites. Furthermore, natural fibre composites and bio-polymer composites are promising candidates for future use in sustainable aircraft design, but these materials are yet to reach their full potential. The sustainability of critical elements [75] used in aircraft design is also thought be an important factor in the future of sustainable aviation.

References

1 Das, R. and Karumbaiah, K.M. (2015). Biodegradable polyester-based blends and composites: manufacturing, properties and applications. In: *Biodegradable Polyesters* (ed. S. Fakirov), 321–340. Wiley-VCH.

2 Zenkert, D. (1997). *An introduction to sandwich construction*, 1e. London: Emas Publishing.

3 Vanderplaats, G.N. (1984). *Numerical Optimization Techniques for Engineering Design: With Application*. New York, NY: McGraw-Hill.

4 Arora, J.S. (1989). *Introduction to Optimum Design*. New York, NY: McGraw-Hill.

5 Das, R., Jones, R., and Xie, Y.-M. (2011). Optimal topology design of industrial structures using an evolutionary algorithm. *Optimization and Engineering* 12 (4): 681–717.

6 Das, R., Jones, R., and Peng, D. (2006). Optimisation of damage tolerant structures using a 3D biological algorithm. *Engineering Failure Analysis* 13 (3): 362–379.

7 Das, R., Jones, R., and Xie, Y.M. (2005). Design of structures for optimal static strength using ESO. *Engineering Failure Analysis* 12 (1): 61–80.

8 Praticò, L., Galos, J., Cestino, E. et al. (2020). Experimental and numerical vibration analysis of plates with curvilinear sub-stiffeners. *Engineering Structures* 209: 109956.

9 Galos, J. and Sutcliffe, M. (2019). Material selection and structural optimization for lightweight truck trailer design. *SAE International Journal of Commercial Vehicles* 12 (4): 281–297.

10 Das, R. and Jones, R. (2009). Development of a 3D biological method for fatigue life based optimisation and its application to structural shape design. *International Journal of Fatigue* 31 (2): 309–321.

11 Das, R. and Jones, R. (2009). Fatigue life enhancement of structures using shape optimisation. *Theoretical and Applied Fracture Mechanics* 52 (3): 165–179.

12 Vogelesang, L.B. and Vlot, A. (2000). Development of fibre metal laminates for advanced aerospace structures. *Journal of Materials Processing Technology* 103 (1): 1–5.

13 Griffith, A.A. (1924)). The theory of rupture. In: *Proc. 1st Int. Congress Appl. Mech.* (eds. C.B. Biezeno and J.M. Burgers). Waltman.

14 Irwin, G.R. (1958). *Handbuch der Physik VI* (ed. Flugge). Springer.

15 Broek, D. (1986). *Elementary Engineering Fracture Mechanics*. 4th rev. Dordrecht: Kluwer Academic.

16 Gallagher, J.P., Giessler, F.J., Berens, A.P., and Engle, R.M. (1984). *USAF Damage Tolerance Design Handbook: Guidelines for the Analysis and Design of Damage Tolerant Aircraft Structures*. Dayton, OH: University of Dayton Research Institute.

17 Das, R. and Jones, R. (2009). Designing cutouts for optimum residual strength in plane structural elements. *International Journal of Fracture* 156 (2): 129–153.

18 Jones, R., Chaperon, P., and Heller, M. (2002). Structural optimisation with fracture strength constraints. *Engineering Fracture Mechanics* 69: 1403–1423.

19 Coffin, M.D. and Tiffany, C.F. (1976). New air force requirements for structural safety, durability and life management. *Journal of Aircraft* 13 (2): 93–98.

20 King, T.T., Hurchalla, J., and Nethaway, D.H. (1985). United States air Force engine damage tolerance requirements. In: *AIAA Paper*, 11. Monterey, New York: AIAA.

21 Das, R. and Jones, R. (2009). Damage tolerance based design optimisation of a fuel flow vent hole in an aircraft structure. *Structural and Multidisciplinary Optimization* 38 (3): 245–265.

22 CES Edupack. (2019). Granta Design Ltd., Cambridge, UK.

23 Starke, E.A. and Staley, J.T. (1996). Application of modern aluminum alloys to aircraft. *Progress in Aerospace Sciences* 32 (2): 131–172.

24 Sinmazçelik, T., Avcu, E., Bora, M.Ö., and Çoban, O. (2011). A review: fibre metal laminates, background, bonding types and applied test methods. *Materials & Design* 32 (7): 3671–3685.

25 Mangalgiri, P.D. (1999). Composite materials for aerospace applications. *Bulletin of Materials Science* 22 (3): 657–664.

26 Alderliesten, R., Rans, C., and Benedictus, R. (2008). The applicability of magnesium based fibre metal laminates in aerospace structures. *Composites Science and Technology* 68 (14): 2983–2993.

27 Cantwell, W. and Morton, J. (1991). The impact resistance of composite materials—a review. *Composites* 22 (5): 347–362.

28 Cortes, P., Cantwell, W., Kuang, K., and Quek, S. (2008). The morphing properties of a smart fiber metal laminate. *Polymer Composites* 29 (11): 1263–1268.

29 Laliberté, J., Straznicky, P.V., and Poon, C. (2002). Numerical modelling of low velocity impact damage in fibre-metal laminates. *International Council of the Aeronautical Sciences 2002 Congress*, Toronto, ON, September, 2002

30 Beumler, T., Pellenkoft, F., Tillich, A. et al. (2006). Airbus costumer benefit from fiber metal laminates. *Airbus Deutschland GmbH* 1: 1–18.

31 Gibson, R.F. (2010). A review of recent research on mechanics of multifunctional composite materials and structures. *Composite Structures* 92 (12): 2793–2810.

32 de Moura, M.F.S.F. and Marques, A.T. (2002). Prediction of low velocity impact damage in carbon–epoxy laminates. *Composites Part A: Applied Science and Manufacturing* 33 (3): 361–368.

33 Koo, J.-M., Choi, J.-H., and Seok, C.-S. (2014). Prediction of post-impact residual strength and fatigue characteristics after impact of CFRP composite structures. *Composites Part B: Engineering* 61: 300–306.

34 Davey, S., Das, R., Cantwell, W., and Kalyanasundaram, S. (2013). Forming studies of carbon fibre composite sheets in dome forming processes. *Composite Structures* 97: 310–316.

35 Mouritz, A.P., Bannister, M.K., Falzon, P.J., and Leong, K.H. (1999). Review of applications for advanced three-dimensional fibre textile composites. *Composites Part A: Applied Science and Manufacturing* 30 (12): 1445–1461.

36 Ma, Y.E. and Hu, H. 2013). Experimental and numerical investigation of fibre-metal laminates during low-velocity impact loading. *13th International Conference on Fracture*, Beijing, China (16–21 June 2013).

37 Mouritz, A.P. and Cox, B.N. (2000). A mechanistic approach to the properties of stitched laminates. *Composites Part A: Applied Science and Manufacturing* 31 (1): 1–27.

38 Mouritz, A.P., Leong, K.H., and Herszberg, I. (1997). A review of the effect of stitching on the in-plane mechanical properties of fibre-reinforced polymer composites. *Composites Part A: Applied Science and Manufacturing* 28 (12): 979–991.

39 Bieniaś, J., Dębski, H., Surowska, B., and Sadowski, T. (2012). Analysis of microstructure damage in carbon/epoxy composites using FEM. *Computational Materials Science* 64: 168–172.

40 Chung, D. (2012). *Carbon Fiber Composites*. Butterworth-Heinemann.

41 Morgan, P. (2005). *Carbon Fibers and their Composites*. CRC press.

42 Asundi, A. and Choi, A.Y.N. (1997). Fiber metal laminates: an advanced material for future aircraft. *Journal of Materials Processing Technology* 63 (1–3): 384–394.

43 Han, K.S. (1983). Compressive fatigue behaviour of a glass fibre-reinforced polyester composite at 300 K and 77 K. *Composites* 14 (2): 145–150.

44 Das, R., Chanda, A., Brechou, J., and Banerjee, A. (2016). 17 - impact behaviour of fibre–metal laminates. In: *Dynamic Deformation, Damage and Fracture in Composite Materials and Structures* (ed. V.V. Silberschmidt), 491–542. Woodhead Publishing.

45 Krishnakumar, S. (1994). Fiber metal laminates—the synthesis of metals and composites. *Material and Manufacturing process* 9 (2): 295–354.

46 Laliberté, J., Poon, C., and Straznicky, P.V. (2002). Numerical modelling of low-velocity impact damage in fibre-metal-laminates. *23rd International Congress of Aeronautical Sciences*, Toronto, Canada (8–13 September, 2002).

47 Alderliesten, R.C. (2005). *Fatigue Crack Propagation and Delamination Growth in Glare*. TU Delft: Delft University of Technology.

48 Marissen, R. (1988). Fatigue crack growth in ARALL: a hybrid aluminium-aramid composite material: crack growth mechanisms and quantitative predictions of the crack growth rates. PhD thesis. Delft University of Technology.

49 Wu, G. and Yang, J.-M. (2005). The mechanical behavior of GLARE laminates for aircraft structures. *JOM* 57 (1): 72–79.

50 Vlot, A. (1993). Impact properties of fibre metal laminates. *Composites Engineering* 3 (10): 911–927.

51 Vlot, A. (1996). Impact loading on fibre metal laminates. *International Journal of Impact Engineering* 18 (3): 291–307.

52 Straznicky, P.V., Laliberté, J.F., Poon, C., and Fahr, A. (2000). Applications of fiber-metal laminates. *Polymer Composites* 21 (4): 558–567.

53 Chai, G.B. and Manikandan, P. (2014). Low velocity impact response of fibre-metal laminates – a review. *Composite Structures* 107: 363–381.

54 Vlot, A. and Gunnink, J.W. (2011). *Fibre Metal Laminates: An Introduction*. Springer Science & Business Media.

55 Afaghi-Khatibi, A., Lawcock, G., Ye, L., and Mai, Y.-W. (2000). On the fracture mechanical behaviour of fibre reinforced metal laminates (FRMLs). *Computer Methods in Applied Mechanics and Engineering* 185 (2–4): 173–190.

56 Vlot, A., Vogelesang, L., and De Vries, T. (1999). Towards application of fibre metal laminates in large aircraft. *Aircraft Engineering and Aerospace Technology* 71 (6): 558–570.

57 Schijve, J. (1994). Fatigue of aircraft materials and structures. *International Journal of Fatigue* 16 (1): 21–32.

58 da Costa, A.A., da Silva, D.F., Travessa, D.N., and Botelho, E.C. (2012). The effect of thermal cycles on the mechanical properties of fiber–metal laminates. *Materials & Design* 42: 434–440.

59 Wang, W.-X., Takao, Y., and Matsubara, T. (2007). Galvanic corrosion-resistant carbon fiber metal laminates. *16th International Conference on Composite Materials*, Kyoto, Japan (8–13 July, 2007).

60 Sadighi, M., Alderliesten, R.C., and Benedictus, R. (2012). Impact resistance of fiber-metal laminates: a review. *International Journal of Impact Engineering* 49: 77–90.

61 Vlot, A. (1993). Low-velocity impact loading: on fibre reinforced aluminium laminates (ARALL and GLARE) and other aircraft sheet materials. PhD thesis. Delft University of Technology.

62 Oliveira, R.C., Gomez, J.G.C., Torres, B.B. et al. (2000). A suitable procedure to choose antimicrobials as controlling agents in fermentations performed by bacteria. *Brazilian Journal of Microbiology* 31: 87–89.

63 Fombuena, V., Bernardi, L., Fenollar, O. et al. (2014). Characterization of green composites from biobased epoxy matrices and bio-fillers derived from seashell wastes. *Materials & Design* 57: 168–174.

64 Ojijo, V. and Ray, S.S. (2014). Nano-biocomposites based on synthetic aliphatic polyesters and nanoclay. *Progress in Materials Science* 62: 1–57.

65 Mohanty, A.K., Khan, M.A., and Hinrichsen, G. (2000). Influence of chemical surface modification on the properties of biodegradable jute fabrics - polyester amide composites. *Composites Part a-Applied Science and Manufacturing* 31 (2): 143–150.

66 Joshi, S.V., Drzal, L.T., Mohanty, A.K., and Arora, S. (2004). Are natural fiber composites environmentally superior to glass fiber reinforced composites? *Composites Part A: Applied Science and Manufacturing* 35 (3): 371–376.

67 Yang, Y., Boom, R., Irion, B. et al. (2012). Recycling of composite materials. *Chemical Engineering and Processing: Process Intensification* 51: 53–68.

68 Galos, J. (2020). Thin-ply composite laminates: a review. *Composite Structures* 236: 111920.

69 Galos, J., Best, A.S., and Mouritz, A.P. (2020). Multifunctional sandwich composites containing embedded lithium-ion polymer batteries under bending loads. *Materials & Design* 185: 108228.

70 Dutta, S., Das, R., and Bhattacharyya, D. (2019). A multi-physics framework model towards coupled fire-structure interaction for flax/PP composite beams. *Composites Part B: Engineering* 157: 207–218.

71 Dutta, S., Kim, N.K., Das, R., and Bhattacharyya, D. (2019). Effects of sample orientation on the fire reaction properties of natural fibre composites. *Composites Part B: Engineering* 157: 195–206.

72 Gargano, A., Das, R., and Mouritz, A.P. (2019). Finite element modelling of the explosive blast response of carbon fibre-polymer laminates. *Composites Part B: Engineering* 177: 107412.

73 Gargano, A., Pingkarawat, K., Pickerd, V. et al. (2018). Effect of seawater immersion on the explosive blast response of a carbon fibre-polymer laminate. *Composites Part A: Applied Science and Manufacturing* 109: 382–391.

74 Galos, J. (in press). Microwave processing of carbon fibre polymer composites: a review. *Polymers and Polymer Composites*. 0967391120903894.

75 Ashby, M.F. (2016). Chapter 1 - background: materials, energy and sustainability. In: *Materials and Sustainable Development* (ed. M.F. Ashby), 1–25. Boston: Butterworth-Heinemann.

10

Low-Emission Propulsive Technologies in Transport Aircraft

Kavindu Ranasinghe[1], Kai Guan[2], Alessandro Gardi[3,4], and Roberto Sabatini[3,4]

[1] Insitec Pty Ltd, Melbourne, Victoria, Australia
[2] RMIT University, School of Engineering, Bundoora, Victoria, Australia
[3] Department of Aerospace Engineering, Khalifa University of Science and Technology, Abu Dhabi, UAE
[4] School of Engineering, RMIT University, Melbourne, Victoria, Australia

10.1 Introduction

Engines are the main source of environmental emissions from aircraft, thereby attracting the attention of researchers worldwide to explore the possible improvements to current technology, since they have a strong potential for moderating these adverse environmental impacts. At present, most airline transport aircraft are equipped with turbofan engines, which have been in widespread use for decades and have evolved considerably in terms of their efficiency and stability.

In recent years, many technological advances have been proposed to improve the efficiency and environmental sustainability of turbofan engines. Due to the physics involved, some of these improvements resulted in trade-offs between the various emissions and other factors, therefore it is essential to formulate a multidisciplinary and multi-objective design optimisation problem. Aerospace industry experts have predicted that by the year 2050, advances in technology will lead to a 40–50% improvement in average fuel efficiency with fuel efficiency taking priority and a 30–40% improvement in fuel efficiency with NO_X reduction taking priority [1].

There are nonetheless several recognised challenges associated with every new technology, with various direct and indirect implications to the design, manufacturing, operation, and maintenance phases. This chapter, originally published in [64, 65], reviews some of the most promising technological advances and outlines the main benefits and challenges identified in the literature to assess their overall viability to advance the turbofan engine technology.

10.2 Turbofan Emissions in Aviation

Turbofan engines generate a variety of gaseous and particulate material in their exhaust during normal operating conditions due to the combustion of fuel. The exhaust comprises about 70% carbon dioxide CO_2 and about 30% water vapour (H_2O) which

Sustainable Aviation Technology and Operations: Research and Innovation Perspectives, First Edition.
Edited by Roberto Sabatini and Alessandro Gardi.
© 2024 John Wiley & Sons Ltd. Published 2024 by John Wiley & Sons Ltd.
Companion Website: www.wiley.com/go/sustainableaviation

are unwanted emissions but are unfortunately unavoidable as long as fossil fuels are used for propulsion [2]. However, the total quantity of exhaust emissions can be reduced by improving the overall efficiency of the aircraft, including improvements to the thermodynamic cycle of the engine as well as aerodynamic efficiency. The rest of the constituents comprise less than 1% of the exhaust and include nitrogen oxides (NO_X), oxides of sulphur (SO_X), carbon monoxide (CO), partially combusted or unburned hydrocarbons (HC), particulate matter (PM), and other trace compounds [3].

Nitrogen oxides (NO_X) are unwanted by-products of combustion that are produced when nitrogen and oxygen combine when air is subjected to extremely high pressure and temperature conditions during the combustion process. Four main types of mechanisms for the formation of nitrogen oxides have been identified in the literature [4], however thermal NO_X was found to be the most dominant form. Sulphur oxides (SO_X) are formed when small quantities of sulphur, inherently present in all petroleum fuel, combine with oxygen from the air during combustion. The formation of sulphur oxides also contributes to the secondary PM formation. Hydrocarbons (HC) are formed as the result of a poor fuel-to-air (FAR) ratio and incomplete combustion. This occurs at low temperature and pressure regions and often leads to the formation of carbon monoxide (CO) due to the lack of oxygen required to complete the combustion reaction and form CO_2. CO can also be produced in high temperature regions due to the dissociation of CO_2 molecules; however this phenomenon is negligible as the CO is later burned out in the air-rich regions downstream of the combustion chamber [2]. PM consist of finely distributed particles of soot that form as a result of incomplete combustion. The characteristics of the PM emissions are largely influenced by the type of engine, the thrust level (power) at which it is operated as well as the properties of the fuel itself [5].

10.2.1 Environmental and Health Impacts of Aircraft Emissions

Figure 10.1 conceptually represents the breakdown of emissions at different flight stages during aircraft operation.

The main contributors to the greenhouse effect, and thereby global warming, are CO_2 and H_2O (in the form of contrails) which trap infrared radiation from the surface of the earth and prevent heat from being emitted into outer space. In particular, contrails and aircraft-induced cirrus clouds are amongst the largest contributors to aviation-induced global warming, possibly with a greater impact compared to CO_2 for aircraft flying at high altitudes [7]. While aircraft emissions produced in the cruise phase cause pollution issues on a global scale, those that are produced in proximity to airport areas have a more direct effect on human health [8]. The products of incomplete combustion, such as hydrocarbons (HC) and PM, are the main contributors to urban smog and can be readily inhaled by humans, leading to increased health risks related to respiratory, cardiovascular, and neurological systems [9]. Additionally, these particles remain in the atmosphere for extended periods of time, absorbing and reflecting sunlight.

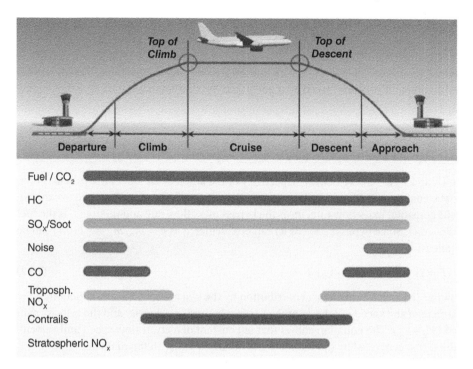

Figure 10.1 Repartition by flight phases of the environmental impacts associated with aircraft operation [6].

Environmental protection organisations are particularly concerned about NO_x emissions from aircraft, mostly because they have a direct impact on the troposphere layer of the atmosphere and ozone depletion [10]. NO_x emissions have an enhanced effect on the environment as they are released in concentrated regions near airports and at higher altitudes in the atmosphere.

Tropospheric NO_x emissions from aircraft trigger a chain reaction of chemical processes in the presence of ultraviolet (UV) light to form ozone (O_3) in the troposphere. The formation of tropospheric O_3 is associated with a positive radiative forcing, which is an indication of a forcing agent's influence on change in global mean surface temperature [11]. Additionally, at this proximity to the ground, ozone is a pollutant which is harmful when breathed by living beings and damages crops, trees, and other vegetation. The production of tropospheric O_3 is magnified due to the fact that this chemical process is self-sufficient, as NO_2 is recycled in the process, with each molecule of NO_2 producing ozone multiple times. The only limiting factor is essentially the amount of volatile organic compounds (VOC), hydrocarbons (HC), and CO in the atmosphere. On the other hand, NO_x emissions from aircraft in the stratosphere destroys O_3 rather than producing it [12], leading to the depletion of the protective ozone layer naturally occurring in the stratosphere and thereby increasing the risk of skin cancer, cataracts, and impaired immune systems in humans from overexposure to UV radiation.

10.3 Increasing Engine Bypass Ratio

The bypass flow of a turbofan engine is the main contributor to the total thrust, and at the same time, contributes to cooling the engine core. Assuming that exhausts are allowed to expand to ambient pressure, the thrust generated by a turbofan engine can be expressed as [13]:

$$
\begin{aligned}
F &= \dot{m}_e V_e - \dot{m}_0 V_0 \\
&= \left(\dot{m}_c + \dot{m}_b + \dot{m}_f\right) V_e - \left(\dot{m}_e + \dot{m}_b\right) V_e = \left(\dot{m}_c + \dot{m}_b\right)\left(V_e - V_0\right) + \dot{m}_f V_e
\end{aligned}
\tag{10.1}
$$

where V_0 and V_e are the velocities respectively of the ambient air and of the exhausts relative to the engine, $\dot{m}_0$ is the mass flow rate of influx air, $\dot{m}_e$ is the mass flow rate of exhaust, $\dot{m}_c$ is the core inlet mass flow rate, $\dot{m}_b$ is the bypass mass flow rate and finally $\dot{m}_f$ is the fuel mass flow rate. Since bypass ratio (BPR) can be calculated as $\beta = \dot{m}_b / \dot{m}_c$, Eq. (10.1) can be rewritten as:

$$
F = (1 + \beta)\, \dot{m}_c \left(V_e - V_0\right) + \dot{m}_f\, V_e
\tag{10.2}
$$

Equation (10.3) gives the thrust contribution by the bypass flow, F_b, as a function of the fan diameter d and speed relative to ambient air, where ρ is air density and the velocity gain is $\Delta V \triangleq (V_e - V_0)$. The equation shows that at constant density, inflow speed and velocity gain the thrust generated by the fan increases quadratically with fan diameter.

$$
F_b = \dot{m}_b \left(V_e - V_0\right) \approx \rho\, A_b\, V_0\, \Delta V = \rho \frac{\pi d^2}{4}\, V_0\, \Delta V
\tag{10.3}
$$

Therefore, larger bypass airflows can greatly contribute to increasing thrust or alternatively to reduce the engine core sizes and required velocity gains at constant thrust.

As evidenced by the equations, larger fans and inlets are beneficial to turbofan engine efficiency as they directly increase the airflow mass ratio and yield higher overall engine BPR. As BPR is related to the propulsive efficiency, increasing it has a direct impact on improving the thrust-specific fuel consumption (TSFC), with added benefits in terms of noise reduction.

Figure 10.2 illustrates the historical trends pertaining to bypass and overall pressure ratios (OPR) of aircraft engines. It is evident that there is an initial increase in BPR in the 1970s due to the introduction of high BPR turbofans. However, this is followed by a long period of time with no significant improvement in BPR, which can be attributed to the several implications and engineering aspects that had to be overcome to increase the BPR of a turbofan engine. On the other hand, the trend in OPR exhibits a continuous growth during this period owing to the improvements in compressor technology, the materials being used, and the introduction of multi-spool engines.

In addition to geometric size limitations for ground clearance and transportation, larger fans and fan cases increase the total empty weight and cruise drag of the aircraft as both aerodynamic form factor and structural loads are worsened. Furthermore, more powerful LP turbines (LPT) are required, further increasing engine weight. Another known issue relates to the tangential velocity of fan blade tips: the overall increase in the magnitude of the vector sum of the incoming flow velocity and the fan blade tip velocity can lead to degradations in aerodynamic efficiency and vibration characteristics of the fan. These,

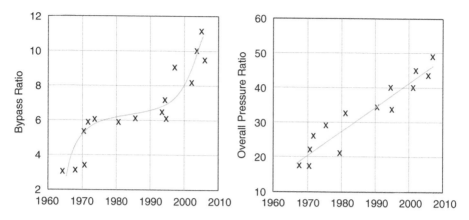

Figure 10.2 Historical trends of BPR and OPR in aircraft engines [14].

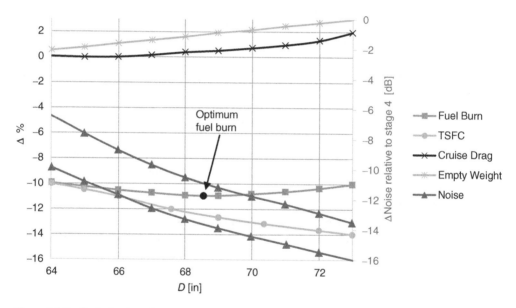

Figure 10.3 Optimization of fan diameter for the CFM LEAP-1B [15].

in turn, can amplify the oscillatory torsional loads on the shaft, degrading its fatigue life characteristics and requiring structural strengthening with its associated weight penalties.

Figure 10.3 shows the variation of weight, drag, noise, TSFC, and fuel burn with increasing fan diameter for the CFM Leading Edge Aviation Propulsion (LEAP)-1B turbofan engine used to power the Boeing 737 MAX [15]. From the figure, it is evident that increasing the fan diameter has both positive and negative impacts on the total fuel burn of a turbofan engine, and it is important for designers to take all these factors into consideration to find the optimum fan diameter to minimise fuel burn for a specific application. In terms of fan blade aerodynamics, the significant drag penalties incurred at high tangential speeds are correlated to the local Mach number M, defined as the ratio between the local tangential velocity as a function of the radial coordinate $v(r) = r\,\omega_{fan}$ and the local speed of sound v_{sound}, as:

$$M(r) = \frac{r\,\omega_{fan}}{v_{sound}} \tag{10.4}$$

The drag divergence Mach number is defined as the Mach number at which the aerodynamic drag affecting an aerofoil begins to increase significantly with respect to a small increase in Mach number. This is primarily due to the formation of shockwaves, which induce flow separation and adverse pressure gradients over the aerofoil. An approximate method for estimating the transonic performance of aerofoils is using the Korn equation:

$$M_{DD} + \frac{C_L}{10} + \frac{t}{c} = \kappa_A \tag{10.5}$$

where M_{DD} is the drag divergence Mach number, c_L is the lift coefficient, t/c is the aerofoil thickness to chord ratio and κ_A is the aerofoil technology factor, with 0.87 assumed for conventional and 0.95 for supercritical aerofoils, respectively. As shown in several studies, the value of M_{DD} for conventional aerofoils is approximately $0.6 \sim 0.7$ [16, 17]. These velocities are easily exceeded when increasing the fan diameter, leading to severe degradations in aerodynamic efficiency. The most restrictive conditions are attained at altitude, as the speed of sound reduces with the square root of the temperature, which decreases linearly with altitude in the troposphere [18]. For instance, an average temperature of $-56.5\,°C$ (216.65 K) and a reference speed of sound of $295\,m\,s^{-1}$ encountered at 11000 m are to be assumed as worst-case conditions. If the maximum tangential Mach number is limited to 0.6 for optimal drag, the blade tip velocity of the fan is $v_r \le 206.5\ m\ s^{-1}$. Thus, assuming a fan diameter (D) of 2.66 m as an example – representative of the General Electric (GE) aviation engine General Electric Next-generation (GEnx)-2B – the optimal angular velocity is $\omega = 155.26\ rad\ s^{-1}$, corresponding to about 1500 RPM. This value is substantially lower than optimal LP shaft rotation speeds.

There are two main strategies being pursued to increasing the fan size and hence the BPR: the use of gearbox technology and the addition of extra compressor and turbine stages. Since the latter approach can easily lead to excessive weight and volumes, a viable strategy is to add an additional shaft. The recent increase in BPR of aircraft engines from the 2000s onwards as shown in Figure 10.2 can be attributed to such strategies.

10.3.1 Geared Turbofan Engines

In conventional turbofans, the fan and LP compressors (LPC) are driven by the same shaft and thus rotate at the same angular velocity. Relatively high rotation speeds are required to maximise the efficiency of compressor stages, but since the LP shaft speed is limited by the restriction on the tangential tip velocity of fan blades, the efficiency of the LPC stages are significantly lower than the HP compressor (HPC) stages. The adoption of a gearbox between fan and LP shaft has been proposed to solve the conflicting optimal rotation speeds of fan and LP compressor and turbine stages, increasing the efficiencies of both and allowing greater BPR to be achieved [19]. The increased efficiency potentially allows for the total number of compressor stages to be reduced, thereby also reducing overall engine weight. As an additional advantage, having the fan run at a low rotation speed reduces the aerodynamic noise generated by the engine and the resulting decrease in structural load and stress provides the opportunity to reduce the blade thickness and thereby reduce the total engine weight [20]. Although the use of gearboxes in turbofan engines was proposed as early as the 1970s, stability and fatigue lifecycle drawbacks discouraged their adoption. Furthermore, the addition of the gearbox has negative impacts on the manufacturing and maintenance processes, with changes required to production lines and additional maintenance personnel training requirements [21]. In addition, with the introduction of new parts and changed sizes, the engine weight distribution is notably affected and this requires new airframe integration studies [22]. Despite these drawbacks, major aircraft engine manufactures such as Pratt and Whitney (P&W) have invested significant resources to improve fan drive gear system (FDGS) technology in order to produce cleaner, quieter, and more efficient high BPR turbofan engines for their customers. Over recent years, significant improvements to FDGS design has enabled greater and more efficient power transmission, higher reliability, and longer lifespans of gear systems in turbofan engines. This section explores the innovative solutions brought forward by the P&W research program to overcome the drawbacks of traditional gear systems in turbofan engines.

10.4 Carbon Fibre Composites

The fan blades of turbofan engines are seeing an ever-increasing use of composite materials as these allow to reduce a considerable amount of weight while maintaining structural integrity. For the next generation of commercial engines, researchers are looking to composites made from carbon fibres coated with carbon nanotubes, called carbon fibre composites (CFRP). In particular, carbon fibre-composite fan blades are 10% lighter than titanium alloy structures of the same strength [23]. However, when considering the strength of composite parts, it is misleading to directly compare properties of carbon fibre versus aluminium or titanium, as these metals are homogeneous and have isotropic properties. By comparison, in a carbon fibre composite, the strength resides along the axis of the fibres, and thus fibre properties and orientation greatly impact mechanical properties. Table 10.1 shows the physical characteristics of a unidirectional lamella of 0° carbon fibres in ideal conditions.

The table highlights the anisotropic behaviour of carbon-fibre composites as the strength in the 0° direction is significantly higher than 90°. Thus, carbon fibres need to be used compositely to withstand shear stress and other non-axial loads. In order to create a composite part, the carbon fibres, which have excellent stiffness properties in tension and compression, need a stable matrix to reside in and preserve their form. Epoxy resin is an excellent plastic with good compressive and shear properties, and is often used to form this matrix, whereby the carbon fibres provide the reinforcement.

It is typical in engineering structural design to measure the benefit of a material in terms of its strength-to-weight ratio and stiffness-to-weight ratio. Therefore, it is also important to consider the density of viable materials to understand their feasibility for different applications. Table 10.2 lists some representative densities of jet engine materials as reviewed by NASA.

It can be noted that aluminium is nearly half as dense as titanium, whereas the density of carbon-fibre composites, which is around 2000 kg m^{-3}, is even lower than aluminium [25].

Table 10.1 Typical mechanical properties of a flat carbon-fibre sheet [24].

Property	Value
Density (kg m^{-3})	1550
0° Tensile strength (MPa)	1400
0° Tensile modulus (GPa)	123
0° Compression strength (MPa)	850
0° Compression modulus (GPa)	100
90° Tensile strength (MPa)	18
90° Tensile modulus (GPa)	8.3
90° Compression strength (MPa)	96
90° Compression modulus (GPa)	8.4

Table 10.2 Representative densities of some commonly considered materials for jet engines.

Material	T Lim (K)	Density (kg m^{-3})
Aluminium alloy	500	2726
Titanium alloy	833	4693
Stainless steel	1111	7633
Nickel alloy	1388	8252
Nickel crystal	1666	8252
Ceramic	1666	2630

Using lightweight composite materials has a direct impact on the fuel efficiency of the engine as the overall weight would be substantially reduced. This also allows room for the fan diameter to be increased to enhance the BPR and thereby further increase engine efficiency. Additionally, when a metallic fan blade fractures, the sharp metallic debris can easily damage other engine components, potentially leading to uncontained failures. Carbon-fibre composite blades, on the other hand, do not have sharp edges when fractured and cannot withstand high temperatures, so their debris will be burned off when passing through the compressor and combustion chamber, thereby protecting engine core components from secondary damage. This allows the engine to meet blade-out test requirements without the manufacturing, assembly, and weight of a separate containment ring. Moreover, because carbon-fibre composites have a better fatigue cycle, the total maintenance efforts could be reduced [26].

Assuming a GEnx engine employing a carbon-fibre composite fan, the overall weight of carbon-fibre composite fan blades is 66% lower than titanium ones but they are 100% stronger [27]. At the same time, because of the increased strength of carbon-fibre composites, the fan blades can be designed thinner and have a larger area than titanium ones. Thus, the GEnx engine achieved a reduction in the count of fan blades from 22 to 18, further reducing engine weight. Furthermore, as the number of blades is reduced, the noise level is also reduced, allowing a further increase to the diameter with no noise penalty [28].

However, due to their limited thermal resistance, CFRP cannot be used to replace core engine components, and this characteristic introduces new risks to the engine. For example, in the event of accidental exposure to high temperatures as a result of an uncontrolled fire, the carbon-fibre composite fan and fan case may incur substantial damage. Carbon fibre composites cannot sustain high temperatures as the mass loss will significantly increase when temperatures are higher than 490 °C for three minutes. Another major concern of using carbon fibre composites is the issue with their recyclability at their end-of-life stage. Most metals used for industrial and manufacturing purposes have a proven track record for recycling and minimising waste, whereas the reclamation and repurposing of CFRP is a relatively novel concept, with further research required into understanding the processes and cost of closing the life-cycle loop of composite parts.

10.4.1 Evolution of Composite Materials in Turbofans

Although CFRP is a relatively new material technology with many refinements and upgrades as more research is carried out over recent decades, the use of composite materials in the aviation industry dates back to the 1950s, when RR used glass fibre/epoxy in its RB108 engine compressor blades and casing. This was followed by a carbon fibre-reinforced epoxy called Hyfil in its RB211-22B turbofan blades in the 1970s. However, RR had to abandon the development of these composite blades, owing to the fact that they failed catastrophically during bird-strike tests [29]. At this stage, designers had a limited understanding of composite materials and were therefore reluctant to use polymer composite materials in safety-critical engine components.

Over the following few decades, advanced research and analysis into the structure and failure mechanisms of composite materials at a micro-scale level was enabled by the development of such techniques as composite fractography and infrared spectroscopy. Consequently, the understanding of damage and failure modes of composites was greatly improved, and the design and manufacture of composites adapted to impede the causes of failure and maximise damage containment [29]. This paved the way for the reintroduction of CFRP in turbofan blades, beginning with GE Aviation's GE90 in 1995, and even the use of CFRP in the front fan case as in the case of the GEnx turbofan in 2006. The driving factor behind the increased use of CFRP in turbofan engines was the desire by engine manufacturers to achieve higher BPRs for their engines, which resulted in better fuel efficiency and weight savings.

In addition to investigation techniques, there were also significant technological advancements with respect to the manufacturing methods used in the production of composite components [30]. For example, the injection of the epoxy resin into the polymer matrix and the curing process was combined into one operation called resin transfer moulding (RTM). Weaving methods that were previously carried out manually were soon replaced by weaving machines able to produce 3D woven textile preforms featuring complex geometry with much more precision and accuracy at a much faster rate of production. Employing these efficient manufacturing methods, CFM International has managed to achieve an extremely high production rate of composite fan blades for the manufacture of their LEAP engines.

After their unsuccessful attempt at introducing composite materials to turbofan blades in the 1970s, RR invested heavily in research into new blade designs incorporating composite materials. The aim was to deliver light fan blades with a strong structural integrity while retaining aerodynamic performance. The outcome of their efforts was the development of the CTi blade, referring to the lightweight carbon composite structure and the aerodynamically efficient titanium leading edge, which was first tested in 2014 on a Trent 1000 'donor' engine with promising results. Currently, RR utilise composite materials in their Trent family of engines, with the Trent XWB having a CFRP fan case, fan track liners, bifurcation linings, and anti-fluid panels. However, CTi fan blade technology is a major step forward for the design of the RR's next generation of turbofan engines, the RR Advance and Ultrafan, allowing the former to save up to 680 kg in total engine weight and achieve 20% better fuel efficiency and reduction in carbon emissions when compared to the early engines in the Trent family [30].

The future of component manufacturing for turbofan engines seems promising with highly automated robots manufacturing composites with complex 3D geometries and high levels of accuracy and precision at a rapid production rate. This provides room for growth into the next stage of manufacturing technology, 'adaptable manufacturing', where the level of automation allows machines to react to changes in manufacturing processes without requiring human intervention.

10.5 Low Emission Combustion Technologies

The design of low emission combustion technologies is a complex multi-disciplinary optimisation (MDO) design challenge, since, although the focus is on the reduction of emissions, there are several other factors and consequences to be considered that could affect other attributes of the aircraft. The fundamental challenge is to increase engine cycle efficiency, while keeping emissions at the lowest possible levels; however, it is important that characteristics such as performance, safety, and operability are not compromised. Some of the design considerations taken into account when reviewing low emission combustion technologies in this section are given below:

- **Combustion efficiency** – how effectively the stored chemical energy in fuel is converted to useful heat energy.
- **Combustion stability** – how vulnerable the process is to small perturbations altering the combustion characteristics.
- **Pressure drop** – the magnitude of the pressure loss due to the increase in temperature.
- **Smooth ignition** – the likelihood of controlled ignition being achieved without auto-ignition/flashback risks.
- **Combustor length/weight** – space and weight savings directly correlated to fuel savings.
- **Temperature distribution** – the ratio of the difference between the average and peak temperatures in the combustion zone.

- **Liner life** – the durability of the inner wall of the combustion chamber.
- **Structural integrity** – the ability to support the required load during operating conditions with an appropriate safety margin.

10.5.1 Rich-burn/Quick-quench/Lean-burn Combustion

The first stage of conventional combustion is the 'rich burn' stage, which involves a fuel-rich environment with stable combustion due to the high concentration of energetic hydrogen and hydrocarbon radical species at a relatively low temperature and concentration of oxygen. Therefore, the amount of nitrogen oxides formed at this stage is very low, but this produces a high concentration of combustible substances such as CO, the unburned hydrocarbons (HC), hydrogen (H_2), and also carbon (soot). These substances cannot be freely released into the atmosphere; therefore, a quench section is employed downstream to dilute the mixture. In conventional combustors, there is a gradual and continuous admission of air to reduce the FAR ratio (ϕ), which results in the rapid formation of NO_X, particularly close to when the reaction achieves its stoichiometric point, as shown in Figure 10.4. Therefore, from the point of view of reducing NO_X emissions, the combustion chamber of the engine should operate far from the stoichiometric point in either air-enriched or depleted zones. The idea of quick-quench is to minimise the time taken to reduce the FAR, and rapidly input the jet air through holes in the walls of the flame tube to dilute the mixture and reach the lean-burn zone, where the low FAR promotes the oxidisation of the unburnt substances. This is the approach taken in rich-burn/quick-quench/lean-burn (RQL) combustion, depicted in Figure 10.5. Ideally, the exhaust gas leaving the combustion chamber should have an acceptable composition of the products of complete combustion (CO_2 and H_2O) along with N_2 and O_2.

The main focus and the problem of such technology is to ensure fast and qualitative mixing of the gas stream in the intermediate stage (quick-mix) in order to prevent the formation of a mixture of stoichiometric composition and minimise NO_X formation. If this process is not achieved effectively, it can lead to a sudden increase in flow temperature with undesirable consequences, both in terms of harmful emissions and damage to structural elements.

Figure 10.4 Formation of nitrogen oxides and working principle of RQL. Adapted from [2].

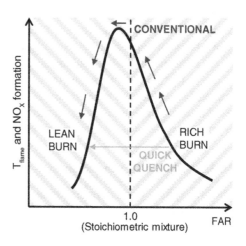

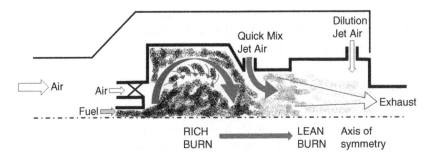

Figure 10.5 Engine cross-section schematic illustrating RQL combustion.

Due to the critical nature of the quick-quench process for RQL combustion, several measures can be taken to ensure its effectiveness [31]:

- **Closer spacing between primary and quench jets** – increasing the proximity of the dilution holes to the primary rich-burn zone where the primary air is injected boosts the strength of mixing and accelerates the process by which the combustion reaction moves from the rich-burn to the lean-burn zone.
- **Minimise time spent by the mixture in the quench zone** – designing the combustor geometry to optimise the volume and distribution of combustor area to reduce the residence time of the mixture in the quench zone.
- **Adding local cooling air** – the NO_x formation rate is highest at gas temperature peaks in the quench zone, therefore injecting air from additional cooling holes downstream of the primary zone leads to further reduction in NO_x.
- **Optimization of jet hole pattern spacing, sizing, and geometry** – these characteristics of the dilution holes play an important role in the structure of the flow field distribution and therefore several research studies and experiments have been carried out to determine the optimum setup.

There is concern that the most widely used RQL technology also has limited potential to further significantly decrease NO_x emissions whilst maintaining the other emissions criteria and satisfying all operability requirements. Nonetheless, it has been refined over the years by the world's largest engine manufacturers for their combustor designs. Pratt & Whitney developed the TALON (Technology for Advanced Low NO_x) family of RQL combustors for commercial aircraft engines with reductions in NO_x emissions with each progressive model, the latest being the TALON X. More detailed information on the design of the TALON family of combustors can be found in the references [32]. Other industrial approaches include GE's Low Emission Combustor (LEC) RQL technology which includes swirling devices to optimise fuel-air mixtures in the different regions, and RR's Phase 5 RQL combustor which was installed in the Trent family of engines.

10.5.2 Double Annular Combustor

The idea behind using annular combustors is to eliminate the separation of the combustion zones and instead have a continuous liner and casing in a ring configuration. The simpler design and smaller size of this combustor greatly reduces the total weight and promotes more uniform combustion, leading to uniform gas exit temperatures. Double annular combustors (DAC) are a variant of this concept with two combustion zones around a ring: the pilot zone and the main zone. The pilot zone (the outer annulus) acts like that of a single annular combustor (SAC), and typically operates at low power settings to raise the FAR so as to increase the combustion efficiency and promote complete combustion. At high power settings, the main zone (the inner annulus) is used as well, increasing air and mass flow through the combustor to achieve lean burn for minimal NO_X and smoke formation. There is also an intermediate setting where part of the main zone is fuelled and ignited. The demand for different fuel supplies to different combustion zones based on power setting demands a more complex fuel injection system.

Similar to the SAC, the radial arrangement enables the stable, efficient, and low-emission combustion to be achieved while shortening the length of the combustion chamber and thereby reducing the weight. The utilisation of separate combustion zones at different power settings also allows more flexibility and control when compared to a SAC, with the potential to have lower NO_X and smoke emissions at all power levels. However, the increase in surface area due to the addition of another annulus requires a larger amount of cooling, and with the injectors arranged in a radial configuration, insufficient cooling could lead to uneven distribution of temperature in 'hot spots'. The radial temperature profile also has negative impacts on the durability of the turbine and its ability to extract work done, thereby reducing the fuel burn efficiency. Another major challenge arising from having the pilot and main stage in parallel with each other is to obtain the desired performance during the intermediate power settings when both the annular chambers are outside their optimum conditions. This provides significant challenges for designers to ensure that the benefits of an additional annulus outweigh the drawbacks in terms of fuel-burn efficiency.

CFM adopted the DAC design for their CFM56 series of engines to power the Airbus A320 and A321 aircraft. GE also extended this technology for large thrust category applications, as is the case in the GE90 series which powers the Boeing 777 aircraft. Emission tests on these engines reveal that the margin between the NO_X emission level and the Committee on Aviation Environmental Protection (CAEP)/6 is generally wider than that of RQL combustors; however the margin tends to decrease with increasing OPR of the engine [33]. The drawbacks of DAC technology, in particular the cooling requirement, seem to limit the OPR that the engine can operate at with acceptable levels of NO_X emissions. Therefore, significant improvements would have to be made to keep this technology viable in the future of engine designs with continually higher OPRs.

10.5.3 Axially Staged Combustors

Similar to DAC, axially staged combustors (ASC) have a pilot and a main zone of combustion with similar working principles. However, instead of a radial configuration, the fuel injection zones are arranged in the axial direction with the main stage downstream of the pilot stage. With this configuration, the pilot zone continuously delivers higher upstream gas temperatures for more reliable and prompter ignition of fuel in the main stage, as well as more stable, efficient, and complete combustion. Due to the arrangement of the stages in an axial manner, NO_X emissions can be kept low because of the short residence time at the high

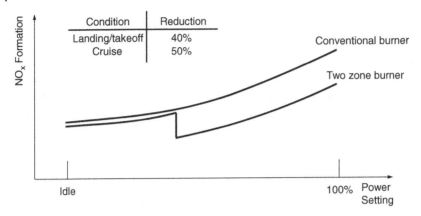

Figure 10.6 NOX emissions vs. power for conventional and axially staged combustors. Adapted from [34].

temperatures experienced at high power settings, as shown in Figure 10.6. The sharp drop in NO_x formation when the power setting is increased can be attributed to the transition from part load to full load when the main stage comes online.

On the other hand, the extremely high downstream temperatures have some negative side effects that need to be considered. The higher the temperature, the more susceptible the fuel becomes to the process of coking, in which a solid residue (coke) is created when fuel undergoes severe oxidative and thermal breakdown at extreme engine temperatures. The separation of the pilot and main fuel injection systems eliminates the cooling effect of the pilot fuel that is present in other two-zone combustors, and also requires separated penetration of the combustor casing, which calls for higher strength and stiffness of the casing material for structural integrity. Furthermore, engine manufacturers have to take extra precautions to ensure the in-line alignment of the pilot and main zone is perfect, otherwise the imperfect flow path would result in substantial pressure losses.

Despite the concept of ASC arising in the same timeframe as DAC in the 1970s, the only major application of this technology is in the Pratt & Whitney IAE V2500 series of two-shaft high-bypass turbofan engines, primarily used to power the Airbus A320 family. Other engine manufacturers have opted to follow different routes when investing resources in the development of combustor technology, thereby the TRL of ASC is still fairly low compared to the other types of combustor discussed here. However, P&W recently carried out further development of this technology, renaming it axially controlled stoichiometry (ACS) under the NASA ERA program, and successfully demonstrated the lowest levels of NO_X emissions tested (88% margin to CAEP/6 LTO NO_X regulations for an advanced N + 2 GTF cycle), shown in Figure 10.7, providing the largest margin for development compared to the other concepts tested [35].

10.5.4 Lean Direct Injection

In lean direct injection (LDI) combustors, fuel is injected directly into the combustion chamber without external premixing. The injector helps to produce a toroidal recirculating

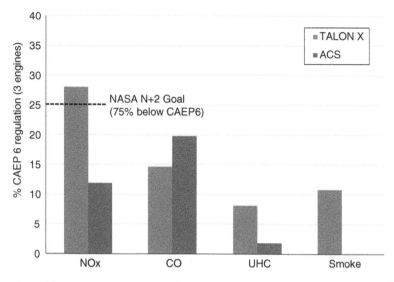

Figure 10.7 Comparison of rig test results of the TALON X and ACS at NASA and UTRC for an advanced N + 2 GTF cycle. Adapted from [35].

air flow with the fuel, which can help the fuel and airflow to mix optimally. Since a large proportion of air is used in this process, the combustion takes place at lower overall FAR compared to conventional combustors, and the absence of a separate dilution zone means that the FAR is kept constant throughout the combustion chamber.

An advantage of direct fuel injection into the combustion zone is that the risk of auto-ignition and flashback is eliminated, and, provided that the fuel-air mixture is well mixed uniformly and rapidly to achieve constant lean burn, an extremely low level of NO_X emissions can be achieved. On the other hand, constant lean burn leads to a lower maximum temperature during combustion reaction, which while significantly improving the liner life, has negative impacts on the stability and efficiency of combustion, leading to the production of CO, HC, and soot.

RR has been investigating LDI concepts of premixed and partially pre-vaporised lean combustion in the framework of the NEWAC (NEW Aero Engine Core) concepts, with the goal of achieving substantial NO_X reduction for high OPR (>30) aero-engine applications. Test results show that the RR LDI combustor was able to achieve a NO_X reduction of 60% versus the CAEP/6 LTO NO_X regulation for an engine with 39 : 1 OPR. Although the TRL for this technology is approaching seven for emissions and operability, further LDI development will include further optimization in terms of weight and system complexity and better understanding of the required fuel staging functionality.

10.5.5 Multipoint Injection Integrated LDI

The high dependency on the quality of mixing performance for LDI technology suggested the use of multiple injection points with several stages of LDI combustion. A study by Heath [36] can be adopted to assess the combustion efficiency of this technology. The study included a multiple LDI combustion consisting of three single LDI modules. In the experiments, the length of the combustor was set as a constant and maintained for all combustions; however, 20 different designs were tested by varying certain parameters of the LDI modules. These parameters include the LDI module diameter X_1 (varying between 1.810 and 3.016 cm), vane angle X_2 (varying from 35.0 to 85.0°), venture angle X_3 (varying from 25.0 to 55.0°), and number of vanes X_4 (ranging from 4 to 8).

The combustor air flow pressure drop (ΔP) is one of the key values that needs to be determined to assess the combustion efficiency. To measure the effect of lean burn technology in reducing NO_X emissions, the EI_{NOx} also needs to be considered. It is very important that the combustion has steady characteristics to make the engine and measured data reliable. Therefore, the non-uniformity U(x) has to be taken into consideration when assessing the results of this study. This can be calculated using Eq. (10.6).

$$U(x) = \frac{1}{n} \sum_{i=1}^{n} \frac{var(x_i)}{\mu_i} \tag{10.6}$$

Where x is the local temperature extracted from each computational solution and μ is the average temperature. At the same time, the average Sauter mean diameter (D_{32}) also needs to be calculated through Eq. (10.7). The σ is fuel surface tension (N m), μ_L is fuel dynamic viscosity (Pa.s), m_L is fuel injector mass flow rate (kg s^{-1}), P_L is injector fuel flow (FF) pressure (Pa), and ρ_A is injector air density (kg m^{-3}). Using these calculations,

$$D_{32} = 2.25\sigma^{0.25} \mu_L^{0.25} \dot{m}_L^{0.25} P_L^{-0.5} \rho_A^{-0.25} \tag{10.7}$$

The exit temperature (T_4) could also be used to assess the lean-burn combustion non-uniformity. The target exit temperature when burning fuel is 1735.5 K, however it was observed that some of the lean-burn models could not reach this temperature, which suggests that an additional source of heat is required to increase the exit temperature up to the target value.

The variation of the combustor air flow pressure drop ΔP with increasing vane angle in designs was also tested. It was found that ΔP varies linearly with the vane angle X_2. When the vane angle decreases, the airflow travels through the combustion chamber in a shorter period of time and the direction of the airflow velocity is more aligned with the combustion chamber. Thus, the pressure drop will become lower, which makes the power drop also lower. However, the models with eight vanes exhibited a large pressure drop when compared to the models with four vanes. It can be deduced that increasing the number of vanes will also cause the pressure drop to increase at the same time. Therefore, the pressure drop is related to both the vane angle and the number of vanes.

When considering the non-uniformity U(x) of every model, it was observed that most of the models outside the mid-range vane angle (56 to 72°) have low non-uniformity, indicating that steady lean-burn combustion could be achieved in these regions. Therefore, it can be inferred that the efficiency of lean burn in MLDI combustion is closely related to the vane angle and the number of vanes. The other two factors could affect the combustion efficiency, but do not have disciplinary characteristics that could be analysed. Therefore, the study only focuses on the vane angle and the number of vanes.

High combustion uniformity supports steady conditions and thus efficiency and repeatability. At the same time, a lower pressure drop could also lead to better combustion efficiency. Lowering EI_{NOx} is the main advantage of using MLDI combustion.

10.5.6 Twin Annular Premixing Swirler Combustion

Evolving primarily from other staged combustion concepts, twin annular premixing swirler (TAPS) technology has many resemblances to conventional SACs, with concentrically mounted pilot and main stages. However, a unique feature of TAPS combustors is the application of premixed combustion, achieved through a multi-swirler arrangement generating two co-annular swirling flow streams produced by the pilot and a main mixer, respectively. The pilot uses a simplex atomizer surrounded by two co-rotating swirlers to spray fuel onto a pre-film lip where it is atomised to make it suitable for engine start-up and low-power operations. At higher power settings, the main mixer, consisting of a cyclone or radial inflow swirler, is utilised. A mixing layer is formed where the flow streams from the two mixers interact with each other, and this helps to stabilise the main flame and ensure clean combustion. Almost all the air entering the combustor is directed to the pilot and main swirlers, with the remaining air being used for cooling the combustor dome and liners. This can be attributed to the elimination of the need for dilution air during the latter stages of combustion, since the fuel-air mixing process is already significantly more thorough due to the TAPS mixers. This leads to better structural integrity of the combustor and shorter overall length due to the absence of dilution holes.

A major advantage of having an internally staged configuration is that the required exit temperature of the TAPS combustor can be easily achieved and maintained, unlike that of conventional DACs. The improved distribution of exit temperature provides longer turbine life as well as lower fuel burn due to better thermal efficiency. The durability could be further improved by using advanced composite materials in the liner to resist higher temperatures, and this is a critical parameter for the next generation of ultra-high OPR engines being developed.

As with other premixed combustion techniques, there is always a risk of auto-ignition and flashback, which has to be handled carefully through rapid and thorough mixing. Therefore the quality and effectiveness of the mixing is of paramount importance to the success of TAPS technology, and this requires detailed research into flow dynamics and heat release analysis using techniques such as computational fluid dynamics (CFD) tools and particle image velocimetry (PIV) [37].

GE has significantly invested in developing TAPS technology for the turbofan engines, beginning with DAC versions in the late 1990s and transitioning into SAC TAPS combustors in the production of their CFM56 7-B engine. Based on the experience gained on fuel staging in these annular combustors, GE developed a TAPS combustor for the GEnx turbofan engine, which was able to achieve ultra-low NO_x emissions (60% and 40% margins relative to CAEP/6 at 35 and 47 OPR respectively [31]) and good combustion efficiency. The success and potential for further emission reduction of TAPS has led to continuous improvement and scaling of the technology, termed TAPS II and TAPS III by GE, and has a very promising future in low emission combustion technology.

10.5.7 Lean Premixed Pre-vaporised Combustors

The underlying working principle of the lean premixed pre-vaporised (LPP) concept is to supply a homogenous fuel-air mixture of low FAR into the combustion zone, and then carry out the combustion phase at an equivalence ratio very close to the lean blowout limit. The probability of NO_X formation is significantly reduced due to the low flame temperature and absence of hot spots in the combustion zone. This also benefits liner durability due to low flame radiation. The LPP split into 3 major areas: the fuel preparation duct for premixing and pre-vaporising, the combustion zone, and the dilution zone.

Inside the premixer, the air passes through axial swirlers and is mixed with an evaporating liquid fuel spray from the fuel injection nozzles. These processes of mixing, fuel drop injection, and evaporation have to be optimised to ensure the suitable conditions for extremely lean burn when the resulting mixture is supplied to the combustion zone. The dilution zone further reduces the FAR in a similar manner to conventional combustors.

In order to optimise the preparation process, it is tempting to increase the residence time of the mixture in the fuel preparation duct; however, there are other implications of this that need to be considered. In addition to a longer duct and thereby larger engine weight, an increase to the residence time also makes the fuel-mixture more susceptible to auto-ignition. Flashback is another major concern of LPP combustors, which occurs when the flame speed is higher than the magnitude of the velocity of the fuel-air mixture exiting the preparation duct. These risks are only amplified at higher engine OPRs, which calls for a new technique or modification of this technology to deal with them for further application in future engines.

10.5.8 Flameless Combustion

Flameless combustion was developed for burners in heating industrial furnaces using preheated combustion air; however its ability to suppress the formation of thermal NO_X even at extremely high temperatures makes it suitable to be adapted for aero-engine applications, especially with the demand for higher OPR engines.

In conventional combustion, the combustible mixture of fuel and air is ignited to develop a flame. The stability and control of combustion is usually determined by the flame stabilisation, where a stable flame front provides a constant and controlled reaction and is also used as an indicator for flame supervision. However, this is not the only way of producing controlled combustion within a combustion chamber. The principle idea behind FC is the elimination of the flame front, which can be achieved in extremely high chamber temperatures (>850 °C) with the help of preheated air. This type of reaction is only able to take place at temperatures above the self-ignition temperature, in a large dispersed volume as opposed to a restricted flame front. There are numerous advantages to this type of reaction, including very low levels of CO and NO_X formation and the elimination of noise. The requirement of flame supervision is also eliminated as there is no danger of the reaction extinguishing, and thereby no risk of explosion. The flameless oxidation mode allows a large reduction of NO_X emissions by avoiding localised peaks of temperature and significantly reducing the concentration of oxygen available for NO_X formation.

As shown in Figure 10.8, the feeding of oxidising air and fuel gas is performed separately at high injection speeds. This creates large internal recirculation flows which, along with the high temperature of combustion products, initiates the flame's mode of combustion and distributes the reaction throughout the volume of the combustion chamber. It is important

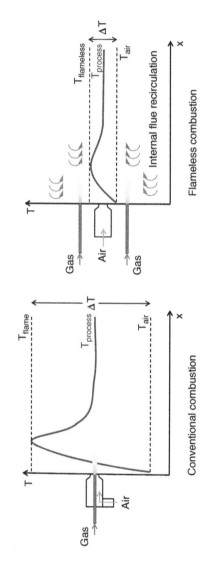

Figure 10.8 Comparison of the two combustion modes. Adapted from [38].

to note the homogeneity in the temperature of the entire process, as opposed to the existence of a peak in the temperature at the reaction zone of conventional combustion.

The use of flameless combustion has been largely restricted to industrial applications, such as power generation and steel and gas manufacturing industries. Unlike these ground-based applications, aero-engine combustors have additional restrictions in terms of volume, weight, and high-pressure ratios, making it difficult to implement flameless combustion. Gas turbine engines also operate at very fuel-lean conditions in order to reduce their specific fuel consumption (SFC), and the combustor liner and turbine blades have a limited resistance to extremely high gas temperatures, with limited space to add heat exchangers within the engine nacelle. This also means a large portion of oxygen remains unconsumed after combustion, making it difficult to effectively lower the oxygen concentration, which is an important factor for the low NO_x emissions. Therefore, there is certainly a gap in the knowledge of applying flameless combustion technology to an aero-engine scenario.

10.6 Casing Treatments

Increasing the pressure ratio of compressor stages can help in reducing the number of the stages and engines' weight. However, increasing the pressure ratio of individual stages equates to an increase of the compressor blade loading. A higher blade loading is more subject to stall, which can compromise the engines' thrust or even lead to catastrophic engine failures [39]. Therefore, the stall margin is a crucial parameter to be observed when investigating evolutionary compressor technologies. One of the factors prompting a compressor stage to stall is the unsteady flow at the tips of the compressor blades. Because of the gaps between the blade tips and the compressor case, the airflow becomes highly unstable in proximity of these gaps. Therefore, carefully designed casing treatments can help control the stability of the air flow at higher stage loadings and prevent the onset of stall [40].

10.6.1 Stall Precursor-suppressed Casing Treatment

The compressor stall precursor wave has been recently identified as a fundamentally important factor and can explain how compressor stall originates. Thus, increasing the stall margin translates into how to mitigate the stall precursor wave completely to avoid compressor stall [41]. The stall precursor-suppressed (SPS) casing treatment is one way that could help to mitigate the precursor wave. With respect to SPS casing treatment, the unsteady tip flow is drawn into an annular backchamber towards the front of the compressor stage rotor. Figure 10.9 illustrates the mechanism of SPS casing treatment.

A study was carried out by Sun et al. (2014) [43] in which a low-speed TA36 fan equipped with controllable bleed valves, which can accurately change the operating conditions near and away from stall, was tested to explore the effect of SPS casing treatment on compressor stability. From this study, it was concluded that the SPS casing treatment helps to improve the compressor stall margin by around 10% by improving its stall point. The pressure efficiency is nearly the same with or without casing treatment, however, because the SPS casing treatment could improve the stall margin, it could indirectly improve the

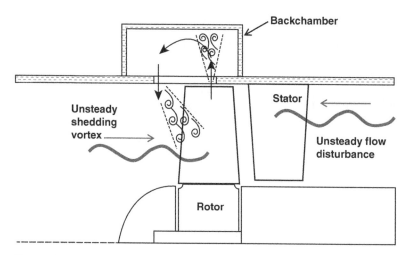

Figure 10.9 SPS casing treatment. Adapted from [42].

compressor efficiency. The compressor blade shape could be designed to sustain more loading without compromising stall margin, and the pressure ratio could be increased by using more efficient compressor blades, thereby indirectly assisting in increasing the compressor efficiency. Furthermore, it could be deduced from the results of the study that with SPS casing treatment, the compressor could support more steady power even at the near stall point. However, because the SPS casing treatment adds extra cases to the compressor casing, the compressor needs to be fully redesigned. At the same time, adding extra case outside the core inlet may affect the fan inlet and cooling system. Thus, the structural integrity of the new compressor case needs to be tested to verify whether the new case is structurally reliable.

10.6.2 Recirculating Casing Treatment

Tip leakage flow (TLF) is another important factor that is known to have negative impacts on the performance and operability of most modern axial flow compressors. Slot-type or groove casing treatments were suggested to provide compressor stability improvements by manipulating the TLF for tip-critical compressors; however there were drawbacks to these strategies in terms of loss in compressor efficiency [44]. A study by Guinet et al. (2014) [45] concluded that the benefits of using casing treatments were more prominent in compressors with higher tip clearance, but on the other hand larger tip clearance leads to the formation of a strong tip leakage vortex which has detrimental effects to compressor stability.

In order to provide compressor stability without compromising for efficiency, a recirculating casing treatment (RCT) was proposed as a potential solution. RCT involves extracting high-pressure air from locations downstream of the compressor and injecting it at the tip of the compressor blades. This tip injection process provides the range extension which is essential to negate the adverse effects of having a large tip clearance. Figure 10.10 illustrates the method by which RCT can be implemented.

A comprehensive study carried out by Wang et al. (2016) included both a numerical and experimental investigation into the effects of RCT in an axial flow compressor.

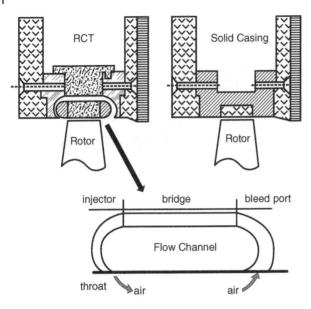

Figure 10.10 Mechanism of RCT. Adapted from [44].

From the results, it is evident that RCT has a positive impact on the stall margin of the compressor, with an improvement in performance and efficiency. The difference in the simulated and experimental pressure as well as the adiabatic efficiency parameter at the near-stall point can be attributed to the non-uniformity of the rotor tip clearance and the incapability of the turbulence model. However, the mass flow rates at the simulated and true stall points are very similar, with both showing a 40% range extension in the stall margin for RCT.

There are clear benefits in terms of stall range extension by using RCT, with no predicted loss in efficiency or pressure increase capability at design speed conditions. However, the accuracy of the design and placement of the injection and bleed ports is an important factor in ensuring the proper recirculation of air, otherwise misalignments of the jets or insufficient fluid injection could have adverse effects on compressor performance.

10.7 Interstage Combustion and Combined Cycle Technologies

When considering fuel efficient technologies in engines, it is important to analyse the conversion of thermal energy into mechanical energy in its thermodynamic cycle. A simple Brayton-Joule cycle of a gas turbine engine is depicted in Figure 10.11. The ideal thermodynamic cycle includes four stages correlating to their respective engine components: two isentropic (compressor and turbine) and two isobaric (combustion chamber and exhaust).

The thermal efficiency η_{th} of the cycle is related to the difference between the heat added (Q) in the combustion chamber (Stage 2–3) and the heat released in the exhaust (Stage 4–1)

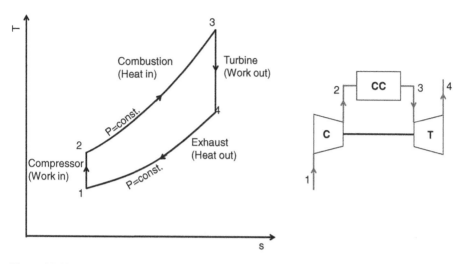

Figure 10.11 Ideal Brayton-Joule cycle for a gas turbine engine.

according to Eq. (10.8).

$$\eta_{th} = \frac{Q_{23} - Q_{41}}{Q_{23}} \tag{10.8}$$

In simple terms, the area enclosed by the curves is an indicator of the thermal efficiency of the entire process, with a larger area correlating to a cycle that is able to achieve a better conversion of thermal energy to useful work done. Improving the efficiency of this conversion using thermodynamic optimization methods is key to improving the overall system performance and reducing pollutant emissions [46]. This section explores technologies that attempt to manipulate the thermodynamic cycle of aircraft gas turbine engines in the hope of achieving better fuel efficiency with minimal loss in thrust and OPR of the engine.

10.7.1 Intercooled and Recuperated Aeroengines

The concept of intercooling is aimed at decreasing the net work input of the compressor required for a given pressure ratio by cooling the air temperature between the outlet of the LP compressor and the inlet of the HP compressor while maintaining constant pressure. This results in an effective increase in the net work output of the engine, albeit at the expense of a reduction in thermal efficiency, as heat is removed from the process [47]. To compensate for this, intercooling must be used in conjunction with a reheating process. This can be easily achieved by introducing a heat exchanging system that utilises the hot exhaust gas from the LP turbine to increase the temperature of the air entering the combustion chamber. The schematic of this combined process and the associated modification to the thermodynamic cycle of the engine is shown in Figure 10.12.

An additional benefit of this cycle is its significantly lower typical OPR, which supports the utilisation of ultra-low NO_x emission combustors, such as DAC and LPP combustors, which otherwise are difficult to implement in high OPR engines. In terms of the aviation industry, however, engine manufacturers are hesitant to incorporate intercooling and recuperation cycles in their engines, mainly due to the added size and weight of the additional

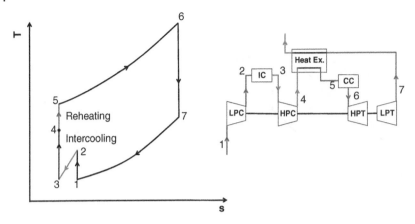

Figure 10.12 Thermodynamic cycle of IRA. Adapted from [48].

components required. Therefore, despite the potential of achieving better fuel consumption and lower emissions in aircraft, the use of this technology has been widely restricted to ground-based gas turbine power plants. In order to successfully implement this technology in aircraft, an innovative design and method of integration that minimises weight and size will be required.

Nevertheless, an EU-funded campaign EEAFAE (Efficient and Environmentally Friendly Aero Engine) has already invested heavily in the testing and validation of the IRA cycle with the help of the CLEAN (Component Validator for Environmentally-Friendly Aero-Engine) demonstrator, due to the potential improvements in fuel consumption and emission reduction that it could achieve. Based on the results of these tests that were completed in 2005, NEWAC (Concepts) have continuously developed the concept further in more detail, taking into account optimum components to utilise, their arrangement, and integration with the engine core. The 'Intercooled Recuperative Core Concept' is included in their list of subprojects that consist of core engine technologies with the best potential to achieve ACARE 2020 objectives. Significant advancements have been made in terms of the associated component technologies of IRA, such as a profiled-tubed recuperator (designed by MTU Aero Engines), centrifugal HP compressor and an advanced LPP combustor, and NEWAC's objective is to investigate and optimise these technologies and address their potential drawbacks in order raise their TRL to a level acceptable for the application of IRA in commercial aircraft.

10.8 Thermofluidic Improvements

The performance of a gas turbine engine is dictated by the flow rate of air through it, thereby any restrictions or losses that inhibit the smooth flow of air will have a direct impact on the aerodynamic efficiency of the engine. Optimising the path of the air flow through a gas turbine engine, in particular by addressing profile, tip clearance and end-wall losses, can contribute to a significant improvement in overall engine performance. Aerodynamic performance benefits can also be gained by optimising the flow characteristics at the engine inlet with the use of morphing engine structures [49], and the engine outlet by manipulating the bypass nozzle after-body design [50]. Several studies have been performed to improve thermofluidic losses in different components of the engine; however, the breadth and width of these studies is beyond the scope of this review. In this section, a top-level overview of some of the notable technologies in this area is presented.

10.9 Integrated Health Monitoring and Engine Management Systems

The thrust control of modern turbofan engines is managed by the full authority digital engine control (FADEC) system on-board the aircraft. Essentially, the FADEC is responsible for interpreting the pilot power request input from the throttle setting or power level angle (PLA) and adjusting the FF to the combustion chamber accordingly, simultaneously ensuring that the engine is able to operate within safety limits. This process is important for the smooth and stable operation of the engines as uncontrolled adjustments to engine parameters could lead to compressor surge, causing significant damage to the engine and potential loss of human life. Figure 10.13 shows a simplified version of the control logic implemented by the FADEC to control engine thrust.

Since the engine's thrust cannot be measured directly, the engine fan speed (N1) or the engine pressure ratio (EPR) can be used as the primarily control variable correlated to the thrust. Other variables that are measured by sensors in the engine include the core compressor spool rotation speed (N2), pressure and temperature measurements at the fan inlet, LP compressor and HP compressor exits, and an exhaust gas temperature (EGT) sensor that measures the gas temperature at the exit of the LP turbine.

Despite providing guaranteed performance and safe operation throughout the engine operating life, current engine control architectures are unable to dynamically adapt to changes in the engine's operating environment, because their design is based on clean

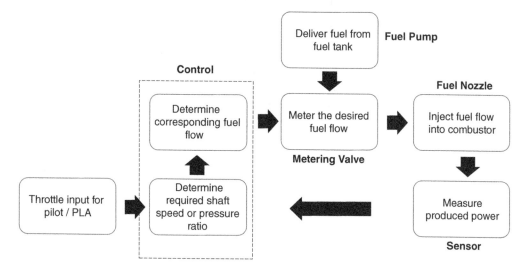

Figure 10.13 Process diagram of simple engine control system. Adapted from [51].

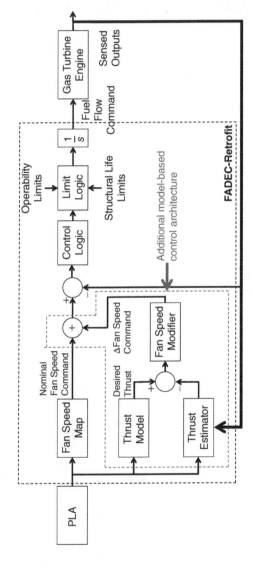

Figure 10.14 Block diagram of modified engine control loop. Adapted from [55].

operating conditions [52]. The set of control gains used in traditional engine control logic is determined using aero-thermal mathematical models of the engine based on mass, energy, and momentum conservation laws [53]. However, during operation, engines are exposed to several physical conditions such as erosion, corrosion, and the build-up of dirt and carbon (soot) in sensitive areas. Consequently, important parameters such as flow capacity, effective nozzle areas, and engine efficiency are affected negatively, thereby altering the performance characteristics of the engine [54]. The problems associated with this gradual deterioration become noticeable over a given period of time, with several pilots stating that they are required to manually make adjustments to the throttle in order to achieve the desired output setting, thereby increasing their workload and stress [55].

In order to implement dynamic adaptation of the control logic, a model-based control architecture which takes data from engine sensors to perform fault diagnostics and recalibration of control gains in real-time has been proposed. An engine monitoring unit (EMU) can be installed in the engine to assist the FADEC with thrust control. The associated modified control architecture is illustrated in Figure 10.14.

The on-board model of the engine adapts to the changes in engine condition and makes adjustments to the fan speed command signal accordingly. This not only alleviates the pilot's workload, but also helps with the maximisation of fuel efficiency, reduction of harmful emissions, and extending engine life. An additional benefit includes better control and operability of the aircraft when flying outside design conditions, for example in an emergency scenario, as the model makes dynamic changes to the command inputs based on real-time data from sensors. However, in order to safely allow full autonomy in intelligent engine management, control, and diagnostics systems, it is essential that the real-time model used to describe the behaviour of the turbofan engine has a high confidence level. Examples of potential models that can be utilised for this application include STORM (Self Tuning On-Board Real-time Model) by P&W and ALPVM (Adaptive Linear Parameter Varying Model) [56]. Aside from real-time applications, engine health monitoring and management systems can also increase the safety and reliability of engine operation in the long term, by early detection of performance degradation, continuous tracking of engine health parameters, and accurate prediction of remaining useful life (RUL) [57].

As of this point, the only factor involved in the control loop that modifies the operating conditions of the engine is the FF. However, this technology could potentially pave the way to other factors, such as variable pitch turbofan blades and variable reduction ratio gearbox technology being involved. The TRL of this technology is still relatively low and would require more development, testing, and validation to reach the level of technical maturity required for commercial use by engine manufacturers.

10.10 Emissions Trends

Unsurprisingly, modern turbofan engines exploiting some of the emerging technologies reviewed in this chapter are consistently more fuel efficient and produce fewer emissions. An analysis was carried out to extrapolate trends in the levels of thrust-specific HC, CO, and NO_x emissions as well as the total fuel burn per unit thrust for an LTO cycle of 584 aircraft engines over the past few decades with certified data taken from the ICAO Aircraft

Engine Emissions Databank Issue 25A (30th May 2018). The Emission Index (EI) of CO and HC (g kg^{-1} of fuel) were taken for the aircraft idle phase and the EI of NO$_X$ (g kg^{-1} of fuel) was taken for the aircraft take-off phase. The amount of fuel consumed in a reference ICAO-defined LTO cycle [58] was used as the most suitable indicator to obtain the thrust specific level of emissions for each engine. The calculation of fuel burned per unit thrust F (kg kN^{-1}) for each aircraft engine is shown in Eq. (10.9).

$$F_{LTO} = \left(\tau_{takeoff} \cdot \frac{FF_{takeoff}}{T_{takeoff}} \right) + \left(\tau_{climb} \cdot \frac{FF_{climb}}{T_{climb}} \right) +$$
$$+ \left(\tau_{descent} \cdot \frac{FF_{descent}}{T_{descent}} \right) + \left(\tau_{taxi} \cdot \frac{FF_{taxi}}{T_{taxi}} \right) \tag{10.9}$$

Where τ is the time (sec), FF is the fuel flow (kg s^{-1}), and T is the thrust (kN) at each corresponding phase. The details of an ICAO-defined reference LTO cycle is given in Table 10.3.

The results of this analysis are plotted in Figures 10.15–10.18. Additionally, the variation of HC, CO, NO$_X$, and CO$_2$ emission indexes as well as the TSFC of these engines with increasing thrust setting are plotted in Figures 10.19–10.22 and are categorised by the decade of introduction to the aircraft engine market. The data points for each thrust setting were obtained from their corresponding flight phase as detailed in Table 10.3. A nonlinear

Table 10.3 Reference Landing and Takeoff (LTO) cycle [58].

Phase	Thrust (T, %)	Time (τ, min)
Takeoff	100	0.7
Climb	85	2.2
Descent	30	4.0
Taxi	7	26

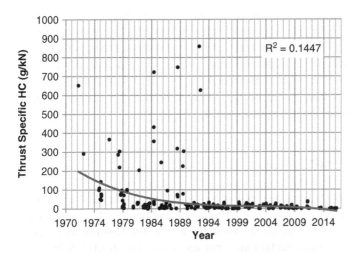

Figure 10.15 Variation of thrust specific HC emissions for a reference LTO cycle of aircraft engines.

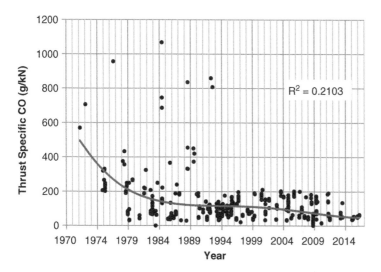

Figure 10.16 Variation of thrust specific CO emissions for a reference LTO cycle of aircraft turbofan engines.

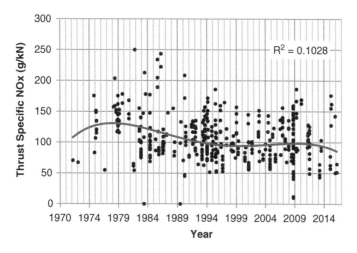

Figure 10.17 Variation of thrust specific NOX emissions for a reference LTO cycle of aircraft turbofan engines.

symbolic regression method was used to obtain the best fit trends of the datasets, using exponential models for HC and CO and second order polynomial models for NO_X, CO_2, and TSFC.

The fitting functions and the associated coefficients of determination R^2 are detailed in Table 10.4.

From Figures 10.15–10.16, it is evident that the general trend of HC and CO emissions from aircraft engines has closely followed a negative exponential decay, with a few exceptions arising from 1985 to 1995. The LTO cycle TSFC depicted in Figures 10.18 and 10.23 has experienced a steady decline since the 1970s. However, while apparently linear, the

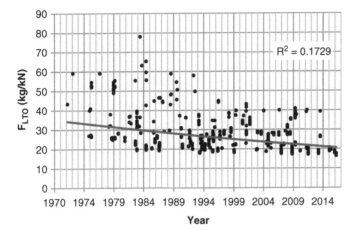

Figure 10.18 Variation of total fuel burn per unit thrust for a reference LTO cycle of aircraft turbofan engines.

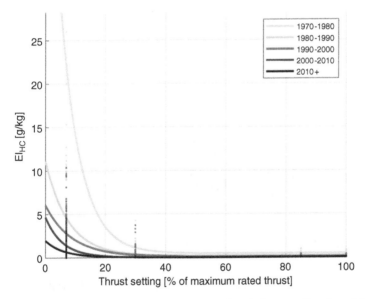

Figure 10.19 Empirical fits of fuel-specific HC emissions as a function of the throttle of aircraft turbofan engines.

trend is more closely captured by a negative exponential fitting (higher R^2), so the trend is flattening out progressively. Due to the significant correlation between fuel consumption and CO_2 emissions, a similar trend is observed in CO_2 emissions in Figure 10.22.

While economic factors (fuel consumption) rather than climate concerns played and still play a prominent role in the decrease of CO_2 emissions, reductions in CO and HC emissions were partly justified by the increase in efficiency and partly by the mitigation of air quality and health hazards posed by these emissions. However, these pollutant-emission reductions were mostly afforded by ordinary technological advancements. Moreover, as

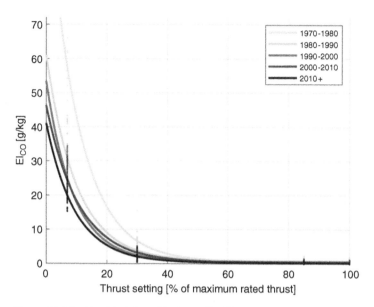

Figure 10.20 Empirical fits of fuel-specific CO emissions as a function of the throttle of aircraft turbofan engines.

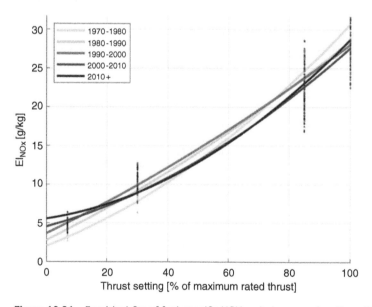

Figure 10.21 Empirical fits of fuel-specific NOX emissions as a function of the throttle of aircraft turbofan engines.

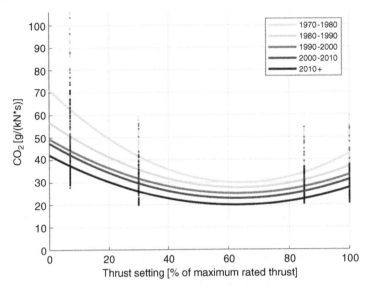

Figure 10.22 Empirical fits of thrust-specific CO_2 emissions as a function of the throttle of aircraft turbofan engines.

Table 10.4 Fitting functions and the associated coefficients of determination of the empirical fits.

Year	EI_{CO}	EI_{HC}	EI_{NO_x}	TSFC
1970–1980	$EI_{CO} =$ $e^{(-9.75\tau + 4.72)}$ $+ 0.570$	$EI_{HC} =$ $e^{(-15\tau + 4.13)} +$ 0.447	$EI_{NO_x} = 13.4\tau^2 +$ $15.2\tau + 2.03$ $R^2 = .9996$	$TSFC =$ $.0319\tau^2 - .0409\tau +$ $.0226$ $R^2 = .9232$
1980–1990	$EI_{CO} =$ $e^{(-9.95\tau + 4.10)}$ $+ 0.769$	$EI_{HC} =$ $e^{(-13\tau + 2.40)} +$ 0.166	$EI_{NO_x} = 6.27\tau^2 +$ $19.5\tau + 2.79$ $R^2 = .9974$	$TSFC =$ $.0228\tau^2 - .0290\tau +$ $.0180$ $R^2 = .9274$
1990–2000	$EI_{CO} =$ $e^{(-11.1\tau + 3.97)}$ $+ 0.541$	$EI_{HC} =$ $e^{(-11.2\tau + 1.79)}$ $+ 0.082$	$EI_{NO_x} = 5.18\tau^2 +$ $19.2\tau + 3.66$ $R^2 = .9910$	$TSFC =$ $.0194\tau^2 - .0245\tau +$ $.0156$ $R^2 = .9142$
2000–2010	$EI_{CO} =$ $e^{(-9.27\tau + 3.83)}$ $+ 0.380$	$EI_{HC} =$ $e^{(-17.2\tau + 1.54)}$ $+ 0.0224$	$EI_{NO_x} = 12.2\tau^2 +$ $10.9\tau + 5.54$ $R^2 = .9799$	$TSFC =$ $.0196\tau^2 - .0247\tau +$ $.0150$ $R^2 = .9086$
2010+	$EI_{CO} =$ $e^{(-10.4\tau + 3.71)}$ $+ 0.293$	$EI_{HC} =$ $e^{(-16.3\tau + 0.666)}$ $+ 0.031$	$EI_{NO_x} = 17.2\tau^2 +$ $5.99\tau + 5.58$ $R^2 = .9602$	$TSFC =$ $.0179\tau^2 - .0224\tau +$ $.0133$ $R^2 = .9064$

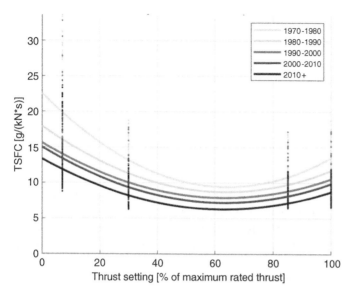

Figure 10.23   Empirical fits of TSFC as a function of the throttle of aircraft turbofan engines.

Figures 10.19 and 10.20 show, these emissions are still very significant in the most critical conditions (low temperature/throttle). Even more noticeable is the historical evolution of specific NO_X emissions shown in Figure 10.17 and 10.21, which follows a completely different trend, with a notable increase up to 1980s and a mostly steady level up to 2010s. This trend is mainly due to the increase in engine core temperatures to achieve better thermal efficiency, which counterbalanced the decreases in NO_X formation afforded by higher BPRs and associated fuel efficiency. Because of this undesirable trend emerging both in aviation and in the road transport sector, concern about NO_X emissions has recently attracted widespread public attention and significant improvements are now mandated, particularly with the development of lean combustion methods.

Advances in low-emissions technologies – particularly including the ones reviewed here – coupled with stricter carbon emission reduction policies may lead to the assumption that a declining trend in emissions could continue. However, Figures 10.15, 10.16 and 10.19, 10.20 very clearly indicate that the magnitude of reduction achieved decreased each decade despite sustained investments. In other words, technological advancements have yielded diminishing returns and will not significantly impact overall emissions in the future. This issue is well acknowledged in the scientific community and was also brought to the attention of policy makers, because those disruptive technologies that could afford a shift of gear will require policy incentives and substantial research funding.

From this point on, changes to other limiting factors, such as type of fuel, propulsion technology, and aircraft configuration, will be required to significantly improve the fuel efficiency and mitigate the environmental impact of commercial transport aircraft. This puts a greater emphasis on research into such areas as alternative aviation fuels [59, 60], intelligent engine control systems and novel propulsion methods, such as hybrid-electric distributed propulsion, in the effort to produce ultra-efficient and environmental-friendly aircraft [6, 61, 62].

10.10.1 Hybrid-Electric Distributed Propulsion

As illustrated in the previous section, the advances in aircraft engine systems to improve efficiency and reduce emissions are beginning to yield lower gains for greater effort, whilst more improvements are still required to achieve sustainability. The next step in the development of aircraft design for improved efficiency is considered to be the integration of the subsystems of the aircraft to gain synergistic benefits through positive interactions between these systems [63]. A proposed method of achieving this is using distributed propulsion, in which the thrust force of an aircraft is distributed about the airframe with multiple smaller engines as opposed to the conventional propulsion configuration of using large engines producing concentrated thrust vectors.

In particular, this concept has a good synergy with emerging hybrid-electric technologies as an alternative to solely using internal combustion engines. Both concepts can combine and provide major benefits in terms of noise reduction, better reliability, shorter take-off and landing distances, better specific fuel consumption, and improved flight stability [6]. Hybrid power systems have been proposed by many to bridge the gap between energy densities of electrochemical cells and liquid chemical fuels used in internal combustion engines.

10.11 Conclusions

Turbofan engines are the most widely adopted aircraft propulsion technology at the moment and they are likely to maintain a dominant role at least for a couple of decades. Significant expectations are placed on advanced turbofan technologies to revert the currently unsustainable impact of aviation on the environment in spite of the steady growth of traffic. The advanced technologies reviewed here improve turbofan engines in different ways, with some addressing weight reductions, pressure, and BPR increases as well as thrust and thermal efficiency increases. Many of the novel turbofan technologies discussed contribute to achieving this objective, however, significant challenges are still to be faced when considering a widespread adoption of these technologies. In addition to the abnormally rapid wearing recently encountered by several production models, the analysis of current trends in aircraft emissions points to the fact that recent advances in turbofan engine technology have yielded diminishing returns despite sustained investment and that the reduction in emissions is progressively flattening. This calls for urgent disruptive technological advances, such as turboelectric propulsion, alternative aviation fuels, and innovative aerodynamic configurations. Due to the significant risk associated with these technologies, effective policy incentives and targeted research funding initiatives will be essential to spur a new generation of more sustainable aircraft technologies that will eventually meet the emission goals set by international organisations.

References

1 Baughcum, S.L., Begin, J.J., Franco, F. et al. (1999). *Aircraft Emissions: Current Inventories and Future Scenarios*. Cambridge: Cambridge University Press, 1999.

2 Wulff, A. and Hourmouziadis, J. (1997). Technology review of Aeroengine pollutant emissions. *Aerospace Science and Technology* 1: 557–572.

3 Federal Aviation Administration (2015). Aviation emissions, impacts & mitigation: a primer.

4 Nicol, D., Malte, P.C., Lai, J. et al. (1992). "NOx sensitivities for gas turbine engines operated on lean-premixed combustion and conventional diffusion flames", *Proceedings of ASME 1992 International Gas Turbine and Aeroengine Congress and Exposition (TurboExpo 1992)*, Cologne, Germany. p. V003T06A012, 1992. DOI: https://doi.org/10.1115/92-GT-115

5 Elser, M., Brem, B.T., Durdina, L. et al. (2019). Chemical composition and radiative properties of nascent particulate matter emitted by an aircraft turbofan burning conventional and alternative fuels. *Atmospheric Chemistry and Physics* 19: 6809–6820. https://doi.org/10.5194/acp-19-6809-2019.

6 Sliwinski, J., Gardi, A., Marino, M., and Sabatini, R. (2017). Hybrid-electric propulsion integration in unmanned aircraft. *Energy* 140: 1407–1416, https://doi.org/10.1016/j.energy.2017.05.183.

7 Lim, Y., Gardi, A., and Sabatini, R. (2015). Modelling and evaluation of aircraft contrails for 4-dimensional trajectory optimisation. *SAE International Journal of Aerospace* V124.

8 Zaporozhets, O. and Synylo, K. (2017). Improvements on aircraft engine emission and emission inventory asesessment inside the airport area. *Energy* 140: 1350–1357.

9 Froines, J. R. Ultrafine particle health effects. http://burningissues.org/car-test/medical_effects/pre-conference_2_froines.pdf.

10 Şöhret, Y., Kıncay, O., and Karakoç, T.H. (2018). An environment-friendly engine selection methodology for aerial vehicles. *International Journal of Green Energy* 15: 145–150. https://doi.org/10.1080/15435075.2017.1324788.

11 Lee, D.S., Pitari, G., Grewe, V. et al. (2010). Transport impacts on atmosphere and climate: Aviation. *Atmospheric Environment* 44: 4678–4734.

12 Penner, E., Lister, D.H., Griggs, D.J. et al. (1999). *Aviation and the Global Atmosphere – a Special Report of IPCC Working Groups I and II*. Cambridge: Cambridge University Press, 1999.

13 Dankanich, A. and Peters, D. (2017). Turbofan engine bypass ratio as a function of thrust and fuel flow. *Mechanical Engineering and Materials Science Independent Study*, 34.

14 Ballal, D.R. and Zelina, J. (2004). Progress in aeroengine technology (1939–2003). *Journal of Aircraft* 41: 43–50. https://doi.org/10.2514/1.562.

15 Tinseth, R. (2011). The right fan size is a balance of all design rconsiderations. Presented at the Northwest Aerospace Alliance 11th Annual Conference, Seattle, WA, USA.

16 Basavaraj, A. (2015). Estimation of wave drag of non-transonic airfoils using Korn equation. *International Journal of Innovative Research in Science, Engineering and Technology* 4: 2119–2126.

17 Mason, W.H. (2006). Transonic aerodynamics of airfoils and wings. In: *Configuration Aerodynamics*.

18 ICAO (1993). Manual of the ICAO standard atosphere.

19 Hughes, C. (2010). *Geared Turbofan Technology*. NASA Ames Research Center.

20 Zhang, J., Zhou, Z., Wei, W., and Deng, Y. (2017). Aerodynamic design of an ultra-low rotating speed geared fan. *Aerospace Science and Technology* 63: 73–81.

21 Pratt & Whitney (2009). *Geared Turbofan: Maintenance Simplified?* vol. 28.

22 Arvai, E. S. (2011). Comparing the new technology narrow-body engines: GTF vs LEAP maintenance costs. http://airinsight.com/2011/11/09/comparing-the-new-technology-narrow-body-engines-gtf-vs-leap-maintenance-costs/#.VXUixGKJhD-

23 Bellini, C. and Carney, J. (2012). The GEnx: Next Generation Aviation. *Mechanical Engineering*.

24 Xiong, J., Ma, L., Vaziri, A. et al. (2012). Mechanical behavior of carbon fiber composite lattice Core Sandwich panels fabricated by laser cutting. *Acta Materialia* 60: 5322–5334.

25 Lee, S.M. (1992). *Handbook of Composite Reinforcements*. Wiley.

26 Marsh, G. (2012). Aero engines lose weight thanks to composites. http://www.reinforcedplastics.com/view/29144/aero-engines-lose-weight-thanks-to-composites-part-1.

27 Mecham, M. (2012). GEnx development emphasizes composites, combustor technology. http://www.aviationweek.com/aw/jsp_includes/articlePrint.jsp?storyID=news/aw041706p1.xml&headLine=GEnx%20Development%20Emphasizes%20Composites,%20Combustor%20Technology.

28 GE Aviation (2014). Thinner, lighter, stiffer, stronger: next gen jet engine fan blades use carbon super material. http://www.gereports.com/post/96107121800/thinner-lighter-stiffer-stronger-next-gen-jet.

29 Greenhalgh, E.S. (2009). 1 - introduction to failure analysis and fractography of polymer composites. In: *Failure Analysis and Fractography of Polymer Composites*, 1–22. Woodhead Publishing.

30 Gardiner, G. (2015). Aeroengine composites, Part 2: CFRPs expand. *Composites World*.

31 Liu, Y., Sun, X., Sethi, V. et al. (2017). Review of modern low emissions combustion technologies for aero gas turbine engines. *Progress in Aerospace Science* 94: 12–45.

32 McKinney, R., Cheung, A., Sowa, W., and Sepulveda, D. (2007). The Pratt & Whitney TALON X low emissions combustor: revolutionary results with evolutionary technology. In: *45th AIAA Aerospace Sciences Meeting and Exhibit*. American Institute of Aeronautics and Astronautics.

33 Mongia, H. C. (2013). N+3 and N+4 generation aeropropulsion engine combustors: Part 2: Medium size rich-dome engines and lean-domes. *ASME Turbo Expo 2013: Turbine Technical Conference and Exposition, GT 2013*, San Antonio, Tx.

34 Koff, B.L. (1993). Aircraft gas turbine emissions challenge. In: *Turbo Expo: Power for Land, Sea, and Air*, vol. 78927, V03CT17A083. American Society of Mechanical Engineers.

35 Lee, C.-M. (2013). NASA project develops next generation low-emissions combustor technologies. In: *51st AIAA Aerospace Sciences Meeting Including the New Horizons Forum and Aerospace Exposition*. American Institute of Aeronautics and Astronautics.

36 Heath, C.M. (2014). Characterization of swirl-venturi lean direct injection designs for Avation gas turbine combustion. *Journal of Propulsion and Power* 30: 1334–1355.

37 Dhanuka, S.K., Temme, J.E., and Driscoll, J. (2011). Unsteady aspects of lean premixed prevaporized gas turbine combustors: flame-flame interactions. *Journal of Propulsion and Power* 27: 631–641.

38 Pepper, C.B., Nascarella, M.A., and Kendall, R.J. (2003). A review of the effects of aircraft noise on wildlife and humans, current control mechanisms, and the need for further study. *Environmental Management* 32: 418–432.

39 Greitzer, E.M., Tan, C.S., and Graf, M.B. (2004). *Internal Flow: Concepts and Applications*. Cambridge University Press, Cambridge, UK.

40 Smith, G.D.J. and Cumpsty, N.A. (1984). Flow phenomena in compressor casing treatment. *Journal of Engineering for Gas Turbines and Power* 106: 532–541.

41 Paduano, J.D., Greitzer, E.M., and Epstein, A.H. (2001). Compression system stability and active control. *Annual Review of Fluid Mechanics* 33: 491–517.

42 Li, F., Li, J., Dong, X. et al. (2017). Influence of SPS casing treatment on axial flow compressor subjected to radial pressure distortion. *Chinese Journal of Aeronautics* 30: 685–697.

43 Sun, X., Sun, D., Liu, X. et al. (2014). Theory of compressor satablility enhancement using novel casing treatment. *Journal of Propulsion and Power* 30.

44 Wang, W., Chu, W., and Zhang, H. (2017). Mechanism study of performance enhancement in a subsonic axial flow compressor with recirculating casing treatment. *Proceedings of the Institution of Mechanical Engineers, Part G: Journal of Aerospace Engineering* 232: 680–693.

45 Guinet, C., Streit, J.A., Kau, H.-P., and Gümmer, V. (2014). Tip gap variation on a transonic rotor in the presence of tip blowing. In: *Turbo Expo: Power for Land, Sea, and Air*, vol. 45608, V02AT37A002. American Society of Mechanical Engineers.

46 Coban, K., Colpan, C.O., and Karakoc, T.H. (2017). Application of thermodynamic laws on a military helicopter engine. *Energy* 140: 1427–1436.

47 Colmenares Quintero, R.F. (2009). *Techno-Economic and Environmental Risk Assessment of Innovative Propulsion Systems for Short-Range Civil Aircraft*. Department Of Power And Propulsion, Cranfield University.

48 Milancej, M. (2011). *Advanced Gas Turbine Cycles: Thermodynamic Study on the Concept of Intercooled Compression Process*. Saarbrücken: VDM Verlag Dr. Müller.

49 Majić, F., Efraimsson, G., and O'Reilly, C.J. (2016). Potential improvement of aerodynamic performance by morphing the nacelle inlet. *Aerospace Science and Technology* 54: 122–131.

50 Goulos, I., Stankowski, T., MacManus, D. et al. (2018). Civil turbofan engine exhaust aerodynamics: impact of bypass nozzle after-body design. *Aerospace Science and Technology* 73: 85–95.

51 Jaw, L.C. and Mattingly, J.D. (2009). *Aircraft Engine Controls: Design, System Analysis, and Health Monitoring*. American Institute of Aeronautics and Astronautics.

52 Mohammadi, E. and Montazeri-Gh, M. (2016). Active fault tolerant control with self-enrichment capability for gas turbine enginesw. *Aerospace Science and Technology* 56: 70–89.

53 Yazar, I., Yavuz, H.S., and Yavuz, A.A. (2017). Comparison of various regression models for predicting compressor and turbine performance parameters. *Energy* 140: 1398–1406.

54 Sogut, M.Z., Yalcin, E., and Karakoc, T.H. (2017). Assessment of degradation effects for an aircraft engine considering exergy analysis. *Energy* 140: 1417–1426.

55 Garg, S. (2012). Intelligent propulsion control and health management. In: *Advances in Intelligent and Autonomous Aerospace Systems*, vol. 241, 201–228. American Institute of Aeronautics and Astronautics, Inc.

56 Lu, F., Qian, J., Huang, J., and Qiu, X. (2017). In-flight adaptive modeling using polynomial LPV approach for turbofan engine dynamic behavior. *Aerospace Science and Technology* 64: 223–236.

57 Yu, J. (2017). Aircraft engine health prognostics based on logistic regression with penalization regularization and state-space-based degradation framework. *Aerospace Science and Technology* 68: 345–361.

58 Graver, B. M. and Frey, H. (2018). Estimation of Air Carrier Emissions at Raleigh-Durham International Airport. Paper 2009-A-486-AWMA, Proceedings, 102nd Annual Conference and Exhibition, Air & Waste Management Association, Detroit, Michigan, June 16–19, 2009.

59 Coban, K., Şöhret, Y., Colpan, C.O., and Karakoç, T.H. (2017). Exergetic and exergoeconomic assessment of a small-scale turbojet fuelled with biodiesel. *Energy* 140: 1358–1367.

60 Yilmaz, N. and Atmanli, A. (2017). Sustainable alternative fuels in aviation. *Energy* 140: 1378–1386.

61 Yin, F., Gangoli Rao, A., Bhat, A., and Chen, M. (2018). Performance assessment of a multi-fuel hybrid engine for future aircraft. *Aerospace Science and Technology* 77: 217–227.

62 Baharozu, E., Soykan, G., and Ozerdem, M.B. (2017). Future aircraft concept in terms of energy efficiency and environmental factors. *Energy* 140: 1368–1377.

63 Kim, H. D. (2010). Distributed propulsion vehicles. Presented at the 27th International Congress of the Aeronautical Sciences.

64 Guan, K. (2015). Opportunities and Challenges of Advanced Turbofan Technologies, MSc Thesis, School of Aerospace, Mechanical and Manufacturing Engineering, RMIT University, Melbourne, Australia.

65 Ranasinghe, K., Guan, K., Gardi, A., and Sabatini, R. (2019). Review of advanced low-emission technologies for sustainable aviation, *Energy*, 188: 115945.

11

Approved Drop-in Biofuels and Prospects for Alternative Aviation Fuels

Graham Dorrington

School of Engineering, RMIT University, Bundoora, Victoria, Australia

11.1 Introduction

As well as modifying aircraft systems, airports, and aircraft operating procedures, another opportunity to bring about 'more sustainable' air transportation is to reduce (or even eliminate) dependence on fuel derived from petroleum, i.e. from crude oil or oil shale/sand. Since the 1950s, nearly all currently certified passenger and cargo-carrying aircraft fitted with gas turbine power plants (including turboprops and turbofans) have used aviation turbine fuel (ATF) produced by petroleum refineries. Currently, the global aviation sector consumes ~220 million tonnes per annum of ATF. After combustion, this global usage results in the net annual atmospheric release of ~700 million tonnes of carbon dioxide (CO_2) as well as other greenhouses gases contributing to human-induced global warming [1, 2]. In effect, fossil carbon that has been essentially trapped within the Earth's crust for ~300 million years is currently being extracted and then released as CO_2 at an unprecedented rate. Collectively, all human industrial and transportation activity has resulted in the CO_2 level of the atmosphere increasing by about 30% over the past 50 years, without any reliable understanding of the possible environmental consequences [3]. Whilst the aviation contribution to this global emission is relatively small compared with the releases caused by coal-fired power stations, there are growing calls to reduce CO_2 emissions in all sectors.

One seemingly obvious way to reduce the aviation component of this emission is utilise biological feedstock to produce aviation biofuel or 'sustainable ATF'. Since plants build themselves by fixing carbon from atmospheric CO_2 via photosynthesis, the use of plant-derived sustainable ATF could in principle result in significantly reduced net carbon emissions. Indeed, a comprehensive study by Stratton et al. [4] concludes that a reduction of more than 50% in life-cycle greenhouse emissions could be achieved. Unfortunately, despite recent efforts in the aviation sector to introduce sustainable ATF, such large reductions have not yet been realised, nor does it seem likely that they are viable in the foreseeable future. Nevertheless, minor yet significant reductions might be achieved within the next decade, provided that commercially viable biochemical production routes can be realised and/or suitable regulations are imposed. More specific reasons for this tentative prognosis are outlined in this chapter, but essentially the main stumbling block is that

Sustainable Aviation Technology and Operations: Research and Innovation Perspectives, First Edition.
Edited by Roberto Sabatini and Alessandro Gardi.
© 2024 John Wiley & Sons Ltd. Published 2024 by John Wiley & Sons Ltd.
Companion Website: www.wiley.com/go/sustainableaviation

all the existing sustainable ATF batches produced to date have been far more expensive than ATF produced from crude oil by well-established petroleum refineries. Successful commercial airlines have also evolved to high levels of efficiency by reducing unnecessary operational costs and it would be difficult for them to justify the uptake of more expensive biofuels unless it is enforced upon them, or the purchasing pressure of passengers becomes more environmentally inclined. Furthermore, to compound this problem, there is no clear evidence that the price of crude oil will rise to problematic high levels, in excess of US$120 per barrel - as was experienced in 2008. On the contrary, over the past decade the price of crude oil has remained at ∼ US$83 per barrel.

To examine the reasons why the introduction of sustainable ATF is likely to be hampered in the near future, it is necessary to consider the overall certification process and the reasons it has evolved. In this chapter, the main focus is on the certification requirements of so-called 'drop-in' ATF within the civil sector, but in the last two sections 'non-drop-in' alternative ATF options are also briefly discussed.

11.2 Currently Approved ATF Production Routes

The International Civil Aviation Organization (ICAO) effectively imposes a general directive on ATF supply [5]. In the civil aviation sector, the most common, approved grades of ATF include:

1. Jet A-1 and Jet A (available in the USA) complying with ASTM D1655 [6].
2. Jet A-1 (or F-35) complying with a UK Defence Standard, DEF STAN 91-91 [7].
3. TS-1 (available in Russia) complying with GOST 10227.
4. Jet Fuel No. 3 (available in China) complying with GB6537

Jet A-1 (or 'Avtur'), is the most commonly available ATF used within the international airline sector. The differences between compliance with ASTM D1655 and DEF STAN 91-91 are negligible, and these documents may be considered as equivalent. ASTM D1655 version 15d [6] calls up 64 other ASTM standards – mainly test methods designed to demonstrate compliance.

To control the introduction of sustainable ATF, an internationally agreed standard specification, 'ATF containing synthesized hydrocarbons', ASTM D7566 [8], was introduced by ASTM International in 2011. At the time of writing, this standard has seven approved annexes that specify production of ATF from seven alternative (non-petroleum) chemical processing routes:

1. Annex A1: Fischer-Tropsch (F-T) Hydroprocessed Synthesised Paraffinic Kerosine (SPK)
2. Annex A2: SPK from Hydroprocessed Esters and Fatty Acids (HEFAs).
3. Annex A3: Synthesised Iso-Paraffins (SIPs) from Hydroprocessed Fermented Sugars.
4. Annex A4: F-T Synthesised Kerosine with Aromatics (SPK/A) Derived by Alkylation of Light Aromatics from Non-Petroleum Sources.
5. Annex A5: Alcohol-to-Jet SPK (ATJ-SPK)
6. Annex A6: Synthesized Kerosine from Hydroprocessed Conversion of Fatty Acid Esters and Fatty Acids
7. Annex A7: SPK from Hydroprocessed Hydrocarbons and HEFAs

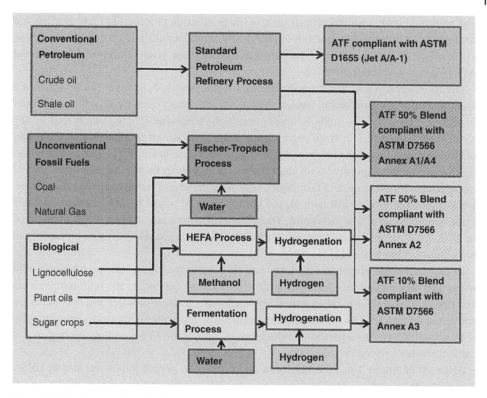

Figure 11.1 Approved production routes for ATF from fossil and bio-carbon sources.

Currently, the main carbon feedstock for the F-T routes (covered by Annex A1 and A4) is natural gas or coal, rather than a renewable biological feedstock, as shown in Figure 11.1. Consequently, ATF produced via the F-T routes does not necessarily result in any net carbon emission reduction and may actually result in net increases compared to the standard petroleum refinery route. Currently, D7566 also requires that ATF batches produced through all the alternative routes are blended with fuel compliant D1566 (i.e. Jet A, or Jet A-1 from conventional petroleum sources). Currently, the maximum allowable blend is 50% and for some production routes it is limited to 10%. The specific reasons for these current blend limits are outlined later in this chapter, but essentially their adoption reflects the first stage of a conservative, safety-motivated development strategy avoiding any possibility of aircraft system failure resulting from the accidental introduction of non-compliant ATF.

The production of SPK from HEFA (Annex A2) depends on the use of bio-oils derived from suitable plants such as jatropha, or switchgrass [4, 9]. The HEFA SPK production process is complex (described in more detail later), but essentially involves trans-esterification of the plant bio-oil using methanol and then chemical reaction with hydrogen in a catalytic chamber at elevated temperature (hydrogenation), in order to produce a final fuel product with a high percentage of paraffins, typically in the c11-C16 range. Despite all the energy and material inputs required at each stage of the production process, life-cycle analysis has demonstrated that the introduction of HEFA SPK would ameliorate greenhouse warming [4]; however, the costs of HEFA SPK currently exceed those of standard petroleum-refined ATF by a factor of about three [9].

The third approved biosynthesis-production route, Annex 3, involves the production a specific synthetic hydrocarbon: farnasane (2, 6, trimethyldodecane, $C_{15}H_{32}$) discussed later in section 11.5.3. This is made by hydroprocessing farnasene ($C_{15}H_{24}$) produced by the fermentation of sugar (e.g. from sugar cane). Again, the current problem is that the prices of farnasene and farnasane far exceed that of petroleum- refined ATF [11]. The latter is also significantly more expensive than the former, since hydrogenation requires three mol of hydrogen gas for every mole of farnasene. Assuming the cost of farnasane could be reduced, a 10% blend (or less) of SIP with D1566 compliant ATF might become more acceptable from a commercial airline viewpoint. The fifth production route, Annex 5, was introduced in 2018. This 'alcohol-to-jet fuel' route involves oligiomerization and hydroprocessing of bio-ethanol or bio-butanol [12].

In the sixth route, Annex 6, free fatty acid from processed waste oils is combined with water and then passed to a "catalytic hydrothermolysis" reactor. At the high temperature and pressure conditions reached within the reactor, the mixture of water and fatty acids are converted into paraffin, isoparaffin, cycloparaffin, and aromatic compounds. Fractionation is subsequenlty used to separate a product that is suitable for ATF blending referred to as a catalytic hydrothermolysis jet (CHJ).

Approval of Annex 7 has recently been fast tracked, to permit the conversion of HEFA derived from an algae feedstock: "hyper-growth" *Botryococcus braunii* which might permit mass production at significantly lower costs.

Currently, Annex 1, 2, 4, 5, and 6 products are limited to 50% blending with ATF certified under specification D1655. Like Annex 3, the biofuels fuel produced via Annex 7 are limited to 10% blending. Therefore, at this juncture, even assuming commercially competitive cost targets could be met, the bio-refinery sector is still unable to supply a compliant, sustainable ATF product independently of the petroleum refinery sector.

To better understand the reasons for this conflicting situation, it is useful to discuss the requirements of ASTM D7566 in more detail and then to outline the reasons why these requirements exist.

11.3 Drop-in ATF Requirements

As might be expected, ASTM D7566 [8] builds on D1655 [6] by calling on the same fuel properties and test methods. A list of some of the fuel property requirements of D7566 is found in Table 11.1. The only major difference between the specifications of Jet A and Jet A-1, is that the former permits a maximum freezing point of $-40\,^\circ$C.

For each annex of D7566, the pre-blend fuel batches must comply with additional requirements. Some of these extra requirements are presented in Tables 11.2 a–f.

Blends complying with these collective requirements are regarded within the aviation sector as 'drop-in' ATFs that may be used 'seamlessly' without any necessary changes to aircraft systems, airport infrastructure, and supply methods. However, it is important to stress that

Table 11.1 Some Jet A-1 requirements specified by ASTM D7566 20c.

Fuel property requirement	Value	ASTM test methods
Minimum net heat of combustion	42.8 MJ kg^{-1}	D4529, D3338, D4809
Density min.-max. range at 15 °C	775–840 kg m^{-3}	D1298, D4052
Minimum aromatic content by volume	8.4%	D6379
Maximum freezing point (Jet A-1)	−47 °C	D5972, D7153, D7154, D2386
Maximum viscosity at −20 °C	8.0 mm^2 s^{-1}	D445, D7042
Minimum flash point	38 °C	D56, D93, D3828
Minimum smoke point	18 mm$^{a)}$	D1322
Maximum naphthalene content by volume	3.0%	D1840
Thermal stability max. filter pressure drop	25 mm Hg at 260 °C	D3421
Maximum wear in fuel lubricity test	0.85 mm	D5001
Maximum total sulphur by mass	0.3%	D1266, D2622, D4294, D5453
Maximum antioxidant additive	24 mg l^{-1}	
Icing inhibitor additive range by volume	0.07–0.15%	D4171

a) This requirement increases to 25 mm, if the naphthalene content is not tested.

their introduction still needs to comply with the 'Standard Practice for Qualification and Approval of New ATFs', ASTM D4054 [13]. This practice requires, for example, potential tests of the new ATF in combustion rigs and ground-based-test turbofan engines, according to an agreed testing programme.

11.4 Reasons Behind ASTM D7566 Property Requirements

11.4.1 Heating Value and Density

One essential requirement of any D7566 compliant ATF blend is the minimum net heat of combustion (or so-called 'lower heating value'), h_f, must exceed a minimum value. D7566 [8] currently invokes the D1655 requirement that $h_f > 42.8$ MJ kg^{-1} (Table 11.1), which effectively ensures that current aircraft designs can achieve a prescribed range with specified minimum fuel reserves. Some Jet A-1 suppliers specify that their product exceeds this minimum with a small margin to avoid the possibility of non-compliance. For example, AIR BP [14] states that its Jet A-1 product has $h_f > 43.15$ MJ kg^{-1} (i.e. a 0.82% margin).

The Bréguet range equation [15] for a turbofan powered aircraft demonstrates that, for a given aircraft gross mass, m_0, fuel mass burn, m_f, and propulsive efficiency, η_p, and lift-to-drag ratio (L/D) the maximum feasible range, R, is directly proportional to h_f,

$$R = \frac{h_f \eta_p \left(\frac{L}{D}\right)}{g_0} \ln\left(\frac{m_0}{m_0 - m_f}\right) \tag{11.1}$$

Table 11.2 Additional requirements as per ASTM D7566.

Batch property requirement	Value	ASTM test methods
(a) Some additional batch requirements ASTM D7566 AnnexA2: HEFA SPK		
Minimum flash point	38 °C	D56, D3828
Density min.-max. range at 15 °C	730–770 kg m^{-3}	D1298, D4052
Maximum freezing point	−40 °C	D5972, D7153, D7154, D2386
Thermal stability max. filter pressure drop	25 mm Hg at 325 °C	D3421
Maximum cycloparaffin mass content	15%	D2425
Maximum aromatic mass content	0.5%	D2425
(b) Some additional batch requirements ASTM D7566 Annex A3, SIP		
Minimum flash point	100 °C	D56, D3828
Density min.-max. range at 15 °C	765–780 kg m^{-3}	D1298, D4052
Maximum freezing point	−60 °C	D5972, D7153, D7154, D2386
Thermal stability max. filter pressure drop	25 mm Hg at 355 °CC	D3421
Maximum olefin mass content	300 mg (Br)/100 g	D2425
Maximum aromatic mass content	0.5%	D2425
(c) Some additional batch requirements ASTM D7566 Annex A4		
Minimum flash point	38 °C	D56, D3828
Density min.-max. range at 15 °C	755–800 kg m^{-3}	D1298, D4052
Maximum freezing point	−40 °C	D5972, D7153, D7154, D2386
Thermal stability max. filter pressure drop	25 mm Hg at 325 °CC	D3421
Maximum aromatic mass content	21.2%	D6379
(d) Some additional batch requirements ASTM D7566 Annex A5		
Minimum flash point	38 °C	D56, D3828
Density min.-max. range at 15 °C	730–770 kg m^{-3}	D1298, D4052
Maximum freezing point	−40 °C	D5972, D7153, D7154, D2386
Thermal stability max. filter pressure drop	25 mm Hg at 325 °CC	D3421
(e) Some additional batch requirements ASTM D7566 Annex A6		
Minimum flash point	38 °CC	D56, D3828
Density min.-max. range at 15 °C	775–840 kg m^{-3}	D1298, D4052
Maximum freezing point	−40 °C	D5972, D7153, D7154, D2386
Thermal stability max. filter pressure drop	25 mm Hg at 325 °CC	D3421

Table 11.2 (Continued)

Batch property requirement	Value	ASTM test methods
(f) Some additional batch requirements ASTM D7566 Annex A7		
Minimum flash point	38 °C	D56, D3828
Density min.-max. range at 15 °C	730–800 kg m^{-3}	D1298, D4052
Maximum freezing point	−40 °C	D5972, D7153, D7154, D2386
Thermal stability max. filter pressure drop	25 mm Hg at 325 °CC	D3421

Consequently the margin in the value of h_f of the supplied ATF directly translates into range reserve: e.g. for an aircraft flying 12 000 km, the range reserve is ∼100 km. This also infers that in principle it would be possible to blend in HEFA SPK at 50% with a slightly lower h_f value. For example, when blending with Jet A-1 supplied by Air BP, the h_f value of the HEFA SPK batch could be reduced to 42.45 MJ kg^{-1}, but no lower. In practice, however, the producer of the HEFA SPK batch would also try to ensure a small margin, to avoid non-compliance.

This requirement also fundamentally restricts the fuel options from a chemical standpoint. The Mendeleyev approximation [16] for net heat release for paraffins in appropriate hydrogen is: carbon mass range is $h_f = 34 + 0.9H$ where H is the percentage value of hydrogen by mass. Therefore, to achieve the required D1655 heating value, H must exceed ∼10%.

Both D1655 and D7566 require that the ATF density must lie between 770 and 840 kg m^{-3} (Tables 11.1 and 11.2a–f). However, the batch specification for the pre-blended (non-petroleum) fuel components permits a slightly lower density range: Annex A2 permits 730–770 kg m^{-3} for HEFA SPK and Annex A3 permits 765–780 kg m^{-3} for SIP. Using Ragozin's approximate hydrocarbon density formula [16], $\rho_f = 1742 - 67H$, implies that H must be in the range 13–15%.

The requirement on maximum density is arguably not as important as that of the minimum density, since aircraft fuel tanks do not have to be completely filled. Indeed, it is hard to identify an engineering reason as to why an ATF with an excessively high density of (say) 850 kg m^{-3} should be rejected. Conversely, it is obviously necessary to ensure the aircraft's maximum available fuel tank volume is not exceeded for a given fuel load.

For some aircraft designs, fuel tank capacity actually limits range instead of fuel mass upload capability. Whilst there is no explicitly stated requirement on the fuel heating value per unit volume, clearly, combining the aforementioned requirements on h_f and density, ρ_f, D7566 requires that $\rho_f\, h_f \sim 33$ GJ m^{-3}. Most civil transport aircraft store fuel in the aircraft wing box. The volume of conventional wing boxes varies roughly in proportion with $S^{3/2}$, where S in the wing planform area. Since the maximum wing lift coefficient and the airfield approach speed are both constrained by operating requirements, to a first approximation, the wing area varies in proportion to the gross weight. Consequently, based on the wing box volume constraint, the maximum feasible fuel load varies roughly as $m_0^{3/2}/\rho_f$. Equation (11.1) demonstrates that (for a given range, fuel heating value, propulsive efficiency, and lift-to-drag ratio) the required fuel load varies in proportion to m_0. For this reason an aircraft such as the Airbus A320 size is more likely to be constrained by wing box volume than a large one, such as the Airbus A380. Indeed, the former uses the central-internal (through- fuselage) wing box for fuel storage and in some configurations employs in-fuselage range extension tanks, whereas the latter only stores fuel in the external wing box. Therefore, to achieve a long range, smaller aircraft would benefit more from denser fuels than larger aircraft. Alternatively, lower density fuels are less problematic for larger aircraft operating over a shorter range.

11.4.2 Freezing Point and Fluidity at Low Temperatures

The requirement for the maximum fuel freezing point of −47 °C for Jet A-1, is a consequence of the fact that all long-range turbofan powered passenger aircraft flights occur in the lower stratosphere. Aircraft tank cooling occurs progressively throughout such flights during ascent and cruise-climb, since the ambient air temperature falls with altitude in the troposphere approximately according to the standard lapse rate of 6.5 K/km. At an altitude of 11 km (the stratopause), the standard atmospheric temperature is −56 °C, but it should be noted that the corresponding stagnation temperature at Mach 0.85, is approximately −25 °C. The temperature reached in the fuel tank depends on the heat transfer rate, but it will become lower as the flight duration increases. On long haul flights (of 12 hours duration, for example), the fuel temperature may drop to a critical allowable limit, but typically it does not approach −47 °C. For domestic flights within the USA (with durations typically less than ~8 hours), Jet A is permitted to have arelaxed freezing point requirement of −40 °C. However, if Jet A is used for long haul international operations, then there is a possibility that the critical limit will be reached and in this case the aircraft would have to reduce altitude to avoid fuel freezing. Like D1566, D7566 allows blended fuel to have either a nominal minimum freezing point of −47 or −40 °C depending on whether the product is considered equivalent to Jet A-1 or Jet A, respectively. The pre-blend fuel batches of F-T SPK, HEFA-SPK, and SPK/A are only required to have a minimum freezing point of −40 °C, but Annex A requires SIP to have a freezing point of −60 °C.

To further ensure sufficient ATF fluidity in cold conditions, D7655 requires the fuel kinematic viscosity to be less than 8 mm^2 s^{-1} at −20 °C and less than 12 mm^2 s^{-1} at −40 °C and since the Hagen-Poiseuille equation implies that the pressure drop in any pipe is proportional to dynamic viscosity (and fuel density is constrained), this requirement effectively ensures the aircraft fuel pumping power requirements cannot be breached by a fuel that is too viscous.

11.4.3 Combustion Cleanliness

ATF is burned at near-stoichiometric conditions in the primary zone of a gas turbine combustor. The primary flow is then diluted with excess air in the secondary zone to reduce its temperature prior to entry into the turbine. This staged flow arrangement also reduces convective heat transfer to the combustor outer walls. In general, paraffins (alkanes) offer the most desirable combustion characteristics and are relatively 'clean' burning, meaning that it produces low soot (or smoke) and does not result in carbon or gum deposits on the combustor walls. Cycloparaffins also offer relatively clean combustion, but it appears that for legacy reasons their content to limited to less than 15% in all four Annexes.

Olefins (alkenes) also display good combustion characteristics, but the tendency of olefins (especially dienes) to form gum deposits/residues [16, 17] when fuel evaporates is an issue of concern. According to D7566 [8] the olefin content is usually restricted to 1% or less, although no actual upper limit is specified in Annex A1, A2, and A4. The batch specification for SIP in Annex A3, however, limits olefin content through test D2710 that measures bromine reactivity. In effect, this limits the farnasene content and ensures the SIP content is almost 100% farnasane.

Both D1655 and D7566 stipulate that the aromatic content of ATF must not exceed 26.5% by volume (Table 11.1). The driver for this requirement is that high aromatic content fuels generally result in soot formation and carbon deposition. Naphthalenes or bicyclic aromatics are especially conducive to producing more soot, which results in high levels of undesired radiation heat transfer to the combustor walls, hence the naphthalene content is restricted to less than 3% [6, 8]. Soot formation is also controlled by a 'smoke point' test (ASTM D1322) which involves measuring the maximum height that a candle-like flame (supplied by fuel though a wick) remains smokeless.

Combustion efficiency and the tendency for soot formation are also influenced by fuel injection, which involves spray atomisation of the ATF into the combustor. Again, the aforementioned constraint on fuel viscosity partly ensures consistency of spray formation, although a lower limit on viscosity is not imposed. Curiously, there is no mention in D7566 of any constraint on liquid surface tension, which is the principle property governing spray droplet size [18]. Jet A-1 has a surface tension of ~28 mNm^{-1}. Any ATF blend that is likely to comply with D4054 combustion tests would likely need a surface tension that does not deviate from this value appreciably.

11.4.4 Fuel System Compatibility

Gas turbines are lubricated by separate turbine oil feeds. However, aircraft fuel systems (especially fuel pumps) rely on ATF to lubricate their parts. Adequate ATF lubricity is assessed via ASTM D5001 by a Ball-On-Cylinder Lubricity Evaluator (BOCLE) test, which involves the measurement of wear caused by a sphere on a cylinder immersed in ATF. The test requires that the resulting wear scar is limited to a diameter of 0.85 mm (Table 11.2). Severe hydrogenation eliminates components that improve lubricity and improves with cycloparaffin and aromatic content. Additives that also inhibit corrosion (such as HiTec™ 580 or Nalco™ 5403) may be employed to achieve/regain acceptable lubricity, although their polar nature adversely effects fuel filtration and, consequently, they are only used in small quantities, e.g. less than 15 g m^{-3} [8].

As well as lubricity, fuel system sealing is an important issue. Aircraft fuel systems contain nitrile rubber O-rings that swell in volume in the presence of aromatic compounds [19, 20]. This swelling determines the nominal sealing properties of O-ring joints: without swelling the joints would leak. ATF produced by petroleum refineries, complaint with ASTM D1655, typically has a 10–20% aromatic content, which ensures that the O-rings swell as intended. However, the introduction of low aromatic F-T SPK would pose a problem since O-ring swell is reduced. For this reason, the 50% blend limit with D1655 compliant fuel was imposed

[8]. The introduction of F-T SPK/A (with aromatics, Annex A4) appears to be a potential way around this problem, hence it is possible that 100% F-T SPK/A may eventually be approved.

Unlike D1655, minimum aromatic content of blended ATF is specifically constrained by D7566 to 8.4% (Table 11.1). To retain the desired nitrile rubber volumetric swelling, aromatics such as Tetralin (1,2,3,4- tetrahydronaphthalene) may be added to aromatic-depleted fuels [e.g. 20], although the deliberate use of such additives will also have the adverse effect of reducing the smoke point of the fuel. In this regard, it is also worth noting that Annex A1 and A3 of D7655 restrict the batch aromatic content of F-T SPK and SIP, respectively, to less than 0.5% prior to blending. In effect, therefore, the 50% blend limit implies that the actual minimum aromatic content of D1655 compliant fuel needs to be in excess of 16% to permit F-T SPK blending.

The eventual replacement of nitrile with fluorocarbon seals [21] appears to be a promising way forward, in order eliminate the paradoxical need for aromatic content. However, whilst fitting production aircraft with fluorocarbon seals is unlikely to pose any major difficulty, retrofitting of existing operational aircraft would be needed before concessions on aromatic content could be agreed with D1655 and D7566.

11.4.5 Flash Point, Thermal, and Oxidation Stability

From a combustion standpoint, high ATF volatility is desirable to ensure good fuel ignitability. However, from a safety viewpoint, the opposite of low volatility and good storage stability is required. Ignition (or flashing) of fuel vapour occurs at certain pressure, temperature, and oxygen concentration conditions. The flash point, as determined by the ASTM D56 test (Table 11.1), is the lowest temperature at which a fuel vapour with air mixture ignites at standard sea level pressure. Both D1655 and D7566 specify a minimum flash point of 38 °C (Table 11.1) is required. In the case of SIP, D7566 Annex A3 specifies a higher minimum flash point of 100 °C. On the other hand, D7566 also states that HEFA-SPK with a lower flash point may be blended with D1655 compliant fuel. The flash point is closely linked to vapour pressure. Fuels with vapour pressures less than that of decane (in the appropriate temperature range) are likely to comply with this requirement.

In ATF feed lines, it is also necessary to avoid chemical decomposition before the fuel reaches the combustor. These feed lines become quite hot, because of the heat transfer from the combustor. Both D1655 and D7566 require ATF to remain thermally stable up to a specified control temperature of 260 °C [6, 8]. However, for the pre-blending batches, Annex A1, A2, and A4 require thermal stability at a control temperature of 325 °C, and Annex A3 calls for 355 °C (without explanation). Compliance is tested via ASTM D3241 by passing the fuel through a heated tube and filter and measuring the filter pressure drop caused by deposit formation. The maximum pressure drop imposed is 25 mmHg in all cases, after a 2.5-hour test duration.

Nearly all current aircraft fuel tanks allow air into the ullage, hence any compliant ATF must also have adequate oxidation stability. Anti-oxidant additives are employed to reduce the risk of oxidation, but added quantities are constrained to lie between 17 mg l^{-1} (minimum) and 24 mg l^{-1} (maximum) for each Annex of D7566.

11.4.6 Sulphur Content and Other Contaminants

Whilst sulphur compounds offer slight improvements to fuel lubricity and possibly increased heating value, the emission of sulphur (and its oxides, etc.) is clearly highly undesirable from an environmental viewpoint. Currently, ASTM D1655 [6] and D7566 [8] restrict total sulphur content to 0.3% by mass (Table 11.1). Past surveys of Jet A-1 composition indicate the actual sulphur content is 321–800 ppm [22] corresponding to up to about 200 000 tonnes of global sulphur emissions per annum. Further reduction of sulphur content, by introduction of ultralow sulphur ATF, is desired by ICAO [22]. In this respect, plant-derived ATF blends have an advantage since they intrinsically low sulphur content, whereas some crude oils have significant sulphur content, and it is expensive to remove. Aside from this environmental desire for this reduction, low sulphur content is required to prevent corrosion of turbine blades by sulphur oxides. Mercaptan (methanethiol) CH_4S content is also already severely restricted to 0.003% [6, 8] since it is known to be reactive with certain elastomers.

Several other fuel contaminants are restricted by D7566, including existent gum to less than 7 mg per 100 ml; metals including phosphorus and halogens to $1 \, mg \, kg^{-1}$; nitrogen to $2 \, mg \, kg^{-1}$; water to $75 \, mg \, kg^{-1}$; electrical conductivity improvers to $5 \, mg \, l^{-1}$. It should also be noted that D1566 calls up a stringent requirement for fatty acid methyl ester (FAME) content to be less than $50 \, mg \, kg^{-1}$, following warnings of unapproved Jet A-1 supply contamination by FAME in 2009 [23].

11.5 Sustainable ATF Production

In previous sections of this chapter the required properties of ATF produced via non-petroleum sources have been discussed, but the actual production processes have not yet been outlined and this is a worthwhile exercise, since it helps to underline some of the problems preventing the immediate introduction of truly sustainable ATF in significant quantities.

11.5.1 Production of Fischer-Tropsch (F-T) SPK

The Fischer-Tropsch (F-T) process was first conceived in Germany around 1944 [24] as a means to convert coal into liquid hydrocarbons, but in principle it can be used to convert any carbon feedstock, including natural gas and plant cellulose. The process essentially involves two stages. Firstly, steam reforming is used to produce a general mixture of carbon monoxide and hydrogen called 'syngas'. For example, combining methane with steam yields three mol of hydrogen for every mole of carbon monoxide, i.e. $CH_4 + H_2O \rightarrow CO + 3H_2$. In the second stage of the process, a catalytic chemical reaction converts the syngas at high pressure into hydrocarbons of various molecular weights according to the following exothermic reaction, $(2n + 1) \cdot H_2 + n \cdot CO \rightarrow C_n H_{(2n+2)} + n \cdot H_2O$ where n is an integer. The F-T process conditions are usually designed to maximise the formation of the desired hydrocarbons. Depending on the catalyst, temperature, and type of process employed, suitable ATF paraffins and olefins with $n > 8$ can be obtained. The obvious downside to this process is that large amounts of power are required to drive the endothermic steam-reforming process.

One of the largest, current operational F-T facilities converting natural gas to ATF, is the Shell Pearl Gas-To-Liquid (GTL) plant in Qatar, which received an investment of ~US$25 billion and currently produces ~140 000 barrels per day of liquid hydrocarbon products, of which ~17% is in the ATF distillate range [24], i.e. ~0.35% of the current global ATF consumption rate. The commercial breakeven point to compete with crude-oil derived ATF is likely to be at least US$65 per barrel. Given that it takes ~4 years to build such plants, it is difficult to conceive that global F-T SPK production will reach a major proportion of the global ATF demand with the next 20 years, unless perhaps the price of crude oil rises well above $65 per barrel and the natural gas price does not rise. As regards more sustainable F-T conversion, there are no plans to use biomass as a significant feedstock at any GTL plant. Consequently, the likelihood of large quantities of sustainable ATF production via the F-T SPK route appears to be unlikely in the near future.

11.5.2 Production of HEFA SPK

Hydroprocessed renewable jet fuel, or HEFA SPK, may be produced from plants that readily yield plant oils, such as camelina, jatropha, algae, or even used cooking oil. Different feedstocks offer different levels of net CO_2 reduction compared to standard Jet A-1. For example, SkyNRG state that carmelina offers a 70% net reduction [25]. Overall environmental life-cycle cost benefits are hard to assess because crops like camelina require water and fertiliser, may alter ground surface albedo, and (in some cases) may displace other food crops. Palm oil plantations have also displaced pristine rain forest in some locations. In terms of yield per hectare, algae appear to be promising, but algae derived oils remain relatively expensive.

The current global production of HEFA SPK is also hard to quantify, since it is closely linked to the associated production of biodiesel. Between 2007 and 2012 the US Department of Defense purchased ~3200 tonnes [9]. In 2013, SkyNRG facilitated the use of ~430 tonnes from JFK airport [25]. From 2015 to 2018, United Airlines intended purchasing ~15 000 tonnes [26].

Development of such production has been hampered by the relatively high cost of HEFA SPK [9]. For example, the US Department of Defence paid ~10 $/litre, for the aforementioned supply.

11.5.3 Production of SIP

The only approved SIP currently specified by Annex 3 of ASTM D7566 is farnasane, which was produced by a Brazilian-based company, Amyris, according to a patented process [27, 28]. The feedstock is sugar (from sugarcane) and the process yields ~30% farnasene, which then has to be hydrogenated. Assuming a global price of sugar of ~US$ 0.35/kg, a minimum base price of farnasene of ~$1 per litre might be reached. However, past prices of farnasane have been much higher [11] with one quote of $9.84 per litre [29] suggesting the need for production scale-up.

Complete turbofan engine tests of 20% farnasane with Jet A-1 [30] have revealed no issues of concern, although, as mentioned earlier, D7566 currently restricts farnasane blending to 10%. The current production rate is low, but ~14 000 tonnes per annum was targeted by Amyris.

11.5.4 Other Synthetic Kerosene Production Routes

The so-called Alcohol-To-Jet (ATJ) route [31], approved by ASTM for blends up to 50% in 2016, appears to be promising at first sight simply since current bio-ethanol production [32] is comparable with global Jet A-1 usage. As well as sugar or corn feedstock, production from waste gas [33], cellulose [34] and lignins [10] may also be commercially possible. If bio-butanol could be produced in similar commercial quantities as corn ethanol [32], then it would be possible to convert the butanol to butene, and then ogliomerize the butene using a patented catalytic process [35].

Fast-pyrolysis of lignocellulose feedstock also appears to be a promising route to achieve high production rates [36].

11.5.5 Case Study: 'Sustainable Mallee Jet Fuel'

In 2013 an industrial-agricultural consortium undertook a project to assess the potential for the conversion of mallee feedstock to jet fuel to supply Perth Airport [37]. The assessment comprised of:

1. A sustainability assessment considered the proposed value chain, including farmer participation.
2. A life cycle assessment to compare mallee jet fuel with standard petroleum derived Jet A-1.

The project was commissioned by Airbus, and involved collaboration with GE, Virgin Australia, Dynamotive Energy Systems (a Canadian-based company specialising in the fast pyrolysis process, which went into receivership), Renewable Oil Corporation, and the Australian Future Farm Cooperative Research Centre. It was one of a number of projects in Airbus's 'European Advanced Biofuel Flighpath' programme [38] aimed at producing 2 million tonnes of sustainable jet fuel by 2020, resulting in a 50% reduction in net greenhouse gas emissions.

Blue mallee is a eucalyptus tree that grows relatively quickly in arid conditions typical of the south western region of Australia, where rainfall is 400–600 mm per annum. It was envisaged the mallee would be planted in rows (or belts) in ~250 000 ha owned by 222 participating farms in the period 2021–2050. The mallee would be cropped as small shrubs by special harvesting equipment and mulched. The mulch would then be transported to an existing fast pyrolysis plant in Katanning (which would have to be scaled-up). The mallee production rate was envisaged to be ~154 000 dry tonnes per annum. The fast pyrolysis process would convert this lignocellulose to bio-oil (~85 000 tonnes per annum) and then the bio-oil would be upgraded by hydrogenation to ~10 000–15 000 tonnes of jet fuel per annum. To bring this into perspective, approximately nine ha of mallee would needed to fuel one flight of an Airbus A330-200 from Perth to Sydney (one way).

The sustainability assessment concluded that the greenhouse gas emissions would be reduced by 40%, assuming the mallee jet fuel could be used 100%. However, if the mallee jet has to be 50% blended with ordinary Jet A-1, as might be expected under current practice, then the green house emission reduction is reduced to 20%. It was found that growing mallee rows would not be detrimental to groundwater depletion and would probably benefit the ecosystem.

The life-cycle cost assessment revealed that the cost of mallee jet fuel was likely to be two to three times higher than ordinary Jet A-1. Unresolved issues included the need to supply low cost hydrogen.

11.6 Past Use of Non-drop-in Alternative Fuels

In the previous sections, the prevailing assumption has been that any future, (more) sustainable ATF must comply with existing or expected ASTM standards, and, in effect, the search is for exact drop-in matches for Jet A/A-1. However, it is important to bear in mind that Jet A/A-1 are not the only ATFs that have been previously employed in the aviation sector. In the 1980s, Tupolev [39] carried out flight-tests on a prototype Tu-155 aircraft first using liquid hydrogen (LH_2). Predictions of the production costs of LH_2 proved to be overly optimistic [40] and hence subsequently the aircraft was fuelled using much cheaper liquefied natural gas (LNG).

The LNG was stored within a $20\,m^3$ cryogenic insulated (vacuum screen) tank-dewar situated at the back of the fuselage and used to supply one NK-8-2 turbofan engine, whilst the other two engines were run on TS-1 (equivalent to Jet A-1). The tests showed that LNG offered a 15% reduction in thrust-specific fuel consumption, and a 17% reduction in CO_2 emissions [40].

In the 1980's a consortium of several other Russian aviation companies, petroleum industries, and universities also proposed a more storable cryogenic fuel as an alternative to TS-1 (the equivalent of Jet A-1), which at that time had a relatively high cost. This new fuel, called 'ACKT', was essentially a mixture of liquid propane (C_3H_8), butane (C_4H_{10}), and pentane (C_5H_{12}), with C_4H_{10} as the dominant component. In order to demonstrate the feasibility of using ACKT and explore its benefits, the Mil Helicopter organisation conducted a development programme [40–42], which resulted in the production of the Mi-8TG prototype using ACKT carried in external pod tanks (see Figure 11.2).

The first flight of the Mi-8TG took place on the 7 September 1987. The pods resulted in an increase in drag coefficient, resulting in a 2% reduction in cruise speed. However, it was found that ACKT offered a 5% lower thrust-specific fuel consumption than TS-1, was significantly cheaper at the time, less toxic, less aggressive to sealants, and did not result in combustion residues in the combustor and turbine blades. The overall operational cost was reduced by 25–30% and the operational range was increased by 5% [40–42].

These examples demonstrate that LNG or C_3H_8/C_4H_{10}-fuelled aircraft are technically feasible, but clearly they would require extensive aircraft and airport system changes.

Figure 11.2 Mi-8TG with external ACKT pods [35–37].

11.7 Basic System Considerations for Alternative Fuels

11.7.1 Aircraft Design

If the fuselage of an aircraft has to be enlarged to cater for the increased fuel volume of a lower density fuel, then there would typically be a reduction in lift-to-drag ratio, as well as an increase in dry mass. The former adverse effect is quite small: if the fuselage contributes about 30% to the overall drag, then a 15% fuselage volume increase would only result in a drag penalty of about 3%. The dry mass increase is appreciable for LH_2 or LNG aircraft, since the storage tank(s) is effectively a heavy cryogenic dewar with thick insulation. In the case of C_4H_{10}, storage (possibly in the wing) would not require any insulation and impose a far lower dry weight penalty.

Unlike Jet A-1, LNG cannot be primarily stored within the limited wing volume of conventional passenger aircraft (to achieve the same range). Several alternative configurations need to be compared. One category has the LNG integrated within the main fuselage, either running above the passenger cabin or bisecting it near the wing, or with two tanks placed at either end of the passenger cabin The other category involves wing-mounted fuel tank pods. The latter introduces a higher drag penalty, but might be considered as an intermediate retro-fit modification option.

11.7.2 Preliminary Performance Comparison

For illustrative purposes, it is useful to compare the approximate relative performance of an LCH_4 aircraft vis-à-vis one fuelled with petroleum-derived Jet A-1. The key performance parameters of primary interest are the relative payload ratio and the relative CO_2 release per unit payload (or passenger) which is proportional to the relative fuel burn-to-payload ratio. The payload ratio of both the LNG and Jet A-1 aircraft are simply given by,

$$m_\lambda/m_0 = 1 - m_{dry}/m_0 - m_f/m_0 \tag{11.2}$$

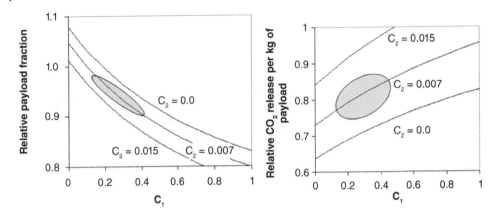

Figure 11.3 Example of preliminary performance evaluation of LNG fuelled aircraft relative to Jet A-1 fuelled baseline, with the following parameters [43].

The dry mass fractions of both aircraft may be expressed as a simple function of the tank volume,

$$m_{dry}/m_0 = \frac{(m_{dry}/m_0)_{avtur}}{(1+c_1)}\left[1+c_1\frac{(m_f/\rho_f)^n}{(m_{JetA}/\rho_{JetA})^n}\right] \tag{11.3}$$

The fuel mass fraction of both aircraft may be approximated by inverting the Breguet equation above,

$$m_f/m_0 = 1 - exp\left[\frac{-Rg_0}{\eta_p h_f(L/D)}\right] \tag{11.4}$$

where the lift-to-drag ratio may also be expressed as a function of tank volume,

$$(L/D)^{-1} = (L/D)_{avtur}^{-1} + c_2\{1-(m_{JetA}/\rho_{JetA})^{2/3}/(m_f/\rho_f)^{2/3}\} \tag{11.5}$$

Figure 11.3 shows that relative performance is dependent on the two unknown parameters c_1 and c_2 (assumed constants) that both effectively dictate the extent of the negative impacts of increased tank mass and size. If $c_1 \cong 0.4$ and $c_2 \cong 0.07$ are achievable, then the relative payload loss of the LNG aircraft may be reduced to about 5–10%, with a CO_2 per passenger-km reduction of around 15–20%. Furthermore, a significant saving would be achieved in direct operating costs (the gain obviously depending on the relative prices of LNG and Jet A-1). These findings are relatively insensitive to changes in operating range and aircraft gross mass

11.8 Future Prospects for Alternative, Non-drop-in Fuels

In principle, it would be possible for an organisation to develop an all-new ATF that is not considered to be drop-in, but is still considered 'fit-for-purpose' by complying with AST D4054 [5]. However, this would require comprehensive engine and aircraft certification trials, as well as the need to introduce all-new fuel supply and bunkering facilities at airports, all of which would be relatively costly.

From an ideal performance standpoint, a fuel with a high heating value and high density is needed by the aviation sector, but unfortunately these two parameters are conflicting. If a moderate reduction in heating value could be tolerated, then denser fuels become viable [44, 45]. On the other hand, the heating values of alkanes rise as the carbon number drops, and the hydrogen-to-carbon ratio increases, but the liquid density falls and the boiling point drops, hence fuels like butane must be stored under pressure or in the sub-cooled condition in insulated tanks, leading to trade-offs between fuel mass and tank mass [43].

Liquid hydrogen (LH$_2$) has been proposed as the ultimate zero-carbon emission ATF solution [e.g. 46, 47], since it has a high energy content ($\sim$120 MJ kg^{-1}). However, the density of liquid hydrogen is relatively low ($\sim$70 kg m^{-3}), leading to designs with relatively large fuel tanks. Whilst the aforementioned trade-off between tank size/mass (with an associated aerodynamic penalty) and fuel mass appears to be favourable, the cryogenic delivery and storage of liquid hydrogen (at $\sim$20 K) also poses significant operational problems.

Prior to any possible introduction of liquid hydrogen later this century, a more likely (intermediate) solution is the possible introduction of LNG, including liquefied methane (LCH$_4$), or liquefied bio-methane fuelled aircraft [40, 45, 46, 48–50]. Methane has a slightly better heat content 1 ($\sim$50 MJ kg^{-1}), compared with Jet A-1. Whilst its liquid density ($\sim$420 kg m^{-3} at $\sim$160 K) is low compared to Jet A-1, it is much denser than LH$_2$ and is far less problematic from a cryogenic storage standpoint. Future aircraft designs with so-called 'blended-wing-bodies' [50] with improved lift-to-drag ratios may also offer increased wing-tank volumes offsetting the disadvantages associated with relatively low fuel density discussed in the previous section. However, to retain a balanced perspective, it is important to emphasise that the capital investment necessary for aircraft development and airport infrastructure changes to permit LNG usage would probably be a formidable barrier to such a radical transformation.

Recently Airbus announced their "ZEROe" initiative, which appears to be focussed on the introduction of LH$_2$ aircraft. Three different passenger aircraft configurations are depicted as design examples: a turboprop configuration with a LH$_2$ tank that is positioned in the aft fuselage; a turbofan configuration also with an aft tank; and a blended wing-body turbofan

Figure 11.4 Conceptual representation of possible future liquid bio-methane aircraft with under-wing fuel pods [51].

configuration. No information is offered as to development costs, including the necessary ground infrastructure changes. Nevertheless, this initiative illustrates the growing recognition of the need for radical transformation.

In summary, the prospects for an all-new, non-drop-in, alternative fuel appear to be poor in the foreseeable future, unless the price of petroleum-refined Jet A-1 rises rapidly, and/or unless strong environmental restrictions on net carbon emissions are imposed (Figure 11.4), in which case radical transformation will need to embraced

11.9 Conclusions

It seems unlikely that sustainable ATF production will reach 10% (as an example) of the current global Jet A-1 consumption within the next decade or so. Nevertheless, a target of about 1% might be achievable, as effectively predicted by the IATA Technology Roadmap [1]. It is difficult to predict what could transpire in the latter half of this century. Advances in biochemical engineering may bring out truly transformational processes that offer low cost 'drop-in' fuels with much reduced net carbon emissions, or it is possible that 'non-drop-in' fuels may be adopted. In arriving at optimum solutions, it is essential that some degree of system change is considered. For example, the relatively simple replacement of nitrile rubber with fluorocarbon seals [21] in fuel systems could permit (synthetic) paraffinic fuels with low aromatic content that have better overall performance than existing Jet A-1. Aside from reduced smoke emissions, low aromatic fuels would probably result in reduced carbon deposition on combustor walls and turbine blades, improving engine lifetime. It therefore seems paradoxical that ASTM D7566 calls for a minimum 8% aromatic content. There also appears to be a good argument for reducing that maximum allowable aromatic content well below 26.5% as is currently permitted by ASTM D1655 and D7566.

If and when the world economy adopts stringent low carbon emission measures, then the aviation sector may be forced to take much more radical transformation measures [52].

References

1 Anon (2014) IATA Technology Roadmap. 4e. http://www.iata.org/whatwedo/environment/Documents/technology-roadmap-2013.pdf.

2 Anon (2015). Facts and Figures, Air Transport Action Group. http://www.atag.org/facts-and-figures.html.

3 Letcher, T. (2014). *Future Energy, Improved, Sustainable and Clean Options for Our Planet*, 2e. London: Elsevier. ISBN 978-0-08-099424-6.

4 Stratton, R.W., Wong, H.M., and Hileman, J.I. (2010), Life Cycle Greenhouse Gas Emissions from Alternative Jet Fuels, PARTNER Project 28 report version 1.2. Massachusetts Institute of Technology.

5 Anon (2012). Manual on Civil Aviation Jet Supply, ICAO, Doc 9977 AN/489. International Civil Aviation Organization, Montreal, Canada. ISBN 978-92-9249-105-5.

6 Anon (2015). Standard Specification for Aviation Turbine Fuels, ASTM D1655-15d. ASTM International, West Conshohocken, PA, USA.

7 Anon (2011). Turbine Fuel Kerosine Type, Jet A-1, NATO code: F-35, Joint Service Designation AVTUR, Defence Standard 91–91, Issue 7, Ministry of Defence.

8 Anon (2020). Standard Specification for Aviation Turbine Fuel Containing Synthesized Hydrocarbons, ASTM D7566-20c, ASTM International, West Conshohocken, PA, USA.

9 Anon (2013). IATA 2013 Report on Alternative Fuels, 8e. International Air Transport Association, Montreal, Canada, ISBN 978-92-9252-269-9.

10 Gnansounou, E. and Dauriat, A. (2005). Ethanol fuel from biomass: a review. *Journal of Scientific and Industrial Research* 64: 809–821.

11 Klein-Marcuschamer, D., Turner, C., Allen, M. et al. (2013). Technoeconomic analysis of renewable aviation fuel from microalgae, Pongamia pinnata, and sugarcane. *Biofuels, Bioproducts, and Biorefining* 7: 416–428.

12 Weiss, K. (2011). Alcohol to Jet, Emerging Through ASTM, ICAO Aviation and Sustainable Alternative Fuels Workshop, Montréal, Canada.

13 Anon (2014). Standard Practice for Qualification of New Aviation Turbine Fuels and Fuel Additives. ASTM D4054–14, ASTM International, West Conshohocken, PA, USA.

14 Anon (2000). Handbook of Products, Air BP, Hemel Hempstead, UK. https://fmv.se/FTP/Drivmedel/M7789-000183%20svensk%20utg%C3%A5va/datablad/M0754-233000_Avgas_100_LL.pdf

15 Küchemann, D. (1978). *The Aerodynamic Design of Aircraft*, 1e. Oxford, UK: Pergamon.

16 Ragozin, N.A. (1961). *Jet Propulsion Fuels"*, (trans. W.E. Jones). Oxford: Pergamon.

17 Hawthorne, W.R. and Olson, W.T. (1960). *Design and Performance of Gas Turbine Power Plants*. Oxford: Oxford University Press.

18 Lefebvre, A.H. and Ballal, D.R. (2010). *Gas Turbine Combustion, Alternative Fuels and Emissions*, 3e. Boca Raton, Florida, USA: CRC Press.

19 Coleman, J.R. and Gallop, L. D.. (1982). Exposure of Nitrile Rubber to Fuels of Varying Aromatics Level, DREO Technical Note 82–86, Defence Research Establishment of Ottawa. Department of National Defence, Canada.

20 Graham, J.L., Striebich, R.C., Myers, K.J. et al. (2006). Swelling of nitrile rubber by selected aromatics blended in a synthetic jet fuel. *Energy and Fuels* 20: 759–765.

21 Ewing, D. (2011) Alternative Seals for Alternative Fuels, SAE Aerospace Engineering magazine, June 8, 2011. See also: Otner, J. Material Compatibility presented in High Biofuel Blends in Aviation (HBBA) study and BioJetMap Workshop. Brussels, Belgium (11 February 2015).

22 Anon (2009). Additional Local Air Quality Benefits from Sustainable Alternative Fuels for Aircraft, ICAO Working Paper, CAAF/09-WP/05, Conference on Aviation and Alternative Fuels, Rio de Janerio, Brazil.

23 Anon (2009) Fuel: Jet Fuel continaing FAME (Fatty Acid Methyl 1 Ester), FAA, Special Airworthiness Information Bulletin, NE-0-25R1.

24 Glebova, O. (2013). Gas to Liquids: Historical Development and Future Prospects, Report NG 80, The Oxford Institute for Energy Studies, University of Oxford.

25 Anon (2014). Evaluation Report, JFK Green Lane Program, SkyNRG. http://skynrg.com/wp-content/uploads/2014/03/SkyNRG-JFK-document_Digitaal.pdf.

26 Anon (2016). Commercial Production of Renewable Fuel is a Reality, Honeywell UOP. http://www.uop.com/processing-solutions/renewables/green-jet-fuel/#commercial-production.

27 Renninger, N., McPhee, D.J. (2008). Fuel Compositions Comprising Farnesane and Far-nesane Derivatives and Method of Making and Using Same, US Patent, US20080083158.

28 Garcia, F. and Marchand, P. (2014). Amyris-Total Biojet Fuel Breakthrough Solu-tion for Aviation, presented at ICAO Workshop, Montréal, Canada (9–10 September 2014). http://www.icao.int/Meetings/EnvironmentalWorkshops/Documents/2014-GreenTechnology/7_Garcia-Marchand_TOTAL.pdf.

29 De Guzman, D. (2014). Turnaround for Amyris. http://greenchemicalsblog.com/2014/12/05/turnaround-for-amyris.

30 Anon (2014). Evaluation of Amyris Direct Sugar to Hydrocarbon (DSHC) Fuel, DOT/FAA/AEE/2014–07. U.S. Department of Transportation, Federal Aviation Adminis-tration, Final Report, Public Release Version.

31 Crago C.L., et al. (2010). Competitness of Brazilian Sugarcane Ethanol Compared to US Corn Ethanol, AAEA Annual Meeting, Denver Colorado.

32 Holmgren, J. (2013). Innovative Use of Industrial Gases to Produce Sustainable Fuels and Chemicals, presented at the Low Carbon Jet Fuel: the Industry Roadmap, Avalon Airshow, Geelong, Australia.

33 Biddy, M. (2012). Biochemical Conversion Processes, presented at the Advanced Bio-Based Jet Fuel Cost of Production Workshop, US Department of Energy. http://www.energy.gov/sites/prod/files/2014/04/f14/biddy_caafi_workshop.pdf.

34 Anon (2015). Biojet Fuel with US Navy. http://www.cobalttech.com/biojetfuel.html.

35 Wright, M.E., Harvey, B.G., Quintana, R.L. (2013) Diesel and Jet Fuels Based on the Oligomerization of Butene. US Patent 8, 395, 007 B2.

36 Liu, C., Wng, H., Karim, A.M. et al. (2014). Catalytic fast pyrolysis of lignocellulosic biomass. *Chemical Society Reviews* 43: 7549–7623.

37 Goss, K., Abadi, A., Crossin, E., Stucley, C., Turnbull, P. (2014). Sustainable Mallee Jet Fuel, Future Farm Industries CRC, ISBN 978-0-9871562-7-3.

38 Anon. (2014). Biofuels for Aviation, European Comission. https://ec.europa.eu/energy/en/topics/biofuels/biofuels-aviation.

39 Andreev, V. and Solozolov (2001). Fuel for flying apparatus of the twenty-first century" (in Russian). *Science and Life* 3: 23–25. See also Tupolev website accessed March 1, 2012: http://www.tupolev.ru/english/Show.asp?SectionID=82.

40 Zhuravlev, V.N. (2014). Comparative Analysis of Russian and Western Approaches to the Problem of Alternative Fuel Usage in Aviation, 9th Congress of the International Council of the Aeronautical Sciences (ICAS), St. Petersburg, Russia.

41 Raznoschikov, V.V. and Yanovskaya, M.L. (2010). An optimisation of aviation con-densed gas fuel composition for cargo aircraft, autogas fuelling complexes including alternative fuels. *Avtogazozapravočnyj Kompleks + Al'ternativnoe Toplivo* 52 (4): 11–14. ISSN-2073-8323.

42 Zaitsev, V., Belov, A. and Samusenko, A. (2006). Liquidified Gas for the Helicopters, ATO - Industry, No 68.

43 Dorrington, G., Baxter, G., Bil, C. et al. (2013). Prospects for liquefied natural gas and other alternative fuels for future civil air transportation. In: *Proceedings of the 15th Aus-tralian International Aerospace Congress (AIAC15)*, Melbourne, Australia, 25-28 February 2013, 116–125.

44 Daggett, D., Hadaller, O., Walther, R. (2006). Alternative Fuels and their Potential Impact on Aviation, NASA/TM—2006-214365.

45 Chuck, C. and Donnelly, J. (2014). The compatibility of potential bioderived fuels with jet A-1 aviation kerosene. *Applied Energy* 118: 83–91. ISSN 0306-2619.

46 Pohl, H.W. (ed.) (1995). *Hydrogen and Other Alternative Fuels for Air and Ground Transportation*. Chichester, UK: Wiley.

47 Verstraete, D. (2015). On the energy efficiency of hydrogen fuelled transport aircraft. *International Journal of Hydrogen Energy* 40 (23): 7388–7394.

48 Carson et al. (1980) Study of Methane Fuel for Subsonic Transport Aircraft, NASA CR 159320, contract NAS-1-15239.

49 Bradley, M.K., Droney, C.K. (2012). Subsonic Ultra Green Aircraft Research Phase II: N+4 Advanced Concept Development, NASA CR-2012-217556.

50 Kawai, R.T. (2013). Benefit Potential for a Cost Efficient Dual Fuel BWB, AIAA paper 2013-0937 in 51st AIAA Aerospace Sciences Meeting including the New Horizons Forum and Aerospace Exposition, Texas, USA.

51 Burston, M., Conroy, T., Spiteri, L., Spiteri, M., Bil, C., Dorrington, G.E. (2013). Conceptual Design of Sustainable Liquid Methane Fuelled Passenger Aircraft, 20th ISPE International Conference on Concurrent Engineering, Melbourne, Australia

52 Anon ZEROe Towards the world's first zero-emission commercial aircraft. https://www.airbus.com/innovation/zero-emission/hydrogen/zeroe.html

Section IV
Research Case Studies

12

Overall Contribution of Wingtip Devices to Improving Aircraft Performance

Nikola Gavrilović[1], Boško Rašuo[2], Vladimir Parezanović[3], George Dulikravich[4], and Jean-Marc Moschetta[1]

[1]*Aerodynamics, Energetics and Propulsion Department, ISAE-SUPAERO, Toulouse, France*
[2]*Faculty of Mechanical Engineering, University of Belgrade, Belgrade, Serbia*
[3]*School of Engineering, RMIT University, Bundoora, Victoria, Australia*
[4]*Florida International University, Miami, Florida, USA*

Aerodynamic drag force breakdown of a typical transport aircraft shows that lift-induced drag can amount to as much as 40% of total drag at cruise conditions and 80–90% of the total drag in the take-off configuration. One way of reducing lift-induced drag is by using wingtip devices. This chapter [1] analyses and compares the impact on commercial aircraft performance of different types of classical winglets versus bio-inspired winglet configurations. The numerical methodology and results are presented and comparative results on the winglet aerodynamic performance are provided. Advanced, multi-objective design optimization software is applied to determine an optimal, one-parameter winglet configuration that simultaneously minimizes drag and maximizes lift and range factor. A bio-inspired winglet design is also compared with a classical winglet in a low-speed flow regime, typical for the operation of unmanned aerial vehicles (UAVs). Here, the bio-inspired winglet has a significant advantage over the classical winglet in mitigating the UAV's wingtip vortex intensity, which is an essential consideration in optimizing the separation distances of both manned and unmanned aerial systems (UAS) for air traffic management (ATM).

12.1 Introduction

Winglets are quasi-vertical extensions of wingtips that improve an aircraft's fuel efficiency and cruising range. Designed as small wing extensions, winglets reduce the aerodynamic drag associated with vortices that develop at the wingtips as the airplane moves through the air. By reducing wingtip drag, the power required is reduced and, therefore, fuel consumption decreases and range and endurance are extended. Fixed-wing flying vehicles of all types and sizes are flying with wingtip devices: from small UAVs via gliders and ultralights to global airliners.

The concept of winglets was originally developed in the late 1800s by British aerodynamicist Frederick W. Lanchester, who patented an idea that a vertical surface (end plate) at the wingtip would reduce drag by controlling wingtip vortices. Unfortunately, the concept never demonstrated its effectiveness in practice because the increase in drag due to skin friction and flow separation outweighed any lift-induced drag benefit. A rapidly increasing oil price in 1973 and a continuous drive for performance improvement made aircraft manufacturers look more closely to improve the operating efficiency of commercial aircraft. R.T. Whitcomb from NASA Langley Research Center was the first to conclude that a sort of vertical wingtip barrier, a near-vertical supplementary wing, would be the most efficient way of reducing wingtip vortices. The initial studies of Whitcomb [2] showed that the improvement is in the order of magnitude of around 5%. Even though manufacturers were slow to adopt a promising invention, winglets have become an indispensable aerodynamic element on every commercial aircraft in the past 30 years [3].

Environmental issues and rising operational costs have forced the industry to improve the efficiency of commercial air transport and this has led to some innovative developments for reducing lift-induced drag. Several different types of wingtip devices have been developed during this quest for efficiency and the selection of wingtip devices depends on the specific situation and airplane type. Commercial passenger aircraft spend most of their operational life in cruise conditions, thus all wingtip shapes need to be examined in those conditions in order to justify their purpose. Winglets allow for significant improvements in the aircraft fuel efficiency, range, stability, and even control and handling [4]. They are traditionally considered to be near-vertical, wing-like surfaces that can extend both above and below the wing tip where they are placed. Some designs have demonstrated impressive results, such as 7% gains in lift-to-drag ratio and a 20% reduction in drag due to lift at cruise conditions [5, 6].

12.2 Winglet Design

Finite span wings generate lift due to pressure imbalance between the bottom surface (high pressure) and the top surface (low pressure). The higher-pressure air under the wing flows around the wingtips and tries to displace the lower pressure air on the top of the wing. The effects of the flow around a wingtip and a winglet are illustrated in Figure 12.1. These phenomena are referred to as wingtip vortices with high velocities and low pressure at their cores. These vortices induce a downward flow known as the downwash, which has the effect of tilting the free-stream velocity to produce a local relative downward wind, reducing the angle of attack that the wing experiences. Since the vector of the lift is also tilted by the same induced angle-of-attack (α_i) creating a new contribution to the drag, which is referred to as the lift-induced drag [7]. Equation of the total drag of a wing is a sum of the parasite drag and the induced drag. In terms of the non-dimensional coefficient of drag it is:

$$C_D = C_{D0} + C_{Di} \tag{12.1}$$

Here, C_{D0} is the drag coefficient at zero lift and is known as the parasite drag coefficient, which is independent of the lift. The second term on the right-hand side of Eq. (12.1) is the lift-induced drag coefficient, C_{Di}, defined as:

$$C_{Di} = \frac{C_L^2}{\pi \lambda e} \tag{12.2}$$

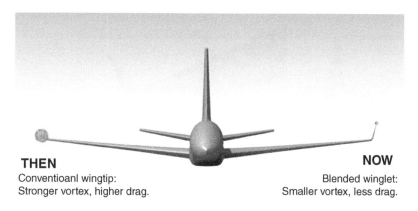

THEN

Conventioanl wingtip:
Stronger vortex, higher drag.

NOW

Blended winglet:
Smaller vortex, less drag.

Figure 12.1 Wingtip vortex for wind with and without winglet.

In Eq. (12.2), C_L is the wing lift coefficient, λ is the wing aspect ratio and e is Oswald efficiency factor – a correction factor that accounts for the difference between the actual wing and an ideal wing, having the same aspect ratio and elliptical lift distribution (i.e. wingspan efficiency).

Winglets increase an aircraft's performance by reducing the recirculation at the tips of the wings, which gives rise to induced drag. The large vortex created by a conventional wingtip (Figure 12.1 – left) is responsible for increased drag and reduced lift, resulting in lower flight efficiency and higher fuel costs. On the other hand, a winglet (Figure 12.1 – right) reduces the wingtip vortex intensity, thus mitigating the losses associated to momentum being translated into the useless (from the standpoint of flight) rotational motion in the plane which is perpendicular to the aircraft's direction. From another perspective it can also be said that winglets operate like a sailboat tacking upwind, producing a forward thrust inside the circulation field of the vortices and reducing their strength. Weaker vortices mean less drag at the wingtips and lift is restored. Improved aircraft efficiency is directly translated to more payload, reduced fuel consumption and a longer cruising range that can allow an airline to expand routes and destinations.

Several different winglet designs are considered in this chapter, as shown in Figure 12.2. A blended winglet attached to the wingtip with a smooth curve instead of a sharp angle is a classical design (by current standards) intended to reduce interference drag at the wing/winglet junction. A sharp interior angle in this region can interact with the boundary layer flow causing a drag-inducing vortex, negating some of the benefits of the winglet. Such blended winglets are used on business jets and sailplanes, where individual buyer preference is an important marketing aspect. An example of the design methodology and optimization of a blended winglet and its airfoil for low-speed airplanes and gliders, using a combination of experimental and numerical techniques is given in [8, 9]. The blended winglet used in the current work is shown in Figure 12.2a.

A wingtip fence, shown in Figure 12.2b, refers to the winglets which include surfaces extending above and below the wingtip. Both surfaces are shorter than or equivalent to a winglet possessing similar aerodynamic benefits. Wingtip fences were the preferred

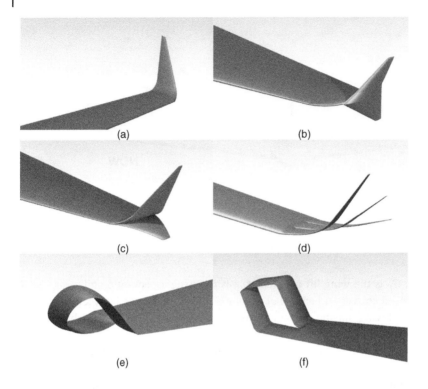

(a)

(b)

(c)

(d)

(e)

(f)

Figure 12.2 Wingtip devices considered: (a) blended winglet, (b) wingtip fence, (c) split-tip winglet, (d) slotted feathers wingtip, (e) spiroid winglet 1, and (f) spiroid winglet 2.

wingtip device of Airbus for many years [10]. The Boeing 737 MAX uses a new type of wingtip device: a split-tip winglet. Resembling a three-way hybrid between a blended winglet, wingtip fence, and raked wingtip, Boeing claims that this new design should deliver an additional 1.5% improvement in fuel economy over the 10–12% improvement already expected from the 737 MAX [11, 12]. A similar shape was analysed in the present study as shown in Figure 12.2c and was recently optimized [13].

Other appealing shapes for winglet design could be adopted by looking at analogous shapes in nature. An example of this is bird wingtip feathers, with their large variety in morphology. In Figure 12.2d, we present a design inspired by the wingtip feathers of an eagle, which are bent up and separated (like the fingers of a spreading hand). The slots also appear to reduce drag by vertical vortex spreading; the drag of a gliding Harris hawk in a wind tunnel increased significantly when the tip slots were removed by clipping the primary feathers [14]. A thorough review of the performance of various implementations of the slotted wingtip feathers may be found in [15]. However, for a variety of reasons, identical copies from nature to man-made technologies are not always feasible. Instead, bionics encompasses a creative conversion into technology that is often based on various steps of abstractions and modifications, i.e. an independent successive construction that is a "new invention," rather than a blueprint of nature. This implementation by bionic abstraction can be improved even further and aesthetically adapted to wings by designing a spiral loop

Figure 12.3 Raked wingtip on a Boeing 787 (left) and foldable winglet on the Boeing 777X (right).

that externally wraps the wing tip extension, referred to as the spiroid winglet [16]. The following Figure 12.2f shows the form that was created after the spiroid model, to develop a technologically more feasible solution by simplifying the shape curvature.

More recently, raked wingtips have been introduced to the Boeing 787 Dreamliner instead of winglets, as shown in Figure 12.3. The raked wingtips are generally found to further reduce the induced drag compared to a blended winglet while benefiting from weight savings. The seamless wing/winglet design of the 787 brings aeroelastic structure technology to a new level; at extreme wing loads, the wingtip of a 787 is displaced upwards by up to 26ft. Another recent wingtip device development is the foldable winglet which is implemented in the new Boeing 777X as means of ensuring compatibility with the existing airport infrastructure. When extended, the winglet increases the span of the wing by 23 ft compared to a regular 777, creating a net fuel efficiency increase of 10%. If we consider the weight efficiency of the folding mechanism of the foldable winglet, and the extreme flexing capacity of the 787's wing structure, future concepts are bound to explore a morphing winglet structure. New materials used in the latter two examples show that morphing structures, capable of adapting to flight conditions as well as ground operation limitations, are steadily becoming technologically feasible. The morphing of the winglet's camber has been shown to be beneficial [17] at flight conditions from low-speed/high load to transonic cruise regime, while a variable-sweep raked wingtip could improve roll rates and fuel efficiency, and mitigate flutter issues for truss-braced wing configurations [18]. Asynchronous deflection of trailing edge tabs of the winglet by morphing is also shown [19] to be beneficial to performance in off-design flight conditions, manoeuvres, and alleviating excessive wing loading. Although these kinds of morphing configurations are not examined in the current paper, one could imagine the application of the optimization methodology, which we present later in this chapter, to be applicable to optimize the preferred states of the morphing structure for various flight conditions.

12.3 Numerical Setup

The computational effort performed in this research consisted of three stages. The work began from a pre-processing stage of geometry setup and grid generation [20]. The geometry of the passenger aircraft model was designed using CATIA V5 [21]. Geometry setup was made using surface design in order to draw a 3D model as shown in Figure 12.1.

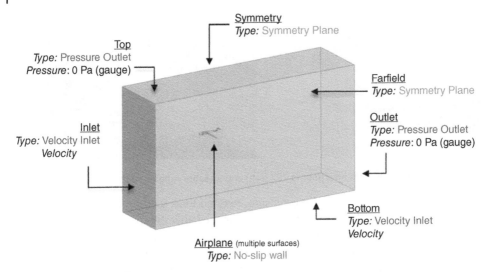

Figure 12.4 Summary of the boundary conditions.

The computational grid was generated by the ICEM program in ANSYS. The second stage was CFD simulation by FLUENT software using a finite volume approach. Finally, the post-processing stage was used where the aerodynamic characteristics of the wing were defined [22].

After specifying the physics continua to be used for the computation, the boundary conditions of the mesh domain are specified. Using the surface parts that were named during the mesh construction process in the ICEM-CFD software, the definition of types, physics conditions and values of the boundary condition is straightforwardly input into the solver software. A diagram describing the different boundary conditions used for simulations with a positive angle of attack is shown in Figure 12.4. Since the reference pressure of the simulation is already fixed at 22 632 Pa (h = 11 000 m), the outlet pressure specification is set to 0 Pa gauge.

An unstructured tetrahedral grid was utilized for computing the flow-field around the 3D configurations. An unstructured grid is appropriate due to the complexity of the model. A typical mesh is made up of approximately eight million elements. After the grinding process, the grid was examined to check its quality by observing the skewness level and abrupt changes in grid cell sizes [23].

The generation of prism mesh layers adjacent to the geometric body of interest is a crucial factor in enhancing the accuracy of the CFD results for a given mesh. By condensing layers of prism mesh extruded from the surface mesh, the boundary layer gradients of the flow over the airplane body can be accurately discretized and computed by the solver. Common practice generally used when conducting CFD simulations involves defining the prism layer heights under the assumption of a flat plate boundary layer thickness. Using the flat-plate boundary layer theory described in [24], the appropriate first cell height can be calculated.

The skin friction coefficient C_f for a flat plate is estimated by the following expression:

$$C_f = \frac{0.026}{Re^{1/7}}$$

The wall-friction velocity u_τ that is defined as a function of the shear stress at the wall can be calculated,

$$u_\tau = \left(\frac{\tau_w}{\rho}\right)^{1/2} \quad where \; \tau_w = C_f \cdot \frac{1}{2}\rho U_\infty^2$$

For the acceptable first layer dimensionless height of $y+ = 1$, the first wall spacing y can be obtained from the following expression,

$$y^+ = \frac{yu_\tau}{\nu}$$

For unstructured mesh generated from ICEM-CFD software, users can specify many variables describing the desired characteristics of the prism layers. One of the features offered in the software is the prism quality control options, where users can specify the minimum prism quality, orthogonal weight, prism fillet ratio, maximum prism angle, prism height over the base, and prism height limit factor. The settings used for the prism meshing and quality control parameters are summarized in Table 12.1. The minimum prism quality is specified very low to encourage more prism to generate. These prism aspect ratios are controlled by the maximum height over base and height limit factor setting, where the program would automatically halt and condense the prism layers down if the prism aspect ratio were to exceed the specified value. Since the maximum height over base setting would abruptly stop growing the prism and cap it off with pyramid cells, as opposed to the height limit factor where the number of prism layers is preserved, the aspect ratio is set lower than the latter as it is preferred to keep all prism layers.

The prism fillet ratio of 0.1 is specified to allow some blending of prisms near sharp and acute corners. Lastly, setting $180°$ in the maximum prism angle function allows prism layers to stay connected around tight corners. This function is found to be especially useful when extruding prism layers near the wing's trailing edge. A mesh cut plane at a main wing cross-section in Figure 12.5 showcases the prism layers and the trailing edge wake refinement.

The numerical simulation by FLUENT solver was performed after the completion of the grid generation and simulated 3D compressible turbulent flow using Navier-Stokes equations. The turbulence model was the realizable k-ε with appropriate solid wall boundary conditions. Solver control parameters and fluid properties were also defined

Table 12.1 Prism mesh parameters and control settings.

Growth law	Exponential
Initial height	2.6×10^{-5} m
Height ratio	1.25
Number of layers	20
Total height	8×10^{-3} m
Min. prism quality	0.0005
Fillet ratio	0.1
Max. prism angle	$180°$
Max. height over base	3
Prism height limit factor	2

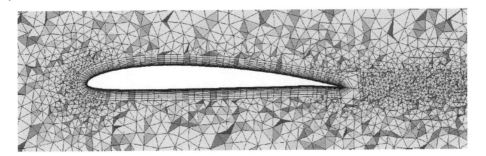

Figure 12.5 Prism mesh at wing's cross-section and the refined wake refinement.

at this stage. FLUENT software provides the user with seven turbulence models [22]. Lower-order turbulence models tend to be less accurate than higher-order ones. The Spalart-Allmaras model is a low Reynolds number one-equation turbulence model that solves for the kinematic eddy viscosity. The last four turbulence models are skipped because they require far too much computing power. Thus, the only options that are left are two-equation turbulence models, the k-ε and k-ω model. Since its introduction, the k-ε model has become the most widely used turbulence model. The key advantages of the k-ε model are its economy, reasonable accuracy, and its applicability to a wide range of flow types. The realizable k-ε model is the best option when dealing with flow containing high-pressure gradients and separation [24].

After all the parameters were specified, the model was initialized. The initializing and iteration processes stopped after the completion of the computations [25]. Computations are carried out on one quad-core processor 3.2 GHz, with 16 GB of RAM, and each case took approximately 12 hours to converge. A typical convergence history of fluent aerodynamic analysis in this work is presented in Figure 12.6.

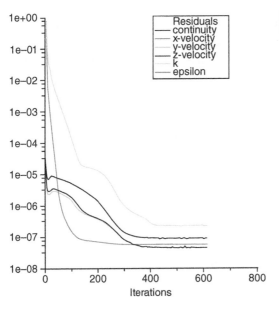

Figure 12.6 Convergence history of fluent simulation.

12.4   Commercial Aircraft Performance Improvement

As a representative aircraft geometry, a typical medium-size commercial aircraft model is shown in Figure 12.7. Such aircraft perform a large number of flights per year; thus fuel consumption is of great importance. The main geometric parameters of the aircraft model are listed in Table 12.2.

The results from the 3D wing with various winglet models were compared to the baseline 3D wing without a wingtip device. We focus on the mean aerodynamic characteristics at cruise conditions, which include drag coefficient, C_D, lift coefficient, C_L, and lift-to-drag ratio, L/D. The simulation was carried out at various angles of attack α, and a Mach number of 0.85. Figures 12.8, 12.9, and 12.10 show the lift, polar and lift-to-drag ratio curves for the clean wing and wing configurations with different winglets.

With the increase of the angle of attack, lift and drag both grow. The increasing lift raises the strength of the wingtip vortices and thus induced drag of the airplane. By using a winglet, we reduce the strength of these vortices and thus induced drag. This performance improvement can be very important for the performance during take-off and landing, where this leads to shorter required runway lengths. By using these aerodynamic structures, the climb performance is also improved by achieving a higher speed of climbing.

Figure 12.7   Clean commercial aircraft configuration.

Table 12.2   Commercial aircraft representative.

Cruise altitude h_{cruise}	11 000 m ($\rho = 0.36$ kg/m³)
Speed of sound c	295 m/s
Cruise speed V_{cruise}	250 m/s (M = 0.85)
Wing area S_{ref}	200 m²
Wingspan b	37.5 m
Mass m_{TOW}	100 000 kg
Lift coefficient (cruise) C_L	0.43
Reynolds number Re	19×10^6

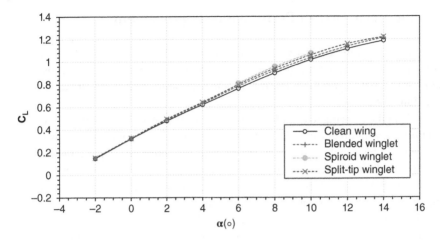

Figure 12.8 Lift curve.

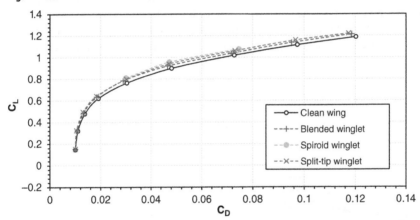

Figure 12.9 Polar curve.

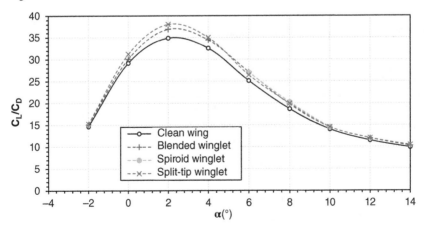

Figure 12.10 Lift-to-drag ratio curve.

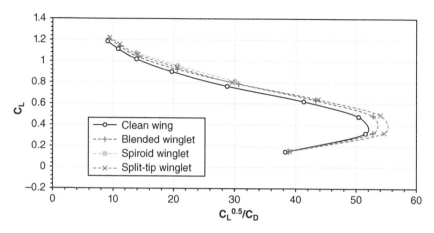

Figure 12.11 Factor of range curve.

The factor that directly affects the range of the aircraft is a relation which is presented in Figure 12.11 for different winglets. It is also interesting to note that the trade-off at the maximum value of the range factor is nearly a 7.1% improvement, which not only affects the range but also fuel consumption as will be shown. The reason why spiroid winglets demonstrated the poorest performance is because of their increased wetted surface and thus larger parasite drag. Also, interference is at much higher levels when using those shapes at wingtips as shown in [13].

Improving performance in cruise flight is essential to increase the value range of an airplane equipped with different winglets in comparison with a clean wing. With reducing drag and increasing lift, range L is significantly increased (Eq. 12.3). All values in Eq. (12.3) are defined for cruise mode [26].

$$L_{aH} = \frac{72}{c} \frac{C_L^{0,5}}{C_D} \sqrt{\frac{2\left(\frac{m_1}{s}\right)}{\rho g}}(1 - \sqrt{1-\zeta}) \tag{12.3}$$

In Eq. (12.3), c is specific fuel consumption, $C_L^{0.5}/C_D$ is a factor of range, m_1 is the mass of an empty aircraft, s is the wing area, ρ is the air density, g is the acceleration of gravity and ζ is the ratio between the mass of the fuel and the mass of empty aircraft.

The resulting ranges obtained with different winglets are given in Figure 12.12, suggesting that the simplified spiroid winglets (spiroid winglet 2) offered the least number of improvements.

We have shown the effects of winglet shapes on aerodynamic coefficients. Now we will demonstrate how they affect fuel consumption. It is assumed that the plane flies 200 routes of up to 4000 km/year. Figure 12.13 shows the amount of fuel consumed for 30 years of use. It can be seen from this that the amount of fuel saved can range from 6000 up to 8000 tonnes, depending on a chosen winglet.

Figure 12.14 shows the number of passengers carried with the same amount of fuel for one year of usage, calculated as shown in [21]. By using winglets, the number of passengers carried increased significantly for the same amount of fuel.

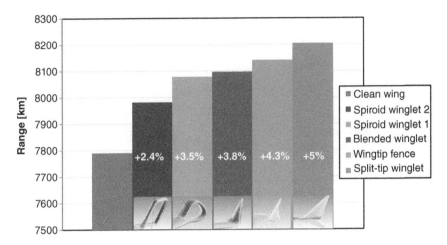

Figure 12.12   Range improvements over clean wing.

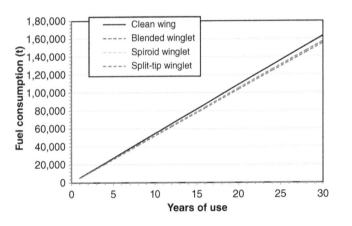

Figure 12.13   Fuel consumption curve.

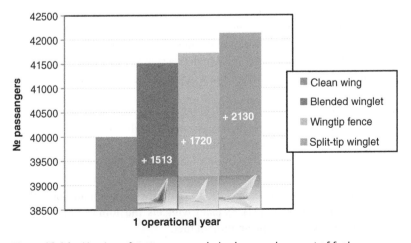

Figure 12.14   Number of passengers carried using equal amount of fuel.

12.5   Winglet Design Optimization

To answer the question as to which is the best aerodynamic shape of the winglet, it is necessary to perform optimization for each of the previously analysed winglets, intending to obtain such a shape that provides the best performance of the aircraft [27, 28]. In the current chapter we present only the optimization of the classic blended winglet design, based on modifying a single geometric parameter. We use only one geometric parameter for optimization which describes the vertical position of the winglet airfoil z_w with respect to the reference of the wingtip airfoil, illustrated in Figure 12.16a. This parameter represents the position of the winglet profile that is connected with the wingtip profile, and by changing its value we modify the length and angle of the winglet with respect to a wingtip. It means that the winglet will be formed by connecting the wingtip and winglet profile that has the freedom to move up and down as described by the parameter of optimization. The vertical position of the winglet profile was in the range of $z_w = \pm 1.6$ m.

Approximately 100 winglet shapes, based on the variation of z_w were generated for the analysis. Each configuration was analysed using ANSYS Fluent software with free-stream Mach number 0.85. The geometry of the clean wing was the same in all of these analyses. The boundary conditions were defined in the same way as in the earlier calculations. Nondominated Sorting Genetic Algorithm (NSGA II) in *modeFRONTIER* optimization software was used to search the response surfaces for the Pareto optimal solutions, similarly to [13, 29, 30]. The three simultaneous optimization objectives were maximizing the coefficient of lift and factor of range, while minimizing the coefficient of drag.

Figure 12.15 shows the result of optimization where each point represents a unique shape of the winglet. It is interesting to notice that winglet shapes that extend from the upper side of the wing tip give higher lift and drag coefficients. On the other hand, winglet shapes that extend from the lower side of the wing tip give lower lift and drag coefficients, which largely corresponds to earlier findings [11]. From a wide range of shapes analysed, the goal is to select the one giving the best aerodynamic characteristics. The value of the factor of the range was chosen in this study as an indicator that defines the most efficient form of the wingtip. Figure 12.16b shows the most efficient shape obtained with optimization and its strong resemblance to the A330 winglet.

Figure 12.15   Optimization result for lift and drag at fixed angle of attack.

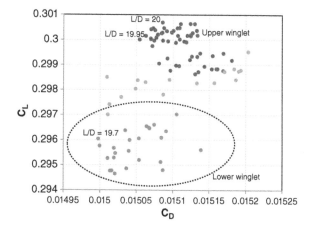

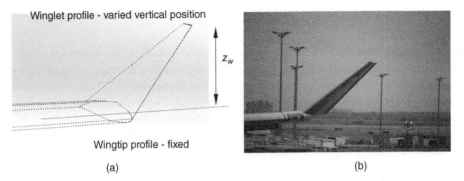

Figure 12.16 The most efficient shape obtained from optimization (a), A330 winglet (b).

12.6 Biologically-Inspired Winglets for UAVs

Due to a huge discrepancy in flight conditions between commercial aircraft and UAVs, a new study has been conducted that is adapted to flight conditions of a small drone. It should be noted here that even a benefit of a few percent is highly valuable as the aircraft is intended to fulfil the mission requirement of having a range of several thousands of kilometres. Two different winglet designs have been studied and compared to clean aircraft performance.

The first design is a classical blended winglet shown in Figure 12.17a, found on various aircraft types, from small UAVs for lateral stability purposes, up to big commercial aircraft for induced drag reduction. The second one is a bio-inspired winglet shown in Figure 12.17b

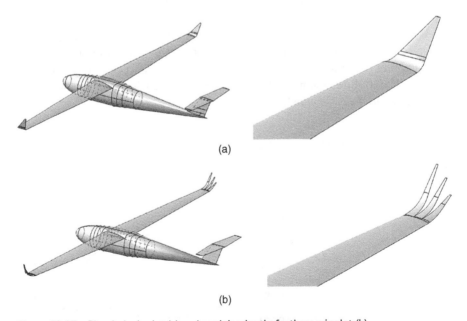

Figure 12.17 Blended winglet (a) and eagle's wingtip feathers winglet (b).

Table 12.3 UAV representative.

Cruise altitude h_{cruise}	100 m ($\rho = 1.2\,\text{kg/m}^3$)
Speed of sound c	340 m/s
Cruise speed V_{cruise}	23 m/s (M = 0.067)
Wing area S_{ref}	0.65 m²
Wingspan b	3.8 m
Mass m_{TOW}	12 kg
Lift coefficient (cruise) C_L	0.57
Reynolds number Re	0.3×10^6

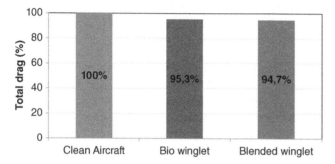

Figure 12.18 UAV total drag reduction with winglets.

that resembles the eagle wing tip feathers. The parameters of a typical representative of a small UAV category are shown in Table 12.3. A detailed winglet optimization procedure for a medium-altitude, long-endurance UAV can be found in [29].

The final comparison between different structures and clean aircraft for cruise operating conditions is presented in Figure 12.18. It can be concluded that winglet structures provide considerable benefits with up to 5% in total drag reduction. Moreover, the addition of winglets usually brings around 4–6% higher lift coefficient for the same angle of attack. That means that the drag can be even further reduced by decreasing the required angle of attack in a cruise for a configuration with winglets. Finally, it should be pointed out that a 5% drag reduction represents a significant achievement; adding 150 km to a total of 3000 km it can be considered as a fuel reserve.

Figure 12.19 illustrates the evolution of the vorticity as a function of distance from a wing trailing edge. Vortex circulation intensity is the main cause for the creation of induced drag. The vortex intensity has been significantly reduced with the use of bioinspired winglets. Since the bioinspired winglet is composed of several "sub-winglets," the area in contact with air is higher than for the blended winglet design. This higher surface in contact with air leads to higher shear drag. Winglets increase the effective aspect ratio of the wing, changing the pattern and magnitude of the vortex. Hence, the kinetic energy is reduced in the circular air flow, which implies a fuel consumption reduction and thus increases the range of the aircraft.

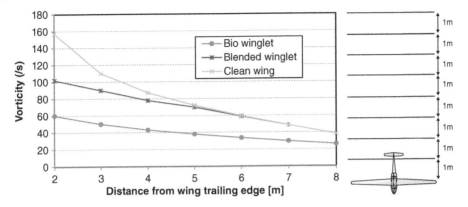

Figure 12.19 Vortex intensity in function of distance from trailing edge.

12.7 Conclusion

In this chapter we have presented an analysis of several wingtip designs, in comparison to a clean wing of a typical mid-range commercial transport aircraft. The performance of the wing with a winglet relative to a clean wing has been studied quantitatively and qualitatively. In terms of performance, lift was improved over the whole range of angles of attack for all winglet configurations, while the total drag was reduced. The lift-to-drag ratio was also improved by up to 15%. On the other hand, some configurations experienced an increased parasitic drag due to a larger wetted surface and have brought smaller improvements in range and fuel economy, as in the case of the spiroid winglets. In general, all of the winglet configurations bring additional benefits in terms of delayed flow separation (wing stall), improved take-off and landing performance, higher climb rates, reduced engine emissions, and reduced intensity of wingtip trailing vortices. The performance disadvantages are mainly due to an increased parasite drag due to an increased wetted surface. From a practical point of view, the disadvantages also include the increased weight due to the device itself and a requirement for a new structural study of the wing.

The blended winglet design was further optimized with respect to a single geometry parameter. The optimization, however, was performed using a Pareto front, multi-objective optimization algorithm targeting the maximization of lift and range factor and the minimization of drag. Of the around 100 different blended winglet geometries, the best individual chosen by the algorithm closely resembles the blended winglet in use on the Airbus A330. This result validates the procedure and suggests that a similar optimization would be useful to optimize the other, more unorthodox winglet designs presented here and elsewhere in order to find a configuration that might be better than the classical designs. In such endeavours, multiple geometry factors should be included, which requires a careful parametrization of each winglet type geometry. Alternatively, it would be interesting to examine if the large variety of winglet designs can be reduced to a unified set of geometric parameters, which could enable an automatic optimization of the design for a specific application.

Finally, in the third part of this chapter, we have presented an analysis of a bio-inspired, slotted feather winglet for use in UAVs. The results have clearly shown large benefits in aerodynamics performance, but also a significant advantage of the bio-inspired design in rapidly dissipating the trailing vortices behind the UAV model. This aspect is significant for the evolution of ATM systems, standards and procedures supporting both manned aircraft and UAS operations. The characterization of vortex wakes and their mitigation by winglets would facilitate the introduction of a unified approach to the management of traffic and airspace [31], with appropriate separation assurance and collision avoidance provisions [32].

Acknowledgments

This chapter is an expanded and updated version of a research article [1] originally published under open access license CC-BY-4.0 in FME Transactions (2015, Belgrade, Serbia).

References

1 Gavrilovic, N.N., Rasuo, B.P., Dulikravich, G.S., and Parezanovic, V.B. (2015). Commercial aircraft performance improvement using winglets. *FME Transactions* 43: 1–8.

2 Whitcomb, R.T. (1976). A design approach and selected wind tunnel results at high subsonic speeds for wing-tip mounted winglets. *NASA Technical Report*, no. L-10908.

3 Chambers, J.R. (2003). Concept to reality: Contributions of NASA Langley Research Center to U.S. Civil Aircraft of the 1990s. NASA SP-2003-4529.

4 Breitsamter, C. (2011). Wake vortex characteristics of transport aircraft. *Progress in Aerospace Science* 47: 89–134.

5 Faye, R., Laprete, R., and Winter, M. (2002). Blended winglets. *Aero, Boeing* 17.

6 Langevin, G.S. and Overbey, P. (2003). To reality: Winglets. NASA Langley Research Center.

7 Anderson, J.D. (2010). *Fundamentals of Aerodynamics*, 5e. McGraw-Hill Science/ Engineering/Math.

8 Maughmer, M. (2003). Design of winglets for high-performance sailplanes. *Journal of Aircraft* 40 (6): 1099–1106.

9 Maughmer, M., Swan, T., and Willits, S. (2002). Design and testing of a winglet airfoil for low-speed aircraft. *Journal of Aircraft* 39 (4): 654–661.

10 Bargsten, C.J. and Gibson, M.T. (2011). NASA Innovation in Aeronautics: Selected Technologies That Have Shaped Modern Aviation. *NASA TM/2011-216987*.

11 Molnar, M. (2012). *Boeing Says Radical New Winglets on 737 MAX Will Save More Fuel*. NYC Aviation.

12 Heerden, A., Guenov, M., and Molina-Cristobal, A. (2019). Evolvability and design reuse in civil jet transport aircraft. *Progress in Aerospace Science, Elsevier* 108: 121–155.

13 Reddy, S.R., Sobieczky, H., Abdoli, A., and Dulikravich, G.S. (2014). Winglets – Multiobjective Optimization of Aerodynamic Shapes. *11th World Congress on Computational Mechanics, 5th European Conference on Computational Mechanics, 6th European Conference on Computational Fluid Dynamics*.

14 Tucker, V.A. (1995). "Drag reduction by wing tip slots in a gliding Harris Hawk," *Parabuteo Unicinctus. Journal of Experimental Biology* 198: 775–781.

15 Cheng, Z.-P., Wu, Y.-M., Xiang, Y. et al. (2021). Benefits comparison of vortex instability and aerodynamic performance from difference split winglet configuration. *Aerospace Science and Technology* 119: 107219.

16 Guerrero, J.E., Maestro, D., and Bottaro, A. (2011). Biomimetic spiroid winglets for lift and drag control. *Comptes Rendus Mécanique, Elsevier* 340 (1–2): 67–80.

17 Eguea, J.P., Bravo-Mosquera, P.D., and Catalano, F.M. (2021). Camber morphing winglet influence on aircraft drag breakdown and tip vortex structure. *Aerospace Science and Technology* 119: 107148.

18 Mallik, W., Kapania, R., and Schetz, J. (2016). Aeroelastic applications of a variable-geometry raked wingtip. *Journal of Aircraft* 54 (1): 1–13.

19 Dimino, I., Andreutti, G., Moens, F. et al. (2021). Integrated design of a morphing winglet for active load control and alleviation of turboprop regional aircraft. *Applied Sciences* 11 (5): 1–27.

20 Saravanan, R. (2012). *Design of Parametric Winglets and Wing tip Devices – a Conceptual Design Approach*. Linkping, Sweden: Lincopiln Studies in Science and Technology.

21 Gavrilovic, N. (2014). Improvement of performance of Commercial Aircraft Using Winglets. M.Sc. Thesis. University of Belgarde, Faculty of Mechanical Engineering.

22 ANSYS Fluent (2013). *14.5 User's Guide*. Canonsburg, PA: ANSYS Inc.

23 Azlin, M.A., Mat Taib, C.F., Kasolang, S., and Muhammad, F.H. (2011). *CFD Analysis of Winglets at Low Subsonic Speed*, vol. I. London, UK: World Congress on Engineering.

24 Versteeg, H.K. and Malalasekera, W. (1995). *An Introduction to Computational Fluid Dynamics: The Finite Volume Method*. Pearson Education Limited.

25 Rasuo, B., Parezanovic, V., and Adzic, M. (2008). *On Aircraft Performance Improvement by Using Winglets*. ICAS.

26 Rasuo, B. (2014). *Flight Mechanics*. Faculty of Mechanical Engineering, University of Belgrade, e-book.

27 Minnella, G., Rodriguez, Y.J. and Ugas, J. (2010). Aerodynamic shape design optimization of winglets. Senior Design Project, *Mechanical and Materials Engineering Dept. Florida Intenrational University*.

28 Soreshjani, H.M. and Jahangirian, A. (2021). An investigation on winglet design with limited computational cost, using an efficient optimization method. *Aerospace Science and Technology* 117: 106957.

29 Panagiotou, P., Kaparos, P., and Yakinthos, K. (2014). Winglet design and optimization for a MALE UAV using CFD. *Aerospace Science and Technology, Elsevier* 39: 190–205.

30 Elham, A. and van Tooren, M. (2014). Winglet multi-objective shape optimization. *Aerospace Science and Technology, Elsevier* 37: 93–109.

31 Gardi, A., Sabatini, R., and Ramasamy, S. (2016). Multi-objective optimisation of aircraft flight trajectories in the ATM and avionics context. *Progress in Aerospace Sciences* 83: 1–36.

32 Ramasamy, S., Sabatini, R., and Gardi, A. (2018). A unified analytical framework for aircraft separation assurance and UAS sense-and-avoid. *Journal of Intelligent & Robotic Systems* 91: 735–754.

13

Integration of Naturally Occurring Materials in Lightweight Aerostructures

Jose Silva[1], Alessandro Gardi[1,2], and Roberto Sabatini[1,2]

[1] School of Engineering, RMIT University, Melbourne, Victoria, Australia
[2] Department of Aerospace Engineering, Khalifa University of Science and Technology, Abu Dhabi, UAE

13.1 Introduction

Over the past few years, we have assisted in raising interest towards the application of natural materials for engineering applications. A natural based material can be defined as a product obtained from natural and renewable resources extracted from agricultural and forestry feedstock, including by-products and residues [1]. These types of materials have followed a pathway parallel to the evolution of mankind, since a myriad of examples of natural occurring materials used with distinct functionalities can be found in many everyday applications, such as wood based products used in civil and naval applications, for example.

In the particular case of air transport, it is interesting to note that the choice of materials used in the early aircraft at the beginning of the twentieth century was a corollary of the existing technology at that time. In fact, from the dawn of aviation history and over nearly three decades most aircraft were built from the lightest materials available, which were dominated by wood based materials for structural purposes, like bamboo, balsa wood, or spruce [2]. One of the best examples of the prevalence of wood in aeronautical construction until WWII is probably the de Havilland DH. 98 *Mosquito*, a British high-speed bomber airplane constructed almost entirely of wood, which granted it the nickname of the *Wooden Wonder* [3]. Due to the shortage of other strategic materials required for the construction of different military gear to be used in the war, designers had to find ingenious solutions from other existing and abundant materials. The fuselage was constructed of balsa wood sandwiched between layers of plywood, creating a remarkably clean-looking aerodynamic configuration and providing strength-to-weight ratios close to the aluminium alloys existing at that time. One important advantage of the wooden design was the readiness and affordability of the related production techniques, most of them common to wood construction used in routine applications.

Another emblematic airplane used in WWII was the Supermarine Spitfire, a British fighter that was the first aircraft reported using a composite material combining both natural and synthetic constituents, namely flax linen roving held in a urea formaldehyde

Sustainable Aviation Technology and Operations: Research and Innovation Perspectives, First Edition.
Edited by Roberto Sabatini and Alessandro Gardi.
© 2024 John Wiley & Sons Ltd. Published 2024 by John Wiley & Sons Ltd.
Companion Website: www.wiley.com/go/sustainableaviation

(phenolic) resin [4]. This airplane also used a cellulose based composite made of paper mixed with adhesives to fabricate the pilot's seat. In fact, by the 1940s cellulose based composites using natural fibres to reinforce polymers were a well-known candidate material for many aeronautical applications in non-critical components, although a rampant interest was being shown for structural applications [5].

The role of natural materials diminished considerably since WWII as a result of the advent of many forms of synthetic materials offering superior mechanical properties. In the 1950–1960s aluminium alloys were the dominant material for aircraft construction due to their superior strength-to-weight ratio and reasonable cost. Nevertheless, aircraft designers kept striving for better performance and improved reliability, which necessarily demands new materials with optimised properties. This triggered a considerable investment in research activities in the 1960s aiming at introducing new synthetic composite materials with enhanced capabilities, which was probably the greatest leap forward in the aeronautical construction in the last century. In particular, fibre-reinforced polymer-based materials (FRPs) started capturing the attention of designers and manufacturers as an alternative material to metallic alloys due to their considerable advantages, namely: higher strength-to-weight and/or stiffness-to-weight ratios; the ability to customise their properties as a result of the proper selection of different lay-up stacking sequences; excellent resistance to fatigue and to corrosion damage; and the possibility of designing parts with complex geometries like double curvature [6, 7].

Typically, the composite parts used in most recent aircraft are made from glass, carbon, or aramid fibres as reinforcing elements, whereas thermoset polymers (predominantly epoxy resins) are frequently used as matrix elements to ensure the best mechanical properties. A good example of how the aeronautical industry has embraced the technology of composite materials is the Boeing 787, a commercial long-range airplane which entered service in 2011. In this aircraft carbon fibre composites (CFRP) are extensively used in primary structures (adding up to 50% of the total weight) allowing for substantial weight savings compared with conventional aluminium alloys.

Notwithstanding their superior mechanical properties, synthetic composites are made of polymers and fibres derived from petroleum, which is a non-renewable commodity. Furthermore, recent concerns about the recyclability of these materials, as a result of their non-biodegradability, sounded the alarm among authorities and decision makers all over the world as the disposal of some of these materials has become a critical issue. By definition, composites are made using two dissimilar materials, which poses some challenges regarding their recyclability or reuse. In most cases, composites end up in landfills, whereas some are incinerated after use [8].

Recently, the European Commission expressed its vision regarding the future of aviation in Europe in the next 3–4 decades through a report [9]. This document points out some priority areas to assist in the development of the aviation sector as one of the most strategic areas in the European agenda. Amongst the many challenges listed in this report, there is a clear goal aiming at protecting the environment and the energy supply, which will lead to a drastic decrease of materials and fuels depending on crude oil. There is also a strong commitment to the design and manufacture of air vehicles with improved recyclability capabilities, which will catalyse the research activities on greener materials, or in other words, with a reduced environmental footprint.

In fact, the climate change has been driving many global initiatives, which have sparked off a *'green fever'* around new recyclable and/or biodegradable materials with potential applications in engineering products. Over the last 15 years, significant efforts have been put in the development and characterisation of 'green composites' which are normally in the form of FRPs incorporating either natural fibres and/or bio-derived polymers. So far, the applications of green composites have been limited to non-critical parts, such as secondary and tertiary structures, panels, fairings, cases, etc. [10]. For this type of application, nature offers a broad range of inexpensive and abundant plant-based fibres possessing many advantages compared to synthetic alternatives: they can be incinerated, are CO_2 neutral when burned, are much less abrasive than inorganic-mineral counterparts to processing machinery, less dangerous for the production employees in case of inhalation, and, because of their hollow and cellular morphology, are lightweight and perform well as acoustic and thermal insulators [8, 10]. These latter features are very interesting from the standpoint of their utilisation in certain parts of transport vehicles to improve the ridding characteristics (e.g. noise and vibration mitigation) and minimise fuel consumption. In fact, the automotive industry was one pioneering sector in terms of the successful integration of green composites in many car parts since the 1990s [1]. Still, and despite the natural charm and euphoria around bio-composites, designers and engineers should bear in mind that these have some pitfalls which cannot be neglected when considering them as candidate materials for structural applications in critical parts.

This case study, originally published in [34], intends to undertake a very concise review of the existing natural materials to assess the viability of their application in aeronautical parts. It should be noted that the authors did not aim for an exhaustive review of all physical and mechanical properties of natural materials, as these aspects have been thoroughly addressed in the literature. Instead, the main objective of this work is to contribute to the identification of possible evolutionary paths for the possible utilisation of more sustainable materials and related technologies in the aviation sector as stipulated by the key international stakeholders for the near future. In order to outline the general characteristics of naturally occurring materials in terms of mechanical strength and stiffness, it is useful to refer to the two Ashby plots in Figure 13.1, developed among others by Dicker et al. [11].

13.2 Composites with Natural Fibres

Over the past few years the development of high-performance composites made from natural resources has been increasing worldwide due to environmental and sustainability concerns. The natural constituents can be used as reinforcing elements (normally in the form of fibres), biopolymers, or both. As the reinforcing fibres have a significant impact in the overall mechanical behaviour of composites, and considering the significant number of scientific literature dedicated to natural fibres, this chapter will only analyse this type of material and therefore will not cover the emerging area of green materials used as matrixes in bio-composites.

Many examples can be found of the use of either plant-based or animal-based fibres as reinforcing elements of FRPs. Figure 13.2 summarises the main categories of the different natural fibres grouped according with their extraction sources. Amongst these, vegetable

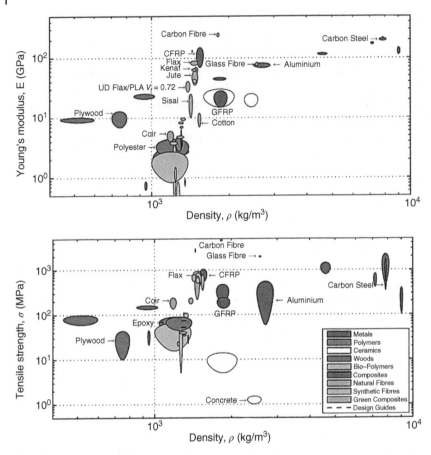

Figure 13.1 Ashby plots of the mechanical stiffness and strength properties of a wide range of materials. Source: Adapted from Dicker et al. [11].

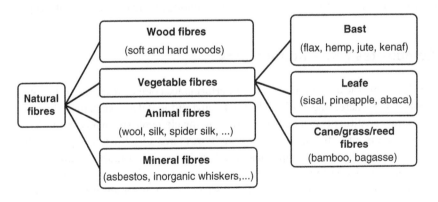

Figure 13.2 Types of natural fibres, depending on their source of extraction.

fibres have taken a leading position in the research and production of green composites over the past decade. The attractiveness of modern agricultural techniques ensuring a high production yield together with very interesting mechanical properties are the key reason for this choice. The cell structure of vegetable fibres is complex, with each fibre being a composite of rigid cellulose microfibrils embedded in a soft lignin and hemicellulose matrix [11]. The theoretical elastic modulus of a cellulose molecule is predicted to be around 138 GPa which is a higher value when compared with glass fibres and within the same range as aramid fibres [12]. However, this high rigidity level cannot be maintained when considering long fibres for practical applications, due either to naturally occurring structural defects in the fibre, defects incurred during its processing phase, or chemical incompatibilities with the polymer matrix [13]. Therefore, we need to be moderately optimistic when considering vegetable fibres as competitors against the conventional synthetic fibres used in FRPs. Another important drawback of natural fibres, which poses some difficulties for their application on an industrial scale, is the considerable variability of their nominal mechanical properties, which are largely influenced by the overall environmental conditions during growth and the traditional methods of preparation of the fibres in which some processing variables are difficult to control.

Nevertheless, and depending on the type of application (and therefore on the design requirements), several works evinced that there might be some cases where the strength (or stiffness) to weight ratios of some natural fibres are in the same range or even higher than some synthetic fibres, namely E-glass fibres. Although glass fibres possess the lowest mechanical properties in the family of engineering composites, they are still extensively used in many applications, particularly in the transport sector (including aeronautics) making it a benchmark material when it comes to measuring the performance of green composites. In the following paragraphs, we will present a short overview of those natural fibres whose properties paint a promising and viable scenario regarding their utilisation in structural components for lightweight applications considering their utilisation as reinforcing constituents of FRPs.

Flax is a bast fibre well known from the textile industry. Interestingly, it was also one of the first natural materials to be considered as a composite constituent in the early days of the aviation history combined with phenolic resin as a construction material for the Spitfire fighter aircraft. Recently, flax fibres have been considered for sporting equipment applications, such as in the reinforcement of snowboards, tennis rackets, racing boats, and bicycle frames [11, 14, 15]. In many cases these fibres are combined with other synthetic reinforcements (carbon fibres for example) to ensure improved mechanical properties as it has been reported that this hybrid formulation provides superior vibration dampening. A significant amount of research has been done at the German Aerospace Center (DLR) on biodegradable plastics and composites using flax fibres. Some of these composites had properties comparable to E-glass/epoxy FRPs and were found to be suitable for a variety of structural applications [8]. One particular potential application where flax fibres proved to be a serious alternative to glass fibres is in components used in aircraft cabins [16]. Researchers from the DLR conducted a life cycle assessment to quantify the environmental impacts of flax fibres compared to glass fibres in a cabin panel made from a phenolic resin prepreg. Results showed that flax had clear advantages in terms of the lower energy consumption throughout its life cycle, including the possibility to recover energy through incineration at the disposal phase.

Belonging to the cannabis family, hemp is another type of bast fibre considered as an alternative to glass reinforcements. Wambua et al. [17] investigated the mechanical behaviour of polypropylene composites reinforced with different natural fibres, including hemp. The authors concluded that hemp based materials led to superior results regarding the tensile strength considering identical volume fractions to the other natural fibres (sisal, kenaf, jute, and coir), whereas their flexural strength and tensile modulus were comparable to that of glass matt composites but with the advantage of having a lower density value (1.48 against 2.55 g cm^{-3}). Another work from Pervaiz and Sain [18] investigated a polypropylene composite reinforced with hemp and found that an energy saving of 50 MJ kg^{-1} (3 kg of CO_2 per kg of material) was possible by replacing a glass fibre component with a fibre weight fraction of 0.3, with a hemp fibre component with a 0.65 fibre weight fraction. Other works showed that PP composites reinforced with hemp fibres exhibit interesting recyclability capabilities as the mechanical properties of this type of material remained well preserved despite the number of reprocessing cycles [19].

Kenaf is also another natural fibre normally considered as a reinforcement element in polymer based composites for many applications due to its superior toughness and high aspect ratio in comparison to other fibres. A single fibre of kenaf can have a tensile strength and elastic modulus as high as 11.9 and 60 GPa, respectively [20]. Further to its interesting mechanical properties, the growth rate of kenaf plants is very attractive (up to 10 cm d^{-1} under optimum ambient conditions), which makes it a very affordable and abundant material [5]. From the viewpoint of energy consumption, the production of 1 kg of kenaf requires 15 MJ of energy, whereas this value rises to 54 MJ to produce 1 kg of glass fibre [20]. For these reasons, kenaf has been used in many different applications with a special relevance in the automobile industry. Toyota, jointly with Araco Corporation, developed a kenaf based composite which is used in some interior components (such as door trims), providing weight savings and better soundproofing characteristics [21].

A comparative study of the mechanical properties of kenaf fibre-reinforced composites and other natural fibres is presented by Akil et al. [20]. These composites were fabricated using polypropylene films, with natural fibre layers randomly spread between them. In general terms, the tensile and flexural strengths are lower than those obtained for the same weight fraction regarding the hemp and flax fibres (around 10% and 20%, respectively) but still providing superior specific properties than E-glass fibres.

Silk is an animal type of fibre that we decided to include in the list of natural materials, as it has a clear potential to act as a reinforcement for composites due to its remarkable properties. Silks are generally defined as protein polymers that are spun into fibres by such animals as silkworms, spiders, scorpions, mites, and flies [22]. Although the absolute stiffness/strength (tensile and flexural) of silk based composites is lower than that of glass-fibre composites, some studies have demonstrated a superior performance in terms of fracture strain capacities and elongation at break, which may be particularly attractive in applications where progressive failure or high compliance is required [23, 24]. For this reason, potential applications of silk fibres may include lightweight, crashworthy, and impact-critical components used in the automotive and aeronautical sectors. It is also worth mentioning that the mechanical performance of silk fibres surpasses most of

Table 13.1 Properties of some natural fibres vs. E-glass.

Type of fibre	Density (g cm⁻³)	Tensile strength (MPa)	Young's modulus (GPa)	Specific strength	Specific modulus
Flax	1.5	345–1500	27–39	230–1000	18–26
Hemp	1.47	550–900	70	374–612	48
Kenaf	1.5–1.6	350–930	40–53	226–600	25–34
Silk (*Bombyx mori*)	1.25–1.35	650–750	16	500–576	12.3
E-glass	2.55	2000	70–73	784	27–28

the natural vegetable fibres in terms of interlaminar shear strength and specific impact strength [24]. As downsides, some issues have been reported regarding the compatibility of silk fibres with some polymeric matrixes and high moisture sensitivity.

The data available in the aforementioned references allowed us to summarise the main mechanical and physical properties of the natural fibres discussed in this chapter in the form of Table 13.1. Average specific stiffness and specific strength are also presented as they are important requirements for the structural design of lightweight structures. It must be stressed out though that these values should be considered as an indicative reference only as they are largely dependent on several factors that might contribute to a significant variability, such as the climate and region influence in the harvesting process and/or the distinct manufacturing processes used to extract the fibres from the plant stem.

Notwithstanding the interesting mechanical properties of natural fibres, there are some drawbacks that hinder their application under some particular conditions. The most probable issue is moisture absorption, as cellulose is a highly hydrophilic molecule and this property is therefore imparted to natural fibres. Water absorption causes fibre swelling which leads to delamination, surface roughening, and a subsequent loss of strength of the material [11]. Symington et al. [25] concluded that moisture plays a significant role in influencing the mechanical properties of some natural fibres exposed to water environments, such as flax and hemp. In this latter case, the degradation level was so intense that the material was unable to be tested, raising concerns about its stability when exposed to various environmental conditions. The moist environment can also facilitate the growth of fungus and bacteria, which leads to rotting [26].

The chemical incompatibility between the hydrophilic natural fibre and hydrophobic polymer matrix is an additional issue responsible for the decrease in the mechanical properties as it leads to poor fibre wetting, producing an inferior interface, and encouraging fibre agglomeration. To overcome this limitation, some treatments have been investigated to modify the fibre surface properties to improve their adhesion with different matrices. The fibre alkali treatment, also referred to as mercerisation, has been utilised with appreciable success [11]. This process improves the capacity for chemical interaction between the matrix and fibres, while allowing for better mechanical interlocking through rougher topography and larger numbers of individual fibrils, having as a side benefit the reduction in water absorption.

13.3   Cork: Nature's Foam

As an example of the many benefits of natural materials to the construction of lightweight structures, this section will be focused on the use of composites with cork. Cork is a natural, renewable, sustainable material extracted from the bark of the oak (*Quercus suber L.*), which is periodically harvested from the tree, usually every 9–12 years. The most intensive cork production is located around the Mediterranean basin and China.

Microscopically, cork may be described as a homogeneous tissue of thin-walled cells, regularly arranged without intercellular space laying under an alveolar structure, analogous to that of a honeycomb [27], which is at the base of its low density (ranging from 100 to 240 kg m^{-3}). Additionally, the cellular morphology provides greater levels of energy absorption under deformation corresponding to cell edge bending or face stretching for low stress levels or progressive cell collapse by elastic buckling or plastic yielding at higher loads [28]. Although the overall mechanical properties of natural cork are relatively worse than other lightweight materials (such as synthetic foams), its combination with adequate binding agents results in cork based agglomerates with competitive specific properties. Besides, its low thermal conductivity plus reasonable compressive strength make it an excellent material for thermal insulation where compressive loads are present.

Further to its well-known usage in non-structural applications in civil construction (e.g. floor panels, thermal insulator, external cladding), recently cork based agglomerates have been considered for applications in the transport sector owing to their excellent vibration/acoustic suppression capabilities and low thermal conductivity [29]. The product range within the transportation industry comprising panels, flooring systems, and anti-vibration barriers can already be found in many components in trains, buses, and boats, including parts with structural capabilities [30]. In the aerospace sector, cork compounds have been used for more than 30 years for the fabrication of thermal protection applications based on ablative heat shields installed in rockets and space vehicles.

One of the most promising forms of applications of cork based composites is in the form of a core material for sandwich type components due to the cumulative benefit of its lightweight, high loss factor, superior energy absorption properties, thermal insulation, and marginal absorption of moisture in the long run. Sargianis et al. [31] suggested that cork agglomerate cores combined with carbon based FRPs facesheets resulted in a virtually noise-free sandwich structure with excellent damping performance providing increased durability under fatigue loading conditions. Also Silva et al. [32] proposed the use of a micro-sandwich cork core as an effective and simple solution to enhance the damping behaviour of high strength/stiffness carbon/epoxy laminates for structural applications. Another recent work [33] aimed at improving the resilience and damage tolerance of properties of CFRP laminates by using the viscoelastic nature of cork. In this case, two types of materials were considered: a sandwich formed by carbon-epoxy facesheets with a cork-epoxy core and a carbon-epoxy laminate with embedded cork granulates. Results were clear about both the shielding effect provided by the cork core on the minimization of the damage extension induced by low energy impact tests (which are representative of many situations within the real operational context of aircraft) and the extension of the flight envelope without weight penalty by increasing the critical flutter speed (as a consequence of the improved damping characteristics of laminates with

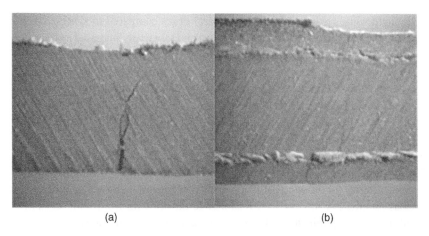

(a)   (b)

Figure 13.3 Extent of damage caused by a low energy (10 J) impact event on a carbon epoxy laminate: (a) without embedded cork layers; (b) with two cork layers embedded close to the surface positions.

embedded cork). A good example of this shielding effect under impact loading is illustrated by the two cross-sections in Figure 13.3, one referring to a conventional carbon-epoxy laminate (Figure 13.3a) and the other (Figure 13.3b) pertaining to an identical laminate but considering the inclusion of two very thin cork layers embedded close to the surface of the material. As can be seen, the damage (in the form of a through thickness crack front) in the cork-based laminate, resulting from a low energy impact loading, has been considerably reduced relative to the conventional material, which is a good indicator of the energy absorption capability provided by natural cork without hindering the low weight requirements of structural composites used in the aerospace sector.

As a general conclusion we can state that cork-based composites can be considered as a serious alternative to synthetic foams or metal honeycombs when the need for a maximisation of the energy absorption capabilities is a major design requirement. Furthermore, the mechanical properties of polymeric foams are strongly influenced by temperature, whereas cork agglomerates can withstand temperatures of up to 200 °C with minimum mass loss and retaining their nominal properties.

13.4 Conclusions

In several industrial applications, naturally occurring materials are of great interest to attain a significant reduction in the use of petroleum-based products and other non-renewable resources, therefore contributing to a more sustainable development. Natural materials are in fact biodegradable, CO_2 neutral, and generally less noxious in case of inhalation. In addition to their environmental benefits, some natural materials can be employed as constituents of high performance, composite materials due to their competitive mechanical properties. In particular, this chapter briefly reviewed the opportunities and challenges arising from the application of selected natural fibres in lightweight composites for the aerospace industry, as well as the potential of using cork based products as an alternative material to synthetic foams in sandwich structures. Based on the existing literature, it was

possible to highlight some clear advantages of natural fibres compared to synthetic reinforcements, particularly their low density, reduced cost and tool wear, increased availability,
and biodegradability. In certain cases, the specific mechanical properties are comparable
or even better than glass fibres, paving the way for possible applications in structural components. The review also addressed cork as an ideally suited core material for sandwich
type components, due to its lightweight, high loss factor, superior energy absorption properties, acoustic, and thermal insulation and marginal absorption of moisture. Nevertheless,
there are still some issues that limit the overall performance of natural materials in composites, such as the interfacial adhesion with the matrix, moisture absorption, and long-term
stability. Significant research is presently underway to overcome these limitations and to
create more opportunities for the successful application of natural materials in lightweight
aerostructures as an effective answer to the challenge put forward from both legislative
authorities and the general public towards a more sustainable aviation industry.

References

1 Koronis, G., Silva, A., and Fontul, M. (2013). Green composites: a review of adequate
materials for automotive applications. *Compos Part B: Eng* 44: 120–127.

2 Cutler, J. (1999). *Understanding Aircraft Structures*, 3e. Oxford: Blackwell Pub.

3 Guttman, R. (2001). De Havilland Mosquito: Britain's wooden wonder. *Aviation History*
11: 42–48.

4 Marsh, G. (1996). Aerospace composites – the story so far. *Reinforced Plastics* 40: 44–48.

5 Baillie, C. (ed.) (2005). *Green Composites: Polymer Composites and the Environment*.
CRC Press.

6 Saúde, J.M.L. and Silva, J.M. (2014). Aircraft industrialization process – a systematic
and holistic approach to ensuring integrated management of the engineering process.
In: *Technology and Manufacturing Process Selection* (eds. E. Henriques, P. Pecas and A.
Silva), 81–104. Springer.

7 Soutis, C. (2005). Carbon fiber reinforced plastics in aircraft construction. *Materials
Science and Engineering A* 412: 171–176.

8 Netravali, A.N. and Chabba, S. (2003). Composites get greener. *Materials Today*: 22–29.

9 European Commission. (2011). Flightpath 2050: Europe's Vision for Aviation - Report
of the High Level Group on Aviation Research, European Commission - Directorate
General for Mobility and Transport, Luxemburg.

10 La Mantia, F.P. and Morreale, M. (2011). Green composites: a brief review. *Composites.
Part A, Applied Science and Manufacturing* 42: 579–588.

11 Dicker, M.P.M., Duckworth, P.F., Baker, A.B. et al. (2014). Green composites: a review of
material attributes and complementary applications. *Composites. Part A, Applied Science
and Manufacturing* 56: 280–289.

12 Nishino, T., Takano, K., and Nakamae, K. (1995). Elastic modulus of the crystalline
regions of cellulose polymorphs. *Journal of Polymer Science: Part B* 33: 1647–1651.

13 Staiger, M.P. and Tucker, N. (2008). Natural-fibre composites in structural applications.
In: *Properties and Performance of Natural-Fibre Composites*, 269–300. Elsevier Inc.

14 Stemergy. (2014). Plain sailing for flax, ed.

15 Lineo-NV. (2014). Applications, ed.

16 Wiedemann, M., Sinapius, M., and Melcher, J. (2012). *Innovation Report 2012: Institute of Composite Structures and Adaptive Systems*. Germany: DLR.

17 Wambua, P., Ivens, J., and Verpoest, I. (2003). Natural fibres: can they replace glass in fibre reinforced plastics? *Composites Science and Technology* 63: 1259–1264.

18 Pervaiz, M. and Sain, M.M. (2003). Carbon storage potential in natural fiber composites. *Resources, Conservation and Recycling* 39: 325–340.

19 Faruk, O., Bledzki, A.K., Fink, H.P., and Sain, M. (2012). Biocomposites reinforced with natural fibers: 2000–2010. *Progress in Polymer Science* 37: 1552–1596.

20 Akil, H.M., Omar, M.F., Mazuki, A.A.M. et al. (2011). Kenaf fiber reinforced composites: a review. *Materials and Design* 32: 4107–4121.

21 Toyota Motor Corporation. (2012). Toyota Boshoku Develops New Automobile Interior Parts Utilizing Plant-based Kenaf Material, 9 Feb 2012. Accessed 17 Sept. https://www.toyota-boshoku.com/global/news/1951.html.

22 Altman, G.H., Diaz, F., Jakuba, C. et al. (2003). Silk-based biomaterials. *Biomaterials* 24: 401–416.

23 Ude, A.U., Eshkoor, R.A., Zulkifili, R. et al. (2014). Bombyx mori silk fibre and its composite: a review of contemporary developments. *Materials and Design* 57: 298–305.

24 Shah, D.U., Porter, D., and Vollrath, F. (2014). Can silk become an effective reinforcing fibre? A property comparison with flax and glass reinforced composites. *Composites Science and Technology* 101: 173–183.

25 Symington, M.C., Banks, W.M., West, O.D., and Pethrick, R.A. (2009). Tensile testing of cellulose based natural fibers for structural composite applications. *Journal of Composite Materials* 43: 1083–1108.

26 Stamboulis, A., Baillie, C.A., Garkhail, S.K. et al. (2000). Environmental durability of flax fibres and their composites based on polypropylene matrix. *Applied Composite Materials* 7: 273–294.

27 Silva, S.P., Sabino, M.A., Fernandas, E.M. et al. (2005). Cork: properties, capabilities and applications. *International Materials Review* 50: 345–365.

28 Gibson, L.J. (2005). Biomechanics of cellular solids. *Journal of Biomechanics* 38: 377–399.

29 Gil, L. (2009). Cork composites: a review. *Maternité* 2: 776–789.

30 Amorim. (2014). Cork Composites, ed.

31 Sargianis, J., Kim, H.I., and Suhr, J. (2012). Natural cork agglomerate employed as an environmentally friendly solution for quiet sandwich composites. *Scientific Reports* 2.

32 Silva, J. M., Gamboa, P. V., Cláudio, R., Nunes, N., and Lopes, J. (2013). Optimization of the damping properties of CFRP laminates with embedded viscoelastic layers, 14th International Conference on Civil, Structural and Environmental Engineering Computing (CC 2013), Cagliari, Italy.

33 Silva, J.M., Nunes, C.Z., Franco, N., and Gamboa, P.V. (2011). Damage tolerant cork based composites for aerospace applications. *Aeronautical Journal* 115: 567–575.

34 Silva, J.M., Sabatini, R., and Gardi, A. (2014). Opportunities Offered by Naturally Occurring Materials in Lightweight Aerostructures Design. In: *Proceedings of Practical Responses to Climate Change. Engineers Australia Convention 2014 (PRCC 2014)*, Melbourne, Australia.

14

Distributed and Hybrid Propulsion: A Tailored Design Methodology

Martin Burston[1], Kavindu Ranasinghe[2], Alessandro Gardi[3,4], Vladimir Parezanovic[3], Rafic Ajaj[3], and Roberto Sabatini[3,4]

[1] *School of Engineering, RMIT University, Melbourne, Victoria, Australia*
[2] *Insitec Pty Ltd, Melbourne, Victoria, Australia*
[3] *Department of Aerospace Engineering, Khalifa University of Science and Technology, Abu Dhabi, UAE*
[4] *School of Engineering, RMIT University, Melbourne, Victoria, Australia*

14.1 Introduction

As extensively discussed in this book, modern transport aircraft have progressively introduced various technological advancements such as composite materials, aerodynamic improvements, and aeroengine advancements, which have already had profound impacts, leading to more than 30% reduction in fuel consumption per available seat kilometre (ASK) over the past decades [1]. However, these approaches are beginning to yield smaller gains for greater effort, whilst the road to achieving carbon neutrality is still long. A promising next evolutionary step in aircraft design is the better integration of multiple aircraft systems to gain synergistic benefits through positive interactions [2], a strategy which can greatly benefit from advanced forms of distributed propulsion (DP).

DP can be defined as the concept of distributing an aircraft's thrust force and associated loads around the airframe [2, 3]. This is in contrast to the conventional approach to aircraft propulsion, whereby a small number of engines produce concentrated thrust vectors. An alternative interpretation to that presented above is that DP occurs when more than two engines are employed in an aircraft design. Gohardani et al. [4] approach the topic from this perspective, reviewing a number of aircraft that employ three or more engines in order to establish historical trends. However, as highlighted by Kim [2], modern DP concepts involve a high degree of integration between the propulsion system and the airframe. The primary objectives of DP configurations are to improve energy efficiency, short take-off and landing (STOL) performance, and reduce noise emissions. The environmental benefits of increased energy efficiency are reduced exhaust gas and particulate emissions, such as nitrogen oxides (NO_X), sulphur oxides (SO_X), carbon dioxide (CO_2), and carbon (soot).

The design of aircraft adopting the DP principles requires a high level of integration between the structure, aerodynamics, and propulsion systems of the aircraft. The objective

Sustainable Aviation Technology and Operations: Research and Innovation Perspectives, First Edition.
Edited by Roberto Sabatini and Alessandro Gardi.
© 2024 John Wiley & Sons Ltd. Published 2024 by John Wiley & Sons Ltd.
Companion Website: www.wiley.com/go/sustainableaviation

is to achieve a high degree of synergy between these systems so that benefits may be obtained from the favourable interactions between them. For these specific reasons, DP strongly benefits from a holistic approach to aircraft design. This chapter, originally published in [83, 84].

14.2 Advanced Distributed Propulsion Configurations

Ever since the interwar period, the distribution of the thrust force across an aircraft has been the fundamental characteristic behind DP configurations. The installation of three or more monolithic engines was a common design solution until the widespread adoption of extended-range twin-engine operations performance standards (ETOPS). More contemporary concepts, here referred to as *advanced distributed propulsion*, represent a paradigm shift compared to conventional DP as they rely on a more modular and optimised architecture rather than multiple monolithic engines. This might be a custom combination of power source, transmission and thrust generation, and the airframe configuration into which the DP system is integrated. Exploring these aspects in detail allows us to appreciate the advantages that advanced DP concepts offer for a more environmentally, economically, and socially sustainable aviation.

Categorising DP configurations is the first necessary step in such an analysis. Various researchers have proposed DP categories, most notably Kim [2] – whose categories are also used by Steiner et al. [3]. The categories developed by Kim provide a broad, generalised overview of DP configurations whilst that described by Gohardani et al. [4] offers a much more comprehensive list of design choices encroaching on conceptual design at an aircraft level. A compromise between these extremes is presented here with the aim of creating an accurate, easy to use categorisation system for DP systems.

Outside this categorisation, but nevertheless notable, is the example of a true DP system using electroaerodynamics (EAD) as a solid-state propulsion system in the case of the Massachusetts Institute of Technology (MIT) flying drone [5]. The propulsive force is created through high-voltage, high-frequency excitation of electrons in the surrounding air (corona discharge), ionising the air and accelerating the ions by way of the Coulomb force. This novel propulsion system can be currently applied only in very limited cases of ultra-light structures and for obtaining very low flight speeds, since the propulsive force is extremely weak. Nevertheless, this experiment is a notable instance in aerospace engineering history, representing the first truly continuous propulsion system with no moving parts. Future research will have to determine if the physics involved might allow such a concept to be scaled up.

14.2.1 Thrust Distribution Methods

The first fundamental categorisation is using the method in which this distribution of thrust is achieved in DP. Kim [2] provides the following classifications:

- Jet flaps;
- Cross-flow fans;
- Multiple discrete engines;
- Distributed multi-fans driven by few engine cores, either gas, gear, or electrically driven.

The first category refers to aircraft where the exhaust jet of the engine (or engines) is ducted through a slot at or near the wing trailing edge and provides spanwise thrust for cruise as well as super-circulation for high lift during take-off and landing [2]. In this chapter, the nomenclature adopted by Gohardani et al. [4], who refer to this method as *distributed exhaust*, will be adopted as it covers a greater breadth of potential applications where redirection of the exhaust jet is employed.

The second category refers to the use of cross-flow fans, which consist of a drum-like rotor with forward curved blades, encased within housing walls with a rectangular inlet and outlet [6]. The key advantage of this system is that it can be extended lengthways and integrated into a wing such that is stretches partially or fully across the wingspan.

The third and fourth categories involve the use of conventional axial-flow fans, but in large numbers and with spanwise distribution. The former makes use of combustion engines, such as turbojet or turbofan engines, whilst the latter has only a small number of these engines and instead relies upon distributing the power generated by these engines to a large number of fans. Since power distribution methods are being considered separately here, these two categories will be combined for the discussion of thrust distribution methods.

Therefore the configurations from Kim [2] have been reduced to three in this section since power transmission methods and power sources are being dealt with as separate topics. These are summarised below:

- Distributed exhaust
- Cross-flow fans
- Multiple discrete fans

14.2.1.1 Distributed Exhaust

The most salient example of thrust distribution is that of ducting or otherwise distributing the exhaust gas produced by the engines around the aircraft. Typically, the exhaust gas is distributed spanwise through slots in the trailing edge of the wing, as proposed by Kuchemann with his jet-wing concept, where the engines are embedded within the wing [7]. The study concluded that there was potential for a propulsive efficiency improvement by using this configuration [8]. This has been investigated by Ko et al. in [9] as well as Liefsson et al. in [10], who derive mathematical formulations to describe this in order to consider the wake-filling properties of this configuration. This refers to the low velocity region directly behind the aircraft being offset by the engine exhaust jet stream, as shown diagrammatically in Figure 14.1. The wake of the aircraft can be filled effectively by ducting most or all of the engine's exhaust jet out through the trailing edge of the wing, or aft of the wing-body when applied to a Blended Wing-Body (BWB) as shown in Figure 14.2.

However, the ducted exhaust configuration is not only promising due to its wake filling properties. Another promising application of this technology is that of the jet flap, where a high-velocity thin jet sheet emanates from a slot at or near the wing trailing edge and provides spanwise thrust for cruise and super-circulation for high lift [2]. The key distinction between jet wing and jet flap configurations is that of purpose – the objective of the former is to increase propulsive efficiency whilst the objective of the latter is to maximise CL_{max} for improved take-off and landing operations. The defining characteristics of the jet flap

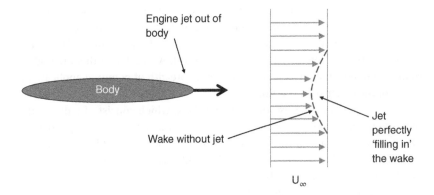

Figure 14.1 A velocity profile of an ideal distributed propulsion body/engine system. Adapted from [10].

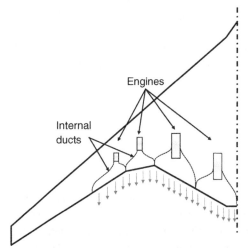

Figure 14.2 The ducted exhaust configuration as applied to a BWB aircraft. Adapted from [10].

configuration are the high deflection angles associated with the jet stream, as well as the (in general) smaller mass flow rates of gas through the trailing edge slot. This method of exhaust distribution, and indeed, thrust distribution, is the most prevalent amongst those aircraft that have so far been built and tested.

Key historical examples of aircraft exhibiting jet flap configurations are the Hunting H.126 aircraft, as shown in Figure 14.3, which was designed to improve take-off and landing performance to allow operation from shorter runways [12]. The aircraft used a system of 16 nozzles positioned along the trailing edge of the wing which blew more than half of the engine's exhaust gas over the upper surface of the flaps. A further 10% of exhaust gas was ducted over the control surfaces to provide adequate control at low airspeeds. Due to this, the aircraft was controllable at airspeeds as low as 51.5 km/h. Despite these benefits, only a single aircraft was ever built.

A more modern example of the application of jet flaps and blown control surfaces is the ShinMaywa US-2 flying boat designed and manufactured in Japan [13]. First flown in 2003, this amphibian is designed with STOL performance in mind and utilises an advanced

Figure 14.3 Hunting H.126 jet flap aircraft [12]. Credits: Tony Hisgett via Wikimedia Commons.

boundary layer control (BLC) system that ducts bleed air from the engines over the flaps and control surfaces. This aircraft, fully loaded to a Maximum Take-Off Weight (MTOW) of 47.7 tons, achieves a take-off distance of 450 m (1476 ft). This aircraft is currently in production.

14.2.1.2 Crossflow Fans

The crossflow, or transverse fan, represents a fundamentally different approach to aircraft propulsion since the fan rotation axis is perpendicular to the direction of motion rather than parallel with it. Crossflow fans can be extended in length so that they produce a uniform inflow and outflow, especially when they are of high aspect ratios [6]. Crossflow fans are an industrial concept common in the heating, ventilation, and air conditioning (HVAC) industry and are used extensively in air conditioners and tower fans. Crossflow fans represent a highly unconventional form of thrust distribution which has seen a resurgence in interest in recent years. The use of the crossflow fan for aircraft propulsion and flow control is not a new concept and was proposed by Dornier in 1962 [14]. Dornier's concept involved a crossflow fan embedded into the wing and was driven by two motors nested inside the fuselage near the wing root.

In aircraft concepts employing crossflow fans, the fans can be driven by internal combustion engines (ICEs), gas-turbine engines, or electric motors. There can be a single fan per wing (if employed on the wing) or multiple fans connected to a common shaft with flexible couplings [15, 16].

The crossflow fan allows the propulsion system to be designed with wake-filling properties just as with distributed exhaust configurations. The crossflow fan also permits high lift coefficients due to active BLC effects caused by the rotor located at the wing trailing edge. These benefits have attracted designers in recent years to consider this unusual propulsive configuration.

An example of an aircraft concept utilising crossflow fans that is currently being actively researched is the FanWing (Figure 14.4). This is being investigated by FanWing Ltd, in partnership with the German Aerospace Research Center (DLR) and the von Karman institute through the distributed Open-rotor AiRcraft (SOAR) project [17]. The FanWing uses a crossflow fan mounted in the wing to provide both DP and augmented wing lift at very low airspeeds [11]. One of the key outcomes of this project is to offer short field performance

Figure 14.4 FanWing cross-flow fan aircraft concept. Source: [11].

similar to vertical take-off and landing (VTOL) aircraft, such as helicopters, at the operating costs of conventional fixed wing aircraft. The safe handling of such an aircraft at low speeds could lead to developments for agricultural applications, such as that depicted in Figure 14.4, as well as many other uses.

14.2.1.3 Multiple Discrete Fans

The third method of thrust distribution is through the use of multiple individual engines and/or fans distributed across the wingspan of the aircraft. The precise number of propulsors required for a configuration to be classified as DP is not clearly defined [2], however it is generally accepted that three or more engines are sufficient [18], provided that the propulsion system is integrated within the airframe, since synergistic design is a key objective of DP. There is no upper limit to the number of individual propulsors, and configurations of up to 100 have been studied [19].

Historically, discrete engines for very large aircraft (such as the An-225, with six engines, or KM ekranoplan, with eight canard-mounted engines [20]) were used primarily for practical reasons, such as the maximum thrust limitations of available engines. These aircraft required many of the largest engines available to achieve their desired performance. In a DP configuration, the choice of multiple propulsors is often driven by other factors, such as increased propulsive efficiency through boundary layer ingestion (BLI) [21–23], reduced noise [24] or improved STOL performance [25]. Distribution of the engines or fans across the wingspan also drives structural mass reductions through bending moment relief.

This method of thrust distribution arguably provides the greatest degree of variability as the number of engines/fans is varied as well as their spanwise and longitudinal locations. Propulsors can be located to best suit each individual aircraft configuration and either embedded within the structure or attached to external pylons. Furthermore, the choice of engine or fan type can be optimised for different phases of flight, for example turbojet engines may be used for high-speed cruising at high altitude, whilst piston-props are used for the other phases of flight, as was the case with the Convair B-36 bomber.

A great deal of research on aircraft concepts utilising multiple discrete fans has been conducted by NASA, with particular attention paid to the BWB aircraft configuration. The NASA N3-X, pictured in Figure 14.5, has been designed to meet the requirements of the NASA fixed-wing project to achieve a mission fuel/energy reduction of 60% by 2025 (the N + 3 timeframe) [27, 28]. The N3-X aircraft concept utilises 14 electric motors driving

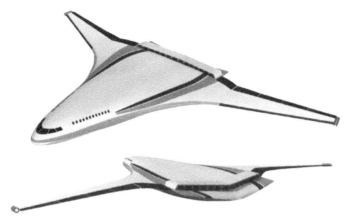

Figure 14.5 NASA N3-X with turboelectric distributed propulsion system [26]. Note propulsors distributed along trailing edge of centre wing/body section and gas turbine generators at wing tips. Credits: NASA.

ducted fans, distributed along the trailing edge of the wing-body. The electrical power for these motors is provided by two gas-turbine generators, one located at each wingtip. This arrangement of electric fans along the trailing edge provides propulsive efficiency benefits through BLI in addition to having a high effective bypass ratio (eBPR) [27]. In combination with the high eBPR, placing the fans along the upper surface of the wing also results in lower fan noise.

The N3-X aircraft addresses requirements for large commercial aircraft of the future. NASA has also recently been researching the multiple discrete fans for smaller, general aviation aircraft, such as the Tecnam P2006, a twin-engine piston aircraft modified to have a number of electric motors distributed along the wing leading edge to provide a blown-wing effect [29]. In this application the electric fans operate at maximum power during landing to increase the lift produced by the wings at low speeds. This allows a reduction in wing area as this is typically sized for a 61-knot stall speed in landing configuration, so that the wing can be designed for cruise conditions and thus drag minimised [29]. The use of wingtip propellers has also been incorporated into this design as discussed in [30] and is part of the NASA leading-edge asynchronous propeller technology program, or LEAPTech, as shown in Figure 14.6.

NASA has been performing research on wingtip mounted propellers and turbines since the 1960s. This has been shown to hold promise for significant improvements in propulsive and overall aircraft efficiency, since the use of a propeller at the wingtip, turning in the direction opposite to that of the wing vortex decreases the wing drag coefficient, increases the maximum lift coefficient, and increases the effective aspect ratio [32]. The benefit is twofold: the wingtip vortex is shifted outboard as previously mentioned, and in addition the propeller advance ratio decreases due to the radial component of the airflow induced by the vortex. This also results in an increase in propulsive efficiency. The location of a propeller at each wingtip was previously impractical due to one-engine inoperative (OEI) cases, however as part of a larger DP system and utilising modern electric motors and power systems may prove feasible in the years to come.

Figure 14.6 NASA LEAPTech Distributed Electric Propulsion concept [31]. Credits: NASA.

14.2.2 Power Transmission Methods

Aircraft propulsion is described in terms of both thrust and power, so it follows that distribution of the thrust force necessitates changes to the way power is transmitted from the source of power to the propellers or fans. As with thrust distribution, there are a number of unique configurations that have been proposed by researchers, each with its own merits. Selection of the power transmission method is performed with respect to the DP system and overall aircraft configuration, and can be grouped into the following categories: mechanical, pneumatic, and electrical.

14.2.2.1 Mechanical Transmission
The simplest and oldest method of transmitting power is through the use of mechanical linkages. These include shafts, gearboxes, belts, or any other form of physical connection between the prime mover (the initial source of motive power) and the propulsors (propellers and fans) which produce thrust.

The use of multiple discrete engines is a case where the prime mover and propulsor are connected as a unit and distributed in large numbers across the span of the aircraft. Figure 14.7 shows an example where 12 turbofan engines are distributed spanwise along the trailing edge of a BWB and are partially embedded within the structure. The integration of the airframe and propulsion system in this configuration results in low-noise STOL operations and efficient high-speed cruising [25].

Alternatively, a number of fans may be used with only a single prime mover (engine core), with the engine connected to the fans via shafts and a common gearbox. The use of this type of transmission system has been studied by Cambridge University and the Massachusetts Institute of Technology (MIT) as a means to minimise noise [24, 31]. This design has a calculated noise level of only 62dBA at the airport perimeter whilst achieving improved fuel efficiency due to the lifting body (BWB) airframe configuration.

So far only axial flow fans have been mentioned with regards to the mechanical transmission of power. Several crossflow fan concepts entailed gas turbine engines at both the wing root and wingtip, powering a number of crossflow fans distributed along and embedded within the wing trailing edge. Flexible shaft couplings are utilised to account for wing

Figure 14.7 Cruise efficient short take-off and landing (CESTOL) BWB aircraft with 12 discrete engines [33]. Credit: NASA.

bending during operation, and the location of these fans results in significant BLI over the wing as well as a blown flap effect, permitting STOL operations [15].

14.2.2.2 Pneumatic Transmission

Another method of transmitting the power from a prime mover is by pneumatic means. This can be achieved through the use of fans with nozzles in the blade tips to allow them to be driven by a high-pressure gas source, such as the compressor bleed air from a gas turbine engine. An example of this is portrayed in Figure 14.8 from a review of aircraft propulsion by NASA in 1970 [34]. This method of power transmission showed promise in the 1970s and 1980s however has not attracted much attention from researchers in recent years.

14.2.2.3 Electrical Transmission

The resurgence of interest in distributive propulsion is partially due to recent technological advancements in electric propulsion, as many recent DP concepts feature electrical power transmission. This is more efficient that the use of multiple combustion engines since these suffer from poor efficiency at reduced scales. The added complications of fuel and lubrication systems brought about by multiple combustion engines is also undesirable [19]. In contrast, electric motors are immune to scaling effects and distributing electrical power to multiple locations on the airframe can be somewhat simpler than doing so mechanically – therefore electric propulsion has acted as an enabler to the DP configuration with the two concepts interacting in a highly synergistic manner.

Electrical power distribution has been studied for use in DP configurations on the scale of large commercial aircraft, such as that depicted in Figure 14.9, as well as at smaller general aviation scales. This is due to efficiency and power-to-weight ratios of electric motors

Figure 14.8 STOL transport aircraft concept with compressor bleed air tip-driven multi-fans along trailing edge [34]. Credits: NASA.

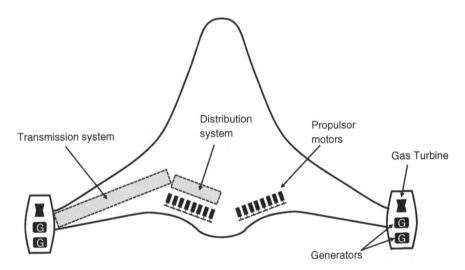

Figure 14.9 Electrical power transmission arrangement of N3-X aircraft. Adapted from [35].

being virtually scale free. The other benefits of electric propulsion include six times higher power-to-weight ratio, three to four times higher efficiency, efficiency across a large range of power settings, as well as others as elaborated by Moore and Fredericks in [36].

There are challenges that remain to be addressed, however, such as the low energy density of batteries. There is much development in this area, though, and significant research has already been conducted on the use of superconducting components in the turboelectric

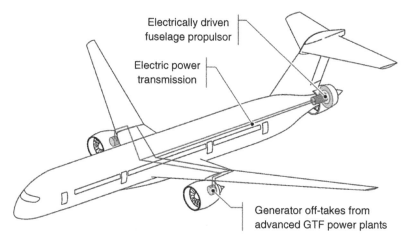

Electrically driven
fuselage propulsor

Electric power
transmission

Generator off-takes from
advanced GTF power plants

Figure 14.10 Isometric view of the propulsive fuselage concept. Source: [37].

distributed propulsion (TeDP) concept N3-X [27, 35]. The diagram presented in Figure 14.10 shows the layout of the electrical power system, of which three electrical power transmission and distribution architectures (DC, hybrid AC-DC, and AC) have been compared by Jones et al. in [35].

This method of power transmission is flexible with regard to the source of electrical power, which may come from an on-board generator or batteries, or some combination of both. The number of unique battery chemistries and fuel/coolant system architectures is vast and the selection of which technologies to use forms an integral part of the design of a DP system employing electrical power transmission.

14.2.3 Power Sources

In the literature a wide variety of power generation architectures and energy storage mediums are proposed, each with their unique benefits and challenges. Here the key power sources and their applications in proposed DP configurations are presented. Conventional aircraft propulsion systems rely upon ICEs to produce power, however the concept of battery powered aircraft is increasingly becoming more popular. The limitations of current battery technology in terms of energy density (energy per unit volume) and specific energy (energy per unit mass) are often augmented by ICEs in hybrid power systems. Therefore, the three key categories are: ICEs, electrochemical cells, and hybrid-electric power systems (HEPSs).

14.2.3.1 Internal Combustion Engines

The use of piston or gas turbine engines in DP configurations in the available literature is common, with many relying upon these as a source of power. The technology is well understood and provides a solid foundation for an often otherwise novel DP configuration. Although the development of combustion engines results in gradual efficiency improvements over time, the greatest area for innovation lies in the fuel used by these engines. Most gas turbine engines (and, increasingly, piston engines – as more diesel engines are certified) run on kerosene-based fuels, such as Jet-A1. However alternative fuels may be used for various reasons, such as liquid hydrogen (LH_2) as is the case with the N3-X concept described in [27]. In this case, hydrogen is proposed as it requires low temperature storage and as a result the fuel itself can be used as a coolant for the superconducting electrical components in the design. Another cryogenic fuel worth consideration is liquid methane [38], as it has a lower cost per unit of energy than petroleum-based fuels and is readily available in many parts of the world.

14.2.3.2 Electrochemical Cells

The use of rechargeable batteries has been considered for DP concepts such as the NASA LEAPTech project, conducted in conjunction with Joby Aviation, a manufacturer of electric motors for unmanned aircraft [29, 30]. Battery chemistries of the lithium polymer family are particularly popular due to their high energy and power densities.

Aside from rechargeable batteries, fuel cells, which consume fuel that requires refilling, have been considered for use in aircraft propulsion systems. As with combustion engines, a variety of fuels are available, however hydrogen fuel cells are the most popular amongst conceptual studies as well as in practice. A design and performance analysis study carried out by Ozbek et al. [39] on a 250 W Proton exchange membrane fuel cell (PEMFC) – Li-Po battery hybrid system for an unmanned aerial vehicle (UAV) showed that the higher energy density can greatly increase flight endurance. Additionally, the study yielded important data insights to the character of the hybrid system under static and dynamic loads during flight.

14.2.3.3 Hybrid-electric Power Systems

Hybrid-electric power systems (HEPSs) have been proposed by many in order to bridge the gap between energy densities of electrochemical cells and liquid chemical fuels used in ICEs [40]. As previously mentioned, the NASA N3-X TeDP concept uses this configuration, and it has also been proposed by Airbus in their E-thrust concept. This concept relies upon a central gas turbine generator embedded within the aft fuselage to provide electrical power to six electric fans, three on each side, mounted along the wing trailing edge close to the wing root.

These two hybrid-electric configurations are both series hybrids, where the prime mover is coupled to a generator and all power generated by the prime mover is converted into electrical power. The alternative to this is a parallel hybrid, where the electric motor and prime mover are connected to the propulsor in parallel via a gearbox, such that one or the other may be used to provide power directly to the propulsor. Both the prime mover and the electric motor may be used together in peak load cases, and in low or no-load cases the motor may be used as a generator to recharge the onboard batteries.

When considering the size and scale of all-electric or HEPSs, it is important to evaluate the gravimetric specific power (power delivered per unit mass), volumetric specific power, and the efficiency of electric motors. In a study by Eqbal et al. [41], one of the critical problems identified in the development of TeDP for remotely piloted aircraft systems (RPAS) was the additional weight of electric motors suitable for distributed propulsion. The importance of finding an optimised system that balances both the propulsion and electric systems is stressed. State-of-the-art electric motors used in aircraft typically exhibit gravimetric specific powers of up to 5 kW/kg and 95% conversion efficiency [37]. With the advent of superconducting material technology [42], which theoretically negates electrical resistance in certain operating conditions, these values are expected to improve up to 15 kW/kg and 99% efficiency respectively, allowing for smaller and more lightweight motors.

Based on recent research trends and market interests, it is reasonable to predict that all-electric and turboelectric motors will increasingly dominate the civilian aircraft sector. In recent studies, TeDP appeared to be the most promising solution for megawatt-class aircraft with 300 or more passengers [43]. TeDP has much potential also in smaller aircraft, provided the existing challenges with respect to design, weight, and material will be overcome with research.

14.2.4 Aircraft Configurations

Since DP architectures involve high levels of integration between the structure, aerodynamics, and propulsion systems of an aircraft design, the aircraft geometric configuration must also be considered.

Future aircraft designs utilising DP architectures typically exhibit a high degree of integration between the airframe and the propulsion system [25]. The objective of these designs is to obtain large energy efficiency improvements through BLI and wake-filling [3]. Certain propulsion configurations have better synergy with particular aircraft geometries – for example, multiple discrete fans with a HEPS are typically BWB designs.

A simplified scheme is adopted here as the actual number of unique aircraft configurations is very large. In lieu of considering all possible wing, fuselage, and tail combinations, three generalised categories are presented here. Using this methodology, DP concepts described in the literature can be grouped efficiently and with reasonable accuracy.

14.2.4.1 Aft-tail (Conventional)

The conventional, aft-tail aircraft (often referred to as tube and wing, or T&W) has been a prime candidate for DP concepts for many years. A large number of discrete fans may be distributed along the wing as shown in Figure 14.6, or crossflow fans may be embedded within the wing. Alternatively, *propulsive fuselage* concepts, such as that shown in Figure 14.10, have been proposed, where BLI is achieved through an aft-mounted fan with annular intake.

A study by Samuelsson and Grönstedt [44] presented a coupled multidisciplinary method for the performance analysis of a turbo-electric, boundary layer-ingesting propulsion system which was aft-fuselage mounted. It was found that the optimal level of ingested drag is 30–57% of the total fuselage drag, which resulted in a net reduction in mission fuel burn of 0.6–3.6%. However, these potential theoretical benefits of this aircraft architecture are greatly offset by losses in the electric transmission system, installation effects, and the difficulties in effectively ingesting the fuselage drag in the outer part of the boundary layer.

14.2.4.2 Canard/Tandem Wing

Very few DP configurations have been proposed which make use of the canard or tail-first aircraft configuration. Recently, Aurora Flight Science's *LightningStrike* design, as shown in Figure 14.11, has been selected by DARPA for the VTOL X-Plane program [45]. This design

Figure 14.11 Aurora flight sciences canard/tandem wing aircraft utilising distributed electric motors [45]. Source: Moses Bunting, http://aurora.aero via Wikimedia Commons.

uses a canard or tandem wing layout with a large number of electric motors sandwiched between the wings in a biplane configuration. Electrical power for the motors is provided by a gas turbine generator located within the fuselage.

14.2.4.3 Tailless

Tailless aircraft include the BWB configuration (Figure 14.12), and have no distinct horizontal stabiliser, or tail, to provide longitudinal stability. These have proven popular amongst DP research, particularly by NASA, which produced a significant body of research in the area of distributed turboelectric propulsion to support development of the N + 3 air transport concepts for the 2025–2030 timeline [27, 33]. It has been discovered that this propulsive architecture marries well with the lifting body aircraft concepts, since a number

Figure 14.12 A BWB aircraft employing BLI. Source: NASA image reproduced from [10].

Table 14.1 Summary of classification factors for distributed aircraft propulsion configurations.

Thrust distribution	Power transmission	Power source	Aircraft configuration
Distributed exhaust	Mechanical	Internal combustion	Aft-tail
Crossflow fans	Pneumatic	Electrochemical	Canard
Discrete fans	Electrical	Hybrid-electric	Tailless

of propulsors distributed along the upper surface of the aft fuselage/centre wing section increases propulsive efficiency by operating in the thick boundary layer created by the fuselage of the BWB. Additionally, there is a benefit gained from re-energising the wake of the aircraft, thus reducing drag.

14.2.5 Distributed Propulsion Configuration Classification

From the configurations discussed a classification system can be developed to aid in the categorisation of DP configurations. A morphology chart is presented in Table 14.1 that shows the categories that have been presented in this section. This morphology chart also provides a conceptual aid for the development of future DP concepts.

By selecting one method from each category one can establish the basic parameters for the majority of DP concepts. Use of the table in this manner results in 81 possible arrangements, however the actual number of useful configurations is slightly lower since some choices are incompatible with others. For example, an electric power transmission system implies the use of either electrochemical cells or a hybrid electric power source.

14.3 Aircraft Design Considerations

Various mathematical models have been proposed to assess the performance of these novel configurations. The interactions between the propulsion system and the aerodynamics of these aircraft are not captured by traditional analysis methods and therefore a more extensive analysis is required. Specifically, traditional performance analysis relies on C_{D0} values that are determined experimentally in a wind tunnel. This testing is usually performed with unpowered models, which yields satisfactory data for conventional airplanes where there is limited interaction between the propulsion system and the aerodynamics but may not be adequate for aircraft employing BLI or other types of integrated propulsion systems. To address this, various authors have proposed novel methods to describe the impact of an integrated propulsion system.

- Steiner et al. [3] make use of the power saving coefficient (PSC) a metric introduced by Smith [46] to quantify BLI benefits at the conceptual design stage. PSC is defined in [3] as the reduction in power required due to BLI compared with the power requirements without BLI.

- Goulos et al. [47] formulated a novel thrust and drag accounting approach for the analysis of integrated airframe-engine systems. This included an integral metric that quantifies the overall performance of coupled airframe-engine configurations based on the unification of various aerodynamic coefficients and engine cycle parameters. Using this methodology, it was shown that a close coupled engine position can result in a net vehicle force difference of almost −0.70% of the nominal standard net trust relative to a representative baseline engine location.
- For distributed exhaust configurations, such as the jet-wing assessed by Ko et al. [9] and subsequently Dippold et al. [48] the wake-filling characteristics of the propulsion system must be accounted for by considering the energy added to the flow by the exhaust stream as well as the energy removed from the flow in the same control volume by the body of the aircraft.
- For crossflow fans, the force in the tangential and radial directions can be estimated using the methods established by Dang and Bushnell [6].
- A sizing procedure proposed by Sgueglia et al. [49] is based on a modified FAST (fixed-wing aircraft sizing tool) code developed originally by ONERA and ISAE-Supaero [50]. The modification includes an evaluation of relevant parameters for the integration of a hybrid propulsion architecture into a classical 'tube and wing' aircraft configuration. The modified code considers the impact of battery, power transmission and digital control cables, ducted fans, electric motors, and hybrid engines.

Ultimately a combination of the methods described above must be used if the broad spectrum of DP concepts is to be analysed. Each unique configuration with respect to Table 14.1 may require different modelling approaches depending upon the combination of characteristics employed in any given design. Development of a modelling methodology that accounts for this, even at a conceptual level, presents significant challenges and has not yet been achieved by the propulsion research establishment.

14.3.1 Conceptual Design

Aircraft design traditionally relies on iterative design processes such as [51], which feed requirements and basic input variables into top level and then iterate to achieve MTOW convergence. Dedicated consideration of DP aspects as part of the conventional aircraft design can greatly enhance the effectiveness of this technology as it allows us to capture its benefits more accurately. This requires evaluating the impacts in terms of fundamental aircraft design parameters, such as weight, lift, and drag coefficients, propulsive efficiency, and effective aspect ratio, among others. Figure 14.13 presents a custom aircraft design methodology which embeds the essential steps to integrate DP systems. The aspects specific to DP are listed in blue. The solution sequence and basic performance equations remain essentially the same, with the exception of modifiers to the fundamental parameters, where applicable.

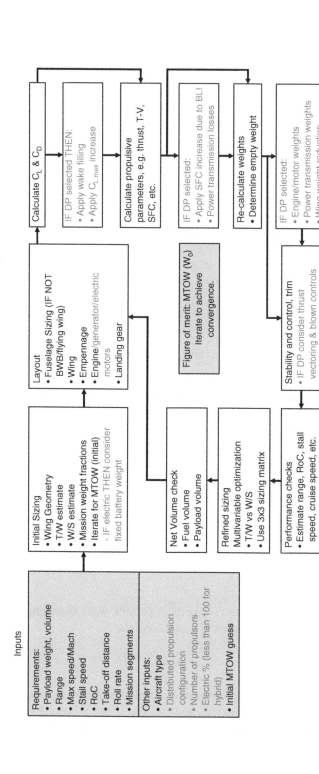

Figure 14.13 Generic model for assessing distributed propulsion configurations.

14.3.2 General Assumptions and Input Variables

In order to undertake the process presented in Figure 14.13, the basic DP configuration should be identified and defined (Table 14.1). Additionally, the following input parameters are required:

- Aircraft geometric properties, such as: wing position (high, low, mid), flap type, flap span as a fraction of wing span, propeller position (pusher or tractor) a wetted area ratio, and fuselage fineness ratio estimates.
- Performance parameters, such as: stall speed, maximum speed in level flight (or maximum Mach number of jet aircraft), design cruise speed, take-off distance, rate of climb (ROC), range, and positive and negative load factors.
- Powerplant parameters, such as: guess of total maximum rated power output (or maximum static thrust for jet aircraft), powerplant dry weight estimate, specific fuel consumption, and engine rpm.
- Advance propulsion parameters such as: number of propulsors and electric propulsion factor (percentage of mission conducted under electric power – for use with hybrid designs).
- Mission type: simple cruise or multiple mission segments, cruise altitude, and payload weight.

To perform conceptual level design and analysis the following assumptions are required:

- Provision of initial estimates for aerodynamic properties such as e, the Oswald efficiency factor (0.8 is reasonable according to [52])
- In the absence of test data, assume a value for propeller efficiency (0.78 is reasonable [52]).

14.3.3 Definition of Aircraft Geometry

Figure 14.14 illustrates the geometry definition convention of BWB aircraft. For the incorporation of DP systems in this configuration, the following parameters must be considered (in addition to the definition of aircraft geometry as would be required for conventional aircraft design):

- Location of propulsors.
- Definition of wake shape parameters (wake displacement, wake momentum areas).
- Control surface locations and location of blown controls.

14.3.4 Weight Estimation

Weight estimation is achieved using conventional methods of aircraft weight estimation found in texts by Roskam [53], Raymer [52], and Torenbeek [54]. Where DP impacts weight estimation specific equations are required to modify the estimated weights.

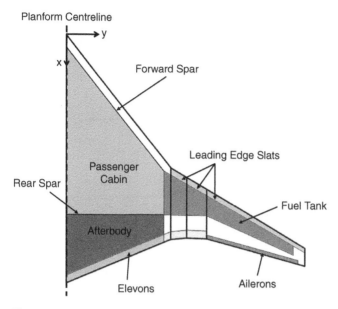

Figure 14.14   Geometry definition convention used by Leifsson et al. Source: Adapted from [14].

Wing structural mass reduction due to bending moment relief provided by wing-mounted engines can be estimated using the following equation from Torenbeek, as used by Ameyugo [19].

$$\frac{\Delta W_{B+S}}{W_{B+S}} = -1.5 \times \sum \frac{\eta_p^{\,2}}{\eta_{cp}} \times \frac{W_e}{MTOW/2}$$

The estimation of engine weight is crucial in the design process. Empirical data is used to estimate the weight of gas turbine engines, such as that from the Virginia Tech engine weight model shown in Figure 14.15.

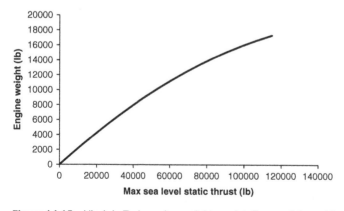

Figure 14.15   Virginia Tech engine weight model. Source: Adapted from [10].

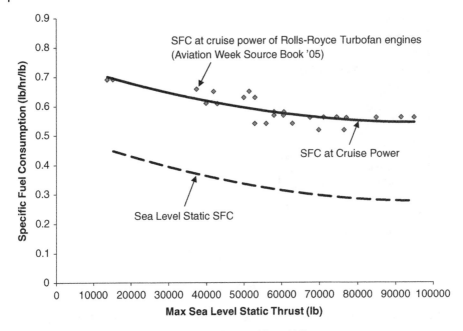

Figure 14.16 SFC at cruise power. Source: Adapted from [10].

For gas turbine engines, empirical data can also be used to provide a benchmark for Specific Fuel Consumption (SFC) as shown in Figure 14.16.

Weight estimation of electric aircraft can be achieved using the range equation for electric aircraft from [40]:

$$R = E^* \cdot \eta_{total} \cdot \frac{1}{g} \cdot \frac{L}{D} \cdot \frac{m_{battery}}{m}$$

where R is range, E^* is the equivalent energy density of the fuel system, η_{total} is the total propulsive efficiency, $\frac{L}{D}$ is the lift to drag ratio, and $\frac{m_{battery}}{m}$ is the battery mass fraction in metric units. Compared to the equation for range for a conventional aircraft which consumes fuel and becomes lighter throughout the mission (the Breguet range equation – in notation consistent with [55]):

$$R = E^* \cdot \eta_{total} \cdot \frac{1}{g} \cdot \frac{L}{D} \cdot ln\left(\frac{1}{1 - \frac{m_{fuel}}{m}}\right)$$

where $\frac{m_{fuel}}{m}$ is the fuel mass fraction of the aircraft. The difference is the term containing the logarithm that results in decreasing fuel consumption as the aircraft burns fuel and becomes lighter. Comparison of the Breguet range equation with the range equation for electric aircraft results in a range that is directly proportional to the mass fraction of batteries to overall aircraft mass, which is illustrated in Figure 14.17.

When using a hybrid configuration both the Breguet range equation and the range equation for electric aircraft must be used. In general, the limitation of E^* of batteries leads to lower range and endurance of hybrid aircraft as compared to their ICE counterparts. Nonetheless, hybrid-electric configurations contribute significantly to minimising exhaust

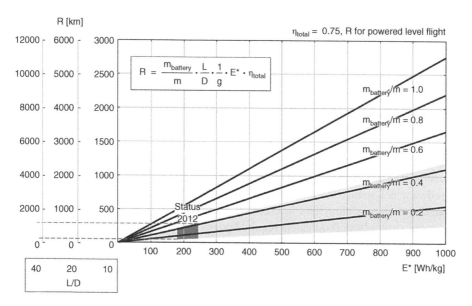

Figure 14.17 Range versus specific energy for different battery mass fractions. Source: From [55].

emissions and show promising complementarities to ICE-only propulsive configurations. Furthermore, optimal battery recharge-depletion profiles could be implemented through the careful design of battery charge cycles, yielding further enhancements in range and endurance [40].

14.3.5 Propulsion System Parameters

Specific fuel consumption of configurations employing BLI can be estimated through the method used in [10], given by Stinton [56]:

$$SFC = \frac{U_\infty}{\kappa_l \eta_P \eta_T}$$

where U_∞ is the free-stream velocity, κ_l is the SFC factor, η_P is Froude propulsive efficiency, and η_T is the engine internal thermal efficiency. κ_l is determined by Stinton to be 4000 ft. hr./s.

Calculating the propulsive efficiency of configurations employing BLI can be performed using the PSC, defined by Smith [46] as:

$$PSC = \frac{P_{NoBLI} - P_{BLI}}{P_{NoBLI}}$$

The PSC has a strong correlation with the ratio of ingested drag to total thrust, D_{ing}/T. Ingested drag represents the viscous drag of the portion of airflow which is ingested by the propulsion system.

A recent study by Uranga et al. [57] proposed a general framework to assess the BLI benefits compared to a baseline engine mount. This framework is applicable to both the conceptual and full-scale high-fidelity analyses of aircraft performance. The required mechanical flow power P_k is related to the dissipative mechanisms as:

$$P_k = \Phi_{jet} + \Phi_{surf} + \Phi_{wake} + \Phi_{vortex}$$

where Φ_{jet} is the dissipation of the jet, Φ_{surf} dissipation due to nacelle surface, Φ_{wake} wake dissipation and Φ_{vortex} trailing vortex dissipation. With BLI, the cruise power requirements can be reduced by around 8% for the same propulsor area, while for an equal propulsor mass flow the obtained benefits are up to 9%. These gains are primarily attributed to the reduced jet dissipation due to BLI. The second in importance is the reduced nacelle wetted surface and lower surface velocities which reduce the overall surface dissipation. Finally, the placement of the engines required to exploit the BLI effect also produces a partial wake filling effect which reduce the wake dissipation. It should be also noted that these gains do not consider other possible system-level benefits, such as the significant potential of weight saving, since a propulsor of equivalent required power can be as much as 40% smaller when exploiting the BLI effect.

On the other hand, when discussing DP comprised of an array of discrete jets, attention should also be turned to the spacing and shape of the exhaust nozzles. An important study performed by NASA [58] in the seventies addressed this issue related to V/STOL aircraft where an array of jet nozzle exhausts was used to provide stability in a hover. The relevant takeaway for our current discussion points to the importance of jet interactions downstream of the exhaust nozzles and their effect on the provided thrust. The NASA report demonstrates that jets in close proximity (broadly similar to what would be used in a DP system) experience a faster decay of the jet dynamic pressure. This effect is present for different scales of the nozzles and persistent for cold and hot jets alike. Furthermore, a 2-D jet nozzle is also more susceptible to the dynamic pressure decay than a round nozzle. This is especially important since an array of jets embedded into the wing trailing edge would most likely use high aspect ratio rectangular nozzle shapes to obtain a higher level of structural integration and strength, as well as to enable the vectorization of thrust in 2-D. Nevertheless, these adverse effects on the produced thrust must be considered.

Finally, the thrust vectorization of the high-aspect ratio nozzles might be an important contribution to performance enhancement of supersonic transport aircraft in such low-speed/high-lift flight regimes as during take-off and landing [59]. Incidentally, a high aspect ratio exhaust outlet requires an overall shorter nozzle design for achieving the same pressure ratio for supersonic jets, contributing to important weight and drag savings.

14.3.6 Aircraft Performance Estimation

The following performance parameters should be estimated:

- Maximum speed in level flight.
- ROC.
- Speed for best ROC.
- Take-off distance or balanced field length.
- Landing distance.
- Missed approach climb rate.

Estimation of these values for DP configurations involves the modification of key variables that impact one of the four fundamental flight forces (lift, weight, thrust, and drag). Weight and thrust have been dealt with already in Sections 3.4 and 3.5 respectively.

In the case of the lift force, in lieu of a more detailed Computational Fluid Dynamics (CFD) analysis such as that conducted by [29], modification of Cl_{max}, the two-dimensional lift coefficient is performed in a similar manner to the simplified method presented by Raymer in [52] to account for flaps. In this method, Cl_{max} is modified through the addition of a factor consistent with the type of flaps employed. In the case of DP, an increase in Cl_{max} of 3.5 is considered acceptable for distributed exhaust configurations based on what is achievable with the Hunting H.126 and Shinmaywa US2.

The drag force is modified by the impacts on induced drag achieved with DP. A useful method of estimating the induced drag of a jet wing, or distributed exhaust configuration, is that developed by Maskell and Spence [60] and applied by Leifsson et al. [10]. This method allows the determination of the induced drag of a wing with a jet blowing out of the trailing edge. The induced drag of a jet wing is:

$$C_{Di_{DP}} = \frac{C_L^2}{\pi AR + 2C_J}$$

where C_L is lift coefficient, AR is aspect ratio, and C_J is the jet momentum flux coefficient, defined as:

$$C_J = \frac{J}{\frac{1}{2}\rho U_\infty^2 S_{ref}}$$

where J is the jet momentum flux, ρ is density, and S_{ref} is the wing planform reference area. The induced drag is then found by calculating the induced drag without the presence of the jet and then corrected using the ratio:

$$\frac{C_{Di_{DP}}}{C_{Di}} = \frac{1}{1 + \frac{2C_J}{\pi AR}}$$

Analysing profile drag, an extensive numerical study on the different aerodynamic integration of DP was recently performed by Lockheed-Martin [61]. They analyse the impact of various DP installation configurations (independent of the power system type) on the scaled lift-to-drag ratio in transonic flight regimes, extremely relevant to the future application to commercial air transport designs. However, the number of parameters which need to be considered for an efficient DP integration is staggering: total propulsive area, its distribution along the wing span, location of the propulsors (upper and lower trailing edge, embedded into the wing, etc.), chordwise location of the nozzle, and the length of the propulsor, etc. An interesting result is that the chordwise and frontal dimensions of the DP should taper off along the span of the wing, otherwise there are additional drag penalties. This particular optimisation could therefore be reasonably adopted only for configurations of distributed exhaust or crossflow fans, and not in the case of multiple discrete fans which would have to be re-sized. Globally, a clear benefit is found for those DP configurations where the embedded propulsor over a short portion of the wingspan yields up to 8% of increase in L/D over a traditional pod-mounted propulsor, for the same propulsor frontal area. However, the authors cautioned that different levels of simulation fidelity (they first explore 2-D cases, then simple and high-fidelity 3-D geometries) yield quite different relative results, which means that this kind of parametric study is very sensitive to the details in geometry and propulsion group flow modelling.

Further exploration of the aerodynamic benefits of DP is needed, particularly in terms of STOL performance and the impact on other aerodynamic and control surfaces (for example the tail section for a classical aircraft configuration).

14.4 Aeroelasticity Considerations

DP systems can have a major effect on the aeroelasticity of the aircraft on which they are installed. Many DP concepts involve the placement of the propulsors along the wingspan, which can significantly alter the mass and stiffness distributions and hence the natural frequencies of the wing structure. This can have a direct effect on the aeroelastic stability of the aircraft. In addition, whirl flutter and interactions of rotating components with the other systems and structures on the air-vehicle must be considered early in the design process [62]. On the other hand, it can be seen from the literature that a large number of studies, concerned with DP, have focused on aircraft with high aspect ratio wings (highly flexible) such as the AeroVironment Helios Prototype and the University of Michigan X-HALE air-vehicles shown in Figure 14.18. Furthermore, DP systems used on high aspect ratio wings can be utilised passively or actively to control the aeroelastic shape of the wing to enhance aerodynamic efficiency, reduce energy consumption, and extend range. In addition, controlling the aeroelastic shape of the wing using DP systems can be used to alleviate gust and manoeuvre loads, resulting in lighter structural weight and in extended life of the airframe structure. Although studying the aeroelasticity of DP systems is still at an early stage, it is evident that the literature can be split into two main themes; aeroelastic stability and aeroelastic control. A similar research trend was observed by Ajaj et al. [63] when studying the aeroelasticity of morphing aircraft. Representative examples on both themes are highlighted in the following subsections.

(a)   (b)

Figure 14.18 Representative examples of highly flexible, high aspect ratio wings equipped with distributed propulsion. (a) AeroVironment Helios Prototype. Source: Credits: NASA. (b) X-HALE. Credits: University of Michigan.

14.4.1 Aeroelastic Stability

The aeroelastic stability theme is mainly concerned with studies conducted to ensure that aircraft with DP satisfy certain aeroelastic requirements/constraints and that such requirements/constraints do not limit the potential benefits of DP. For example, Amoozgar et al. [64] studied the aeroelasticity of an electric aircraft (similar to the NASA X-57 Maxwell) using two types of distributed electric propulsion motors. On each side of the wing, six high-lift motors were distributed and one cruise propulsor was attached to the tip of the wing. An aeroelastic model of the wing was developed and validated. They concluded that the tip propulsor had a major effect on the flutter speed and frequency of the wing, whereas the high-lift motors had little effect on the aeroelastic stability. Moreover, they illustrated that the propulsor angular momentum could alter the stability characteristics of the wing for higher thrust values. Similarly, Cravana et al. [65] conducted numerical and experimental investigations to understand the aeroelastic behaviour of high aspect ratio wings with distributed electric propulsion systems. Several configurations of a flexible wing plus engine propellers were considered in order to investigate the influence of motor position and spinning propellers on modal frequencies of the wing. In their study, they considered cases of a clean wing, a wing with pods only, and a wing with pods plus spinning propellers with different numbers of pods/propellers. The study demonstrated that edgewise and torsional frequencies are significantly affected by mass/loads distribution, as well as the number of stores and their spanwise locations. ONERA assessed the aeroelastic stability aspects associated with using DP on a transport aircraft concept (Dragon: Distributed fans Research Aircraft with electric Generators) by considering two different configurations of DP at a range of flight conditions [66]. They did not find critical aeroelastic issues in cruise flight except at low altitude and high speed, which required increasing the bending stiffness of the wing.

On the other hand, An [67] developed a gradient-based optimisation framework for aircraft with distributed electric propulsion using structural and aeroelastic constraints. The doublet lattice method was used for surface aerodynamics coupled with an actuator disk to model the effect of a propeller (one-way coupling). The structural analysis was performed using the Toolkit for the Analysis of Composite Structures (TACS). Different configurations of distributed electric propulsion were studied to minimise structural weight subject to structural and aeroelastic constraints. The author concluded that larger aspect ratio wings coupled with a large number of distributed electric motors lead to heavier wings that are usually more vulnerable to aeroelastic instabilities and structural failures, thus necessitating comprehensive multidisciplinary design optimisation studies. Finally, Massey et al. [68] conducted an aeroelastic analysis of a wing equipped with distributed electric propulsion. The wing used was similar to that of the early version of the X-57 Maxwell flight demonstrator. They used linear aerodynamic theory, based on the doublet lattice method, and the nonlinear aerodynamic solver based on FUN3D Reynolds Averaged Navier-Stokes, as illustrated in Figure 14.19. For the wing configuration studied, no dynamic aeroelastic issues were reported. However, static divergence occurred in the first bending mode at very high dynamic pressure (around three times the flutter clearance condition). They also reported that flow separation might exist at the wing-pylon junctions, which requires careful consideration to identify possible aeroelastic excitations.

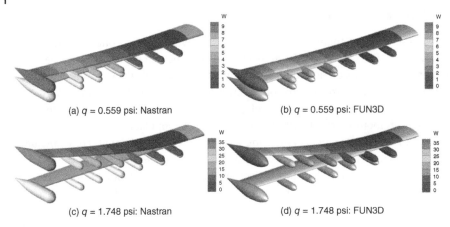

(a) q = 0.559 psi: Nastran   (b) q = 0.559 psi: FUN3D

(c) q = 1.748 psi: Nastran   (d) q = 1.748 psi: FUN3D

Figure 14.19 Static aeroelastic position coloured by vertical displacement in inches (rigid wing surface shown in grey) [68].

14.4.2 Aeroelastic Control

The aeroelastic control theme mainly includes studies that utilised DP (actively or passively) to enhance flight characteristics and/or control flight loads. For example, Nguyen et al. [69] presented an aeroelastic wing shape control concept using DP to increase the aerodynamic efficiency on the NASA Generic Transport Model. Aeroelastic wing shape control was achieved using propulsors distributed along the wingspan to provide aircraft flight control and change the wing shape. The concept exploited synergistic interactions between lightweight materials, electric DP, and active aeroelastic control. They developed an integrated framework that included a propulsion performance model, aerodynamic model, and finite element model, coupled with an automated geometry deformation tool that generated new deformed aircraft wing geometry using the output of the finite element model as shown in Figure 14.20. They concluded that reducing the wing structural stiffness resulted in an improvement in the lift-to-drag ratio (L/D) when compared to a rigid wing design across a representative mission profile consisting of a minimum fuel climb, cruise, and continuous

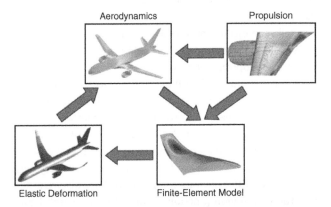

Aerodynamics   Propulsion

Elastic Deformation   Finite-Element Model

Figure 14.20 An integrated aero-propulsive-elastic analysis framework [70].

descent. Overall, a 2% increase in the cruise range was achieved with the DP aircraft when the wing torsional stiffness is reduced by a factor of 2. This improvement was associated with a thrust distribution that had 50% more thrust on the inboard propulsors and 50% less thrust on the outboard propulsors as compared to the uniform thrust distribution. The reduction in fuel burn was mainly achieved by actively tailoring the spanwise lift distribution using DP to change the wing shape throughout the flight envelope. Similarly, Nguyen et al. [70] utilised DP devices to twist the wing during flight and control its aeroelastic shape whilst maintaining aeroelastic stability. The propulsion devices were distributed along the inboard and outboard sections of the wing. A flight controller was used to independently control the thrust of the inboard and outboard propulsion devices to significantly change the flight dynamics. For instance, thrust of the outboard propulsion devices was varied to change the wing twist. They also investigated differentially varying the thrust on each side wing (starboard and port) to change yaw and other aspects of the flight mechanics during various stages of the flight mission. They noted that each side of the wing could use one or more generators to provide the required power for the propulsion devices.

14.5 Digital Control of Distributed Propulsion Systems

Digital control in conventional aircraft involves the use of an on-board, full-authority digital engine control (FADEC) system which interprets the pilot's power request input and adjusts the fuel flow to the engines accordingly. This form of thrust modulation allows the pilot to either increase or decrease the forward driving force acting on the aircraft. The emergence of DP systems introduces the possibility of an additional responsibility for the propulsion system of an aircraft: to perform vehicle control. The concept of propulsion-controlled aircraft (PCA) was, in fact, originally put forward in the 1990s, however it faced a number of technical and certification challenges [71]. The idea behind the concept of PCA was to utilise the thrust force vector applied at a distance from the aircraft's centre of gravity by the individual aeroengines to produce a moment and an associated angular acceleration. The use of DP systems makes this concept much more feasible, as the higher number of propulsors positioned across the aircraft allow for an increased influence on the forces and moments acting on the aircraft. Additionally, this concept favours the use of electric propulsors, which are commonly associated with DP systems, as they exhibit significantly shorter time intervals between a desired throttle input and the resulting change in thrust output.

There is limited literature published regarding the theoretical development of aircraft dynamics models to enable flight controllers that take advantage of DP [71]. However, there are records of DP-enabled aircraft successfully demonstrating the use of propulsion for flight control during flight tests, such as the NASA Greased Lightning (GL-10) [72] and the Aurora LightningStrike [45]. If the use of DP as a means to perform vehicle control were to entirely replace traditional control surfaces, the question arises of how the stability and control standards for airworthiness can be met, particularly in the case of such advanced aircraft configurations as BWB that are known to have stability issues. Traditional stability analysis techniques, such as those presented in Roskam [53], Raymer [52], and Torenbeek [54], focus on a set of relevant flight conditions, including sudden onset of crosswind gust and failure of a single engine, to demonstrate airworthiness in line with Part 25

Standards of Federal Aviation Regulations (FAR). The adverse effects of these conditions are attenuated in the case where there are a large number of propulsors due to the intrinsic redundancy in such a propulsion system. It is widely accepted in the literature that DP would be instrumental in resolving the stability issues of BWB configurations [73], however, control bandwidth challenges need to be addressed by way of stability augmentation techniques, such as thrust vectoring.

The increased level of complexity due to interaction between DP systems and the aerodynamics of advanced aircraft configurations demand a more robust and intelligent digital engine control system with adaptive thrust modulation and thrust vectoring capabilities in order to provide adequate stability and control during flight. Furthermore, there is greater emphasis on ensuring a higher level of fault tolerance and integrity in the propulsion system. This specific responsibility is best handled by an integrated vehicle health management (IVHM) system that interacts with the digital engine control system in real-time during flight. The IVHM system takes into account degradations in engine performance, predicts imminent faults in engine components or sensors, and recommends appropriate restorative actions. In doing so, it augments the safety, reliability, and integrity of the entire DP system with both reactive and predictive capabilities of responding to faults and degradations in performance. The IVHM system employs numerous model-based and data-driven reasoning techniques, powered by artificial intelligence (AI) and machine learning, to make inferences regarding the current and future (predicted) state-of-health of the engines. This includes digital twin technology as well as diagnostic and prognostic algorithms. Figure 14.21 illustrates the concept of this combined intelligent, digital engine control system with health management capabilities.

The pilot provides control inputs which are fed into the closed loop control of the digital engine control system. These are in the form of adjustments to the throttle lever position and, in the case of PCA, adjustments to pitch, roll, and yaw moments of the aircraft. The control inputs are interpreted and mapped to corresponding response commands that are sent to the engines, which in turn alter their operating parameters. For example, the fuel flow rate into the combustion chambers of the engines can be adjusted to increase or decrease the fan speed, and in thrust-vectoring enabled propulsion systems, the angle of thrust can be adjusted to achieve different moments about the aircraft's centre of gravity. The same engine control inputs are sent to digital twins of each engine within the engine health management system. These are dynamic virtual replicas of the physical engines which mimic their behaviour. It is important that the digital twins accurately mirror the life and experience of their physical counterparts in order to provide high fidelity simulations of their responses to given inputs and external conditions. The estimated engine parameters resulting from these models are then compared against the measured sensor outputs by diagnostic and prognostic algorithms. Some key parameters that need to be monitored to ascertain the state-of-health of standard combustion engines include temperature and pressure measurements at various locations along the gas path, fan and core speeds, fuel flow ratio, fuel-to-air ratio, and bypass ratio. In the case of turbo-electric propulsion, such parameters as voltage, current, battery temperature, and battery state-of-charge need to be considered. Furthermore, as digital engine control systems work primarily based on sensor signals, it is also important to perform timely and effective condition assessments of the engine sensors themselves using intelligent fault diagnosis schemes [74]. The AI-based diagnostic

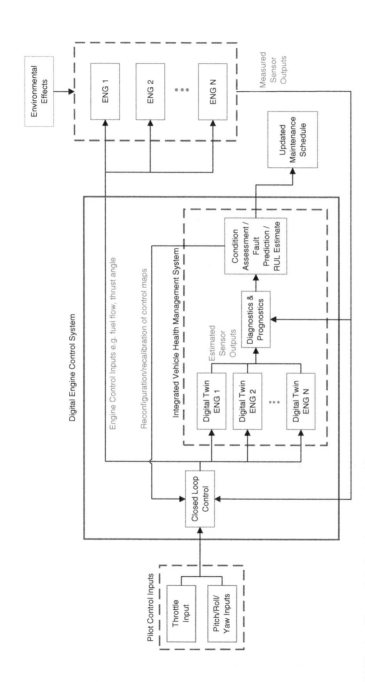

Figure 14.21 Intelligent digital engine control architecture.

and prognostic algorithms make comparisons between the estimated and measured sensor parameters, analyse their variations and trends, and subsequently form an assessment of the state-of-health of the engines. Thereby, they can detect or predict any faults or degradations in performance that have occurred or could potentially occur, and subsequently provide an estimate of remaining useful life (RUL). If an incipient fault is detected, a signal is sent back to the closed loop control to perform the appropriate reconfiguration or recalibration actions of the control architecture in order to prevent a loss in performance, avert a failure event, or, at the very least, provide the option of graceful degradation of performance. This feature introduces predictive integrity into the entire digital engine control architecture. For example, the harsh environmental conditions that engines are exposed to during flight alter important engine operating parameters, such as flow capacity and effective nozzle areas over time. The IVHM system would be able to detect these degraded conditions early on and subsequently provide the closed loop control with corrections to its control maps to account for these degradations [75]. This capability of the digital engine control system to dynamically adapt to different operating conditions is an important feature with the potential to improve aircraft performance in challenging flight conditions and increase aviation safety in the event of severe damage or unforeseen failures [76]. Moreover, constantly operating at optimal performance settings reduces the amount of fuel burn while increasing the lifespan of engine components, thereby contributing to environmental sustainability of air transport. In addition to these operational benefits, the engine health management system can optimise the maintenance schedule for the multiple engines in DP systems, thereby greatly reducing maintenance costs and minimising downtime.

Overall, incorporating engine health management capabilities into the digital engine control system of DP configurations provides an advanced solution that combines increased flight safety and mission optimisation as well as environmental sustainability as part of the same design, thereby enhancing both the cases for airworthiness certification and environmental regulations compliance [77–79].

14.6 Benefits and Challenges

There are a number of desirable effects resulting from the use of DP. Many of these can be attributed to the fact that many DP configurations feature the wing operating in the wake of several propellers distributed across it, leading to wake re-energisation. Aside from the reduction in induced drag obtained from this effect [10, 60], several studies indicate that the efficiency of propulsion is greatly improved if part or all of the propulsive fluid originates from the wake of the aircraft being propelled [46, 56]. Moreover, wingtip propellers provide significant improvements to propulsive efficiency in addition to increasing the effective aspect ratio of the wing. If the propellers or fans are aft-mounted, they can ingest the boundary layer of the aircraft and fill its wake with the exhaust jet, thereby reducing cruise energy consumption and improving overall cruise efficiency [46, 56]. A study by Kirner et al. [80] showed a 5.3% fuel-burn saving of a DP BWB configuration relative to the BWB turbofan baseline aircraft with 2% BLI duct total pressure loss. This fuel burn saving was increased to 7% if the duct total pressure loss was decreased to 1%, which highlights the importance of low loss inlet and outlet ducting in DP fan installation.

The increased dynamic pressure across the wing at low speeds in DP configurations has a profound effect on the maximum lift coefficient [29], which can be further improved by the ducting of bleed air through jet flaps in some configurations. Jet flaps allow for a reduced wing area sizing that matches the optimal area for maximum range at cruise conditions. In this way, stall speed, take-off and landing performance requirements can still be met while providing the ride quality and cruise drag benefits of smaller wing size. In fact, a higher maximum lift coefficient leads to better STOL performance, permitting mission profiles that would have otherwise been impossible. In addition, the use of blown control surfaces contribute to flight control augmentation, which enhances controllability and contributes to safe flight at low speeds. Flight safety is also improved due to the inherent engine redundancy in DP configurations, meaning that an engine-out scenario is not as critical to thrust availability and aircraft controllability as compared to conventional propulsion [10].

Another major benefit of DP is the significant noise reduction that can be achieved through the use of over-wing mounted propulsors, especially in a BWB configuration. This provides a more comfortable flight experience for commercial aircraft passengers as well as a reduction in noise pollution for communities and wildlife residing within the proximity of airports. Studies on silent fuel-efficient aircraft considered the noise level of conceptual BWB aircraft with an embedded boundary-layer ingesting DP system [24, 28]. A noise level of 62 dBA at the airport perimeter was calculated, corresponding to the background noise level of a well-populated area, making the aircraft imperceptible to the human ear on take-off and landing.

Structural benefits can also be realised through the implementation of DP, owing to the load redistribution provided by the engines which has the potential to alleviate gust load/flutter problems [10]. Additionally, distributing the engines along the wing aids in counteracting the lift loads during cruising, providing inertia relief and resulting in a lower wing weight, with a maximum achievable wing weight reduction of 3.5% [19].

In addition to the aforementioned benefits, the implementation of DP configurations in aircraft presents several challenges that need to be addressed. Firstly, there is limited experimental data for the majority of DP concepts presented in the literature, making accurate performance predictions challenging. A significant increase in the number of experimental studies is required to provide data to validate the performance benefits of DP. In particular, the high levels of propulsion-airframe integration create complex interfaces and interactions that have not yet been modelled in the literature.

DP configurations that use podded engines represent a significant challenge, due to Reynolds number and nacelle length scaling effects, which result in increased friction drag caused by a larger number of engines. For a mid-sized, long-range subsonic aircraft, this could result in a 6.5% increase in total drag. Additionally, the possibility of shock formation between engines makes podded engine configurations undesirable [19].

The complexity of the control systems required for DP configurations is generally higher than that of conventional aircraft. Concepts such as thrust vectoring are often associated with DP in lieu of conventional aircraft control surfaces [19]. This is due to the fact that having a large number of engines reduces the safety issues associated with thrust vectoring. The use of thrust vectoring to achieve pitch, roll, and yaw motion provides an opportunity to eliminate the need of the horizontal and vertical tail, resulting in significant take-off weight reduction. However, limitations of the latest technologies, including engine spool-up times and controllability in low power situations, as well as weight and performance penalties associated with thrust vectoring prove to be significant challenges in its implementation.

Lastly, regulatory compliance/certification along with the structure of the aerospace industry represents another significant hurdle for novel configurations such as DP, however environmental pressure and the increasing importance of sustainable aviation technologies could eventually lead to their implementation in aircraft.

14.7 Research Opportunities

DP configurations are varied and have potential applications ranging from general aviation and UAVs to transport category aircraft. The use of crossflow fans in agricultural aircraft has been considered in the FanWing project [11]. Aurora Flight Sciences considers the use of distributed hybrid-electric aircraft in a canard configuration for a UAV with VTOL capability [45]. The NASA LEAPtech aircraft uses multiple discrete electric motors in a general aviation aircraft with improved cruise efficiency [36]. Multiple studies have considered the use of various DP concepts in conjunction with the BWB airframe. DP principles may be applied to an aircraft of any size and there appears to be few applications or scale for which it has not been considered.

To date, research efforts have focussed primarily on large commercial aircraft, although more recent work has addressed light aircraft configurations, such as the Joby S2 or FanWing projects. At the smallest scale, the body of research regarding implementation of DP in UAVs is limited, despite these being the most accessible platforms to test, as shown by NASA Langley's VTOL tilt-wing. Studying the application of DP to UAV design could result in systems that require lower launch energy without sacrificing endurance.

The studies on hybrid electric DP consider primarily series hybrid configurations where the prime mover (e.g. gas turbine) is sized for take-off power requirements and reliance on batteries is low. Airbus is considering battery power for take-off in a concept aircraft [81], however this is not studied in depth in peer-reviewed literature. The optimisation of battery weight and overall propulsion system weight is required to understand the benefits and drawbacks of such a configuration.

The DP configuration has also not been specifically studied in combination with alternative fuels. Studies by Kim et al. into liquid hydrogen use as a cryogenic coolant for superconducting components in electric motors and generators. However, only a small amount is required for this purpose whilst the remainder of the energy required for flight comes from conventional jet fuel (kerosene). Combinations with other alternatives, including liquid methane, proposed separately in NASA N + 4 advanced concept development [82], or with drop-in alternatives, such as biofuels, are not considered.

14.8 Conclusions

New propulsion concepts exploiting distributed and discrete configurations are promising in terms of advancing the sustainability, resilience, and flexibility of atmospheric flight, as well as facilitating reduction in noise and fuel consumption, while enhancing safety and reliability. Distributed propulsion systems are nowadays classified according to a combination of factors including the method of thrust distribution, the type of power transmission, the nature of the power source, and the structural configuration/type of aircraft. A critical review was conducted to assess the benefits and challenges of integrating advanced DP systems in contemporary airframes to increase aircraft performance by taking advantage of the positive interactions between the propulsion system and the aerodynamics of the aircraft. Some of the most promising aspects of contemporary DP concepts, such as improvements to propulsive efficiency due to wake re-energisation, better STOL performance due to increased dynamic pressure across the wing at low speeds, noise reduction benefits and structural benefits, were discussed. Due to the significant interdependencies between the airframe, propulsion system, and flight controls, the design of advanced DP systems is particularly sensitive to the combination of technologies used. Of the various combinations discussed in this review, several achieved good performance characteristics in certain areas, such as aerodynamic manoeuvrability (including stall speed), noise footprint reduction, drag, and fuel consumption reduction. Some highlights include the Hunting H.126 jet flap aircraft with a measured stall speed of 28 kts at a MTOW of 10 740 lbs, the MIT/Cambridge SAX, with a predicted noise level of 62 dBA at the airfield boundary, and the NASA N3-X with an estimated 72% energy usage reduction compared to the B777-200LR. The large variety of aero-structural and power plant configurations highlights the fact that contemporary aircraft design is a multi-disciplinary and multi-objective optimisation process. In this context, advanced DP systems introduce new variables and interactions between aircraft systems, increasing the complexity of the design process and therefore requiring proper handling starting from the conceptual design phase. Furthermore, as advanced DP systems can play a more integral role in the stability augmentation and control aspect of the aircraft, greater emphasis is placed on advanced digital engine control systems capable of dynamic thrust modulation and intelligent management of engine resources and health. This not only aids in mission optimisation but also enhances the cases for airworthiness certification and environmental compliance in novel aircraft configurations incorporating advanced DP systems.

Advanced DP concepts can significantly impact the mass and stiffness distributions of the aircraft. Additionally, they might produce new kinds of interactions between rotating components and other airframe systems and structures. This can have a significant effect on the aeroelastic stability boundaries and the flight envelope. Detailed modelling and analysis are required to capture these interactions and assess their influence. Furthermore, advanced DP systems can be used, passively or actively, to control the aeroelastic shape of the wing to improve flight characteristics and to control flight loads. Novel flight control strategies are required to facilitate such aeroelastic control and to exploit the potential benefits in reducing energy consumption.

Significant challenges that need to be addressed to guarantee the widespread adoption and long-term success of advanced DP technology include a significant increase in complexity of the digital engine control systems required both for stability and manoeuvrability, greater emphasis fault-tolerant hardware/software components design, and regulatory compliance (including the evolution of present-day certification standards). Therefore, further research and technology maturation is required on several fronts, also addressing the scalability of current DP concepts, which would ultimately support the successful integration of DP technologies into a variety of aircraft types and operational fruition in multiple environments.

References

1 Airbus (2015). Global Market Forecast 2015–2034. http://www.airbus.com/company/market/forecast (accessed 15 March 2023).

2 Kim, H.D. (2010). Distributed propulsion vehicles. Presented at the 27th International Congress of the Aeronautical Sciences.

3 Steiner, H.J., Seitz, A., Wieczorek, K., (2012). Multi-disciplinary design and feasibility study of distributed propulsion systems. *28th Congress of the International Council of the Aeronautical Sciences 2012, ICAS 2012*, Brisbane, 403–414.

4 Gohardani, A.S., Doulgeris, G., and Singh, R. (2011). Challenges of future aircraft propulsion: a review of distributed propulsion technology and its potential application for the all electric commercial aircraft. *Progress in Aerospace Sciences* 47: 369–391.

5 Xu, H., He, Y., Strobel, K.L. et al. (2018). Flight of an aeroplane with solid-state propulsion. *Nature* 563: 532–535.

6 Dang, T.Q. and Bushnell, P.R. (2009). Aerodynamics of cross-flow fans and their application to aircraft propulsion and flow control. *Progress in Aerospace Sciences* 45: 1–29.

7 Attinello, J.S. (1957). The Jet Wing. *IAS 25th Annual Meeting*.

8 Kuchemann, D. (2012). *The Aerodynamic Design of Aircraft*. New York: Pergamon Press.

9 Ko, A., Schetz, J.A., and Mason, W.H. (2003). *Assessment of the Potential Advantages of Distributed-Propulsion for Aircraft*. International Society for Air Breathing Engines.

10 Leifsson, L., Ko, A., Mason, W.H. et al. (2013). Multidisciplinary design optimization of blended-wing-body transport aircraft with distributed propulsion. *Aerospace Science and Technology* 25: 16–28.

11 Seyfang, G.R. (2012). FanWing – developments and applications. *28th Congress of the International Council of the Aeronautical Sciences 2012, ICAS 2012*, 238–246.

12 Royal Airforce Museum. (2016). Hunting H126. www.rafmuseum.org.uk/research/collections/hunting-h126 (accessed 15 March 2023).

13 ShinMaywa Industries Ltd. (2016). By Land, Sea, or Air US-2. www.shinmaywa.co.jp/aircraft/english/us2 (accessed 15 March 2023).

14 Peter, D. (1962). Multiple drive for aircraft having wings provided with transverse flow blowers. US Patent, Issue 1962.

15 Hancock, J. P. (1980). Test of a high efficiency transverse fan. AIAA Paper.

16 J. F. Kummer) (2012). Allred III, Jimmie B. (Skaneateles, NY, US), Cross-flow Fan Propulsion System. United States Patent, Issue 2012.

17 DLR. (2016). The SOAR Project - diStributed Open-rotor AiRcraft. http://www.soar-project.eu (accessed 15 March 2023).

18 Gohardani, A.S. (2013). A synergistic glance at the prospects of distributed propulsion technology and the electric aircraft concept for future unmanned air vehicles and commercial/military aviation. *Progress in Aerospace Sciences* 57: 25–70.

19 Ameyugo, G., Taylor, M., and Singh, R. (2006). Distributed propulsion feasibility studies. Presented at the 25th International Congress of the Aeronautical Sciences.

20 Rozhdestvensky, K.V. (2006). Wing-in-ground effect vehicles. *Progress in Aerospace Sciences* 42: 211–283.

21 Felder, J.L., Kim, H.D., Brown, G.V., and Chu, J. (2011). An examination of the effect of boundary layer ingestion on turboelectric distributed propulsion systems. *49th AIAA Aerospace Sciences Meeting Including the New Horizons Forum and Aerospace Exposition.*

22 Laskaridis, P., Pachidis, V., and Pilidis, P. (2014). Opportunities and challenges for distributed propulsion and boundary layer ingestion. *Aircraft Engineering and Aerospace Technology* 86: 451–458.

23 Arntz, A. and Atinault, O. (2015). Exergy-based performance assessment of a blended wing-body with boundary-layer ingestion. *AIAA Journal* 53: 3766–3776.

24 Hileman, J.I., Spakovszky, Z.S., Drela, M. et al. (2010). Airframe design for silent fuel-efficient aircraft. *Journal of Aircraft* 47: 956–969.

25 Kim, H. D., Berton, J.J., and Jones, S.M. (2006). Low noise cruise efficient short take-off and landing transport vehicle study. *Collection of Technical Papers - 6th AIAA Aviation Technology, Integration, and Operations Conference*, 362–372.

26 Felder, J.L. (2014). NASA N3-X with turboelectric distributed propulsion. *Proc. Disruptive Green Propulsion Technologies Conference*, 1–18.

27 Kim, H.D., Felder, J.L., Tong, M.T., and Armstrong, M. (2013). Revolutionary Aeropropulsion Concept for Sustainable Aviation: Turboelectric Distributed Propulsion. Presented at the International Symposium on Air Breathing Engines.

28 Hathaway, M., Del Rosario, R., and Madavan, N.K. (2013). NASA fixed wing project propulsion research and technology development activities to reduce thrust specific energy consumption. *49th AIAA/ASME/SAE/ASEE Joint Propulsion Conference.*

29 Stoll, A.M., Bevirt, J., Moore, M.D. et al. (2014) Drag reduction through distributed electric propulsion. *14th AIAA Aviation Technology, Integration, and Operations Conference.* American Institute of Aeronautics and Astronautics.

30 Borer, N.K., Moore, M.D., and Turnbull, A.R. (2014). Tradespace exploration of distributed propulsors for advanced on-demand mobility concepts. *AIAA AVIATION 2014 -14th AIAA Aviation Technology, Integration, and Operations Conference.*

31 De La Rosa Blanco, E., Hall, C.A., and Crichton, D. (2007). Challenges in the Silent Aircraft engine design. *Collection of Technical Papers – 45th AIAA Aerospace Sciences Meeting*, 5418–5437.

32 Snyder, J.M. and Zumwalt, G.W. (1969). Effects of wingtip-mounted propellers on wing lift and induced drag. *Journal of Aircraft* 6: 392–397.

33 Felder, J.L., Kim, H.D., and Brown, G.V. (2009). *Turboelectric Distributed Propulsion Engine Cycle Analysis for Hybrid-Wing-Body Aircraft.* American Institute of Aeronautics and Astronautics.

34 Lewis Research Center (1970). *NASA SP-259: Aircraft Propulsion.* Cleveland, Ohio: NASA Lewis Research Center.

35 Jones, C.E., Norman, P.J., Galloway, S.J. et al. (2016). Comparison of candidate architectures for future distributed propulsion aircraft. *IEEE Transactions on Applied Superconductivity* 26.

36 Moore, M.D. and Fredericks, B. (2014). Misconceptions of electric propulsion aircraft and their emergent aviation markets. *52nd AIAA Aerospace Sciences Meeting – AIAA Science and Technology Forum and Exposition, SciTech 2014.*

37 Pornet, C. (2018). Conceptual Design Methods for Sizing and Performance of Hybrid-Electric Transport Aircraft. PhD Thesis. Technical University of Munich http://mediatum.ub.tum.de?id=1399547 (accessed 15 March 2023).

38 M. Burston, T. Conroy, L. Spiteri et al. (2014). Misconceptions of electric propulsion aircraft and their emergent aviation markets. *52nd AIAA Aerospace Sciences Meeting - AIAA Science and Technology Forum and Exposition, SciTech.*

39 Ozbek, E., Yalin, G., Karaoglan, M.U. et al. (2021). Architecture design and performance analysis of a hybrid hydrogen fuel cell system for unmanned aerial vehicle. *International Journal of Hydrogen Energy.*

40 Sliwinski, J., Gardi, A., Marino, M., and Sabatini, R. (2017). Hybrid-electric propulsion integration in unmanned aircraft. *Energy* 140: 1407–1416.

41 Eqbal, M., Fernando, N., Marino, M., and Wild, G. (2021). Development of a turbo electric distribution system for remotely piloted aircraft systems. *Journal of Aerospace Technology and Management* 13.

42 Vepa, R. (2018). Modeling and dynamics of HTS motors for aircraft electric propulsion. *Aerospace* 5: 21.

43 Alrashed, M., Nikolaidis, T., Pilidis, P., and Jafari, S. (2021). Utilisation of turboelectric distribution propulsion in commercial aviation: a review on NASA's TeDP concept. *Chinese Journal of Aeronautics* 34: 48–65.

44 Samuelsson, S. and Grönstedt, T. (2021). Performance analysis of turbo-electric propulsion system with fuselage boundary layer ingestion. *Aerospace Science and Technology* 109: 106412.

45 Aurora Flight Sciences. (2016). DARPA Selects Aurora to Build VTOL X-Plane Technology Demonstrator. https://www.multivu.com/players/English/7617851-aurora-flight-sciences-vtol-xplane-darpa (accessed 15 March 2023).

46 Smith, L.H. (1993). Wake ingestion propulsion benefit. *Journal of Propulsion and Power* 9: 74–82.

47 Goulos, I., Otter, J., Tejero, F. et al. (2021). Civil turbofan propulsion aerodynamics: thrust-drag accounting and impact of engine installation position. *Aerospace Science and Technology* 106533.

48 Dippold, V., Hosder, S., and Schetz, J.A. (2004). Analysis of jet-wing distributed propulsion from thick wing trailing edges. *AIAA Paper* 7938–7950.

49 Sgueglia, A., Schmollgruber, P., Bartoli, N. et al. (2018). Exploration and sizing of a large passenger aircraft with distributed ducted electric fans. *AIAA Aerospace Sciences Meeting.*

50 Schmollgruber, P., Bedouet, J., Sgueglia, A. et al., Use of a certification constraints module for aircraft design activities. *17th AIAA Aviation Technology, Integration, and Operations Conference, 2017.*

51 Kirner, R., Raffaelli, L., Rolt, A. et al. (2015). An assessment of distributed propulsion: advanced propulsion system architectures for conventional aircraft configurations. *Aerospace Science and Technology* 46: 42–50.

52 Raymer, D.P. (2006). *Aircraft Design: A Conceptual Approach*, 4e. AIAA.

53 Roskam, J. (1985). *Airplane Design Vol. 1–7*. DARcorporation.

54 Torenbeek, E. (1982). *Synthesis of Subsonic Airplane Design*. Springer-Verlag.

55 Hepperle, M.J. (2012). Electric Flight – Potential and Limitations. *AVT-209 Workshop on Energy Efficient Technologies and Concepts Operation*, Instituto Superior Tecnico and Academia da Força Aérea, Lisbon, Portugal

56 Stinton, D. (1998). *The Anatomy of the Airplane*, 2e. Reston, VA: American Institute of Aeronautics and Astronautics.

57 Uranga, A., Drela, M., Hall, D.K., and Greitzer, E.M. (2018). Analysis of the aerodynamic benefit from boundary layer ingestion for transport aircraft. *AIAA Journal* 56: 4271–4281.

58 Margason, R.J. (1970). *Review of Propulsion-Induced Effects on Aerodynamics of Jet/Stol Aircraft*. Washington, DC, USA: NASA.

59 Sehra, A.K. and Whitlow, W. Jr., (2004). Propulsion and power for 21st century aviation. *Progress in Aerospace Sciences* 40: 199–235.

60 Maskell, E.C. and Spence, D.A. (1959). A theory of the jet flap in three dimensions. *Proceedings of the Royal Society of London* 251 (1266): 407–425.

61 Wick, A.T., Hooker, J.R., Hardin, C.J., and Zeune, C.H. (2015). Integrated aerodynamic benefits of distributed propulsion. *53rd AIAA Aerospace Sciences Meeting*.

62 Heeg, J., Stanford, B.K., Wieseman, C.D. et al. (2018). Status report on aeroelasticity in the vehicle development for X-57 maxwell. *36th AIAA Applied Aerodynamics Conference*.

63 Ajaj, R.M., Parancheerivilakkathil, M.S., Amoozgar, M. et al. (2021). Recent developments in the aeroelasticity of morphing aircraft. *Progress in Aerospace Sciences* 120.

64 Amoozgar, M., Friswell, M.I., Fazelzadeh, S.A. et al. (2021). Aeroelastic stability analysis of electric aircraft wings with distributed electric propulsors. *Aerospace* 8.

65 Cravana, A., Manfreda, G., Cestino, E. et al. (2017). Aeroelastic behaviour of flexible wings carrying distributed electric propulsion systems. *SAE AeroTech Congress and Exhibition, AEROTECH 2017*, vol. 2017-September, 2017.

66 Schmollgruber, P., Döll, C., Hermetz, J. et al. (2019). Multidisciplinary exploration of dragon: an ONERA hybrid electric distributed propulsion concept. *AIAA Scitech Forum 2019*.

67 An, S. (2015). Aeroelastic Design of a Lightweight Distributed Electric Propulsion Aircraft with Flutter and Strength Requirements. MSc Thesis, Georgia Institute of Technology, GA, USA.

68 Massey, S.J., Stanford, B.K., Wieseman, C.D., and Heeg, J. (2017). Aeroelastic analysis of a distributed electric propulsion wing. *58th AIAA/ASCE/AHS/ASC Structures, Structural Dynamics, and Materials Conference*.

69 Nguyen, N.T., Reynolds, K., Ting, E., and Nguyen, N. (2018). Distributed propulsion aircraft with aeroelastic wing shaping control for improved aerodynamic efficiency. *Journal of Aircraft* 55: 1122–1140.

70 Nguyen, N.T., Reynolds, K., and Ting, E., (2017). Aeroelastic wing shaping using distributed propulsion. US Patent US 9,751,614 B1 (45), Issue, 5 Sept. 2017.

71 Kim, H.D., Perry, A.T., and Ansell, P.J. (2018). A review of distributed electric propulsion concepts for air vehicle technology. Presented at the *2018 AIAA/IEEE Electric Aircraft Technologies Symposium*.

72 McSwain, R.G., Glaab, L.J., Theodore, C.R. et al. (2017). Greased lightning (gl-10) performance flight research: Flight data report.

73 Chen, Z., Zhang, M., Chen, Y. et al. (2019). Assessment on critical technologies for conceptual design of blended-wing-body civil aircraft. *Chinese Journal of Aeronautics* 32: 1797–1827.

74 Li, H., Gou, L., Zheng, H., and Li, H. (2021). Intelligent fault diagnosis of aeroengine sensors using improved pattern gradient spectrum entropy. *International Journal of Aerospace Engineering* 2021.

75 Ranasinghe, K., Guan, K., Gardi, A., and Sabatini, R. (2019). Review of advanced low-emission technologies for sustainable aviation. *Energy* 188.

76 Xargay, E., Hovakimyan, N., Dobrokhodov, V. et al. Adaptive control in flight. In: *Advances in Intelligent and Autonomous Aerospace Systems*, 129–172.

77 Sabatini, R., Roy, A., Blasch, E. et al. (2020). Avionics systems panel research and innovation perspectives. *IEEE Aerospace and Electronic Systems Magazine* 35: 58–72.

78 Batuwangala, E., Kistan, T., Gardi, A., and Sabatini, R. (2017). Certification challenges for next generation avionics and air traffic management systems. *IEEE Aerospace and Electronic Systems Magazine* 33: 44–53.

79 Lim, Y., Bassien-Capsa, V., Ramasamy, S. et al. (2017). Commercial airline single-pilot operations: system design and pathways to certification. *IEEE Aerospace and Electronic Systems Magazine* 32: 4–21.

80 Kirner, R., Raffaelli, L., Rolt, A. et al. (2016). An assessment of distributed propulsion: Part B – Advanced propulsion system architectures for blended wing body aircraft configurations. *Aerospace Science and Technology* 50: 212–219.

81 Airbus (2014). E-Thrust Brochure. http://www.airbusgroup.com (accessed 15 March 2023).

82 Bradley, M.K. and Droney, C.K. (2011). *Subsonic Ultra Green Aircraft Research: Phase I Final Report*. National Aeronautics and Space Administration.

83 Burston, M. (2016). Opportunities and Challenges of Distributed Propulsion Configurations, MEng Thesis, School of Aerospace, Mechanical and Manufacturing Engineering, RMIT University, Melbourne, Australia.

84 Burston, M., Ranasinghe, K., Gardi, A. et al. (2022). Design principles and digital control of advanced distributed propulsion systems. *Energy* 241: 122788.

15

Integration of Hybrid-Electric Propulsion Systems in Small Unmanned Aircraft

Jacob Sliwinski[1], Alessandro Gardi[2,3], Matthew Marino[1], and Roberto Sabatini[2,3]

[1] School of Engineering, RMIT University, Bundoora, Victoria, Australia
[2] Department of Aerospace Engineering, Khalifa University of Science and Technology, Abu Dhabi, UAE
[3] School of Engineering, RMIT University, Melbourne, Victoria, Australia

15.1   Introduction

Conventional transport aircraft propulsion technology revolves around the use of petroleum-based internal combustion engines (ICEs), in the form of either aviation turbine fuel (ATF) or aviation gasoline (AvGas). Despite their widespread adoption, these fossil fuels have a significant impact on the environment, both in terms of pollutants, such as carbon monoxide (CO), nitrogen oxides (NO_X), and unburnt hydrocarbons (UHCs), and in terms of greenhouse emissions, the principal one of which is carbon dioxide (CO_2). Improvements are therefore currently being pursued in alternative fuels and hybrid-electric technologies [1]. Some non-mainstream propulsion technologies are expected to mitigate considerably the environmental impacts in aviation [2]. The integration of hybrid-electric technologies exploits the synergies with distributed propulsion configurations and is therefore one of the most promising approaches to introduce new more sustainable forms of propulsion [3–7]. Distributed propulsion is based on subdividing the desired thrust across a number of smaller engines with benefits in terms of noise reduction, higher reliability, shorter take-off and landing distances, better specific fuel consumption, and improved flight stability [3, 5, 7]. From a research and development perspective, small-size, remotely piloted aircraft systems (RPAS) are seen as a stepping-stone, enabling experimental scaling studies for larger aircraft applications. This case study, originally published in [37, 38], focusses on hybrid RPAS, employing a combination of ICE and electrical propulsion systems. Medium-large blended-wing-body (BWB) transport aircraft integrating distributed hybrid-electric propulsion technology are assumed as the long-term target for this research as they can combine most of the benefits offered by all these emerging technologies. Various current BWB concepts are proposing hybrid-electric distributed propulsion configurations, and the optimisation of size and number of ICE systems and electric motors is an area of active research.

An electric aircraft concept was considered in conjunction with research projects on solar-powered aircraft undertaken by NASA and AeroVironement Inc. in the early 1970s. Among the various emerging design concepts, solar powered aircraft, such as the

Sustainable Aviation Technology and Operations: Research and Innovation Perspectives, First Edition.
Edited by Roberto Sabatini and Alessandro Gardi.
© 2024 John Wiley & Sons Ltd. Published 2024 by John Wiley & Sons Ltd.
Companion Website: www.wiley.com/go/sustainableaviation

Helios prototype, typically featured a multi-propulsor arrangement consisting of brushless electric motors [7]. Electric aircraft research highlights potential for future commercial exploitation, yet encounters some fundamental technological limitations. Weight considerations are among the key challenges, due to components such as batteries, electric motors, generators, and converters, introducing significant penalties while still not achieving the typical range and endurance performances of ICE propulsion. Hybrid electrical propulsion can nevertheless help increase system reliability and power distribution/quality. One notable approach is the use of renewable energy sources by implementing suitable harvesting technologies. These may include solar radiation and wind speed, for instance exploiting solar photovoltaic (PV) array, fuel cell (FC) stack and battery with a provision for onsite hydrogen (H_2) generation, by using an electrolyser and H_2 tank. Harvesting energy in the most efficient way possible and optimising the propulsion configuration and mission profile are among the major difficulties. Energy harvesting from renewable sources, as opposed to relying on just one power source (ICE), can enhance endurance and range performances, increase safety, and decrease carbon emission. This chapter focusses on piston engines and electric motors for light aviation use. Range and endurance can be determined for both piston propeller and jet aircraft via the *Breguet equation*. Estimations of range and endurance for electrical aircraft are less established, but consideration of the electrical energy storage nature suggests equating power delivered by the battery (accounting for losses due to propeller motor) and the motor controller, to the power required to overcome drag as an acceptable method to produce initial performance estimates. Subsequently, the actual behaviour of the battery and its effective capacity/energy density can be analysed in a more comprehensive design trade-off study. This chapter will focus on a hybrid-electric propulsion system (HEPS) retrofit for RPAS. Critical parameters for HEPS considered are flight endurance, range, and velocity required to perform such RPAS missions as intelligence, surveillance, and reconnaissance (ISR).

15.1.1 Range and Endurance Performance of ICE Powered Aircraft

The *United States Air Force Institute of Technology* has developed a number of HEPS concepts [8]. One particular concept considered an ICE engine to be activated in the cruise flight phase and provide primary power. In this case, the range can be correlated accurately to fuel burn. Fuel flow, symbolised by $\dot{m}_f$, becomes the reference metric. The amount of fuel available m_f is also required to assess potential energy. If we assume a steady level condition (cruise), range becomes a function of fuel flow, fuel available and velocity, as outlined in Eq. (15.1). This equation assumes that the weight of the aircraft remains constant; however, this is not accurate. Fuel is burned and thus weight is reduced through the flight and must be taken into consideration. The reduction in weight also has an effect on the amount of lift and drag generated as the overall weight of the aircraft reduces so will the amount of lift required for steady level flight. Therefore, a modified form of the Breguet equation can be adopted for propeller driven aircraft [9]. This modified empirical relationship is shown in Eq. (15.2), where W_1 and W_0 are the initial weight and zero fuel weight on the aircraft respectively, L/D is represented as the lift-to-drag ratio, and C_{SFC} is the power specific fuel consumption as defined in Eq. (15.3), as the rate of fuel consumption per unit shaft power [8–10]. These equations do not explicitly account for density; however, this is taken into account when the L/D ratio is numerically approximated. Furthermore, Eq. (15.2) does not account for wind speeds.

$$R = \frac{m_f}{\dot{m}_f} V_\infty \tag{15.1}$$

$$R = \frac{\eta_{prop}}{C_{SFC}} \frac{L}{D} \log \frac{W_0}{W_I} \tag{15.2}$$

$$C_{SFC} = -\dot{m}_f / P \tag{15.3}$$

Following Eqs. (15.1)–(15.3), to maximise the range for reciprocating engines, design efforts are aimed at:

Highest propeller efficiency η_{prop}.
Lowest fuel consumption, C_{SFC}.
Highest W_0/W_1 ratio, equating to largest fuel weight W_f since $W_0 = W_1 + W_f$.

Anderson and Payne also present a Breguet endurance equation derived to maximise loiter time, as shown in Eq. (15.4), where t_{end} is the endurance time. Similar to the range equation, the endurance equation accounts for fuel burn during loiter, suggesting to maximise operations at $[c_L^{3/2}/c_D]_{MAX}$ to enhance endurance performance.

$$t_{end} = \frac{\eta_{prop}}{C_{SFC}} \cdot \sqrt{2\rho_\infty S} \cdot \frac{c_L^{\frac{3}{2}}}{c_D} \cdot \left(\frac{1}{\sqrt{W_1}} - \frac{1}{\sqrt{W_0}} \right) \tag{15.4}$$

As a result, the maximum endurance for propeller driven airplanes powered by reciprocating engines can be achieved in the following conditions:

Highest propeller efficiency, η_{prop}.
Lowest fuel consumption, C_{SFC}.
Highest W_0/W_1 ratio.
Aerodynamic condition $\frac{c_L^{\frac{3}{2}}}{c_D}$.
Flight at sea level, where density maximum.

15.1.2 Range and Endurance Performances of Battery Powered Aircraft

In case of battery powered aircraft, the mass remains constant throughout the flight, hence simplifying the range equation. The range of an aircraft is defined by flight speed and time:

$$R = V_\infty \cdot t \tag{15.5}$$

Specifying endurance for battery powered RPAS, by simply adapting Eq. (15.4) is not possible for battery powered aircraft, due to singularity of the solution in the case of constant weight (i.e. $W_1 = W_0$). Similar considerations apply to the range Eq. (15.2). This reinstates the original range strategy suggested by Payne as appropriate based on battery capacity, E_{BAT}, and on the average current and voltage associated with the battery drain, respectively $\bar{I}$ and $\bar{V}$ [8, 10]. The battery capacity can be further elaborated as the product of the mass of the battery and of the specific energy of the battery. Consequently, the maximum endurance conditions will correspond to the velocity of minimum power required. In particular, the endurance t_{MAX}, measured in units of time, can be calculated as a function of mass $(m_{battery.})$ and mass specific energy, E^* of the battery as:

$$t_{MAX} = E_{BAT}/\overline{IV} = \frac{m_{battery} \cdot E^*}{P_{Drain}} \tag{15.6}$$

Introducing this flight time into the range Eq. (15.5) yields:

$$R = V_\infty \frac{m_{battery} \cdot E^*}{P_{Drained}} \tag{15.7}$$

The power drained from the battery is related to the propulsive power required by the aircraft:

$$P_{Drained} = \frac{P_{aircraft}}{\eta_{total}} \tag{15.8}$$

Whereby the power required from the aircraft is related to its weight, L/D ratio and flight speed:

$$P_{aircraft} = D_{aircraft} \cdot V_\infty = \frac{m\,g}{\left(\frac{L}{D}\right)} V_\infty \tag{15.9}$$

From Eq. (15.9), the battery power is modified becoming:

$$P_{Drained} = \frac{m\,g}{\left(\frac{L}{D}\right) \cdot \eta_{total}} V_\infty \tag{15.10}$$

Which can be inserted into Eq. (15.7)

$$R = V_\infty \frac{m_{battery}\, E^*}{\frac{m \cdot g}{\left(\frac{L}{D}\right) \eta_{total}} V_\infty} \tag{15.11}$$

Simplifying Eq. (15.11) yields the final range equation for battery electric operated aircraft

$$R = E^* \eta_{total} \frac{1}{g} \frac{L}{D} \frac{m_{battery}}{m} \tag{15.12}$$

Similar to the Breguet range equation for the ICE case, range depends significantly on the lift-to-drag ratio (aerodynamic efficiency) and total propulsive efficiency η_{total}. Conversely, the specific energy appears as a new term. In order to maximise range, the following parameters should be maximised:

Battery mass-to-total mass ratio $\frac{m_{battery}}{m}$.
Battery energy density.
Lift-to-drag ratio.
Total system efficiency η_{total}.
Minimise the total aircraft mass.

15.2 Hybrid-Electric Aircraft Configurations

A promising solution to mitigate the disadvantages of each propulsion technology taken individually is to combine both technologies, integrating an ICE with an electric motor

(EM), thus implementing a HEPS. Hybrid propulsion has been an intense area of research within the automotive industry, where hybridisation yielded significant increases in fuel efficiency, which have sparked a growing interest in exploring this approach in aviation. Heavily focused on the automotive scene, the research in hybrid propulsion technologies for aircraft is a relatively new research area with focused developments on theoretical control methods for hybrid propulsion systems and their effects on aircraft [3]. In most instances, the primary production of power is from an ICE that is supplemented by some form of electric energy through either batteries, fuel cells, or solar panels. The electrical systems can provide excess power when needed for take-off and climb flight phases. This chapter explores the use of HEPS specifically implementing a battery and fossil fuel cell combination. In this configuration the secondary energy source usually offers a lower specific energy (energy/mass) than fossil fuel, although it allows additional capabilities in exchange for the added weight penalty. Current trends demonstrate improved fuel economy, such as in the case of the Diamond DA-36 E-Star aircraft [11]. Due to the diminished reliance on fossil fuels, the hybrid propulsion concept offers promising solutions in shaping a more sustainable aviation future.

15.2.1 Energy Storage Technology

Energy storage remains one of the fundamental technologies that are essential to support hybrid propulsion for larger aircraft. Fossil fuels are capable of storing energy in liquid form with favourable specific energy to volume ratios. Liquid fuel also has the advantage of being stored in complex and sealed geometries, such as the aircraft wings. Alternative energy storage solutions are being actively investigated in the search for innovative ways to increase potential energy by unit mass and unit volume. Research into energy harvesting in flight is also showing promise to increase overall energy capacity [12].

Various sources of energy have been explored for hybrid aircraft including active generators, solar panels, fuel cells, and batteries. Generator systems rely on fossil fuels to generate kinetic energy to turn a generator for electricity production. While this concept does not mitigate the dependency on fossil fuels, it supports greater flexibility with respect to the propulsion configuration. Similar concepts can be applied to aircraft and can be designed in various configurations discussed in subsequent sections. AeroVironment's High Altitude Long Endurance (HALE) Global Observer is fuelled on liquid hydrogen, by using a fuel cell that provides power to its eight electric engines [13–16]. The airframe of the Global Observer achieved its maiden flight in August 2010, with the first hydrogen powered flight taking place in January 2011. Engine and automobile manufacturer, Rolls Royce, has also taken an interest in matching its turbine engines to electric generators for a variety of RPAS applications [15]. Photovoltaic solar panels are another popular and considered option. Photovoltaic electricity generation is applicable to large wing aircraft, such as HALE vehicles, as large wing surface areas provide space for solar panels. High altitude flight maximises solar exposure due to the lack of tropospheric weather and increased levels of solar radiation. Two prime examples considered are the Pathfinder and Helios aircraft, propelled by 14 engines using solar energy stored in lithium ion batteries. The Helios prototype is one successful example achieving flight at design cruise speeds for 40 minutes and reaching a maximum altitude of 96 863 ft, which set the record for sustained flight by a winged aircraft [17].

Hydrogen fuel cells are another possibility for hybridisation, with many current trends utilising a full fuel cell system rather than a hybrid configuration. Fuel cells are thermodynamically efficient, however, require pressurised hydrogen as the main fuel source before reacting with oxygen to created electricity. Hydrogen is stored in a pressurised vessel, which allows significantly high energy per unit mass properties, but a relatively low energy per unit volume. This results in large storage systems requiring significant aircraft design modifications. Aircraft with a hydrogen-based system have been explored with some entities adapting fuel cell technology by modifying the existing electric aircraft to carry fuel cells. The AeroVironment Puma AE aircraft was a successful platform that incorporated fuel cell technology and resulted in a 300% increase in aircraft endurance from three to nine hours [12].

Batteries, on the other hand, are the mainstream electrical energy storage technology. The rapid evolution in battery technology has been prompted by both traditional and emerging consumer electronics applications, spanning smart phones to electric cars. Table 15.1 lists the representative specific energies achieved by various state-of-the-art battery technologies. Innovative battery technologies that show promising potential to advance the maximum specific energies include lithium-air, lithium-sulphate, zinc-air, aluminium-air, magnesium ions, and graphene [18].

Due to the low specific energy of batteries, remote-controlled aircraft suffer from poor endurance and range when compared to their ICE driven counterparts. Nonetheless, despite the poor energy densities and specific energies, in recent years there have been several attempts in the development of battery based hybrid designs with different sizes. In 2009, German aircraft builder Flight Design demonstrated a light, manned, sport aircraft where a 40 hp electric motor provided approximately five minutes of boost power to a 115-hp ICE [19]. Since electric motors can deliver the additional thrust required during take-off operations, the ICE could be reduced in size, reducing the overall aircraft weight. In 2011, Siemens AG, Diamond aircraft and EADS took flight in the world's first manned hybrid-electric series configuration. The aircraft propeller was driven by a 70-kW electric motor running on batteries with the primary source being a 30-kW combustion engine generator. This configuration is conceptually similar to submarine hybrid-propulsion technology where the ICE engine runs at its most efficient RPM to charge the batteries when needed. The flight test campaign demonstrated that the series configuration achieved significant fuel savings, since the series arrangement allowed the engines to run at its ideal

Table 15.1 Current battery technology.

Battery technology	Energy density
Nickel-cadmium (NiCd)	$50 \sim 150 \, \text{Wh kg}^{-1}$ 1.2 V per cell
Nickel-metal-hydride (NiMH)	$140 \sim 300 \, \text{Wh l}^{-1}$ 1.2 V per cell
Lithium iron (LiFe)	$\sim 220 \, \text{Wh l}^{-1}$
Lithium ions (Li-Ion)	$250 \sim 676 \, \text{Wh l}^{-1}$
Lithium polymer (LiPo)	$250 - 730 \, \text{Wh l}^{-1}$

speed. It also demonstrated less noise on take-off operation when in the electrical phase. The most promising power sources for HEPS aircraft will be discussed in more depth in the following sections.

15.2.2 Inverter and Controller Technology

An inverter is required to change direct current (DC) electricity to alternating current (AC) electricity at a certain frequency and phase. This conversion is necessary when AC electric motors are used for electric propulsion. An ideal process would convert power without any losses; however, the conversion efficiency from DC to AC is commonly above 95% [20].

The HEPS controller, on the other hand, is an integral component of a hybrid-electric remotely piloted aircraft systems (HERPAS). It carries out the required control functions for all components featured in a HEPS such as the ICE, EM, generator, battery, current inversion, and transmission (CVT). The controller must integrate individual components to achieve maximum efficiency. Different control strategies can be implemented for a system via different control methods. Two main categories specified are intelligent controllers and rule-based controllers [3].

15.2.3 Rechargeable Batteries

Energy density (energy stored per unit volume) and specific energy (energy stored per unit mass) are the key figures to evaluate the performance of various alternative energy sources for aviation, and are particularly important in assessing battery technology. A comparison between the energy densities and specific energy of various energy storage technologies is well researched [12, 21]. State-of-the-art rechargeable batteries can be geometrically designed and scaled in accordance with their design requirements. The comparison between button cells to large lead acid automotive batteries highlights their impressive scalability. Current energy storage technologies include alkaline, lead acid, nickel cadmium (Ni-Cd), nickel metal hydride (Ni-MH), lithium-ion (Li-ion), lithium-ion polymer (li-ion polymer), and lithium sulphur (Li-S). The best candidates for RPAS in terms of specific energy are Li-ion, Li-ion polymer, and Li-S based on their relatively higher energy densities. Li-ion polymer share similar properties as Li-ion batteries, with the additional benefits of lighter weights and custom geometries. Based on these attributes, these kinds of batteries are widely used in consumer technology. Although the technology behind Li-ion and Li-ion polymer battery types have not yet reached full maturity when compared to other energy storage technologies available today, they are the most frequently used and selected, as seen in the solar-powered electric RPAS listed in Table 15.2. The energy densities of Li-ion and Li-ion polymer batteries for solar RPAS vary between $522 \sim 864 \, \text{kJ} \, \text{kg}^{-1}$. Even though Li-ion and polymer have the highest energy densities and a low loss of charge when not in use, their figures are still two orders of magnitude lower than hydrocarbon fuels ($\sim 43 \, \text{MJ} \, \text{kg}^{-1}$). The demand for advanced rechargeable batteries has nevertheless increased significantly. Under these circumstances, a lithium sulphur (Li-S) battery has been developed by Sion Power since 1994 [21]. A Li-S battery has a theoretical specific capacity of $1600 \, \text{mA} \, \text{h} \, \text{g}^{-1}$, and a specific energy value of $9360 \, \text{kJ} \, \text{kg}^{-1}$ ($2600 \, \text{W} \, \text{h} \, \text{kg}^{-1}$) assuming a complete reaction is made between lithium and sulphur formulating Li_2S [22]. These attributes suggest a promising battery technology for aircraft

Table 15.2 Specific energy of Li-ion and Li-ion polymer batteries applied in solar-powered aircraft.

Model name	Year	Type of battery	Specific energy (kJ kg^{-1})
Solar Impulse HB-SIA	2009	Li-ion polymer battery, attained drastic weight penalties of 400 kg (1/4 of the aircraft mass)	864
Sunrise II	1985	Li-ion polymer battery	522
Xihe	2009	Li-ion polymer battery	705
Green flight challenge: Taurus G4	2011	Li-polymer pack, consisting of 88 series-connected cells arrangement (battery mass ~500 kg)	648
So long	2005	120 Sanyo 18 650 Li-ion polymer battery 5.6 kg	770
Skysailor	2004	8× E-tec 1 200 cell batteries arranged in series, six in parallel, weight of 1.2 kg	619

Source: Adapted from [12].

Table 15.3 Zephyr 7, equipped with Li-S battery.

Name	Year	Type of battery	Specific energy (kJ kg^{-1})
Zephyr 7 solar powered high-altitude long endurance UAS	2010	576 cells Sion Li-S battery pack (12 in series and 48 in parallel)	1800–2160 (achievable), 1260 (demonstrated)

Source: Adapted from [12].

use. The first use of Li-S battery was demonstrated on the Zephyr 7, which flew over 336 h in Yuma, Arizona in July 2010, reaching an altitude of 21.6 km. Table 15.3 gives an overview of aircraft using Li-S battery technology as their main source of energy for propulsion. Battery type and energy density is also given. An important advantage of the Li-S battery is its suitability to operate at very low temperatures. In order to achieve this measure, introducing an advanced electronic control system will help assist and maintain battery temperature during flight.

15.2.4 Future Energy Storage Technology for Aircraft

Energy storage solutions for transport vehicles are a subject of very active research. Table 15.4 outlines some emerging energy storage technologies for aircraft propulsion applications.

Metal air, sodium sulphur, and zinc bromine development have shown promising advantages, however the specific energy density has not exceeded lithium based batteries so far [18]. Advanced flywheel technology uses a mechanical concept of storing power in rotation kinetic energy. This kinetic energy can be transferred to a shaft and propeller using a gearbox. The mechanical energy losses will be significant using this approach and a clutch and gearing system is required. There are also issues with the gyroscopic and procession effects of high-energy rotating objects on board an aircraft. One promising technology emerging from research is the lithium air battery in which lithium metal is used as the oxidating element, yielding a theoretical energy density of 11 680 Wh kg^{-1} [23]. Realistic expectations

Table 15.4 Emerging energy storage technologies.

Battery technology	Energy density
Lithium air	$1700\,Wh\,kg^{-1}$ (expected) [23]
Super capacitors	$0.1-5\,Wh\,kg^{-1}$
Metal air	$110-420\,Wh\,kg^{-1}$
Sodium sulfur	$150-240\,Wh\,kg^{-1}$
Zinc bromine	$37\,Wh\,kg^{-1}$
Advanced flywheel	$30-100\,Wh\,kg^{-1}$

are 14.5% of the theoretical maximum, which has been deemed achievable through further research and development. The high energy density is only made possible by using air and the anode and electrolyte significantly reducing the overall weight of the cell [24]. Research and development into lithium-air battery technology is ongoing, identifying the most optimum battery composition using carbon or graphene based cathodes. Material experimentation of catalyst composition is also another area of research to improve relative electrical efficiency and cycle life [24].

15.2.5 Fuel Cell Technology

Fuel cells (FC) technology is based on the production of electrical energy through a controlled chemical reaction of hydrogen fuel and oxygen. Fuel cells have many advantageous properties compared to rechargeable batteries, such as higher efficiency and reduced carbon footprint. There are several types of fuel cells, each depending on a different electrolyte type used. Examples include phosphoric acid fuel cell (PAFC), molten carbonate fuel cell (MCFC), and solid oxide fuel cell (SOFC). The oxidising agent can be sourced from air intakes, much like ICE engines. Hydrogen exists as the stored fuel on board the aircraft. Fuel cell technology has potential usage in the aerospace industry. Invested interest in fuel cell technology is evident with the 2014 reported investment of approximately US$850 million per year due to its multipurpose usage across the transportation industry [25, 26]. Fuel cells for aircraft application are commonly coupled with energy harvesting technology. An example of this is detailed in Table 15.5 where solar panels on the external surfaces harvest energy from the sun to increase endurance. In particular, regenerative fuel cell energy based storage systems hold the key in achieving energy densities above $400\,Wh\,kg^{-1}$, however the efficiency of a hydrogen fuel cell system remains between 25% and 58% [18]. The Helios was the first successful flight-tested, solar-powered aircraft equipped with fuel cells. A significant disadvantage of fuel cell aircraft application is the inability to distribute the system across the aircraft, as opposed to batteries or liquid fuels. The Helios approach involves the installation of a few heavy fuel cell pods located at the centreline of the wing with two high-pressure hydrogen tanks located near the wing tips. This restricts structural flexibility associated with the large masses of the fuel cell system. The three-point mass effects and introducing substantial complexity in aircraft flight dynamics, contributing to persistent high dihedral wing flex during operation.

Table 15.5 Solar-powered aircraft equipped with regenerative fuel cell storage systems.

Name	Year	Type of battery	Specific energy (kJ kg^{-1})
Proposed in Mars aircraft	1990	Regenerative FC, utilising reactants of gaseous hydrogen and oxygen	1584
Heliplat	2004	FC consisting of FC array, hydrolyser, hydrogen, oxygen, and water tanks. Gases are stored at high pressure 120 bar, operating during the night feeding the FC which supply the motor; the water feeds the electrolyser completing the cycle	1980
Helios	2003	Uses a PEM fuel cell and electrolyser stacks, light weight gas, and water storage tanks.	180

Source: Adapted from [12].

15.2.6 Comparison Between Rechargeable Batteries and Fuel Cells

Solar-powered aircraft predominantly use rechargeable batteries of the Li-ion and Li-ion polymer types, however their energy density is not high enough to support the HALE flight concept. Fuel cells demonstrate promising characteristics achieving energy densities of up to 1980 kJ kg^{-1}, as shown in Table 15.5. Although the energy density is high, their relative output per second is low as energy cannot be released at high wattage or at wattages of current battery technology. A hybrid approach, where excess and quick release power can be accessed, would be a solution to the high power requirements of take-off and climb. The fuel cells concentrated mass also raises concern as aerodynamic and structure design must accommodate the point mass of the fuel cell. The storage of hydrogen is also another complexity to aircraft design as pressurised cylindrical tanks must be used and strategically placed on the aircraft. For light aviation, fuel cells offer a more direct and efficient means in converting fuel to electrical power: up to 50–60% in systems without heat recovery cycles, minimised sub-components, and increased performance. Concerning existing surveillance and reconnaissance RPAS applications, batteries are known to be a viable and preferred option with capabilities offering significant noise reduction, ideal for stealth objectives.

15.2.7 HEPS Mission Profiles

Three baseline HEPS mission profiles are available: electric-only, charge-depleting, and charge-sustaining. In electric-only operations, no battery recharge is carried out during the mission. In charge-sustaining operation, the ICE tops up the energy during periods of low propulsive demand along the mission, so that on average a certain battery charge level is maintained. As an example, the motor can intermittently provide alternate boost power during take-off, climb, and cruising. When battery levels deplete, the ICE would be engaged and the motor will switch roles in being used as a generator to recharge the batteries [27]. This strategy is commonly adopted in hybrid automotive vehicles, such as the *Volt*, enhancing performance of the limited electric range of the vehicle. A careful design of charge and discharge phases along the mission is crucial in aerospace applications to prevent premature degradation or failures leading to safety-critical situations. In the charge-depletion strategy, a certain amount of battery recharge is performed throughout the mission, but the average battery charge state will be depleted progressively during the mission. This strategy is commonly implemented in plug-in vehicles, such as electric scooters. The distinctive power regimes in aircraft flight missions are take-off and landing, cruise, cruise with regeneration, and endurance. In the take-off and climb phases, the propulsion requirements are fulfilled by the ICE up to its ideal operating point, beyond which the electric motor provides additional power. In the case of serial HEPS, this strategy is commonly referred to as the torque split method. In cruise operation, the ICE is operated at its maximum efficiency regime above the power required for cruise flight and the excess power is transferred to the electric motor acting as a generator, recharging the batteries. When cruising without regeneration, the ICE provides only the power required to fly the aircraft.

15.2.8 Energy Harvesting

Notwithstanding the HEPS battery recharge strategies exploiting ICE, battery depletion remains the limiting factor for endurance and range performance. Although ICE has been successfully adopted in HEPS aircraft to recharge batteries throughout the mission, other forms of battery charging using environmental energy sources have been investigated. The most traditional is solar energy harvesting, exploiting photovoltaic panels on the wings and fuselage. More advanced energy harvesting concepts include the exploitation of naturally occurring vibrations caused by the propeller and by turbulence through piezoelectric generators [28]. Recent research explores passive methods of energy gain through dynamic soaring flight to maximise range. This can be achieved by developing kinetic energy in flight exploiting the wind velocity profile of the atmosphere [29]. Temporary excesses in airspeed can be electrically harvested through a small wind turbine. Urban soaring is also explored as high-rise structures provide updraught flow profiles where stationary flight can be achieved to significantly increase endurance and electrically harvest flow energy using a small wind turbine [30]. Much of this research is inspired by flying animals, as their efficiency in travelling long distances with minimal energy expenditure is of interest for both small and large aircraft.

15.3 Reference Platform and Integration Case Study

In our HEPS retrofit methodology development, the Aerosonde RPAS, developed by Textron Systems, was selected as the reference platform due to its operational versatility and because a variety of performance and flight dynamics data are available for numerical simulation studies. The fundamental HEPS design goals were set to prioritise endurance over range and reliability. The Aerosonde specifications and aerodynamic parameters were used to determine the flight dynamics and power requirement characteristics necessary to estimate range and endurance performance. The ICE used in the pre-existing Aerosonde model is the Modified Enya R120 4C piston engine, capable of a maximum power of 2.1 hp [31]. The propeller efficiency (η) was assumed to be 0.9 to account for blade tip slipstream, friction, compressibility, and other losses [9].

Table 15.6 Plettenberg HP320/30 electric motor specifications.

Plettenberg HP320/30	
V (V)	29.7
η	0.9
I (A)	46
P_{in} (W)	1366.2
P_{out} (W)	1229.58

Source: From [32].

15.3.1 Electric Motor Model

The EM model was selected based on commercially off the shelf (COTS) availability and power requirements, satisfying conditions of *electric-only* phase in generating at least 1112 W or greater. Based on this, the Plettenberg HP320/30 was selected as, based on manufacturing data, its key properties matched the electric motor requirements, as detailed in Table 15.6 [32].

The selected electric motor recommends a constant voltage input of 29.7 V. This value assures longer lifetime and optimal power consumption. In a more detailed analysis, the motor torque profile should be considered instead. Accurate measurements of torque are difficult, though, and the mechanical characteristics of the motor change with the operating temperature and load regimes. Therefore, accurate estimates of parameters, such as speed, torque, power, and efficiency of the motor, are difficult. An efficient planning of the torque has nonetheless significant implications on energy efficiency, though, and should be pursued [12].

15.3.2 Battery Model

The main function of the battery is to provide the required current and voltage to the on-board avionics as well as to the electric motor. Additionally, the battery needs to switch to charging when the ICE is generating excess power to that required at a specified altitude. The discharging characteristics of state-of-the-art batteries are modelled as:

$$V = V_0 - K\frac{Q}{Q - i \cdot t} + Ae^{-B \cdot i \cdot t} \tag{15.13}$$

Where V is the battery output voltage (V), V_0 is the battery constant voltage (V), K the polarity voltage (V), Q represents the maximum battery capacity; A the exponential voltage coefficient, B the exponential capacity coefficient (Ah^{-1}), i the discharge current (measured in A), and t representing time (seconds) [3]. The battery charging characteristics can also be modelled as per Eq. (15.15), with the alteration of adopting negative current to indicate charging instead of discharging. Another important characteristic that needs to be considered when modelling batteries is their state of charge (SOC), which is approximated by

Table 15.7 Battery characteristics as provided by manufacturer [33].

Thunder power Li-Po air battery 8-cell pack 29.6 V	
Mass per pack (kg)	0.93
Dimensions (mm)	$74 \times 43 \times 136$
Maximum capacity (mAh)	5000
Operating tension (V)	29.6
Max continuous current (A)	125
Burst current (A)	250
Maximum charge current (A)	25
Assumed η_{BATT}	0.9

equation:

$$SOC = 100 \left(1 - \frac{Q \int_0^t i \cdot t}{Q} \right) \tag{15.14}$$

This equation represents the exact derivation of the SOC as discussed in the generator section [3]. The battery selection for the case study was driven by electrical compatibility requirements (voltage and current) to power the Plettenberg EM, as well as COTS availability [33]. The battery pack selected as reference for our case study consists in the Thunder Power Li-Po Air Battery 5000 mAh 8-cell pack. Table 15.7 outlines the key characteristics of the selected battery.

The effects of adding more battery packs on maximum range and endurance performances will be examined in the case study. On average, an Li-Po battery pack can last up to 300–500 cycles before finally depleting [33].

15.3.3 Generator Model

For aerospace applications, it is desirable to resort to the capability of electric motors to be functionally reversed and act as generators when necessary. The main function of the generator is to supply charging current to the battery when there is extra torque available in the propulsion system (sufficient values from the ICE). As with the EM model, the generator only functions an output when it is activated. In terms of generator modelling, a basic approach was considered, governed by the following equations:

$$t_{discharge} = \frac{\eta \cdot e_{bat}}{i_{discharge}} \tag{15.15}$$

$$t_{charge} = \frac{(1 - \eta_{gen}) \, e_{bat} + e_{bat}}{Max \, i_{charge}} \tag{15.16}$$

Where in Eq. (15.13), e_{bat} is the energy density of the battery, with η being the efficiency of the battery (given by manufacturing data). In Eq. (15.14), the charge time, measured in seconds, is analogous to the same as the discharge time, however, in replace of efficiency of the battery, the η_{gen}, was added, to account for generator losses operating on average

60% efficiencies. The discharge time, gives an approximate solution, for the endurance of the Battery as stated in Eq. (15.6). From these equations, calculations were carried out attaining a rough estimate in the duration of time required for the generator to reach maximum charge [34].

15.3.4 Aircraft Performance

The estimation of an aircraft's performance and the determination of its stability and control properties crucially rely on the knowledge of its flight dynamics characteristics. Key data including drag polar and power available for levelled flight were therefore determined based on the available dynamics models to estimate the range and endurance performance characteristics. An earlier variant of the Aerosonde RPAS was adopted as a reference platform in our numerical evaluation. For this aircraft, the lift coefficient (c_L) was available as a cubic polynomial fit of the angle of attack as:

$$c_L = c_{L0} + \left(c_{Lmax} - c_{L0}\right) c_{L,\alpha}\, \alpha \left[1 - \frac{\alpha^2}{3\alpha_{max}^2}\right] \tag{15.17}$$

where $c_{L0} = 0.23$ is the lift coefficient at zero angle of attack, $c_{Lmax} = 1.3$ is the lift coefficient at aerodynamic stall, $c_{L,\alpha} = 5.61$ is the first derivative with respect to the angle of attack and $\alpha_{max} = 15°$ is the stall angle of attack. This curve is plotted in Figure 15.1. The drag coefficient (c_D), on the other hand, was available as a full quadratic fit of the angle of attack as:

$$c_D = c_{D\alpha_0} + c_{D,\alpha}\, \alpha + c_{D,\alpha^2}\alpha^2 \tag{15.18}$$

where $c_{D\alpha_0} = 0.0434$ is the drag coefficient at null angle of attack, while $c_{D,\alpha}$ and c_{D,α^2} are respectively the first and second order fit coefficients with respect to the angle of attack.

Figure 15.2 shows the resulting drag polar of the adopted aircraft model. The graph depicts the lift coefficient C_L against the drag coefficient C_D. Although the data plotted in

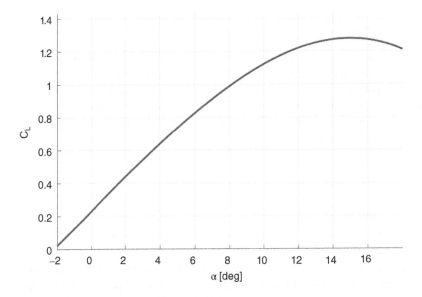

Figure 15.1 Adopted lift coefficient model.

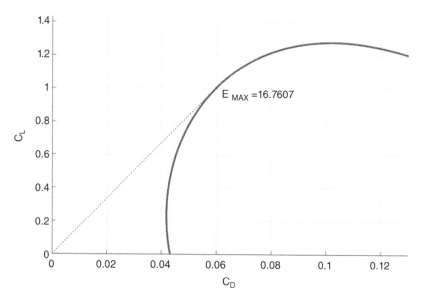

Figure 15.2 Drag polar as retrieved from the available aircraft dynamics model.

Figures 15.1 and 15.2 is based on higher order models and allows us to infer an approximate behaviour beyond the stall, in most calculations it is sufficient to approximate the drag coefficient with a conventional parabolic fit of the lift coefficient:

$$C_D = C_{D0} + \frac{C_L^2}{\pi e AR} \tag{15.19}$$

where e is the Oswald efficiency number, AR is the aircraft's aspect ratio, and C_{D0} as the parasite drag coefficient. From Figure 15.3, it can be inferred that the maximum aerodynamic efficiency E, which corresponds to the maximum ratio of Lift (L) over Drag (D), equates to

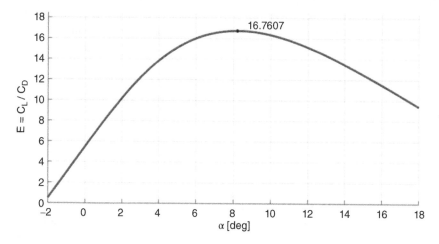

Figure 15.3 Aerodynamic efficiency as a function of the angle of attack.

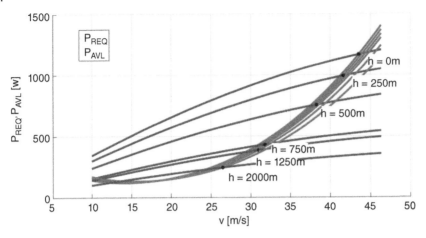

Figure 15.4 Power required and power available as a function of true airspeed.

16.76 and is achieved at approximately 8.2° of angle of attack (α). The relatively gentle slope of the graph ensures good manoeuvrability at a variety of angles of attack.

Assuming the aircraft is travelling in steady level flight conditions at a given altitude and given velocity, the powerplant must produce a net thrust equal to the drag, while the lift must balance the weight. The thrust required (T_R) can therefore be simply written as:

$$T_R = \frac{W}{\frac{C_L}{C_D}} = \frac{W}{\frac{L}{D}}$$

(15.20)

Concerning power required (P_R) the following equation can be used:

$$P_{R,0} = \sqrt{\frac{2W^3 C_D^2}{\rho_0 S C_L^3}}$$

(15.21)

With the subscript of 0 specifying velocity and power components at *sea level*, with in relations to a change in density, ρ, as being the only parameter that will affect results. Conversely, the thrust in a propeller-driven aircraft is a function of the power available at the shaft from either a reciprocating engine or an electric motor. In both classes, the thrust decreases with increasing altitudes due to the diminishing density of the air on which the propeller exerts its aerodynamic forces. Additionally, reciprocating engines experience a decay in shaft horsepower (SHP) due to the decreasing pressure at the air intake. The power required and power available plots at various altitudes are plotted in Figure 15.4.

15.4 Results and Discussion

The Breguet Equation (15.2) described in a previous section provides a widespread approximate methodology to calculate the range for traditional ICE powered aircraft. Equation (15.4) is used to determine the endurance at the calculated range. Table 15.8 details the results, which are separately depicted in Figures 15.5 and 15.6.

Table 15.8 Range and endurance for the ICE case.

m_{fuel} (kg)	Range (nmi)	Endurance (h)
0.35	32.8	1
1.28	114.1	3.3
2.21	188	5.4
3.14	255.7	7.1
4.07	318.2	8.7
5	376.2	10.1

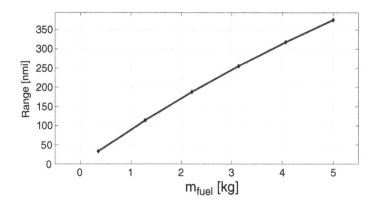

Figure 15.5 Range of the ICE only model.

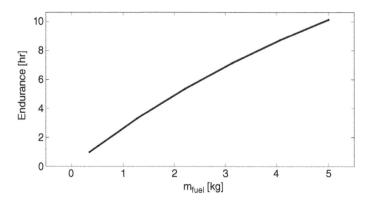

Figure 15.6 Endurance of the ICE only model.

15.4.1 Range and Endurance for Electric Only

In examining the electric-only propulsive configuration, no change in weight is observed since no fuel consumption is considered. Instead of the fuel components, an electric motor and battery packs were considered. Table 15.9 details the results, which are separately

Table 15.9 Range and endurance for the electrical-only case.

Number of battery packs	m_{Batt} (kg)	Range (nmi)	Endurance (h)
1	0.930	29.5	0.6
2	1.86	59	1.2
3	2.79	88.5	1.7
4	3.72	118	2.3
5	4.65	147.5	2.9

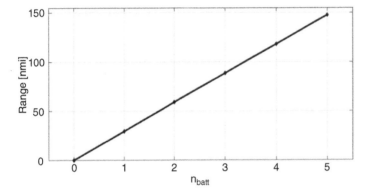

Figure 15.7 Range of the electric only model.

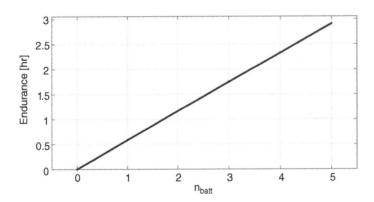

Figure 15.8 Endurance of the electric only model.

depicted in Figures 15.7 and 15.8. Both range and endurance performances were lower on the entirely electrical model in comparison with the ICE-only model, consistent with the significantly lower specific energies.

The hybridisation method was utilised considering range and endurance contributions by both technologies and summing them up. Parameters were examined, altering the number of battery packs, whose mass was subtracted from the mass of fuel. Table 15.10 details

Table 15.10 Range and endurance for the hybrid case.

Number of battery packs	m_{fuel} (kg)	Range (nmi)	Endurance (h)
0	5	376.2	10.1
1	4.07	347.7	9.3
2	3.14	314.7	8.3
3	2.21	276.5	7.1
4	1.28	232.1	5.7
5	0.35	180.3	3.9

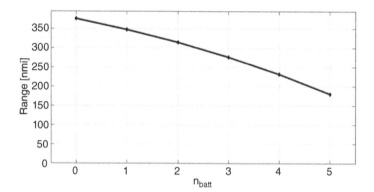

Figure 15.9 Range of the hybrid model.

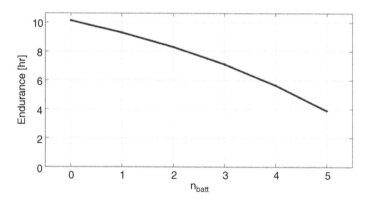

Figure 15.10 Endurance of the hybrid model.

the results, which are separately depicted in Figures 15.9 and 15.10. It was observed that an approximate hybridisation of 50% fuel and 50% electric (roughly equating to three battery packs) achieved a range of approximately 276.5 nmi range and 7.1 hours of endurance. In comparison, the range and endurance for the traditional ICE model still dominated, achieving 376.2 nmi and 10.1 hours of flight.

Table 15.11 Range and endurance for the hybrid case.

Number of battery packs	CO_2 emission (kg)	Breakeven number of missions	Savings at battery end-of-life
0	16.5	0	$0
3	8.45	109	$390
5	0	182	$650

By applying our numerical estimation models, we calculated that in order to achieve unaltered range performance, the battery capacity should be increased to approximately 12 000 mAh at equal weight (corresponding to roughly 2.4 times the specific energy of the selected Li-Po battery). Conversely, in order to achieve unaltered endurance performance, the battery capacity should be increased to approximately 16 000 mAh at equal weight (corresponding to roughly 3.2 times the specific energy of the selected Li-Po battery). These results are promising as they show that it will not be necessary to attain the specific energy of hydrocarbon fuel to match its performance.

Even though range and endurance do not outperform the traditional ICE model, significant cost and fuel savings as well as emission reductions can result from the HEPS retrofit. As an initial estimate, assuming the average chemical formula of AvGas 100 to be C_8H_{15}, the amount of carbon dioxide (CO_2) emitted in the atmosphere by an ICE can be approximated by [35]:

$$CO_2 \ [kg] = 3.16 \ m_{fuel} \ [kg] \tag{15.22}$$

As mentioned before, CO_2 is only one of the various pollutant emissions from ICE. Other emissions associated with hydrocarbon fuels include carbon monoxide (CO), unburned hydro-carbons (HC), sulphur oxides (SO_X), and nitrogen oxides (NO_X). Estimating these emissions is more complex than in the case of CO_2 as they manifest notable dependencies on the throttle setting and altitude, for which some empirical models are available [36]. In terms of costs, the unit price of the selected battery pack is around $200, while the current average price of AvGas per litre can be assumed to be around $1.1 l^{-1}. Selecting cases of ICE-only, a hybrid 3-pack configuration, and an electric only, a comparison was made and the results are shown in Table 15.11. Based on these results, implementing more battery packs yields economic savings and offers a more environmentally friendly solution, emitting less CO_2. Conversely, if range and endurance performance are to be preferred to economic savings, a lower number of battery packs and an appropriate recharge profile should be implemented.

15.5 Conclusions and Future Work

This paper investigated the integration of hybrid-electric propulsion technology in a small RPAS. The overall aim was to develop a design retrofit methodology with potential scale-up to medium and large transport aircraft. A current RPAS model was considered and different parameters of range, endurance, and fuel performances were evaluated in representative conditions for an ICE only, electric only, and hybrid propulsion cases. Consistent with the energy-based estimations, range was higher for the ICE case when compared to the electric and hybrid cases, with the specific energy of the batteries being the main limitation. Hybrid-electric configurations can nonetheless contribute to minimising exhaust emissions and show promising complementarities to ICE-only propulsive configurations. For instance, energy harvesting, research on which is well underway, is a major advance that crucially relies on hybrid-electric technologies for practical deployment. On the other hand, advances in fuel cell technology will potentially meet more- and all-electric aerospace vehicle needs. Future research will address design optimisations, aiming to overcome the challenges in battery energy densities, weight penalties, and power-plant integration issues by focussing on battery charge-depleting profiles, distributed propulsion, and unconventional aircraft configurations. In particular, by considering carefully designed battery charge cycles and allowing for ICE downsizing, optimal battery recharge-depletion profiles could be implemented, yielding further enhancements in range and endurance performance. Finally, RPAS hybridisation offers a valuable stepping stone in pursuing design scale-ups, supporting research in unconventional transport aircraft design configurations and distributed hybrid propulsion systems.

References

1 Seresinhe, R., Sabatini, R., and Lawson, C.P. (2013). Environmental impact assessment, on the operation of conventional and more electric large commercial aircraft. *SAE International Journal of Aerospace* 6: 56–64.

2 Williamson, M. (2014). Air power: the rise of electric aircraft. *Engineering and Technology* 9: 77–79.

3 Hung, J.Y. and Gonzalez, L.F. (2012). On parallel hybrid-electric propulsion system for unmanned aerial vehicles. *Progress in Aerospace Sciences* 51: 1–17.

4 Khandelwal, B., Karakurt, A., Sekaran, P.R. et al. (2013). Hydrogen powered aircraft: the future of air transport. *Progress in Aerospace Science* 60: 45–59.

5 Sehra, A.K. and Whitlow, W. Jr., (2004). Propulsion and power for 21st century aviation. *Progress in Aerospace Science* 40: 199–235.

6 Qin, N., Vavalle, A., Le Moigne, A. et al. (2004). Aerodynamic considerations of blended wing body aircraft. *Progress in Aerospace Science* 40: 321–343.

7 Gohardani, A.S. (2013). A synergistic glance at the prospects of distributed propulsion technology and the electric aircraft concept for future unmanned air vehicles and commercial/military aviation. *Progress in Aerospace Science* 57: 25–70.

8 Ausserer, J. K. (2012). Integration, testing, and validation of a small hybrid-electric remotely-piloted aircraft, MSc, Dept. of Aeronautical and Astronautical Engineering, Air Force Institute of Technology, Wright-Patterson Air Force Base, OI, USA.

9 Anderson, J.D. (2005). *Introduction to Flight*. Boston, MA, USA: McGraw-Hill.

10 Payne, J.P. (2002). *Flight Test Handbook*, 4e. Colorado Springs, CO/CA, USA: United States Air Force (USAF) Academy/Edwards AFB.

11 Whitfield, B. (2011) EADS/Diamond Unveil Electric Hybrid, *Flying Magazine*, vol. June 2011.

12 Gao, X.Z., Hou, Z.X., Guo, Z., and Chen, X.Q. (2015). Reviews of methods to extract and store energy for solar-powered aircraft. *Renewable and Sustainable Energy Reviews* 44: 96–108.

13 Liebeck, R. H. (2003). Blended Wing Body design challenges, *AIAA\ICAS International Air and Space Symposium and Exposition: The Next 100 Years,* Dayton, OH.

14 Liebeck, R.H. (2004). Design of the blended wing body subsonic transport. *Journal of Aircraft* 41: 10–25.

15 Rolls-Royce and Airbus Group. (2013). E-Thrust - Electrical distributed propulsion system concept for lower fuel consumption, fewer emissions and less noise.

16 R. H. Liebeck, M. A. Page, and B. K. Rawdon, (1998). "Blended-Wing-Body subsonic commercial transport," *36th AIAA Aerospace Sciences Meeting and Exhibit, 1998.*

17 Kalsi, S.S., Weeber, K., Takesue, H. et al. (2004). Development status of rotating machines employing superconducting field windings. *Proceedings of the IEEE* 92: 1688–1703.

18 Mahlia, T.M.I., Saktisahdan, T.J., Jannifar, A. et al. (2014). A review of available methods and development on energy storage; technology update. *Renewable and Sustainable Energy Reviews* 33: 532–545.

19 M. Trancossi, A. Dumas, P. Stewart, and D. Vucinic, (2014). "Increasing aeronautic electric propulsion performances by cogeneration and heat recovery," *SAE 2014 Aerospace Systems and Technology Conference, ASTC 2014,* vol. 2014-September.

20 P. C. Vratny, P. Forsbach, A. Seitz, and M. Hornung, (2014). "Investigation of universally electric propulsion systems for transport aircraft," 29th Congress of the International Council of the Aeronautical Sciences, ICAS 2014.

21 Y. Mikhaylik, I. Kovalev, R. Schock, K. Kumaresan, J. Xu, and J. Affinito, (2014). "High Energy Rechargeable Li-S Cells for EV Application. Status, Challenges and Solutions," ed: SIONPOWER Lithium Sulfur rechargeable battery.

22 Jeon, B.H., Yeon, J.H., Kim, K.M., and Chung, I.J. (2002). Preparation and electrochemical properties of lithium-sulfur polymer batteries. *Journal of Power Sources* 109: 89–97.

23 Girishkumar, G., McCloskey, B., Luntz, A. et al. (2010). Lithium– air battery: promise and challenges. *The Journal of Physical Chemistry Letters* 1: 2193–2203.

24 Rahman, M.A., Wang, X., and Wen, C. (2014). A review of high energy density lithium–air battery technology. *Journal of Applied Electrochemistry* 44: 5–22.

25 Sharaf, O.Z. and Orhan, M.F. (2014). An overview of fuel cell technology: fundamentals and applications. *Renewable and Sustainable Energy Reviews* 32: 810–853.

26 DoE, U. (2015). *Fuel Cell Technologies Market Report 2014.* US Department of Energy.

27 J. L. Felder, H. D. Kim, and G. V. Brown, (2009). "Turboelectric distributed propulsion engine cycle analysis for hybrid-wing-body aircraft," *47th AIAA Aerospace Sciences Meeting including the New Horizons Forum and Aerospace Exposition*, Orlando, FL.

28 Erturk, A., Renno, J.M., and Inman, D.J. (2008). Modeling of piezoelectric energy harvesting from an L-shaped beam-mass structure with an application to UAVs. *Journal of Intelligent Material Systems and Structures* 20 (5): 529–544.

29 Barnes, J. P. (2015). Aircraft Energy Gain From an Atmosphere in Motion: Dynamic Soaring and Regen-electric Flight Compared.

30 Fisher, A., Marino, M., Clothier, R. et al. (2015). Emulating avian orographic soaring with a small autonomous glider. *Bioinspiration & Biomimetics* 11: 016002.

31 Enya. (2017). Technical data sheet: R120-4C 4 stroke cycle engine, ed: Enya.

32 Elektromotren Plettenberg. (2015). Technical sheet: Plettenberg HP 320-30 motor, ed: Elektromotren Plettenberg.

33 Hobby King. (2015). Turnigy nano-tech 5000 mAh 6S 45~90C Lipo Pack.

34 MPower UK. (2015). Battery chargers and charging methods, ed: MPower UK.

35 Hupe, J., Ferrier, B., Thrasher, T. et al. (2013). *ICAO Environmental Report 2013: Destination Green – Aviation and Climate Change*. Montreal, Canada: ICAO Environmental Branch.

36 Gardi, A., Sabatini, R., and Ramasamy, S. (2016). Multi-objective optimisation of aircraft flight trajectories in the ATM and avionics context. *Progress in Aerospace Science* 83: 1–36.

37 Sliwinski, J. (2015). Hybrid Propulsion Technologies for Small Unmanned Aerial Vehicles, BEng Final Report, RMIT University, School of Aerospace, Mechanical and Manufacturing Engineering, Melbourne, VIC 3000, Australia.

38 Sliwinski, J., Gardi, A., Marino, M., and Sabatini, R. (2017). Hybrid-electric propulsion integration in unmanned aircraft. *Energy* 140: 1407–1416.

16

Benefits and Challenges of Liquid Hydrogen Fuels for Commercial Transport Aircraft

Stephen Rondinelli[1], Alessandro Gardi[1,2], and Roberto Sabatini[1,2]

[1] *School of Engineering, RMIT University, Melbourne, Victoria, Australia*
[2] *Department of Aerospace Engineering, Khalifa University of Science and Technology, Abu Dhabi, UAE*

16.1 Introduction

The highly dynamic evolution in air transport demand is prompting the aviation industry to pursue ever-improving economic, environmental, and social performances. A major challenge is to establish and develop the long-term future of the sector beyond 2050. This will likely include systematic changes to the design, manufacture, and operation of aircraft, particularly in terms of innovative air vehicle designs, propulsion, and fuel technologies, air traffic management (ATM) strategies, and evolutions in certification standards. The average annual growth rate of passenger and cargo traffic over the next two decades is estimated at 4.1%, and this is a major driver of change in aviation. This steady increase is mainly due to the estimated 3.2% annual growth in worldwide gross domestic product (GDP) over the next 20 years [1]. Furthermore, after 2042 it is expected that coal will be the only fossil fuel available [2], eliciting research into alternative fuels and towards a more sustainable development. One promising alternative relies on the use of hydrogen (H_2) as the main energy source for commercial transport aircraft due to its negligible environmental impacts. Combustion processes utilising H_2 only produce water (H_2O) and reduced amounts of nitrogen oxides (NO_X) as emissions [3]. However, great challenges and safety implications for the design and operation of aircraft and airports are associated with the introduction of hydrogen fuels. This chapter, initially published in [39], reviews the key concepts that have been explored throughout the past decades and assesses the viability and potential performance of cryogenic fuels in commercial transport aircraft.

16.1.1 Historical Overview

Hydrogen (H_2) was commonly featured in the eighteenth century in the design of gas balloons, such as the Charlière Hydrogen Balloon in 1783 [4]. In the nineteenth century Ferdinand von Zeppelin utilised hydrogen for buoyancy in his rigid frame airships in conjunction with gasoline propulsion systems. The twentieth century saw a substantial exploitation

Sustainable Aviation Technology and Operations: Research and Innovation Perspectives, First Edition.
Edited by Roberto Sabatini and Alessandro Gardi.
© 2024 John Wiley & Sons Ltd. Published 2024 by John Wiley & Sons Ltd.
Companion Website: www.wiley.com/go/sustainableaviation

of H_2 propellants in space propulsion systems [5]. Russia experimented with H_2 fuel for aviation in a customised TU-155 aircraft, running one engine on H_2 [3]. Russia successively partnered with Germany in 1991 in a joint programme to develop a 200-passenger aircraft with a predicted range of 500 nautical miles [6, 7]. Both the Airbus 310 and the TU-204 airframes were evaluated as a reference platform. This cooperation led to a design placing the H_2 tanks on top of the aircraft fuselage and wings. Meanwhile, the United States (US) National Aeronautics and Space Administration (NASA) was also developing its own design, featuring twin spherical tanks [3]. This configuration limited the surface-to-volume ratio and allowed for 400 passengers travelling at Mach 0.85 for 5500 nautical miles. The twenty-first century saw 35 aviation industry partners come together under the guidance of Airbus Deutschland to undertake a project known as 'Cryoplane'. The project was funded by the European Commission and was aimed at initiating progress towards H_2-fuelled aircraft. Over 25 months of study followed, fostering political and industrial support for introducing H_2 in aviation [3]. The project assessed the fundamental metrics associated with the introduction of H_2. Safety standards of H_2 were also evaluated and compared to those of conventional jet fuels, highlighting the additional needs for H_2 handling. However, the overall safety levels that could be achieved with H_2 were determined to be well on par with those of conventional jet fuels [8].

16.1.2 Production of Hydrogen

The energy usage and the pollutant emissions associated with the production must be considered when evaluating the environmental impact of any alternative energy source. Similarly, all the costs associated with start-up, production, maintenance, and operations must be accurately estimated due to their major role in the success of any fuel technology [9]. Many economically and environmentally sustainable H_2 production strategies have been proposed. The most probable and realistic sources of H_2 include fossil fuels, such as gas and coal, and renewable sources, such as water, biomass, wind, solar, or hydropower. Depending on these sources, various technologies and processes are proposed for H_2 production. These include photolytic, biological, electrolytic, thermo-chemical, and chemical processes [10].

16.2 Aircraft Design

Hydrogen fuels prompt challenges for designers in terms of mass and volume requirements, as well as for fuel management and storage on board aircraft. As shown in Figure 16.1, the high volume-to-energy characteristics of liquid hydrogen (LH_2) require its carrier to load a larger volume of fuel than conventional jet fuel aircraft. The design of a successful hydrogen aircraft is therefore mainly centred on optimising the tank configuration to carry the required amounts of LH_2. Tank configurations can be distinguished as either non-integral or integral. Non-integral configurations are external to the fuselage of the aircraft and involve fuel tanks mounted either on the airframe, above or under the wing. Non-integral tanks must be able to cope with the aerodynamic and inertial loads, in addition to the fuel containment loads [9].

On the other hand, integral tanks are located inside the fuselage; hence their shape and dimensions are interdependent with the fuselage design. Integral tanks are not required

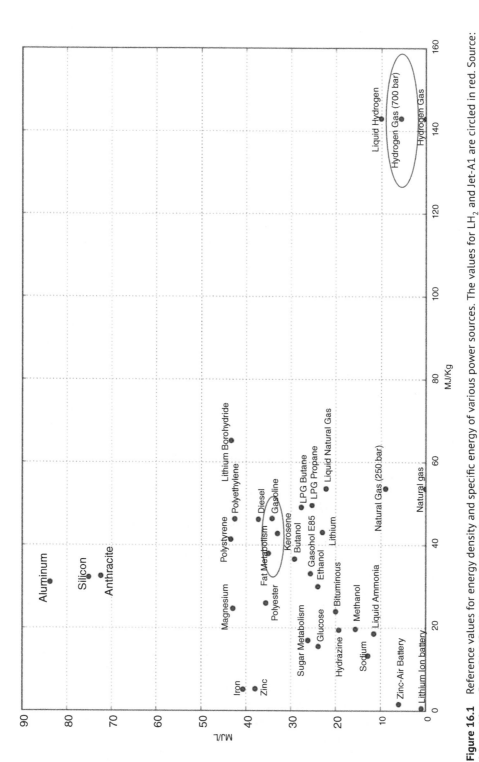

Figure 16.1 Reference values for energy density and specific energy of various power sources. The values for LH₂ and Jet-A1 are circled in red. Source: Adapted from Scott Dial [Public domain], via Wikimedia Commons.

to withstand aerodynamic loads, and conversely may contribute to the structural integrity of the fuselage by improving the resistance to bending and shear forces. Due to the larger volume of cabin available and to reasons further discussed below, integral tanks represent a more promising configuration for wide-body aircraft [9]. The Cryoplane project leaned towards an integral design for LH_2-fuelled aircraft, mainly due to the cryogenic temperatures required for LH_2 containment [11] and to the capacity required for long-haul flights [9]. The length and width of the fuselage both increase to accommodate integral LH_2 tanks [11]. The elimination of wing tanks detracts the associated shear stress and bending moment alleviations. In order to compensate for this, an approximate increase of 37% in the wing structure is required, leading to an overall weight increase of 6% [11] to support and affix the integral LH_2 tanks. However, this can enhance the overall safety, as the tanks are further protected by the supportive and rigid structure of the fuselage [4]. The increased drag and the boil-off issues should also be considered as they impact range and operating costs. For all these reasons, a significant part of the Cryoplane project addressed the various possible tank designs. Streamlined tanks along the fuselage and the wing were also considered. These configurations improved the overall volume exploitation and attainable tank capacities, but the LH_2 containment loads significantly impacted the structural weights. Therefore, spherical and cylindrical designs were preferred [12]. The spherical tank design minimises surface-to-volume ratio and hence reduces the passive heat transfer across the tank wall, minimising the boil-off rate. For these reasons, spherical or quasi-spherical tanks have been frequently adopted in space launchers and vehicle. However, these shapes require a larger frontal area for the same volume in comparison to a cylindrical tank design [13]. A cylindrical tank also provides greater volumetric efficiencies by maximising space usage within the fuselage [4]. However, fuel pressure loads are significantly inhomogeneous in a purely cylindrical tank. The ideal compromise is therefore a cylindrical tank with its bases shaped into a semi-spherical design, minimising the drawbacks of both cylindrical and spherical shapes while combining their advantages [9]. For an LH_2 regional airliner, several designs are possible. One layout incorporates a single tank at the rear of the fuselage, which offers the greatest benefits in terms of weight metrics. However, this design might lead to frequent weight and balance issues, which may in turn require increments to the weight and dimensions of tail planes. In the second considered layout, LH_2 tanks may be positioned in both the aft and front of the fuselage. However, this poses problems in terms of crew access to and from the flight deck, which may be resolved by designing a passageway. Lastly, the LH_2 tanks may be configured along the top of the fuselage above passengers in conjunction with a tank in the aft of the fuselage, impacting upon luggage storage [14]. Long-range aircraft, on the other hand, traditionally feature a wide-body design with multiple aisles in the passenger cabin. In conjunction with the ongoing shift towards larger long-range aircraft, some researchers investigated the possibility of exploiting the increased fuselage cross section, such as in the case of a tri-storey aircraft with LH_2 tanks located in the aft and front of the fuselage [4, 14]. For this design, the forward tank contains approximately 40% of the total fuel in order to satisfy weight and balance requirements. Furthermore, compared to a conventional fuel aircraft, such as the Airbus A380 and Boeing 747, the size metrics vary considerably. For instance in the case of a tri-storey LH_2 aircraft, the fuselage diameter should be extended up to 8.5 m in comparison to the smaller 7.14 and 6.1 m airframe widths of the A380 and B747-8 respectively [14].

16.2.1 Tank Structure and Materials

A major factor for achieving the certification of LH_2 tank designs is the insulation, which upholds the safety standards of kerosene-based, conventional aircraft. In this domain, major developments in the adopted materials are aimed at alleviating the boil-off of hydrogen. This occurs when cryogenic conditions are compromised due to the inward heat transfer [9]. An effective insulation in LH_2 tanks will reduce the boil-off rate of LH_2, increasing operational efficiencies and improving safety. Three types of insulation have been recommended [9], and in particular:

- Multilayer insulation (MLI): this layout consists of up to 100 layers of insulating material such as polyester of glass fibre alternated with metal layers for fuel containment and radiative shielding, arranged perpendicularly to the heat transfer direction. The outside of the inner layer consists of a reflective foil to minimise radiative transfer. The effectiveness of this type of insulation is dependent on factors including the composition and pressure of the fuel gas phase. MLI does not operate effectively in pressure greater than 0.001 mbar [12]. It is also quite susceptible to manufacturing faults during production and is a heavy form of insulation [9].
- Vacuum insulation: this layout involves a pumping system to maintain the vacuum within the tank walls. Such a system must ensure that air does not come into contact with the vacuum walls as its freezing would cause a seizure of the walls [15]. The vacuum insulation is also vulnerable to the external ambient pressure, as the walls may not withstand pressure spikes and fail. Therefore, further structural strength must be introduced through stiffeners, which increase the mass of the tank and the inward conductive heat transfer [16]. The vacuum insulation offers a promising alternative to the multilayer design, having the highest potential in terms of minimising mass lost during boil off. However, it involves heavier structures and costs [17].
- Foam insulation: this layout involves an insulating foam introduced between an inner and an outer tank walls. The outer wall can consist of a thin metal sheet, which protects the foam and assists the structure in maintaining structural integrity. The insulating foam has certain favourable characteristics like low thermal conductivity, whilst maintaining a low density [18]. Foam insulation provides acceptable boil-off rates, tank weight, and size characteristics. Foam insulation also represents a much cheaper option to that of multi-layer or vacuum insulation. In comparison to a vacuum system, the rate of failure of a foam system is also lower [9].

16.2.2 Blended Wing Body

Given the predicted timeframes for the widespread adoption of LH_2 fuels and innovative propulsion systems in aviation, NASA and other governmental, academic, and industrial R&D entities have also extended the study to encompass more innovative aircraft configurations. A particularly notable configuration is the blended wing body (BWB), offering higher aerodynamic and payload efficiencies, greater airframe volumes, higher flexibility in propulsive configurations, and reduced noise footprints. BWB is particularly expected to enhance the technological feasibility of hydrogen propulsion systems, thanks to improved volumetric efficiencies and operational capabilities in terms of passenger and freight movements, while yielding considerable environmental benefits in comparison to conventional aircraft [19]. NASA's model for a clean commercial aircraft of the future is based upon the 'Quiet Green Transport' concept, implementing a hydrogen fuel cell-powered BWB that eliminates hydrocarbon and carbon/sulphur oxides (CO_X/SO_X) emissions. The hydrogen used is contained in insulated integral tanks located inside the airframe. With hydrogen fuel cells, NO_X emissions are also entirely eliminated, along with a significant reduction in noise emissions. The BWB concept features top mounted ducted fans, improving noise shielding as well as aerodynamic efficiency. The fuel-cell based electric propulsion typically involves a higher number of smaller engines to generate the desired amount of thrust, leading to higher frequency noise with smaller amplitudes [19]. The airframe also shows advances in terms of noise mitigation through the opti-misation of gaps and edges in aerodynamic control surfaces. This is achieved through continuous mould-line technology [19]. NASA also explored the potential for operational improvements related to this aircraft. An important contribution for the reduction of noise footprint is to increase the final approach slope angle by 9° (from 3 to 12°). This increases the altitude of arrival traffic on final approach. NASA has further suggested a reduction in cruise altitude for its BWB aircraft, which is more environmentally sustainable as conventional cruise altitudes in the upper troposphere provide ideal conditions for contrail formation from the H_2O exhausts [19].

16.2.3 Systems Impacts

In order to accommodate LH_2, the present propulsion technologies will need to be partially redesigned. This will particularly affect sub-systems, including the fuel lines and combustion chamber. An LH_2-fuelled auxiliary power unit (APU) will also be proposed. This would eliminate CO_2 emissions on the ground when external power sources cannot be accessed. During engine start up, ambient air contamination within the fuel lines poses the risk of flashback, which may be prevented by flushing with an inert gas, such as nitrogen. Similarly, fuel lines should also be flushed upon shut down [9, 20]. From the thermodynamic perspective, it is beneficial to pre-heat LH_2 before it enters the combustion chamber. This can be achieved in a heat exchanger, which can capture heat from the hot parts of the engine (i.e. turbine, exhaust, and combustion chamber), thereby also improving the thermal efficiency and longevity of the engine. An electrical heater may be used to heat the fuel when the engines are still cold. Furthermore, a tailored metering system will also be required to provide LH_2 to the engine in line with the throttle set by the flight crew [9, 20]. The combustion of hydrogen in aircraft engines raises complications beyond that of simple fuel to air mixing [21]. The use of LH_2 in commercial transport aircraft requires redesigning the conventional combustors in order to attain optimal efficiency [20]. Use of hydrogen in conventional kerosene combustors would lead to excessive NO_X emissions, due to undesirable increase in temperature during combustion process [22]. Studies to reduce such effects on board LH_2 aircraft have been undertaken, with an emphasis on improving combustion efficiency, noise, and flame stability. Current efforts have identified lean direct injection (LDI) and micro-mix concepts as potential combustors. These two concepts are similar in methodology and have both proven to be viable. Both aim to reduce the presence of large flames in order to minimise NO_X emissions, whilst reducing flashback. This is achieved through altering and increasing the mix intensity since NO_X is dependent on residence time and temperature [9].

- *LDI:*
  Marek et al. [23] conducted several experiments aimed at evaluating NO_X emissions and combustion performance. The LDI system used featured quick mixing and multiple injection points, as well as high velocities and reduced mixing times in order to combat flashback. Results from these experiments demonstrated the capability of hydrogen to attain same NO_X levels of modern advanced kerosene LDI combustors [23].
- *Micro-mix combustion:*
  If managed correctly, the *micro-mix* or *miniaturised* diffusive combustion process of hydrogen can produce fewer NO_X emissions than that of conventional kerosene combustion [24]. In this layout, the number of local mixing zones between fuel and air are increased in comparison to conventional kerosene burner designs, thereby improving the mixing intensity while reducing its scale. Therefore, the micro-mix combustion process involves thousands of miniature diffusion flames, thereby reducing the likelihood of flashback [20]. Dahl and Suttrop [20] examined the effects of micro-mix combustion on a modified KHD T215 gas turbine engine on an Airbus 320. Their study highlighted the possibility of precisely metering hydrogen, while maintaining engine control during conditions similar to that of a kerosene fuelled aircraft engine. Their configuration also demonstrated hydrogen's ability under a micro-mix system to produce less nitrogen oxides than kerosene combustion, yet still adhering to a diffusive burning process that demonstrates a reduced risk of flashback and engine failure. Safe engine start-up and ignition procedures were also demonstrated, with reduced risk of excess pressure and heat transfer. Furthermore, this technology has a potential for APU as well [20].

16.3 Hydrogen Aircraft Operations

LH_2-fuelled aircraft pose exciting prospects for the aviation industry by not only eliminating CO_2 emissions but also in terms of potential reductions in operational costs for airlines [3]. However, in order to accurately evaluate the potential benefits of LH_2 aircraft, an in-depth analysis of all factors is required, going beyond the scope of current knowledge. This includes, but is not limited to, the traditional aircraft performance metrics, such as payload and range capabilities. All direct and indirect operating costs associated with such aircraft have to be considered, including logistics and maintenance implications. The past decade has seen new historical records in aviation fuel prices [1]. In particular, data has shown that fuel costs can contribute to over 30% of operating expenses for airlines, becoming the greatest direct operating cost [25]. The most effective strategy to mitigate this financial exposure is by reducing dependence on conventional fossil fuels. Historical evidence suggests that even without subsidies, alternative fuels will outcompete the fuel they are directly contending with [26]. The adoption of fuels such as LH_2 in aviation may hold the key to reducing fuel-related operating costs [11]. Benefits of the adoption of LH_2 based fuels are also associated with its excellent energy specific fuel consumption (ESFC) in comparison to conventional aircraft fuels. The high ESFC of LH_2 allows for lighter engines, potentially leading to indirect savings of up to 3% in energy consumption. Similar results have been highlighted by Verstraete [11] and in several other papers, corroborating the energy efficiency of LH_2 engines. The adoption of LH_2 fuel may lead to a 30% reduction in gross weight due to the lower mass of LH_2 in comparison to kerosene. On the other hand, although the operating empty weight (OEW) of both aircraft would be similar, a long-range LH_2 fuelled aircraft is likely to be about 7 m longer. Coupled with a double deck fuselage and smaller wing size, a comparable LH_2 aircraft will see a reduction of approximately 15% in its average cruise lift-to-drag ratio. However, this increase in drag will be counteracted by an 11% improvement in terms of energy usage [11], thereby reducing direct operating costs. Savings in direct operating costs are expected to be diluted in the early stages due to the predicted rise in aircraft purchase price, maintenance, and servicing in comparison to conventional aircraft [11]. Airfreight has a growing importance in the profitability of airline routes and in order to benefit from this, there must be a sufficient residual cargo capacity in addition to passenger luggage requirements. In comparison with conventional kerosene fuelled aircraft, LH_2 aircraft will have to resort to reducing the volumetric capacity for payload. This economic disadvantage may be outweighed by extended range capabilities. The weight and energy advantages of LH_2 aircraft allow them to fly at greater distances compared to conventional kerosene aircraft, as represented in Figure 16.2 [11].

16.4   Airport Design and Operations

Airports will necessarily need to evolve to host regular operations of H_2 powered aircraft, ensuring the required maintenance and support services are available. The integration of LH_2 fuel systems may involve a dedicated and integrated logistics and supply chain that can meet the LH_2 demands of aircraft operators. This can be achieved by having an onsite hydrogen production facility, or the adoption of infrastructure that can securely and safely house the reserves of LH_2 locally. For LH_2 fuel systems to be economically viable, there needs to be a reasonable demand by local aircraft operators and, conversely, multiple airports will have to be equipped with an LH_2 supply in order for this fuel to be commercially viable for regular airline operations. During the implementation stage of hydrogen fuel systems at airports, these airports will have to seamlessly accommodate a hydrogen-fuel infrastructure in conjunction with conventional kerosene delivery systems. For these reasons, it is expected that the larger airports will be the first to adopt such infrastructures, as the first LH_2 aircraft will likely be of the long-range transport category [27]. Modern international route structures are mainly based on the hub and spoke model, where feeder flights are flown into a central location or hub and passengers can benefit from a significant number of flight connections. This network structure is particularly well suited for long range LH_2 aircraft, allowing hubs to be major supply and maintenance centres for LH_2 aircraft. Future network planning will need to consider the regions that are supported by stronger hydrogen production capacities. Within these regions, the largest airports are likely to be the first integrators of LH_2 technologies. Based on departure movements and location, certain cities stand out as plausible LH_2 adopters. These include Chicago, Los Angeles, and Ontario in the United States, Tokyo and Osaka in Japan, and Amsterdam in Europe [28]. The optimal fuel delivery systems for LH_2 aircraft of the future would likely involve onsite production.

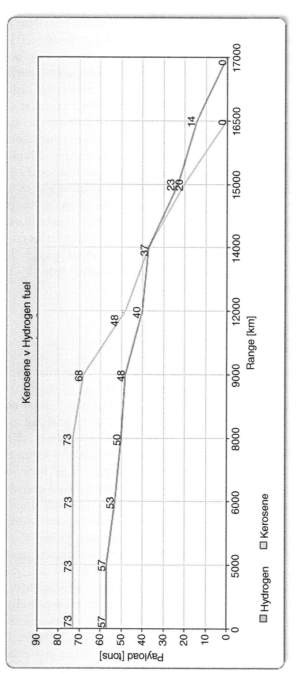

Figure 16.2 Payload vs. range curve of LH$_2$ and kerosene-fuelled aircraft. Source: Data from [11].

Careful consideration should be taken when locating the storage tanks [27, 29]. The piping will need adequate insulation in order to transport the liquid hydrogen at −253° and may consist of three pipes, satisfying the requirements to transfer the LH_2, collect the boiled off H_2, and allow for redundancy [27, 30]. Airports themselves also contribute approximately 30 million tons or 5% of the total air pollution within the aviation industry [31]. Contributing factors include aircraft, passengers, freight, and airside/landside vehicle movements [27]. The widespread adoption of hydrogen fuels for ground vehicles too will provide great environmental benefits to airports, not just restricted to the elimination of carbon emissions. Currently, fuel delivery systems for kerosene aircraft are usually large and in some instances can be quite complex, typically involving a tank area or *fuel farm* within reasonable distance from the apron. These tanks usually provide a fuel supply for one to three days and their fuel is supplied to airport users via trucks or a system of underground pipes. Typically, large airports utilise underground piping to ease congestion, however smaller or regional airports may utilise fuel tankers for simplicity [27, 30]. A fuel distribution system based on fossil fuel-powered trucks results in emissions throughout the whole process. Any proposition for the installation of distribution lines should include entrenched yet open plans, which allow for the adequate vent of potentially dangerous hydrogen gas [29]. Further improvements or reductions in airport-related aircraft emissions may come about through the adoption of LH_2 powered APU, which could also contribute to aircraft weight reduction by eliminating the need for generators within the engine assembly [28].

16.5 Safety

Aircraft using hydrogen historically gained a reputation for being a dangerous endeavour. This was largely brought about by the Hindenburg disaster, but the flammable cloth of the containment bag back then is vastly different to the highly insulated and structurally sound ergonomic tanks proposed for modern LH_2 applications [32]. Most recent in-depth studies highlight hydrogen as a safer alternative to conventional kerosene fuels [9]. In the event of an aircraft crash, liquid hydrogen is more likely to result in a safer outcome than kerosene. Due to the rigidity of LH_2 tanks, they are less likely to rupture. Also, because of buoyancy, hydrogen dissipates quickly, and its fires have lower temperatures and intensity [32]. Unlike kerosene, hydrogen cannot contaminate water or soil. Hydrogen in its liquid form is much safer than in its gaseous state due to the lower pressure in storage tanks, which reduces the likelihood of fatigue-induced structural failures [29]. However, hydrogen's ability as a gas to seep through containment lines or tanks, unlike air or other gases, creates challenges in identifying leaks. Hydrogen can even engrain itself in such solid materials as polymers through permeation, demanding careful consideration when selecting hydrogen containment materials [29, 33]. Like most of the fuels, hydrogen poses a flammability hazard. In its gaseous state, hydrogen has a greater potential to mix with air or kerosene fumes and form a dangerous detonating mixture. However, the temperature of a hydrogen flame is about a tenth of that of a hydrocarbon fuelled one. This not only reduces the extent of possible damage caused during a major accident, but also allows rescue authorities to get closer to the heat source. In the event of a leak, hydrogen in reasonable quantities can cause asphyxiation. Although hydrogen is not corrosive or poisonous, its cryogenic temperatures could injure a person upon contact [29]. A local liquefier at airports requires careful design considerations. Components such as pumps, connections, and accessories for LH_2 require specific arrangements due to the cryogenic conditions they experience [34, 35]. Personnel in contact with such systems require specialist training, as contact with any cryogenically cooled metals can result in injury. Airports must implement technologies, procedures, and policies for safe and economical handling of LH_2. Risk mitigation should be in the form of suitable location of ground LH_2 resources, by enforcing certain minimum safety distances. Likewise on aircraft, tank shape is an important aspect in terms of upholding safety. Though there is not much difference in terms of operation of either a cylindrical or spherical tank, there is a greater associated risk of encountering manufacturing faults of spherical tanks due to its manufacturing complexity. Cylindrical tanks also offer more efficient use of capital resources through vertical installation. As with any LH_2 structure, the fuel tanks require protection from external elements. This may come about in the form of partial submersion of LH_2 tanks underground. Further efforts should be made to intensify security measures, implementing strict scrutiny, and restriction in the access to LH_2 reserves [29].

16.6 Environmental Gains Estimation

In line with all the factors described above, we introduce a quantitative analysis of the potential environmental and economic benefits associated with the adoption of liquid hydrogen (LH_2) fuels for aviation. As a practical reference, we assume conventional aircraft currently adopted in both regional, medium-range, and long-range commercial airline flights, evaluating their hypothetical retrofit for conversion to LH_2 fuel. For this analysis, we consider the Breguet range equation in its traditional form:

$$R = (\eta^L/_D) \frac{h_C}{g} \ln \frac{m_i}{m_f} \tag{16.1}$$

where:

R = range
η = overall propulsive efficiency
$^L/_D$ = nominal lift-to-drag ratio
h_C = lower combustion heat
m_i = initial mass
m_f = final mass
g = gravity.

We also define $K = (\eta^L/_D)$. The comparison is based on the following assumptions:

o Aircraft are listed by their International Civil Aviation Organisation (ICAO) code, which are:
o E190: Embraer E-190.
o A320: Airbus A320-200.
o B738: Boeing 737-800.
o A333: Airbus A330-300.
o B788: Boeing 787-800.

- o B77W: Boeing 777-300ER.
- o A388: Airbus A380-800.
- o The assumed characteristics for each aircraft are meant to represent an average of the advertised or published ones.
- o Aircraft are configured for maximum range, therefore commencing the flight at maximum take-off weight ($MTOW$) and loaded with maximum fuel ($m_{MaxFuel}$) and a partial payload (m_{PL}).
- o K_{Jet-A1} is derived from the advertised aircraft performance by means of the rearranged Breguet range equation.
- o K_{LH_2} is calculated as 85% of K_{Jet-A1}, to represent the increased drag associated with the additional LH_2 tank volumes, in line with the findings documented in section 2.
- o The OEW of the hydrogen aircraft is increased by 6% to represent the additional structural mass required for the LH_2 tank, in line with the findings documented in section 2.
- o The chemical composition of Jet-A1 is approximated as 99.7% in mass of $C_{11}H_{22}$, with a sulphur content of 0.15% in mass, corresponding to half of the maximum regulatory threshold (0.3% in mass). The residual mass fraction of fuel includes all other species.
- o The density of Jet-A1 is 840 kg m^{-3}, whereas the density of cryogenic LH_2 is 70.8 kg m^{-3}.
- o The specific energy of Jet-A1 is 42.8 MJ kg^{-1}, whereas the specific energy of LH_2 is 142 MJ kg^{-1}.
- o The chemical composition of Jet-A1 emissions is calculated by assuming that 1% of the carbon content is processed into CO and 0.5% of it becomes unburnt hydrocarbons (HC).
- o In order to economically capture their adverse effects on the environment, emission charges are hypothetically set to: 20\$t^{-1} for CO_2; 200\$ t^{-1} for CO and SO$_X$; 2000\$ t^{-1} for HC; 10\$ t^{-1} for H_2O. The carbon dioxide charge is very closely related to the average value from a number of nations presently adopting carbon taxation schemes. The remaining figures are meant to represent an educated guess correlated to the noxious potential of the various substances to the environment and living beings.

The assumptions above can be summarised by the following equations:

$$\rho_{C_{11}H_{22}} = 840\frac{kg}{m^3}; \quad \rho_{LH_2} = 70.8\frac{kg}{m^3}; \quad e_{C_{11}H_{22}} = 42.8\frac{MJ}{kg}; \quad e_{LH_2} = 142\frac{MJ}{kg} \tag{16.2}$$

$$K_{LH_2} = 85\% \cdot K_{Jet-A1} \tag{16.3}$$

$$OEW_{LH_2} = 106\%OEW_{C_{11}H_{22}} \tag{16.4}$$

$$m_{Payload} = MTOW - m_{MaxFuel} \tag{16.5}$$

m_{LH_2} calculated as per Breguet imposing the same range and using the modified K (16.6)

$$m_{C_{11}H_{22}} = 99.7\%m_{Fuel}; \quad m_S = 0.15\%m_{Fuel} \tag{16.7}$$

$$98.5\%m_{C_{11}H_{22}} : \quad C_{11}H_{22} + 33\,O_2 \rightarrow 11\,CO_2 + 11\,H_2O \tag{16.8}$$

$$1.0\%m_{C_{11}H_{22}} : \quad C_{11}H_{22} + 22\,O_2 \rightarrow 11\,CO + 11\,H_2O \tag{16.9}$$

$$0.5\% m_{C_{11}H_{22}} : \quad C_{11}H_{22} \qquad\qquad \rightarrow C_{11}H_{22} \qquad\qquad (16.10)$$

$$Charge_{CO_2} \, [\$] = 20 \, m_{CO_2} \, [t]; \quad Charge_{CO/SO_X} \, [\$] = 200 \, m_{CO/SO_X} \, [t] \qquad (16.11)$$

$$Charge_{HC} \, [\$] = 2000 \, m_{HC} \, [t]; \quad Charge_{H_2O} \, [\$] = 10 \, m_{H_2O} \, [t] \qquad (16.12)$$

16.7 Results and Discussion

Table 16.1 summarises the assumed aircraft characteristics, the estimated Jet-A1 gaseous emissions, and the corresponding hypothetical charges. Table 16.2 presents the results of the analysis based on the assumptions in terms of changes in weight, volume, and economic savings.

The tables highlight the considerably lower LH_2 mass required to cover the same range with the same payload compared to the Jet-A1 mass (i.e. $m_{Jet-A1} = MTOW - m_{PL}$), despite worse aerodynamic and structural performances (evident when comparing K_{LH_2} from Table 16.2 with K_{Jet-A1} in Table 16.1). Additionally, the equivalent fuselage length taken by the hypothetical integral LH_2 tank retrofit is calculated for reference. As a fraction of the total fuselage length, the value is quite significant for long haul aircraft ($37.1 \sim 46.1\%$), compared to short/medium range aircraft ($30.3 \sim 34.6\%$).

Further research will extend and integrate the analysis considering the environmental gains associated with enhanced flight trajectories and operations [36]. Particular consideration will be given to identifying the combined benefits and additional challenges associated with the adoption of hydrogen fuels in advanced aircraft configurations [37] and in more electric aircraft configurations [38].

Table 16.1 Assumed aircraft characteristics and calculated emissions for Jet-A1 fuel.

Aircraft model	E190	A320	B738	A333	B788	B77W	A388
Range (nmi)	2400	2950	3060	5550	7850	7930	8500
Total length (m)	36.2	37.5	39.5	63.7	56.7	73.9	72.7
Hydraulic diameter of fuselage (m)	3.15	4.04	3.76	5.64	5.87	6.2	7.75
Approximate K_{Jet-A1}	3.6	4.4	4.3	5.7	6.9	6.2	6.0
OEW (t)	28.1	42.6	41.4	124.5	118	167.8	276.8
Payload (t)	11	16.2	17	30	23	37	40
MTOW (t)	51.8	78	79	233	228	351.5	575
Generated CO_2 (t)	39.4	59.5	63.9	243.4	269.7	454.8	800.4
Generated CO (t)	0.1	0.2	0.2	0.8	0.9	1.5	2.6
Generated SO_X (t)	0.0	0.1	0.1	0.2	0.3	0.4	0.8
Generated HC (t)	0.1	0.1	0.1	0.4	0.4	0.7	1.3
Generated H_2O (t0	16.3	24.6	26.4	100.5	111.4	187.8	330.5
Total pollution charges (US$)	**1110**	**1678**	**1800**	**6861**	**7604**	**12 822**	**22 567**

Table 16.2 Results of the analysis.

Aircraft model	E190	A320	B738	A333	B788	B77W	A388
Assumed K_{LH_2}	3.1	3.7	3.6	4.8	5.8	5.3	5.1
TOW of corresponding hypothetical LH_2-powered aircraft (t)	45.9	69.1	69.2	192.7	181.5	270.0	429.0
Total LH_2 mass (t)	6.8	10.3	10.8	38.2	40.5	65.2	112.2
Total LH_2 volume (m^3)	96.3	145.7	151.9	538.2	569.9	918.8	1580.8
Equivalent fuselage length (m)	12.4	11.4	13.7	21.5	21.1	30.4	33.5
Fraction of the total length (%)	34.1	30.3	34.6	33.8	37.1	41.2	46.1
Weight savings (%)	11	11	12	17	20	23	25
Generated H_2O (t)	61.0	92.3	96.2	340.9	360.9	581.9	1001.1
Total environmental charge (US$)	610	923	962	3409	3609	5819	10011
Total savings per flight (US$)	**500**	**755**	**838**	**3452**	**3995**	**7003**	**12556**

16.8 Conclusions

This chapter described the benefits and challenges associated with the introduction of hydrogen fuels in aviation, specifically targeting liquid hydrogen in cryogenic tanks. The major works on this subject have been reviewed and the key numerical findings mentioned. A simplified numerical this chapter for the estimation of environmental gains in realistic operational conditions was also introduced. The results highlight some of the remarkable economic and environmental benefits associated with hydrogen fuels, even after considering the lower aerodynamic efficiency and higher structural mass. The lower volumetric efficiency and the challenges associated with the production and supply are nonetheless substantial, and will require significant technological and political initiatives to achieve a widespread adoption of liquid hydrogen fuels in aviation.

References

1 Boeing, (2014). "Current Market Outlook 2014–2033", Boeing Commercial Airplanes, Seattle, WA, USA, Market Analysis.

2 Singh, B.R. and Singh, O. (2012). Global trends of fossil fuel reserves and climate change in the 21st century. In: *Fossil Fuel and the Environment* (ed. S. Khan), 167–193. InTechOpen.

3 Contreras, A., Yiğit, S., Özay, K., and Veziroğlu, T.N. (1997). Hydrogen as aviation fuel: a comparison with hydrocarbon fuels. *International Journal of Hydrogen Energy* 22: 1053–1060.

4 Brewer, G.D. (1991). *Hydrogen Aircraft Technology*. CRC Press.

5 Kocer, K. (1994). *Consideration of Liquid Hydrogen for Commercial Aircraft*. Miami, FL, USA: University of Miami.

6 Leonorovitz, J. (1990). Soviets seek international interest in alternative-fuel-powered transport. *Aviation Week and Space Technology (New York)* 132: 112.

7 Pohl, H.W. and Malychev, V.V. (1997). Hydrogen in future civil aviation. *International Journal of Hydrogen Energy* 22: 1061–1069.

8 Cryoplane Project, (2003). "Liquid Hydrogen Fuelled Aircraft - System Analysis", Cryoplane Project, Hamburg, Germany, Final Technical Report G4RD-CT-2000-00192.

9 Khandelwal, B., Karakurt, A., Sekaran, P.R. et al. (2013). Hydrogen powered aircraft: the future of air transport. *Progress in Aerospace Science* 60: 45–59.

10 International Energy Agency, (2006). "Hydrogen production and storage - R&D priorities and gaps", International Energy Agency, Paris, France.

11 Verstraete, D. (2013). Long range transport aircraft using hydrogen fuel. *International Journal of Hydrogen Energy* 38: 14824–14831.

12 Allideris, L. and Janin, F. (2002). "Fuel system components - mechanical tank design trade-off", Cryoplane project Task technical report 3.6.2.2.

13 Mital, S.K., Gyekenyesi, J.Z., Arnold, S.M., Sullivan, R.M., Manderscheid, J.M., and Murthy, P.L.N. (2006). "Review of current state of the art and key design issues with potential solutions for liquid hydrogen cryogenic storage tank structures for aircraft applications", NASA John H. Glenn Research Center, Cleveland, OH, USA, Technical Memorandum NASA/TM—2006-214346.

14 Verstraete, D., Hendrick, P., Pilidis, P., and Ramsden, K. (2010). Hydrogen fuel tanks for subsonic transport aircraft. *International Journal of Hydrogen Energy* 35: 11085–11098.

15 Colozza, A.J. (2002). "Hydrogen storage for aircraft applications overview", NASA John H. Glenn Research Center, Cleveland, OH, USA, Final technical report NASA/CR-2002-211867.

16 Millis, M.G., Tornabene, R.T., Jurns, J.M., Guynn, M.D., Tomsik, T.M., and Van Overbeke, T.J. (2009). "Hydrogen fuel system design trades for high-altitude long-endurance remotely-operated aircraft", NASA John H. Glenn Research Center, Cleveland, OH, USA, Technical Memorandum NASA/TM—2009-215521.

17 Wilkins, F. (2002). *National Hydrogen Energy Roadmap - Production, Delivery, Storage, Conversion, Applications, Public Education and Outreach*. Washington, DC, USA: US Department of Energy.

18 Cumalioglu, I. (2005). "Modeling and simulation of a high pressure hydrogen storage tank with dynamic all," MSc thesis, Texas Tech University.

19 Guynn, M.D., Freeh, J.E., and Olson, E.D. (2004). "Evaluation of a hydrogen fuel cell powered blended-wing-body aircraft concept for reduced noise and emissions", NASA Langley Research Center, Hampton, VA, USA, Technical memorandum NASNTM-2004-2 12989.

20 Dahl, G. and Suttrop, F. (1998). Engine control and low-nox combustion for hydrogen fuelled aircraft gas turbines. *International Journal of Hydrogen Energy* 23: 695–704.

21 Juste, G.L. (2006). Hydrogen injection as additional fuel in gas turbine combustor. Evaluation of effects. *International Journal of Hydrogen Energy* 31: 2112–2121.

22 Dahl, G. and Suttrop, F. (2002). "Combustion chamber and emissions, the micro mix hydrogen combustor technology", Cryoplane project Task Technical Report 4.4-5A.

23 Marek, C.J., Smith, T.D., and Kundu, K. (2005). "Low emission hydrogen combustors for gas turbines using Lean Direct Injection," *41st AIAA/ASME/SAE/ASEE Joint Propulsion Conference and Exhibit*, Tucson, AZ, USA.

24 Heywood, J.B. and Mikus, T. (1973). *Parameters Controlling Nitric Oxide Emissions from Gas Turbine Combustors*. Cornell University.

25 IATA, (2014). "Fact sheet: fuel", International Air Transport Associacion (IATA).

26 Price, R.O. (1991). Liquid hydrogen-an alternative aviation fuel? *International Journal of Hydrogen Energy* 16: 557–562.

27 Janic, M. (2010). Is liquid hydrogen a solution for mitigating air pollution by airports? *International Journal of Hydrogen Energy* 35: 2190–2202.

28 Stiller, C. and Schmidt, P. (2010). "Airport Liquid Hydrogen Infrastructure for Aircraft Auxiliary Power Units," presented at the 18th World Hydrogen Energy Conference (WHEC 2010), Essen, Germany.

29 Schmidtchen, U., Behrend, E., Pohl, H.W., and Rostek, N. (1997). Hydrogen aircraft and airport safety. *Renewable and Sustainable Energy Reviews* 1: 239–269.

30 Korycinski, P.F. (1978). Air terminals and liquid hydrogen commercial air transports. *International Journal of Hydrogen Energy* 3: 231–250.

31 Cherry, J. (2008). "Airports - Actions on Climate Change," A. C. I. (ACI), Ed., ed. Montreal, Canada.

32 Brewer, G.D. (1983). An assessment of the safety of hydrogen-fueled aircraft. *Journal of Aircraft* 20: 935–939.

33 Schmidtchen, U., Gradt, T., Börner, H., and Behrend, E. (1994). Temperature behaviour of permeation of helium through Vespel and Torlon. *Cryogenics* 34: 105–109.

34 Jones, L., Wuschke, C., and Fahidy, T.Z. (1983). Model of a cryogenic liquid-hydrogen pipeline for an airport ground distribution system. *International Journal of Hydrogen Energy* 8: 623–630.

35 Brewer, G.D. (1976). "LH2 airport requirements study", Lockheed-California Company, Burbank, CA, USA, NASA Contractor Report NASA/CR-2700.

36 Gardi, A., Sabatini, R., and Ramasamy, S. (2016). Multi-objective optimisation of aircraft flight trajectories in the ATM and avionics context. *Progress in Aerospace Sciences* 83: 1–36.

37 Marino, M. and Sabatini, R. (2014). Advanced lightweight aircraft design configurations for green operations, *Practical Responses to Climate Change, Engineers Australia Convention 2014 (PRCC 2014)*, Melbourne, Australia.

38 Seresinhe, R., Sabatini, R., and Lawson, C.P. (2013). Environmental impact assessment, on the operation of conventional and more electric large commercial aircraft. *SAE International Journal of Aerospace* 6: 56–64.

39 Rondinelli, S., Sabatini, R., and Gardi, A. (2014). Challenges and Benefits offered by Liquid Hydrogen Fuels in Commercial Aviation. In: *Proceedings of Practical Responses to Climate Change. Engineers Australia Convention 2014 (PRCC 2014)*, Melbourne, Australia.

17

Multi-Objective Trajectory Optimisation Algorithms for Avionics and ATM Systems

Alessandro Gardi[1,2], Roberto Sabatini[1,2], and Trevor Kistan[3]

[1] Department of Aerospace Engineering, Khalifa University of Science and Technology, Abu Dhabi, UAE
[2] School of Engineering, RMIT University, Melbourne, Victoria, Australia
[3] THALES Australia, Melbourne, Victoria, Australia

17.1   Introduction

Innovative air traffic management (ATM) and avionics technologies are expected to deliver most of the operational performance enhancements over the next few years and are therefore at the core of major aviation modernisation initiatives, including the Single European Sky Air Traffic Management Research (SESAR) in Europe and the Next Generation Air Transportation System (NextGen) in the United States [1–4]. These programs fundamentally aim to establish a more predictable and flexible management of airspace and airport resources through higher levels of information sharing, automation, and more accurate navigation, thereby supporting a harmonised evolution of ATM into a more efficient, resilient, integrated, and collaborative system [3, 4]. Among the various operational enhancements targeted by these initiatives, the introduction of 4-dimensional trajectories (4DT) and of the associated trajectory-based operations (TBOs) paradigm represents an evolutionary shift of ATM, allowing both human operators and machines to collaboratively resolve conflicts and improve overall traffic efficiency [3, 5]. Capitalising on the significant opportunities introduced by TBO, the aim of the research activities described in this chapter is to develop automated 4DT planning, negotiation and validation (4-PNV) software functionalities to be integrated in avionics flight management systems (FMSs) and ATM systems to assist human operators in managing dense traffic and mitigate unforeseen perturbations in dynamic (4D) airspace. This research focusses on the online contexts (i.e. when the flight mission is active), which pose stringent time requirements and introduce the need to efficiently resolve situations that could not be predicted in the offline context (i.e. before the beginning of the mission). These uncertainties are typically due to adverse weather and inaccurate winds aloft forecasts, but also include unexpected aircraft turn-around issues and other disruptions. The focus of this chapter is on the 4DT planning process, as state-of-the-art flight planning methods are still based on optimised vertical planning techniques (i.e. altitude and airspeed) initially developed in the 1970s [6–9] and on a preliminary lateral path selection based on offline wind forecasts, with

Sustainable Aviation Technology and Operations: Research and Innovation Perspectives, First Edition.
Edited by Roberto Sabatini and Alessandro Gardi.
© 2024 John Wiley & Sons Ltd. Published 2024 by John Wiley & Sons Ltd.
Companion Website: www.wiley.com/go/sustainableaviation

limited consideration for weather and airspace blockage issues, as further detailed in [10]. Although well-proven, these approaches are very restrictive in the 4D-TBO context. Additionally, the limited initial optimality only captures a reduced set of operational and economic factors (i.e. fuel costs and time costs) and is progressively compromised when online interventions are performed as part of ATM and air traffic flow management (ATFM) processes. The integration of more comprehensive and versatile 4DT optimisation functionalities in avionics FMS and ground-based ATM systems not only mitigates the degradation of optimality, but also allows us to account for more complex criteria and constraints, in larger amounts, thereby supporting enhancements in efficiency and flexibility of online air traffic operations [10, 11]. To account for multiple performance criteria and constraints, the online 4DT planning process proposed here exploits a custom multi-objective trajectory optimisation (MOTO) algorithm based on optimal control and on commonly adopted aircraft models. In addition to providing significantly more detail on the implementation aspects, this case study, originally published in [63, 64], extends the research presented in [12, 13] by introducing and evaluating the complete 4DT post-processing stage as well as a novel multi-objective articulation of preference. The chapter is structured as follows: Section 2 presents the concept of operations and the top-level 4-PNV working processes, while Section 3 describes the 4DT planning process based on a custom MOTO algorithm, with particular focus on mathematical formulation, solution technique, model set, weather data, articulation of preferences, and operational realisation. Section 4 presents the simulation-based 4-PNV verification activities performed and a discussion of key results, while Section 5 presents the conclusions and outlines the needs for future research.

17.2   4-PNV Concept and Processes

The 4DT concept, which emerged as part of the advances in ATFM and automation-assisted guidance [14–18], is one of the fundamental ATM enhancements and is at the very core of the TBO paradigm currently under development. The TBO concept entails a progressive transition away from legacy flight plans into ever-updating 4DT defining the aircraft's flight path in three spatial dimensions (i.e. latitude, longitude, and altitude) and time from origin to destination, as well as on the associated precise estimation and correction of current and predicted traffic states [5]. This supports a more flexible management of the traffic and an optimal synchronisation of the network, which in turn are expected to greatly enhance operational safety, efficiency, and environmental sustainability. In the TBO operational scenario, each flight is allocated a 4DT contract, which evolves from an initial reference business trajectory through a collaborative decision-making (CDM) process involving the air navigation service providers (ANSPs), the aircraft operator and the relevant airport services and management companies. Although such 4DT is progressively outlined up to months before the actual flight takes place in the so-called *offline* timeframe (before the flight is commenced), increased efficiencies and higher throughputs are obtained in a TBO context by optimally managing 4DT contracts in the *online* timeframes, exploiting suitable airborne and ground-based decision support systems (DSSs), as unforeseen perturbations are only detectable in the online timeframe and therefore it is at this stage that they have to be resolved [19, 20]. Compared to legacy ATM interventions, these online 4DT management processes are no longer limited at tactical/short-term deconfliction but instead look at a much broader picture, aiming at optimizing the operational efficiency for the entire flight and over wide geographic areas. Clearly, this requires a significantly higher level of information sharing, which is being realised as part of the system-wide information management (SWIM) framework. In order to optimally exploit these rapidly growing amounts of data, an increase in automation support and a transition away from the centralised command and control-oriented ATM paradigm towards more distributed/collaborative air-ground planning are essential. Emerging ground-based and airborne DSS concepts for TBO therefore aim to support the human ATM operators and flight crews in planning, negotiating, reviewing, and validating 4DT intents in a safe and timely manner, while exploiting the substantially increased levels of airspace, weather and traffic information that will be available [21–26]. Eurocontrol's Program for Harmonised Air Traffic Management Research in Europe (PHARE) proposed a possible solution for single-attempt negotiation in the TBO context [27]. Building upon this approach, the 4-PNV concept of operation has been developed by introducing more specific negotiation and validation loops for transactions initiated by the aircraft and ATM DSS [13].

With no loss of generality, Figure 17.1 illustrates the 4-PNV concept of operation in a terminal airspace structure. For completeness, the control area (CTA), the terminal manoeuvring area (TMA) and airways typically surrounding major airports are also depicted. To support distributed online CDM processes, next generation avionics FMS (NG-FMS) and ground-based ATM DSS are networked by Next Generation Aeronautical Data Links (NG-ADLs). Similar ATM DSS are situated within area control centres (ACCs), terminal control stations (TCSs) and airport control towers (TWRs). All these ATM units are, in turn, networked by SWIM and collaborate in pursuing an optimal exploitation of airspace and airport resources. The airborne and ground-based systems also integrate online 4-PNV functionalities, ensuring both signal-in-space and system-level interoperability.

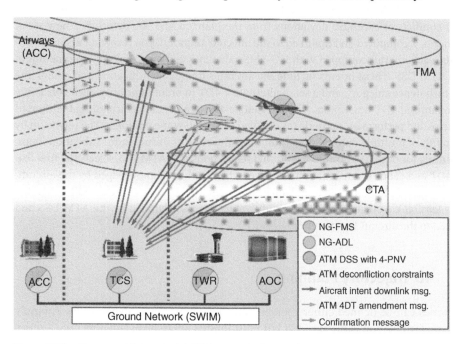

Figure 17.1 Air-ground integrated 4-PNV concept of operations.

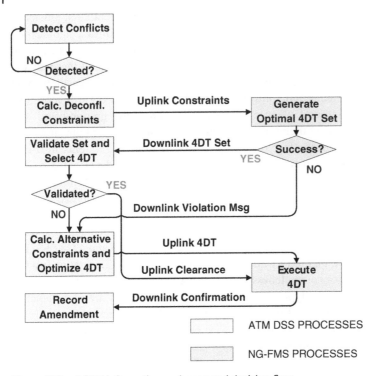

Figure 17.2   4-PNV information exchange and decision flow.

Figure 17.2 represents the 4-PNV information exchange and decision flow. The DSS of the designated ATM unit (the TCS in this example) tracks all aircraft within its jurisdiction and, if conflicts or excessive deviations from the planned 4DT are detected, a new set of constraints is computed and uplinked to the affected aircraft. Upon receiving updated constraints, each aircraft equipped with NG-FMS calculates and downlinks an optimal intent to the ATM DSS. These intents include the aircraft's unique identification and model, the wake-turbulence category, and a set of alternative 4DT in order of priority, each consisting of standardised 4DT descriptors. The alternative 4DT are generated according to different objectives and performance criteria (as detailed in the following sections). The provision by the aircraft of multiple optimal 4DT minimises the risks of unsuccessful validation loops. Accounting for other traffic and operational restrictions, the ATM DSS selects the first feasible 4DT and sends a confirmation message, which is then acknowledged by the aircraft. If, on the other hand, the validation loop is unsuccessful (i.e. none of the aircraft-generated 4DT is feasible), a single conflict-free and flyable 4DT is calculated by the ATM DSS and uplinked to the aircraft for execution.

17.3   4D Trajectory Planning

Trajectories can be defined either geometrically, kinematically, or dynamically. A purely geometric formulation is the simplest but most limited alternative, entailing only a series of points in space and providing neither information on the velocity and attitude of the aircraft, nor on the time at which these points will be crossed, making this formulation unsuitable for 4D TBO. To ensure the flyability of geometrically-defined trajectories at any possible ground speed, very large turn radii need to be conservatively assumed, resulting in an inefficient exploitation of the available airspace. In the kinematic formulation, velocity information is introduced. Current RNP standards [28, 29] can be considered to some extent a practical implementation of the kinematic trajectory concept, as further detailed in subsection G. The dynamic formulation, on the other hand, is the most rigorous and complete description of a vehicle's trajectory from the physical principles and supports the most comprehensive evaluation of operational, economic, and environmental performances, as well as the widest array of possible constraints and optimality criteria. It is also important to note that from a more complex trajectory model it is always possible to extract the simpler ones, but not vice versa. Hence, from a dynamically-defined trajectory it is possible to extract the kinematic and geometric descriptions, but from a kinematic or geometric trajectory it is not possible to univocally derive the dynamic one, nor guarantee its existence. For all these reasons and to support further work, the MOTO algorithm proposed here adopts a dynamically defined trajectory, leading to an optimal control problem (OCP) formulation, which is very well-suited for 4D TBO. Alternative formulations either derived from the OCP or custom-developed can be preferred depending on the length of the trajectory being optimised, as the size and complexity of the OCP can become unmanageable [10].

17.4 Multi-Objective 4DT Optimisation Algorithm

The proposed MOTO-based 4DT planning algorithm for TBO is outlined in Figure 17.3. As depicted in the figure and further detailed in the following subsections, the trajectory optimisation process receives a combined objective as input from an *a priori* weighted product multi-objective articulation of preferences process (subsection F), as well as adequate deconfliction constraints from a sequencing and spacing algorithm. A mathematically optimal trajectory is then generated using pseudospectral methods (subsection C). This optimal trajectory is subsequently treated by suitably defined control input smoothing and operational trajectory realisation processes (subsection G) to obtain an automatic flight control systems (AFCSs) flyable and concisely described 4DT. These two elements, depicted on the right-hand side of the figure, are the focus of this chapter and are described in detail, together with all other main components. To address the various optimality criteria and operational constraints, the MOTO suite ideally comprises a number of models including the aircraft's flight dynamics and propulsion model, an operational costs model, emissions models, an atmosphere and weather model (subsections D and E) and, as necessary, an airspace model, a contrails model, and a noise model.

17.4.1 Mathematical Formulation

The OCP formulation is based on a scalar time $t \in [t_0; t_f]$ and on vectors of time-dependent state variables $\boldsymbol{x}(t) \in \mathbb{R}^n$, control variables $\boldsymbol{u}(t) \in \mathbb{R}^m$ and system parameters $\boldsymbol{p} \in \mathbb{R}^q$. Adopting the OCP formulation, the trajectory optimisation problem can be analytically stated as follows [30, 31]:

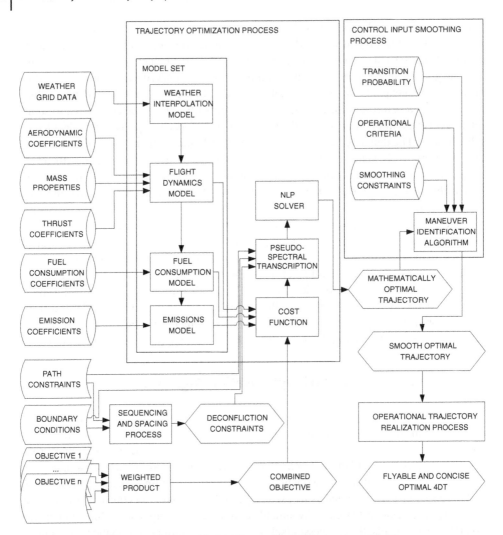

Figure 17.3 Block diagram of the multi-objective optimal 4DT planning algorithm.

Determine the states $x(t) \in \mathbb{R}^n$, the controls $u(t) \in \mathbb{R}^m$, the parameters $p \in \mathbb{R}^q$, the initial time $t_0 \in \mathbb{R}$ and the final time $t_f \in \mathbb{R} \mid t_f > t_0$ that optimise the performance index(es):

$$J = \Phi\left[x\left(t_f\right), u\left(t_f\right), p\right] + \int_{t_0}^{t_f} \Psi\left[x(t), u(t), p\right] dt \tag{17.1}$$

subject to the dynamic constraints:

$$\dot{x}(t) = f\left[x(t), u(t), t, p\right] \tag{17.2}$$

to the path constraints:

$$C_{min} \leq C\left[x(t), u(t), t; p\right] \leq C_{max} \tag{17.3}$$

and to the boundary conditions:

$$\mathscr{B}_{min} \leq \mathscr{B}\left[\mathbf{x}\left(t_0, t_f\right), \mathbf{u}\left(t_0, t_f\right); \mathbf{p}\right] \leq \mathscr{B}_{max} \tag{17.4}$$

where Φ is the Mayer term and Ψ is the Lagrange term of the cost function expressed in a Bolza form.

The optimisation is classified as *single-objective* when an individual performance index J is introduced and *multi-objective* when two or multiple conflicting performance indexes J_i are defined, so that the attainment of a better J_k leads to a worse $J_h, \{h, k \in [1; n_J], h \neq k\}$ so that a number of trade-off solutions can be calculated.

17.4.2 OCP Solution Methods

Two mainstream strategies were identified for the numerical solution of OCP, namely *direct methods* and *indirect methods* [30–33]. Direct methods involve the discretisation of the infinite-dimensional OCP into a finite-dimensional non-linear programming (NLP) problem and are further classified into direct shooting and collocation methods. Direct shooting methods employ user-defined analytical functions to parameterize only the control variables for the entire time domain, while collocation methods adopt piecewise polynomial functions to parameterize states and control. The most computationally efficient collocation methods adopt linearly independent (orthogonal) polynomial functions and are called pseudospectral. In pseudospectral methods, the evaluation of state and control vectors is performed at discrete collocation points across the problem domain using suitably selected orthogonal (spectral) interpolating functions [31, 34]. The computational resources required for a direct solution of OCP can be significant, but this issue is normally mitigated by exploiting the high sparsity of Jacobian and Hessian matrixes, comprising a majority of null terms.

Based on the literature, an OCP formulation was adopted and the pseudospectral solution method was selected for the MOTO software tool. This widely used method employs orthogonal collocation and implicit integration based on Gaussian quadrature, where collocation is performed at the Legendre-Gauss-Radau points. Publicly available pseudospectral optimal control solvers were chosen due to their suitability for aerospace applications. The user can typically define a number of parameters used in the optimisation process, including the minimum and maximum numbers of mesh discretisation intervals and of collocation points in each interval, other quadrature mesh characteristics, the maximum number of NLP iterations and of mesh refinements (if adaptive meshing schemes are available), and the numerical differentiation scheme. The suitability of these and similar techniques for online trajectory planning has been demonstrated in recent research [11, 35–39]. The pseudospectral transcription process of the general OCP formulated in subsection *B* commences with the introduction of the following transformation:

$$\tau = \frac{2}{t_f - t_0}t - \frac{t_f + t_0}{t_f - t_0}, \quad t \in \left[t_0, t_f\right] \tag{17.5}$$

The solution process now requires finding the state and control trajectories $x(\tau)$ and $u(\tau)$ respectively, in the interval $\tau \in [-1, 1]$, as well as the initial and final times, t_0 and t_f respectively. Subdividing the continuous time domain in $N + 1$ intervals τ_k, the original continuous functions $f(t)$ can be approximated by a unique polynomial of degree N whose evaluations at the intervals correspond to the values of the original function. For instance, Lagrange polynomials can be adopted, given by [40]:

$$f(\tau) = \sum_{k=0}^{N} f(\tau_k) L_k^N(\tau) \tag{17.6}$$

$$L_k^N(\tau) = \prod_{\substack{k \neq l \\ 0 \leq l \leq N}} \frac{(\tau - \tau_k)}{(\tau_l - \tau_k)} \tag{17.7}$$

These are mutually orthogonal over the interval $(-1, 1)$ with respect to a weight function $w \neq 0$ [41]:

$$\int_{-1}^{1} L_k^N(\tau) L_m^N(\tau) \, w \, d\tau = 0 \text{ when } m \neq k \tag{17.8}$$

In the Legendre pseudospectral approximation, the state and control trajectories $x(\tau)$ and $u(\tau)$ are approximated by N^{th} order Lagrange polynomials $x^N(\tau)$ and $u^N(\tau)$ based on interpolation at Legendre-Gauss-Lobatto (LGL) nodes τ_k:

$$x(\tau) \approx x^N(\tau) = \sum_{k=0}^{N} x(\tau_k) L_k^N(\tau) \tag{17.9}$$

$$u(\tau) \approx u^N(\tau) = \sum_{k=0}^{N} x(\tau_k) L_k^N(\tau) \tag{17.10}$$

where $x^N(\tau)$ and $u^N(\tau)$ are Lagrange interpolating polynomials. The derivative of the state vector is approximated as follows:

$$\dot{x}(\tau_k) \approx \dot{x}^N(\tau_k) = \sum_{i=0}^{N} D_{ki} x(\tau_i), i = 0, 1, \ldots, N \tag{17.11}$$

where D is the $(N+1) \times (N+1)$ differentiation matrix corresponding to the LGL nodes. The objective function of the OCP is therefore approximated as follows [40]:

$$J \approx \varphi\left[x^N(1), t_f\right] + \frac{t_f - t_0}{2} \sum_{k}^{N} L\left[x^N(\tau_k), u^N(\tau_k), \tau_k\right] \omega_k \tag{17.12}$$

where the weights ω_k are defined at the LGL nodes.

The resulting finite-dimensional NLP problem is then solved by iterative algorithms. The initial steps adopted to implement an iterative NLP solution algorithm for multiple variables involve the adoption of an n-dimensional Taylor series expansion of the finite-dimensional NLP function $F(x)$ to the third term as in:

$$F\left(x^{(k+1)}\right) \cong F\left(x^{(k)}\right) + \nabla_x^T F\left(x^{(k)}\right) \cdot s^{(k)} + \frac{1}{2} s^{(k)^T} H\left(x^{(k)}\right) s^{(k)} \tag{17.13}$$

where:

$$s^{(k)} \triangleq x^{(k+1)} - x^{(k)} \tag{17.14}$$

$$\nabla_x F(x) = \left[\frac{\partial F(x)}{\partial x_1}; \ldots \frac{\partial F(x)}{\partial x_n}\right]^T \tag{17.15}$$

$$H\left(F\left(x\right)\right) = \begin{bmatrix} \dfrac{\partial^2 F\left(x\right)}{\partial x_1^2} & \cdots & \dfrac{\partial^2 F\left(x\right)}{\partial x_1 \partial x_n} \\ \vdots & \ddots & \vdots \\ \dfrac{\partial^2 F\left(x\right)}{\partial x_n \partial x_1} & \cdots & \dfrac{\partial^2 F\left(x\right)}{\partial x_n^2} \end{bmatrix} \tag{17.16}$$

An iterative NLP solution can thus be formulated such that the search direction at step k based on the n-dimensional Newton method is written:

$$s^{(k)} = -\left[H^{-1}\nabla_x F\left(x\right)\right]_{x^{(k)}} \tag{17.17}$$

Due to the very large size of the formulated NLP problem, opportune expedients (such as Broyden-based recursive update strategies) are introduced in most available NLP solvers to prevent the need to recompute the complete matrixes at each iteration, as detailed in [10]. For this reason and thanks to additional source code-based and compilation-time optimisations, software toolsets available in the public domain offer remarkable computational efficiencies and are the ones typically preferred. The Interior Point OPTimizer (IPOPT) toolbox is therefore adopted in our computer-based 4-PNV prototype. Further details on the NLP solver development are given in [42, 43].

In our proposed MOTO implementation, opportune provisions are introduced to prevent the occurrence of ill-posed problems, which would not converge on a feasible numerical solution. All these provisions have been extensively tested and refined, so that significant evidence supporting their validity was accumulated during the MOTO development. In particular, the length of the slant path (d_{slant}) to be optimised must be between 10 and 50 nmi and the altitudes (z_f) and time constraints (t_f) set on the final waypoint shall be within the achievable set, as calculated using conventional heuristics: $t_{Min} \leq t_f \leq t_{Max}$ and $v_{Z_{Min}} t_{Min} \leq z_0 - z_f \leq v_{Z_{Max}} t_{Min}$, where $t_{Min} = \frac{d_{slant}}{v_{Max}}$, $t_{Max} = 2\frac{d_{slant}}{v_{Min}}$, with $\left\{v_{Z_{Min}}, v_{Z_{Max}}\right\}$ specific to each aircraft. The minimum and maximum trajectory lengths are conservatively restricted compared to the distances (5–70 nmi) at which the developed pseudospectral-based MOTO algorithm consistently converges to a feasible trajectory in the desired computation time. The minimum trajectory length constraint also ensures that any combination of initial, final, and connecting track angles is achievable with minimal manoeuvring complexity.

To expedite the 4DT planning stage, the number of intervals and of collocation points in each interval is predefined and constraints are introduced on their maximum increases allowed during the iterative solution. Moreover, the maximum number of IPOPT iterations and of pseudospectral mesh refinements is also limited to terminate slowly- or non-converging runs. Any unfeasible solution is then detected by the post-processing stages detailed below (subsection G) and discarded. Details are provided in Table 17.1.

17.4.3 Transport Aircraft Model Set

Flight dynamics models for MOTO can be based on conventional point-mass or rigid-body models, which are characterised, respectively, by three and six degrees-of-freedom (3-DoF/6-DoF). Due to the presence of short-period modes promoting numerical instabilities, 6-DoF are seldom adopted, unless the timeframe is very small (i.e. $t_f < 15$ seconds) or the particular application demands their usage. Consequently, OCP involving transport

Table 17.1 Numerical solution parameters.

Parameter	Initial value	Maximum value
Number of intervals	$\dfrac{d_{Slant}}{2\,v_{Max}}$	$\dfrac{d_{Slant}}{v_{Min}}$
Number of collocation points in each interval	2	3
Number of NLP iterations	–	500
Numerical NLP tolerance	–	10^{-4}

aircraft on medium-long timeframes typically rely on point-mass models. For numerical implementations, dynamics are expressed as a set of differential algebraic equations (DAEs). A general formulation of 3-DoF flight dynamics DAE is:

$$\begin{cases} \dot{v} = g\,(T/W - D/W - sin\gamma) \\[2mm] \dot{\gamma} = \dfrac{g}{v} \cdot [Ncos\mu - cos\gamma] \\[2mm] \dot{\chi} = \dfrac{g}{v} \cdot \dfrac{Nsin\mu}{cos\gamma} \\[2mm] \dot{\varphi} = \dfrac{vcos\gamma\,cos\chi + v_{w_\varphi}}{R_E + z} \\[2mm] \dot{\lambda} = \dfrac{vcos\gamma\,sin\chi + v_{w_\lambda}}{\left(R_E + z\right)cos\varphi} \\[2mm] \dot{z} = vsin\gamma + v_{w_z} \\[2mm] \dot{m} = -FF \end{cases} \tag{17.18}$$

where the state vector consists of: True Air Speed (TAS) v [m s^{-1}], flight path angle γ [rad]; track angle (clockwise from North) χ [rad]; geodetic latitude φ [rad]; geodetic longitude λ [rad]; altitude z [m]; aircraft mass m [kg]; whereas the control vector includes: thrust force T [N]; load factor N []; bank angle μ [rad]. Other variables and parameters include: aircraft weight $W = m\,g$ and aerodynamic drag D [N]; wind velocity v_w in its three scalar components [m s^{-1}]; gravitational acceleration g [m s^{-2}]; local Earth radius R_E [m]; fuel flow FF [kg s^{-1}]. This set of 3-DoF DAE assumes variable aircraft mass, constant and longitudinal thrust vector orientation, constant vertical gravity and considers the effects of winds. The aerodynamic drag is modelled as

$$D = \frac{1}{2}\rho v^2 S \left(C_{D0} + C_{D2}\,C_L^2\right) \tag{17.19}$$

where $\rho = \rho(\varphi, \lambda, z, t)$ is the local air density retrieved from weather input data grid or a weather model, S is the reference wing surface, C_{D0} and C_{D2} are the parabolic drag coefficients typically available from aircraft performance databases such as Eurocontrol's Base of Aircraft Data (BADA). The lift coefficient C_L can be calculated from:

$$N\,W = \frac{1}{2}\rho v^2 S\,C_L \tag{17.20}$$

The thrust force magnitude is expressed in terms of a normalised throttle coefficient τ as well as maximum and zero-throttle thrust terms T_{MAX} and T_0 dependent on the air density (or on pressure altitude and temperature) as in:

$$T(t) = \tau(t) \cdot [T_{MAX}(\rho) - T_0(\rho)] + T_0(\rho) \tag{17.21}$$

For instance, for turbofan aircraft the following empirical expressions were adopted in the development of BADA to determine the climb thrust, which closely match the maximum continuous thrust (T_{MAX}) conditions in all flight phases excluding take-off and the fuel flow FF [44]:

$$T_{MAX} = C_{T1}\left(1 - \frac{H_P}{C_{T2}} + C_{T3}H_P^2\right) \cdot [1 - C_{T5}(\Delta T - C_{T4})] \tag{17.22}$$

$$FF = max \begin{bmatrix} \tau\, T_{MAX}\, C_{f1}\left(1 + \frac{v_{TAS}}{C_{f2}}\right) \\ C_{f3}\left(1 - \frac{H_P}{C_{f4}}\right) \end{bmatrix} \tag{17.23}$$

where H_P is the geopotential pressure altitude in feet, ΔT is the deviation from the standard atmosphere temperature in kelvin, v_{TAS} is the true airspeed. $C_{T1}...C_{T5}$, $C_{f1}...C_{f4}$ are the empirical thrust and fuel flow coefficients, which are also supplied as part of BADA [44].

To simplify the numerical implementation and particularly the 4DT post-processing stage described later, an alternative formulation of 3-DoF dynamics adopting the geometric trajectory curvature (κ) in place of the bank angle (μ) as the lateral control variable is proposed here. The principal advantage is that the geometric trajectory curvature is the inverse of the local turn radius, which can thus be directly computed at any point from the optimisation results. Additionally, while the turn radius features a mathematical singularity in the admissible domain (i.e. $R \to \infty$ for straight segments), the curvature does not. Adopting the curvature instead of the bank angle, the flight path angle and track angle DAE can be written as:

$$\dot{\gamma} = \frac{g}{v} \cdot \left[\left(N^2 - \frac{\kappa^2 v^4 \cos^2\gamma}{g^2}\right)^{\frac{1}{2}} - \cos\gamma\right] \tag{17.24}$$

$$\dot{\chi} = \kappa \cdot v \tag{17.25}$$

where constraints on the minimum turn radius (i.e. max κ) consistent with the operational envelope of commercial transport aircraft are sufficient to ensure that N never drops below the value required to sustain the turn (which would yield imaginary solutions).

Estimates of pollutant emissions can be calculated as a function of the fuel flow, pollutant-specific emission indexes (EI) are adopted so that:

$$GP = \int_{t_o}^{t_F} FF(t)\, EI_{GP}\, dt \tag{17.26}$$

where the acronym GP represents a generic pollutant. While carbon dioxide (CO_2) emissions are characterised by an approximately constant emission index of $EI_{CO_2} = 3.16\,\left[Kg_{CO_2}/Kg_{Jet-A1}\right]$ [45], other emissions have notable dependencies on the thrust setting (τ). An empirical model for carbon monoxide (CO) and unburned hydrocarbons (HC) emission indexes ($EI_{CO/HC}$) in [g/Kg$_{Jet-A1}$] at mean sea level based on nonlinear fit of experimental data from the International Civil Aviation Organization (ICAO) emissions databank is [13]:

$$EI_{CO/HC}(\tau) = c_1 + exp\left(-c_2\tau + c_3\right) \tag{17.27}$$

where the fitting parameters $c_{1,2,3}$ accounting for the emissions of 165 operating turbofan engine models are $c = \{0.556, 10.208, 4.068\}$ for CO and $c = \{0.083, 13.202, 1.967\}$ for HC [13]. The nitrogen oxides (NO_X) emission index [g/Kg $_{Jet\text{-}A1}$] based on the curve fitting of 177 operating aircraft engines models is [13]:

$$EI_{NO_X}(\tau) = 7.32\,\tau^2 + 17.07\,\tau + 3.53 \tag{17.28}$$

In case the mitigation of radiative forcing associated with condensation trails (contrails) is one of the environmental objectives to be considered, the MOTO suite can include a model such as the one presented in [46, 47].

17.4.4 Weather Input Data

The weather module currently adopted processes the global data available from the United States National Climatic Data Center of the National Oceanic and Atmospheric Administration (NOAA) and selectively interpolates the required information on a local structured 4D grid. In particular, the 4D weather grid data encompasses wind (latitude and longitude components), pressure, temperature, and humidity and is extracted from the Global Forecast System (GFS), collected on a 0.25° latitude and longitude resolution grid, updated every 6 hours (4 times daily), including projections of up to 180 hours in 3-hour intervals. The GFS is operationally classified into the Weather Planning and Decision Service (WPDS) that, as recommended by Radio Technical Commission for Aeronautics (RTCA) standards, should be used for offline and strategic online planning duties targeting time horizons greater than 20 minutes [48–51]. A more detailed discussion on weather data and services for ATM DSS is included in [52]. To simplify the numerical processing of weather data in the MOTO algorithm, the gridded data for wind and air density (calculated from temperature and pressure) was interpolated adopting custom polynomial functions in 3D space and time.

WPDS data such as the GFS have been deemed sufficient by RTCA for flight planning and offline/strategic online trajectory optimisation applications within airborne avionics and ground-based ATM DSS, as the main factors driving the lateral/vertical planning are the 4D wind and density/temperature/pressure fields. Nevertheless, a significant gap is represented by the lack of information regarding weather cells and other adverse phenomena in these macroscopic models. Consequently, as acknowledged by the ICAO's Aviation Systems Block Upgrade roadmap, further improved meteorological services are required to support advanced functionalities, such as those involved in 4D-TBO. Therefore, this is an area of major active research [52–57], as also acknowledged by the RTCA [51].

17.4.5 Multi-Objective Optimality

When introducing multiple performance indexes as in (17.1), the optimisation process leads to a set of optimal solutions called *Pareto* frontier and suitable decision criteria (or articulation of preferences) are therefore necessary to identify a unique solution. Preferences can be articulated either *a priori* (i.e. beforehand) or *a posteriori* (i.e. afterwards). A priori methods entail strategies to identify a combined objective to be supplied to the OCP solution algorithm, while a posteriori methods assist the selection of the final solution from an already calculated Pareto set [10, 58]. The most commonly used a posteriori articulation is the weighted sum, which was used in initial 4-PNV developments but was later discarded as the intrinsic susceptibility to bad scaling and normalisation issues prevented its reliable use. A very promising alternative for our 4-PNV implementation is the weighted product. Due to its mathematical nature, this a posteriori multi-objective optimality criterion is significantly more robust to normalisation and scaling than the weighted sum, so this approach can prove useful when the range of the objective function is unknown or unbounded. Following this approach, the combined performance index is defined as a product of all the objective functions, to the power of the assigned weight w_i, as in:

$$\tilde{J} = \prod_{i=1}^{n_J} \left[\frac{J_i}{k_i} \right]^{w_i} \tag{17.29}$$

where $0 < k_i < \min(J_i)$ is an optional scaling factor that can be introduced to prevent numerical saturation associated with the large magnitude range of the resulting combined objective, which might exceed the double precision limits of the computation platform. This limitation was considered and, in our implementation, the values of individual costs/performances were scaled to remain within two orders of magnitude of a baseline value. Another limitation of the weighted product is that the computational complexity of the iterative solution increases when more than two highly nonlinear objective functions are introduced, but our numerical verification activities suggested that this is not of particular concern up to four conflicting objectives. Clearly, in order to attain a non-trivial solution, it is strictly necessary that $J_i > 0, \forall i$ at all times. The performance weightings are a multi-objective generalisation of the cost index (CI) and can be varied dynamically along the different phases of the flight.

17.4.6 Control Input Smoothing and Operational Trajectory Realisation

Two sequential processes were introduced to transpose the mathematically optimal trajectory, which involves a discretised continuous and piece-wise smooth (CPWS) series of system states (i.e. longitude, latitude, altitude, airspeed, flight path angle, track angle, and mass), into an operationally flyable 4DT. The first of the two steps consists of a control input smoothing (CIS). The second step is named operational trajectory realisation (OTR). The aim of these two processes is to generate a 4DT that is characterised by a reduced sequence of standardised 4DT descriptors so as to facilitate system processing and data-link exchange for air-ground negotiation and validation. Additionally, it only involves manoeuvres that can be flown by current generation AFCS (in terms of control input time series), while still retaining most of the optimality of the original mathematical optimum. While the CPWS nature of state variables is ensured by the dynamic constraints (17.2), no smoothness is in principle imposed on the three control inputs τ, κ, N so the OCP solutions can involve step or impulse changes in the control inputs, which cannot be executed by real aircraft. Furthermore, simultaneous variations of multiple control input parameters are also not desirable when considering an implementation with state-of-the-art AFCS. The

CIS process is therefore tasked to recompute the control inputs required to fly the optimal trajectory while meeting user-defined constraints on the rate (or frequency) of change in the control inputs and ensuring that only conventional flight manoeuvres are followed, including: straight and level flight, level turns, straight climb and descent, turning climb and descent. The CIS module is also designed to detect any unfeasible numerical solution entailing jumps or mathematical singularities in the variables. Unfeasible trajectories are flagged as such and discarded from further processing. The parallel computation of multiple optimal trajectories ensures that these indeed very rare cases do not compromise the overall 4-PNV process. To implement the desired smoothing, with respect to the conventional 3-DoF formulation (17.18), the following restrictions are introduced, respectively, on bank angle rate, load factor rate and throttle rate to capture the limitations associated with conventional medium-large transport aircraft [59]:

$$|\dot{\mu}| \leq \dot{\mu}_{MAX} = 15°/s \tag{17.30}$$

$$|\dot{N}| \leq \dot{N}_{MAX} = 0.25 \, s^{-1} \tag{17.31}$$

$$|\dot{\tau}| \leq \dot{\tau}_{MAX} = 0.2 \, s^{-1} \tag{17.32}$$

These smoothing constraints can be combined with the control input constraints specified in the MOTO path constraints, which for a medium-large transport aircraft are:

$$-30.0° \leq \mu \leq 30.0° \tag{17.33}$$

$$0.75 \leq N \leq 1.5 \tag{17.34}$$

$$0 \leq \tau \leq 1 \tag{17.35}$$

Combining (17.30) with (17.33), it is possible to determine the maximum frequency of change in bank angle:

$$f_{\delta\mu} \leq \frac{\dot{\mu}_{MAX}}{\Delta\mu} = 0.5 \, Hz \tag{17.36}$$

where the maximum bank angle change used is $\Delta\mu = 25°$, corresponding to the difference between the maximum bank angle and the horizontal level attitude. Similarly, combining (17.31) with (17.34), it is possible to calculate the maximum frequency of change in load factor:

$$f_{\delta N} \leq \frac{\dot{N}_{MAX}}{\Delta N} = 0.5 \, Hz \tag{17.37}$$

where the maximum load factor change used is $\Delta N = 0.25$, corresponding to the maximum difference from the horizontal level condition. Analogously, combining (17.32) with (17.35), it is possible to determine the maximum frequency of change in throttle:

$$f_{\delta\tau} \leq \frac{\dot{\tau}_{MAX}}{\Delta\tau} = 0.2 \, Hz \tag{17.38}$$

where the maximum change in throttle setting adopted is $\Delta\tau = 1$. These smoothness constraints can either be implemented by including (17.30)–(17.32) as part of the dynamic constraints and adopting $\dot{\tau}$, $\dot{\mu}$, $\dot{N}$ as new control inputs (in an a priori smoothing approach), or by exploiting computationally efficient low-pass filters as part of the CIS process to apply

(17.36)–(17.38) as cut-off frequencies. Both approaches were successfully demonstrated in our study and, although including the control smoothness constraints in the OCP leads to a slightly improved optimality of the solutions, the CIS post-processing has a distinctive advantage in terms of computational speed and simplicity.

The 4DT resulting from the CIS process still consists of a high number of only fly-over waypoints, and the combination and frequency of control input changes is still beyond the capabilities of human pilots and current AFCS. It is therefore desirable to derive a 4DT involving a suitable combination of both fly-over and fly-by waypoints as this reduces the overall number of waypoints, especially in the case of long constant radius turns. The purpose of the OTR process is therefore to transpose the operationally smooth trajectory generated by the CIS process into a concisely described RNP 4DT that can be efficiently processed by the DSS and that can be easily amended by the human operator if necessary. All output data from CIS algorithms are therefore reprocessed and information is extracted, in particular, in terms of turn radius and flight path angle for each and every segment. Multiple segments in the original trajectory are then clustered together in either level/climbing/descending fly-by or fly-over legs. The clustering is performed when the variation of turn radius or flight path angle are limited in a 6.4% interval relative to the feasible range. Whenever this threshold is exceeded, the cluster is split and a new leg is introduced. Constant radius turn clusters are then redefined by a fly-by 4D waypoint. Particular conditions are introduced to ensure that the extrapolated fly-by turns meet the specifications set for RNP 4DT descriptors as specified in RTCA DO-236C and DO-229D [28, 29]. In particular, the fly-by transition is defined by the following equations:

$$R = \frac{1}{\kappa} = \frac{GS^2}{g\tan\mu} \tag{17.39}$$

$$Y = R\tan\left(0.5\,\delta\chi\right) \tag{17.40}$$

$$GS = |\vec{v}_{TAS} + \vec{v}_{wind}| = \begin{cases} 500\,kts\ (low\ altitude) \\ 750\,kts\ (high\ altitude) \end{cases} \quad unless\ actual\ Ground\ Speed\ (GS)\ is\ used \tag{17.41}$$

$$\mu = \begin{cases} \min\left[0.5\,\delta\chi;\ 23.0°\right]\ (low\ altitude) \\ \qquad 5.0° \qquad\quad (high\ altitude) \end{cases} \tag{17.42}$$

$$if\ Y > 20.0\,nmi \quad \Rightarrow \quad \begin{cases} Y = 20.0\,nmi \\ R = \frac{20.0}{\tan(0.5\,\delta\chi)} \end{cases} \tag{17.43}$$

The geometry of the fixed radius transition is therefore defined only by the track change $\delta\chi$ and the radius $R = 1/\kappa$. Based on these, it is possible to calculate the Lead Distance Y (from the turn initiation to the waypoint) and the Abeam Distance X (between the waypoint and the arc point abeam the fly-by waypoint) as per Eq. (17.40) and the following equation:

$$X = R \cdot \left(\frac{1}{\cos\delta\chi/2} - 1\right) \tag{17.44}$$

As the MOTO implementation presented here mainly addresses terminal sequencing and spacing, the criteria for the 'Feeder, Departure and Missed Approach' context from RTCA DO-229D [28] were adopted.

17.5 Numerical Verification

The numerical verification of the online tactical 4DT planning algorithm presented here focuses on the TMA, which is one of the most relevant online traffic management scenarios. The TMA is the key airspace structure around major airports or multiple airports, and 4-PNV functionalities can offer significant benefits in this context as many climbing and descending aircraft intersect in a short period of time and less than 50 nautical miles (nmi) from the airport, generating multiple potential conflicts. Particular emphasis is placed on the terminal sequencing and spacing of inbound traffic, as carried out by the arrival manager (AMAN), because it demonstrates the multi-aircraft 4DT planning capability and is an area of significant potential efficiency improvements. The AMAN aims to maximise the operational efficiency (i.e. minimising fuel and time costs) while resolving any conflict and establishing the required minimum spacing between aircraft on the final approach. The AMAN operational objectives and requirements closely match the ones associated with several other ATM duties, therefore the numerical verification presented here ideally captures most operational contexts. To accomplish the terminal sequencing and spacing of inbound traffic, the ATM DSS identifies the best arrival sequence among the feasible set considering the required spacing at the final approach fix and subsequently allocates the time slots to each aircraft. The minimum spacing on the approach is calculated based on the following equations [60]:

$$T_{i,j} = max \left[\frac{r + s_{i,j}}{v_j} - \frac{r}{v_i}, \ o_i \right] \text{ when } v_i > v_j \tag{17.45}$$

$$T_{i,j} = max \left[\frac{s_{i,j}}{v_j}, \ o_i \right] \text{ when } v_i \leq v_j \tag{17.46}$$

where $T_{i,j}$ is the separation interval between aircraft i and j, v_i and v_j are the ground speeds of the leading and trailing aircraft respectively, o_i is the runway occupancy time of the leading aircraft on landing, $s_{i,j}$ is the minimum longitudinal separation distance due to wake turbulence and r is the minimum lateral (radar) separation, assumed to be 4 nmi. Adopting the wake turbulence separation minima from the applicable regulations (see for example [61]) and adopting realistic runway occupancy times, the calculated spacing typically ranges between 60 and 330 seconds. The sequencing algorithm is a conventional limited set sorting, detailed in [18]. To realistically simulate the AMAN traffic management context, the sequencing and spacing algorithm calculated the deconfliction constraints in the form of required time of arrival (RTA) at the FAF. The capabilities of the ATM DSS to perform the optimal sequencing and spacing and of the OTR process to generate concise 4DT consisting of a limited number of overfly and fly-by waypoints were presented in [52]. The multi-objective optimality and the CIS process performances are therefore targeted in this simulation-based verification.

17.5.1 Results and Discussion

The 4DT planning verification case study involves four aircraft with different initial positions inbound on the same final approach segment. To study the behaviour of the MOTO algorithm, the same make and model were selected for the four aircraft and the initial

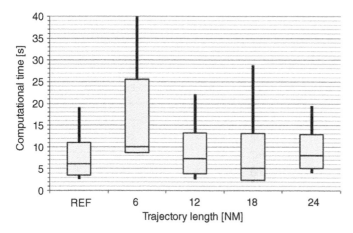

Figure 17.4 Box-and-whisker plot of the typical optimal 4DT computation times as a function of the total 4DT length from the entry point to the FAF.

positions were defined at progressive distances and bearings from the final approach fix (FAF). In particular, the distances were 6, 12, 18, and 24 nmi, whereas the bearings measured relative to the final approach track were 20, 40, 60, and 80. Various aircraft were successfully tested resulting in similar algorithm performance, hence this section presents the results for the Airbus A320 only. The initial track angle was set to the opposite of the final approach track and the initial flight path angle was set to zero. A 3-DoF initializer algorithm was implemented to calculate the initial airspeed and altitude conditions for steady flight. The statistics for 4DT computation time are depicted in Figure 17.4. These results were obtained running the 4DT planning algorithm in a MATLAB simulation environment on a computer equipped with an Intel Core i7-4790 processor and 8 gigabytes of random access memory (RAM). With the exception of the reference (REF) 4DT, calculated using fuel costs only, these results consider all the cost functions elements described in the previous section. The number of discretisation intervals was set to the distance to the FAF divided by twice the maximum TAS, resulting in approximately three seconds duration each. The results indicate that the 4DT planning algorithm yields best performances with intermediate distances (i.e. 12 and 18 nmi) from the FAF and this is likely due to the increased manoeuvre complexity at shorter distances and higher number of intervals in longer trajectories. Various combinations of weightings are used in the weighted product (17.29) to generate multiple Pareto-optimal trajectories, addressing the conflicting objectives of CO/HC emissions from (17.27) and NO_X emissions from (17.28) as well as fuel costs. The weightings adopted and the resulting emissions and fuel-burn performance of aircraft number 4 are listed in Table 17.2. Despite the relatively short distance covered, significantly different performances are observed, with trajectory number 4 frequently representing the best compromise, as it equally accounts for CO_X, HC, NO_X emissions, and fuel consumption, but this is case-dependent. Finer Pareto fronts can be generated as needed by introducing intermediate weightings within the range. The spatial coordinates of optimal 4DT number 4 are depicted for all four aircraft in Figure 17.5. A representation of all 4DT for each aircraft is not included here as the spatial differences are typically too small to be

Table 17.2 Overview of the weightings adopted and performance achieved in the terminal area case study.

	Adopted weightings			Resulting performance		
4DT no.	w_{Fuel}	w_{CO} and w_{HC}	w_{NOx}	CO Emission (g)	NO_x Emission(g)	Fuel burn(kg)
1	1	0	4.000	911.7	205.4	39.4
2	1	0.667	3.333	907.5	205.5	39.3
3	1	1.333	2.667	577.1	235.5	39
4	1	2.000	2.000	410.1	264.9	39.3
5	1	2.667	1.333	151.5	1156.3	94.8
6	1	3.333	0.667	192.6	3633.3	237.3
7	1	4.000	0	172.4	3425	209.8

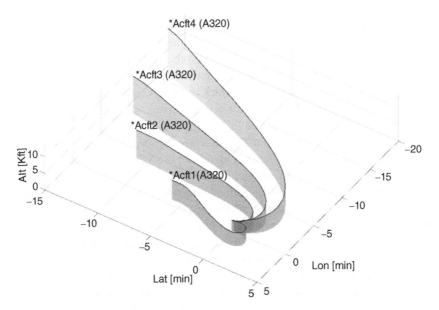

Figure 17.5 Optimal 4DT intent number 4 plotted for all four aircraft.

noticed and do not represent the actual differences, which normally involve the other state variables and especially the control variables.

Figure 17.6 provides a detailed outline of the optimal 4DT intent number 4 for the third aircraft (i.e. starting at 18 nmi from FAF). The figure particularly represents the original CPWS optimal trajectory and the standardised descriptors for lateral 4DT path as calculated by the OTR post-processing module, which are overfly 4D waypoints (points and segments in magenta) and fly-by 4D waypoints (points and segments in red), described in section III-G Eqs. (17.39)–(17.44).

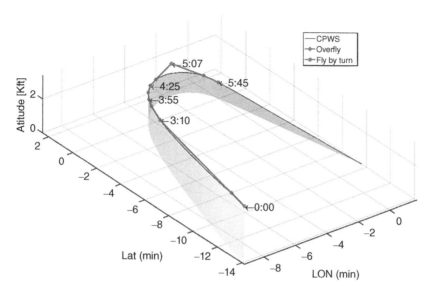

Figure 17.6 Optimal 4DT number 4 for aircraft number 2: details of the operational trajectory realisation (times are in minutes and seconds relative to the initial point of the trajectory).

As highlighted by the figure, the OTR procedure reduces mathematically optimal CPWS trajectories consisting of between 30 to several hundred discrete segments down to a number of fly-by and overflying 4D waypoints consistently below 20. These results meet the set design requirements for tactical online data-link negotiation of the 4DT. Finally, as a reference, the time histories of control inputs and state variables associated with intent number 4 of aircraft number 3 (18 nmi) are depicted in Figure 17.7.

As previously discussed, the CPWS nature of the state variables is already ensured by the dynamic constraints. All abrupt control input variations, on the other hand, are processed via the CIS algorithm, thereby fulfilling the smoothing constraints (17.31)–(17.33).

17.6 Conclusion

A suitable set of 4-PNV functionalities was developed for integration in future avionics and ATM systems. The 4DT planning process exploits custom optimal control-based MOTO algorithms, which can natively accommodate time constraints, offering significant benefits for 4D TBO traffic management. Custom post-processing steps were introduced to successfully translate the mathematical optimum into smooth guidance commands that can be executed by current autopilots using a limited number of 4D trajectory descriptors. A suitable multi-objective articulation of preferences based on the weighted product was proposed, which allows us to formulate the desired operational, economic, and environmental criteria and supports the generation of intents having very different fuel and emission performances. These functionalities support 4D TBO and particularly address the strategic and tactical online timeframes, as the complete 4-PNV process can fulfil the time requirements for online tactical routing/rerouting tasks in representative TMA traffic conditions. Ongoing research is assessing the losses in optimality introduced by the 4DT post-processing

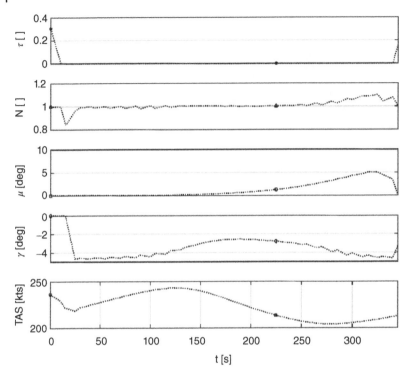

Figure 17.7 Time histories of the control inputs and state variables associated with 4DT number 4 of aircraft number 2.

stages and preliminary results are very encouraging. In future research, the 4-PNV will be extended to incorporate suitable integrity monitoring and augmentation strategies, including both predictive and reactive features similarly to what has been proposed for satellite navigation in [62]. Research is now focusing on the system's human-machine interfaces and interactions for 4D TBO, particularly by exploring the benefits offered by adaptive interfaces. Ongoing research is also addressing the certification challenges of integrated avionics and ATM systems for TBO [18].

References

1 H. Erzberger, (2004) "Transforming the NAS: the next generation air traffic control system", NASA Ames Research Center, Moffett Field, CA, USA NASA/TP-2004-212801.

2 Ky, P. and Miaillier, B. (2006). SESAR: towards the new generation of air traffic management systems in Europe. *Journal of Air Traffic Control* 48.

3 Federal Aviation Administration (FAA), (2013). "NextGen Implementation Plan", Washington DC, USA.

4 SESAR, (2015). "European ATM Master Plan - The roadmap for delivering high performance aviation for Europe", SESAR, Brussels.

5 SESAR (2007). "The ATM Target Concept - D3", SESAR Definition Phase DLM-0612-001-02-00.

6 Barman, J.F. and Erzberger, H. (1976). Fixed-range optimum trajectories for short-haul aircraft. *Journal of Aircraft* 13: 748–754.

7 Sorensen, J.A., Morello, S.A., and Erzberger, H. (1979). Application of trajectory optimization principles to minimize aircraft operating costs. In: *18th IEEE Conference on Decision and Control* 1: 415–421.

8 H. Q. Lee and H. Erzberger, (1980). "Algorithm for fixed-range optimal trajectories", National Aeronautics and Space Administration (NASA), Ames Research Center, Moffett Field, CA, USA NASA Technical Paper TP-1565.

9 Erzberger, H. and Lee, H. (1980). Constrained optimum trajectories with specified range. *Journal of Guidance, Control, and Dynamics* 3: 78–85.

10 Gardi, A., Sabatini, R., and Ramasamy, S. (2016). Multi-objective optimisation of aircraft flight trajectories in the ATM and avionics context. *Progress in Aerospace Science* 83: 1–36.

11 Villarroel, J. and Rodrigues, L. (2016). An optimal control framework for the climb and descent economy modes of flight management systems. *IEEE--Transactions on Aerospace & Electronic Systems* 52: 1227–1240.

12 A. Gardi, R. Sabatini, S. Ramasamy, and K. de Ridder, (2013). "4-Dimensional trajectory negotiation and validation system for the next generation air traffic management," *AIAA Guidance, Navigation, and Control Conference, GNC 2013*, Boston, MA, USA.

13 A. Gardi, R. Sabatini, T. Kistan, Y. Lim, and S. Ramasamy, (2015). "4-Dimensional trajectory functionalities for air traffic management systems," *Integrated Communication, Navigation and Surveillance Conference (ICNS 2015)*, Herndon, VA, USA.

14 Erzberger, H. and Chapel, J.D. (1984). Concepts and algorithms for terminal-area traffic management. In: *1984 American Control Conference*, 166–173. New York, NY/San Diego, CA: IEEE.

15 Bradbury, J.N. (1991). ICAO and future air navigation systems. In: *Automation and Systems Issues in Air Traffic Control* (eds. J.A. Wise, V.D. Hopkin and M.L. Smith), 79–99. Springer.

16 D. H. Williams and S. M. Green, (1991). "Airborne four-dimensional flight management in a time-based air traffic control environment", NASA Langley Research Center, Hampton, VA, USA NASA Technical Memorandum TM-4249.

17 Visser, H.G. (1992). Terminal area traffic management. *Progress in Aerospace Science* 28: 323–368.

18 Kistan, T., Gardi, A., Sabatini, R. et al. (2017). An evolutionary outlook of air traffic flow management techniques. *Progress in Aerospace Science* 88: 15–42.

19 B. Korn and A. Kuenz, (2006). "4D FMS for increasing efficiency of TMA operations," *25th IEEE/AIAA Digital Avionics Systems Conference (DASC 2006)*, Portland, OR, USA.

20 Korn, B., Helmke, H., and Kuenz, A. (2007). 4D trajectory management in the extended TMA: Coupling AMAN and 4D FMS for optimized approach trajectories. In: *25th Congress of the International Council of the Aeronautical Sciences (ICAS 2006)*, 4103–4112. Hamburg, Germany: ICAS.

21 T. Nikoleris, H. Erzberger, R. A. Paielli, and Y.-C. Chu, (2014) "Autonomous system for air traffic control in terminal airspace," *14th AIAA Aviation Technology, Integration, and Operations Conference (ATIO 2014)*, Atlanta, GA, USA.

22 Erzberger, H., Lauderdale, T.A., and Chu, Y.C. (2012). Automated conflict resolution, arrival management, and weather avoidance for air traffic management. *Proceedings of the Institution of Mechanical Engineers, Part G: Journal of Aerospace Engineering* 226: 930–949.

23 Azzopardi, M.A. and Whidborne, J.F. (2011). Computational air traffic management. In: *30th Digital Avionics Systems Conference (DASC 2011)*, 1B61–1B622. Seattle, WA: IEEE.

24 E. Mueller and S. Lozito, (2008). "Flight deck procedural guidelines for datalink trajectory negotiations," *the Joint 26th International Congress of Aeronautical Sciences and 8th Aviation Technology, Integration, and Operations conference (ICAS+ATIO 2008)*, Anchorage, AK, USA.

25 E. Mueller, (2007). "Experimental evaluation of an integrated datalink and automation-based strategic trajectory concept," *AIAA Aviation Technology, Integration and Operations conference 2007 (ATIO2007)*, Belfast, Northern Ireland.

26 J. E. Robinson III, and D. R. Isaacson, (2000). "A concurrent sequencing and deconfliction algorithm for terminal area air traffic control," *AIAA Guidance, Navigation and Control Conference 2000 (GNC 2000)*, Denver, CO, USA.

27 I. A. B. Wilson, (1998). "Trajectory negotiation in a multi-sector environment", Programme for Harmonised Air Traffic Management Research in EUROCONTROL (PHARE), Bruxelles, Belgium.

28 RTCA, (2013). "Minimum operational performance standards for global positioning system/satellite-based augmentation system airborne equipment", Radio Technical Commission for Aeronautics Inc. (RTCA), Washington DC, USA RTCA DO-229D.

29 RTCA, (2013). "Minimum aviation system performance standards: required navigation performance for area navigation", Radio Technical Committee for Aeronautics Inc. (RTCA), Washington DC, USA RTCA DO-236C.

30 Betts, J.T. (1998). Survey of numerical methods for trajectory optimization. *Journal of Guidance, Control, and Dynamics* 21: 193–207.

31 Rao, A.V. (2010). Survey of numerical methods for optimal control. *Advances in the Astronautical Sciences* 135: 497–528.

32 von Stryk, O. and Bulirsch, R. (1992). Direct and indirect methods for trajectory optimization. *Annals of Operations Research* 37: 357–373.

33 Visser, H.G. (1994). *A 4D Trajectory Optimization and Guidance Technique for Terminal Area Traffic Management*. Delft University of Technology.

34 Rao, A.V., Benson, D.A., Darby, C. et al. (2010). Algorithm 902: GPOPS, A MATLAB software for solving multiple-phase optimal control problems using the gauss pseudospectral method. *ACM Transactions on Mathematical Software* 37 (2): 22-1–22-39.

35 Soler, M., Olivares, A., and Staffetti, E. (2010). Hybrid Optimal Control Approach to Commercial Aircraft Trajectory Planning. *Journal of Guidance, Control, and Dynamics* 33: 985–991.

36 Bousson, K. and Machado, P. (2010). 4D Flight trajectory optimization based on pseudospectral methods. *World Academy of Science Engineering and Technology* 70: 551–557.

37 Soler, M., Olivares, A., and Staffetti, E. (2014). Multiphase optimal control framework for commercial aircraft four-dimensional flight-planning problems. *Journal of Aircraft* 52: 274–286.

38 Rodionova, O., Sbihi, M., Delahaye, D., and Mongeau, M. (2014). North atlantic aircraft trajectory optimization. *IEEE Transactions on Intelligent Transportation Systems* 15: 2202–2212.

39 Park, S.G. and Clarke, J.P. (2015). Optimal control based vertical trajectory determination for continuous descent arrival procedures. *Journal of Aircraft* 52: 1469–1480.

40 Chircop, K., Zammit-Mangion, D., and Sabatini, R. (2012). Bi-objective pseudospectral optimal control techniques for aircraft trajectory optimisation. In: *28th Congress of the International Council of the Aeronautical Sciences 2012, ICAS 2012*, 3546–3555. Brisbane: ICAS.

41 Canuto, C., Hussaini, M.Y., Quarteroni, A., and Zang, T.A. (2007). *Spectral Methods - Fundamentals in Single Domains*. Heidelberg: Springer-Verlag Berlin Heidelberg.

42 Betts, J.T. (2010). *Practical methods for optimal control and estimation using nonlinear programming*, 2e, vol. 19. SIAM.

43 Bertsekas, D.P. (1999). *Nonlinear Programming*, 2e. Belmont, MA: Athena Scientific.

44 Eurocontrol Experimental Centre (EEC), (2013). "User manual for the base of aircraft data (BADA) revision 3.11", Brétigny-sur-Orge, France Technical/Scientific Report No. 13/04/16-01.

45 J. Hupe, B. Ferrier, T. Thrasher, C. Mustapha, N. Dickson, T. Tanaka, *et al.*, (2013). "ICAO environmental report 2013: destination green - aviation and climate change", ICAO Environmental Branch, Montreal, Canada.

46 Sridhar, B., Ng, H., and Chen, N. (2011). Aircraft Trajectory Optimization and Contrails Avoidance in the Presence of Winds. *Journal of Guidance, Control, and Dynamics* 34: 1577–1584.

47 Lim, Y., Gardi, A., and Sabatini, R. (2015). Modelling and evaluation of aircraft contrails for 4-dimensional trajectory optimisation. *SAE International Journal of Aerospace* 8: 248–259.

48 ICAO, (2010). "Annex 3 to the convention on international civil aviation - meteorological service for international air navigation", The International Civil Aviation Organization (ICAO), Montreal, Canada.

49 RTCA, (2007). "RTCA DO-308: operational services and environment definition (OSED) for aeronautical information services (AIS) and meteorological (MET) data link services," ed: SC-206.

50 RTCA, (2010). "RTCA DO-324: safety and performance requirements (SPR) for aeronautical information services (AIS) and meteorological (MET) data link services," ed: SC-206.

51 RTCA, (2012). "RTCA DO-340: Concept of Use (CONUSE) for Aeronautical Information Services (AIS) and Meteorological (MET) Data Link Services," ed: SC-206.

52 A. Gardi, Y. Lim, T. Kistan, and R. Sabatini, (2016). "Planning and negotiation of optimised 4D trajectories in strategic and tactical re-routing operations," *30th Congress of the International Council of the Aeronautical Sciences, ICAS 2016*, Daejeon, Korea.

53 Joint Planning and Development Office (JPDO), (2008). "Four-dimensional weather functional requirements for NextGen air traffic management", NextGen Joint Planning and Development Office (JPDO).

54 K. Claypool, A.-H. Sanh, M. Schultz, M. Win, E. Mandel, M. Simons, *et al.*, (2010). "4-Dimensional weather data cube web feature service reference implementation (WFSRI) requirements", MIT Lincoln Laboratory - National Oceanic and Atmospheric Administration (NOAA) - National Center for Atmospheric Research (NCAR).

55 T. G. Reynolds, Y. Glina, S. W. Troxel, and M. D. McPartland, (2013). "Wind information requirements for NextGen applications - Phase 1: 4D-trajectory based operations (4D-TBO)", MIT Lincoln Laboratory, Lexington (MA), USA Project Report ATC-399.

56 T. G. Reynolds, P. M. Lamey, M. D. McPartland, M. J. Sandberg, T. L. Teller, S. W. Troxel, *et al.*, (2014). "Wind information requirements for NextGen applications - Phase 2 final report: framework refinement and application to four-dimensional trajectory based operations (4D-TBO) and interval management (IM)", MIT Lincoln Laboratory, Lexington (MA), USA Project Report ATC-418.

57 C. Edwards, M. D. McPartland, T. G. Reynolds, M. J. Sandberg, T. L. Teller, and S. W. Troxel, (2014). "Wind information requirements for NextGen Applications - Phase 3 final report", MIT Lincoln Laboratory, Lexington (MA), USA Project Report ATC-422.

58 Marler, R.T. and Arora, J.S. (2004). Survey of multi-objective optimization methods for engineering. *Structural and Multidisciplinary Optimization* 26: 369–395.

59 Airbus Industrie, (1993). "A320 Electrical Flight Control System", Airbus Industrie,, Blagnac, France AI/EV-0 472.0395/93.

60 De Neufville, R., Odoni, A., Belobaba, P., and Reynolds, T. (2013). *Airport Systems - Planning, Design and Management*, 2nde. New York: McGraw-Hill.

61 UK Aeronautical Information Services, (2014). "Wake turbulence", NATS Aeronautical Information Circular P 3/2014.

62 Sabatini, R., Moore, T., and Hill, C. (2012). Avionics-based integrity augmentation system for mission- and safety-critical GNSS applications. In: *25th International Technical Meeting of the Satellite Division of the Institute of Navigation 2012, ION GNSS 2012*, 743–763. Nashville, TN: ION.

63 Gardi, A. (2017). A Novel Air Traffic Management Decision Support System – Multi-objective 4-Dimensional Trajectory Optimisation for Intent-Based Operations in Dynamic Airspace, PhD Thesis, RMIT University, School of Engineering, Melbourne, Australia.

64 Gardi, A., Sabatini, R., and Kistan, T. (2019). Multiobjective 4D Trajectory Optimization for Integrated Avionics and Air Traffic Management Systems. *IEEE Transactions on Aerospace and Electronic Systems* 55: 170–181.

18

Energy-Optimal 4D Guidance and Control for Terminal Descent Operations

Yixiang Lim[1], Alessandro Gardi[2,3], and Roberto Sabatini[2,3]

[1]*Agency for Science, Technology and Research (ASTAR), Singapore*
[2]*Department of Aerospace Engineering, Khalifa University of Science and Technology, Abu Dhabi, UAE*
[3]*School of Engineering, RMIT University, Melbourne, Victoria, Australia*

18.1   Introduction

During the descent phase, an aircraft transits from quasi-cruise to approach conditions, while dissipating both kinetic and potential energy as it loses altitude and velocity. With a shift towards trajectory-based operations (TBOs), aircraft flying the descent phase will need to meet strict time and space constraints in the form of 4-dimensional (4D) waypoints. The descent phase is also characterised by a significant number of tasks happening in rapid sequence, posing great challenges for the introduction of single-pilot and unmanned transport aircraft. Compared with legacy paradigms, TBO offer a safer and more efficient exploitation of the terminal airspace capacity [1]. This case study, originally published in [15, 16], presents a guidance and control strategy to support the use of 4D trajectories (4DTs) in descent operations. The guidance process determines an energy-optimal trajectory to be flown along the descent phase by an aircraft operating under instrument flight conditions. The guidance process calculates regular updates to the active 4DT, performing replanning when significant deviations from the previously defined trajectory are detected. That notwithstanding, due to the complex mathematical nature of the guidance problem, trajectory recalculations should be limited to updating the current trajectory every 15–30 seconds. Consequently, deviations from the descent trajectory predefined as part of the guidance process are typically addressed by a suitable control method. For this, we propose the adoption of two control modules separately determining the attitude corrections (addressing cross-track and vertical deviations) and the time and energy (T&E) corrections (addressing along-track and airspeed deviations). Additionally, this chapter also addresses the possible need for human pilot involvement and intelligent system adaptation. As such, it is part of a larger research project tackling the development of cognitive human machine interfaces and interactions (HMI²) in aviation [2]. In particular, the lack of established HMI² formats and functions for 4DT following was identified as a major limitation to be addressed, which creates significant potential synergies with collision avoidance [3, 4].

Sustainable Aviation Technology and Operations: Research and Innovation Perspectives, First Edition.
Edited by Roberto Sabatini and Alessandro Gardi.
© 2024 John Wiley & Sons Ltd. Published 2024 by John Wiley & Sons Ltd.
Companion Website: www.wiley.com/go/sustainableaviation

The chapter is structured as follows. Section II introduces the problem statement, the overall guidance and control architecture, and the energy-based problem formulation. Section III describes the guidance strategy implemented. The control strategy is subsequently presented in Section IV. The verification case studies are presented and discussed in Section V. Finally, the proposed 4DT HMI[2] prototype is introduced in Section VI.

18.2   4D Trajectory Management

The management of the 4DT requires the pilot to perform a number of tasks:

- Ensure that the lateral and vertical position of the aircraft stay within the bounds of the nominal trajectory, as specified by required navigation performance (RNP).
- Ensure that the energy of the aircraft does not deviate too significantly from the nominal energy profile. This ensures that the future aircraft position will not exceed the RNP buffer at a later stage.
- Ensure that the aircraft crosses the specified 4D waypoints along the nominal trajectory within the required time tolerances.
- Recalculate a new 4DT in the event of excessive cross-track, vertical, energy or time deviations from the nominal profile.

To assist the pilot in managing the time and the vertical position of the aircraft, an energy management approach was adopted. Energy state models have been frequently used in the past, and particularly in the guidance domain [5–7] and are commonly investigated [8, 9]. As seen in Figure 18.1, during the descent phase, the aircraft sheds energy at a controlled rate as it loses altitude and velocity. The specific energy, e, of the aircraft is given as the sum of potential and kinetic energies:

$$e = g \cdot h + v^2/2 \tag{18.1}$$

where g is the gravitational acceleration, h is the altitude and v is the velocity. Equation (18.1) can be linearized to obtain:

$$e_0 - \Delta e = g \cdot (h_0 - \Delta h) + (v_0 - \Delta v)^2/2$$
$$= g \cdot (h_0 - \Delta h) + (v_0^2 - 2v_0 \cdot \Delta v)/2 \tag{18.2}$$

where e_0, h_0, and v_0 are the energy, altitude and velocity of the nominal trajectory, Δe, Δh, and Δv are perturbations from the nominal trajectory, and assuming $\Delta v^2 \approx 0$. Differentiation of Equation (18.2) gives the change in energy flux $\Delta \dot{e}$ due to perturbations in h and v:

$$d/dt \left(e_0 - \Delta e\right) = -g \cdot \left(\Delta \dot{h}\right) - v_0 \cdot \Delta \dot{v} \tag{18.3}$$

$$\Delta \dot{e} = g \cdot \gamma_0 \cdot \Delta v + v_0 \cdot \Delta \dot{v} \tag{18.4}$$

Since $\Delta \dot{h}$ can be linearized as $\Delta \dot{h} = \Delta v \cdot \sin(\gamma_0) + v_0 \cdot \sin(\Delta \gamma) \approx \Delta v \cdot \gamma_0$, assuming small γ_0 and $\Delta \gamma = 0$ (i.e. no change in the flight path angle).

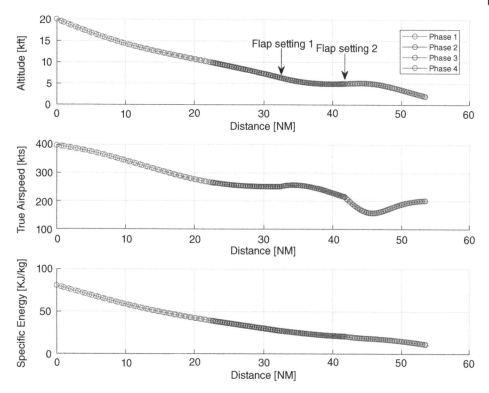

Figure 18.1 Altitude, velocity and specific energy of the descending aircraft as a function of along-track distance.

The energy flux is determined by the thrust and drag of the aircraft, which are primarily managed by the pilot through controlling the throttle and spoilers (or equivalently, speed brakes), as approximated by:

$$\dot{e} = v \cdot (T{-}D)/m$$
$$= v \cdot \left[\eta \cdot T_{\max} - q \cdot C_{D0} \cdot (1 + \delta) \right]/m \tag{18.5}$$

where v is the true airspeed (TAS, in m/s), T is the aircraft thrust and D is the aircraft drag (both in Newtons), m is the aircraft mass (in kg), T_{max} is the maximum continuous thrust (MCT), q is the dynamic pressure, C_{D0} is the zero-lift drag, $\delta = C_{Dsplr} / C_{D0}$ is the ratio between the aircraft zero-lift drag and the spoiler drag and η is the throttle. Linearising this equation about η and δ gives the change in energy flux due to changes in η and δ:

$$\left(\dot{e}_0{-}\Delta\dot{e} \right) = \left(v_0/m \right) \cdot \left[\left(\eta_0{-}\Delta\eta \right) \cdot T_{max} - q \cdot C_{D0} \cdot \left(1 + \delta_0{-}\Delta\delta \right) \right]$$
$$= \left(v_0/m \right) \cdot \left[\eta_0 \cdot T_{max} - q \cdot C_{D0} \cdot \left(1 + \delta_0 \right) \right] - \left(v_0/m \right) \cdot \left[\Delta\eta \cdot T_{max} - q \cdot C_{D0} \cdot \left(\Delta\delta \right) \right] \tag{18.6}$$

$$\Delta\dot{e} = \left(v/m \right) \cdot \left[\Delta\eta \cdot T_{max} - q \cdot C_{D0} \cdot \left(\Delta\delta \right) \right] \tag{18.7}$$

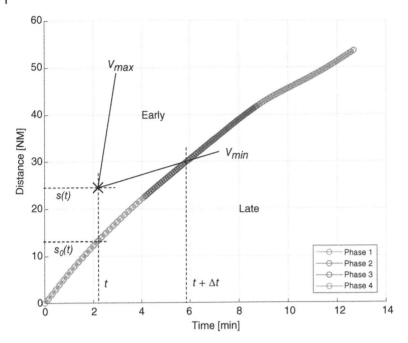

Figure 18.2 Along-track distance as a function of time, used to determine the time taken for the aircraft to rejoin the nominal trajectory.

where τ_0 and δ_0 are the nominal throttle and spoiler controls values, and $\Delta\eta$ and $\Delta\delta$ are their respective deviations. Equations (18.4, 18.7) can be equated to obtain:

$$g \cdot \gamma_0 \cdot \Delta v + v_0 \cdot \Delta \dot{v} = (v_0/m) \cdot \left[\Delta\eta \cdot T_{max} - q \cdot C_{D0} \cdot (\Delta\delta) \right] \tag{18.8}$$

The along-track distance is represented by a coloured curve in Figure 18.2 as a function of time. The figure clarifies the role of along-track deviations from the nominal trajectory, which in 4D TBO are treated as estimated time of arrival (ETA) deviations. In particular, at time t, if the aircraft, denoted by the 'x', lies above the curve, it is flying ahead of the nominal 4DT (early); whereas if the aircraft position lies beneath the curve, the aircraft is flying behind the nominal 4DT (late). The aircraft correct these deviations by adjusting its velocity to re-join the nominal trajectory at a later time $t + \Delta t$. Assuming a step-change in velocity, the shortest time taken to re-join the nominal trajectory is given by $(\Delta t)_{min}$:

$$(\Delta t)_{min} = \left[s(t) - s_0(t) \right] / \left[v_0(t) - v_{min} \right], \qquad \text{if early} \tag{18.9}$$

$$(\Delta t)_{min} = \left[s(t) - s_0(t) \right] / \left[v_0(t) - v_{max} \right], \qquad \text{if late} \tag{18.10}$$

The block diagram in Figure 18.3 illustrates our proposed strategy for managing a 4D descent profile. An outer guidance loop provides updates to the 4DT at 15–30 second intervals. The inner control comprises attitude and T&E control processes, which allow the 4DT to be flown manually (i.e. through the 4DT HMI energy director) or automatically (i.e. through the attitude and T&E controllers). The manual and automatic control modes provide a greater degree of operational flexibility. For instance, autopilot and autothrottle do not have sufficient control authority and are therefore frequently unable to cope with strong winds, windshear, and gusts, requiring the pilot to assume manual control

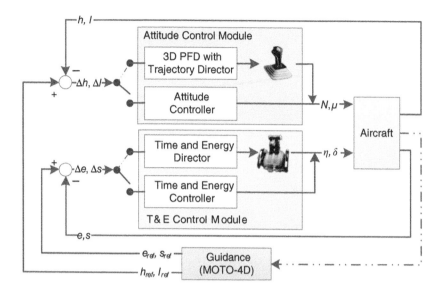

Figure 18.3 Guidance and control architecture for an energy-based 4DT descent.

in these instances. Support for manual control is also essential in the unlikely case of autopilot/autothrottle failure. We therefore introduce some customised HMI[2] formats supporting pilot awareness of the aircraft 3D position and energy state by presenting suitable throttle and spoiler recommendations.

18.3 Guidance Strategy

A variety of trajectory optimisation methods have been proposed and evaluated for the guidance process of transport aircraft [10]. Multi-objective trajectory optimisation (MOTO) methods allow the minimisation of conflicting objectives, such as flight time, fuel consumption, and aircraft emissions. MOTO can also natively support 4D-TBO by capturing the mathematical nature of the guidance problem. However, MOTO does not natively produce a concisely described 4DT, hence operational trajectory realisation (OTR) processes were proposed [10]. MOTO can also be computationally expensive due to the nonlinear dynamics and a large number of state variables being modelled. Therefore, an efficient strategy is to restrict the MOTO process to the vertical profile only, and implement a different technique for the lateral path planning. This is the approach adopted in our case study, as it was greatly beneficial in limiting the computational time, allowing it to be consistently completed in under five seconds. The lateral guidance algorithm determines the ground track using a purely geometric method based on conventional design principles. Given initial and final waypoints and headings, as well as the aircraft turn radius, the algorithm calculates the intermediate waypoints, segments, and fly-by turns required to achieve the final waypoint and heading. An example of the lateral guidance algorithm's result is shown in Figure 18.4.

The vertical component of the trajectory is then formulated as a multi-phase optimal control problem (MP-OCP) subject to multiple path constraints and integrating multiple

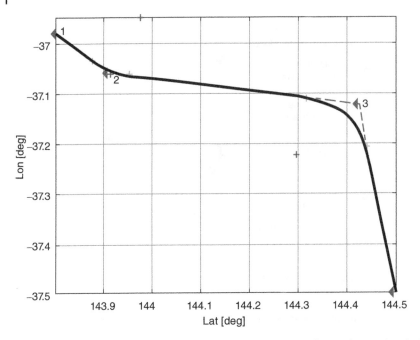

Figure 18.4 Lateral guidance algorithm computes the intermediate trajectory (marked by waypoints 2 and 3) given the initial and final position and headings, marked by waypoints 1 and 4, respectively.

objectives. The MP-OCP uses a reduced-order three degree of freedom (3DOF) model as dynamic constraint, omitting dependencies on the lateral dynamics of the aircraft. In the adopted model, a spatial dimension s is used, which represents the curvilinear coordinate along the ground track. The solution to this MP-OCP involves optimal time histories of state and control variables. The reduced-order 3DOF point mass model of the fixed-wing aircraft flight dynamics is:

$$\dot{x} = \begin{cases} \dot{v} = \dfrac{T - D}{m} - g \cdot \sin(\gamma) \\[2mm] \dot{\gamma} = \dfrac{g}{v} \cdot [N \cdot \cos(\gamma)] \\[2mm] \dot{s} = v \cdot \cos(\gamma) \\[2mm] \dot{z} = v \cdot \sin(\gamma) \\[2mm] \dot{m} = -FF \end{cases} \qquad (18.11)$$

where v is the true airspeed (in m/s), γ is the flight path angle (in radians), s is the distance along the ground track (in metres), z is the altitude (in metres), m is the aircraft mass (in kg), N is the load factor (dimensionless) and FF is the fuel flow (in kg/s). The control variables are $\boldsymbol{u} = \{\alpha, \eta, \delta splr\}$, which respectively denote the angle of attack (in radians), throttle (dimensionless), and spoiler angle (in radians). Although this model ignores the bank angle, we assume that the residual authority in load factor is sufficient to maintain horizontal flight when bank angle effects are considered.

The thrust T, drag D and fuel flow FF are obtained using aircraft performance models in Eurocontrol's Base of Aircraft Data (BADA) 3.9, expressed as:

$$T = \eta \cdot T_{max} \tag{18.12}$$

$$D = q \cdot S \cdot \left[C_{D0} + \left(C_{D2} \cdot C_L{}^2\right) + C_{Dsplr}\right] \tag{18.13}$$

$$FF = C_{F1} \cdot \left(1 + v/C_{F2}\right) \tag{18.14}$$

$$T_{max} = \left[C_{T1} \cdot \left(1 - z_{ft}/C_{T2} + C_{T3} \cdot z_{ft}{}^2\right) \cdot \left(1 - C_{T5} \cdot \left(\Delta T - C_{T4}\right)\right)\right] \tag{18.15}$$

$$C_L = N \cdot m \cdot g \cdot \cos(\gamma) / (q \cdot S) \tag{18.16}$$

$$C_{Dsplr} = \delta_{splr} \cdot \left(\partial C_{D0splr}/\partial \delta_{splr}\right) \tag{18.17}$$

$$q = \frac{1}{2} \cdot \rho \cdot v^2 \tag{18.18}$$

where S, C_{D0}, and C_{D2} are the wing span, zero-lift drag and induced drag, respectively, C_{F1} and C_{F2} are the fuel flow coefficients, and C_{T1}, C_{T2}, C_{T3}, C_{T4}, and C_{T5} are the thrust coefficients, all obtained from empirical models in BADA. q is the dynamic pressure, ΔT is the temperature deviation from the standard atmosphere and $\partial C_{D0splr}/\partial \delta_{splr}$ is spoiler deflection drag slope, which is dependent on the aircraft and can vary between 0.01 and 0.1. The load factor N is given by:

$$N = q \cdot S \left[(\alpha - \alpha_0) \cdot C_{L\alpha} + \alpha_F \cdot C_{LF}\right] / (m \cdot g) \tag{18.19}$$

$$C_{L\alpha} = 2\pi / (1 + 2/AR) \tag{18.20}$$

where $C_{L\alpha}$ is the aircraft lift curve slope, C_{LF} is the lift curve slope of the aircraft flaps, α_0 is the zero-lift angle of the aircraft, α_F is the flap angle and AR is the aircraft aspect ratio. The descent and approach were modelled with four phases requiring the following conditions:

- Phase 1: descent from initial altitude to 10 000 ft without regulatory airspeed constraints.
- Phase 2: descent below 10 000 ft, where the indicated airspeed (IAS) is limited to a maximum of 250 kts.
- Phase 3: further descent with flap setting 1 (slats + typically 5 degrees of flaps).
- Phase 4: final descent up to final approach fix (FAF) with flap setting 2 (typically 15 degrees of flaps).

The conditions above were introduced by parameterising a number of variables (e.g. flap angle α_F) in the dynamics model, as well as imposing bounds on airspeed, altitude, flight path angle, and time in the problem formulation. To minimise the workload of the pilot and to increase the flyability of the trajectory, the control variables $\boldsymbol{u} = \{\alpha, \eta, \delta splr\}$ were kept constant over each phase and only allowed to change during a change of phase. Finally, the cost function J was designed as a weighted product for minimising the use of spoilers and throttle, as well as the total weight of fuel used:

$$J = (1 + \Sigma_{i=4} \, \delta_{splr,Phase[i]})^{w\_splr} \cdot (1 + \Sigma_{i=4} \, \tau_{,Phase(i)})^{w\_\eta}$$
$$(m_0 - m_F)^{w\_fuel} \tag{18.21}$$

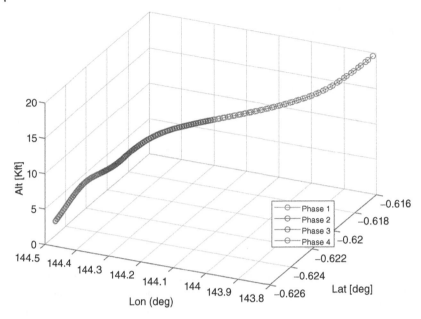

Figure 18.5 Optimal 4D trajectory, generated by MOTO algorithm.

Figure 18.5 illustrates a 4D trajectory generated by the MOTO algorithm using pseudospectral-based optimal control software. The algorithm determines the optimal state and control variables, as well as the phase changes that satisfy the problem constraints and minimise the cost function.

18.4 Control Strategy

The control strategy is separately decomposed into attitude and T&E control. Attitude control aims at correcting cross-track or vertical deviations from the nominal trajectory using the load factor N and bank angle μ, while T&E control aims at correcting deviations in the nominal energy and along track distance using the throttle and spoiler controls, η and δ. The T&E controller determines the corrections to be made to η and δ based on the aircraft's deviations in energy Δe and along-track distance Δs. Typically, η and δ are not used simultaneously as the former is used to increase thrust whereas the latter is used to increase drag. Thus, only one of these two controls should be active at any one time, that is, spoilers should only be used at idle throttle and throttle should only be used when spoilers are retracted. The conceptual control strategy is summarised in Table 18.1.

If the aircraft is behind (late) and is at a low energy state, the throttle is used to increase the aircraft's kinetic energy; conversely, if the aircraft is ahead (early) and at a high energy state, spoilers are used to decrease the aircraft's kinetic energy. The remaining two cases depend on the specific proportion between the energy and along-track deviations.

Due to its implementation simplicity and widespread adoption, a proportional-integral-derivative (PID) controller is used to determine the required throttle and spoiler inputs. The mathematical form of the control function is given by:

$$\beta_e(t) = K_{p,e} \cdot \Delta e(t) + K_{i,s} \cdot \int_t \Delta e(\tau) \, d\tau + K_{d,e} \cdot \Delta \dot{e}(t) \tag{18.22}$$

$$\beta_s(t) = K_{p,s} \cdot \Delta s(t) + K_{i,s} \cdot \int_t \Delta s(\tau) \, d\tau + K_{d,s} \cdot \Delta \dot{s}(t) \tag{18.23}$$

where β_e and β_s are the separate PID corrections to energy and along-track distance. The final correction $\beta(t)$ is given as a weighted average of the two corrections:

$$\beta(t) = \left[w_e \cdot \beta_e(t) + w_s \cdot \beta_s(t) \right] / \left(w_e + w_s \right) \tag{18.24}$$

The sign of $\beta(t)$ determines if the throttle or spoilers should be used, while its magnitude determines the amount of throttle or spoiler to apply:

$$\eta(t) = \beta(t) \, , \text{if } \beta(t) > 0 \tag{18.25}$$

$$\delta(t) = -\beta(t) \, , \text{if } \beta(t) < 0 \tag{18.26}$$

Limits are imposed on the controls η and δ, such that $0 < \eta < 1$ and $0 < \delta < C_{Dsplr,max} / C_{D0}$.

18.5 Case Study

The T&E control module was evaluated in a set of representative case studies. The aircraft model considered for these cases is an Airbus A320 with a mass of 64 000 kg and the associated 3DOF model coefficients were retrieved from BADA 3.9. The PID controller was manually tuned to provide robust performance with proportionate and derivative gain terms only for energy $K_{p,e} = 5 \times 10^{-4}$ and $K_{d,e} = 1 \times 10^{-4}$, as well as gain terms for the along-track distance $K_{p,s} = 2 \times 10^{-3}$ and $K_{d,s} = 2 \times 10^{-2}$.

Four cases were studied, each corresponding to one of the four strategies described in Table 18.1. The first 300 seconds of the controller's throttle and spoiler responses to deviations in the initial velocity and along-track distance were studied, with the aircraft's altitude assumed to be following the nominal profile. For all these case studies, the nominal trajectory used is the trajectory shown in Figure 18.2: the aircraft starts from an initial altitude of 20 000 ft at 397 kts true airspeed (TAS) and descends to a final altitude of 2000 ft at 200 kts TAS. The first 300 seconds of the controller's response to minor offsets in the initial velocity and along-track distances were investigated.

Table 18.1 Time and energy control strategy.

	$\Delta s = s_0 - s > 0$(behind)	$\Delta s = s_0 - s < 0$(ahead)
$\Delta e = e_0 - e > 0$ (low energy - slow)	Increase thrust	Return to the along-time profile, then rejoin the energy profile.
$\Delta e = e_0 - e < 0$ (high energy - fast)	Return to the along-time profile, then rejoin the energy profile.	Increase drag

18.5.1 Low Energy (Slow) and Behind

The first simulation case study is shown in Figure 18.6 and starts with the aircraft two nautical miles behind the nominal profile at a lower initial velocity of 377 kts (i.e. 20 kts slower than the initial nominal velocity). The controller initially applies full throttle with zero spoilers, which allows the aircraft to temporarily maintain a state of higher (kinetic) energy to cover a larger distance and rejoin the optimal profile.

18.5.2 High Energy (Fast) and Ahead

The second simulation case study is shown in Figure 18.7 and starts with the aircraft two nautical miles ahead of the nominal profile at a higher initial velocity of 417 kts (i.e. 20 kts faster than the initial nominal velocity). The controller initially applies full spoilers with zero throttle, which allows the aircraft to temporarily maintain a state of lower (kinetic) energy to cover a smaller distance and rejoin the optimal profile. Minor throttle corrections are then made to help the aircraft reach the nominal energy profile.

18.5.3 High Energy (Fast) and Behind

The third simulation case study is shown in Figure 18.8 and starts with the aircraft two nautical miles behind the nominal profile at a higher initial velocity of 417 kts (i.e. 20 kts faster than the initial nominal velocity). The controller follows the same strategy used in the first case study by initially applying full throttle followed by minor throttle corrections.

18.5.4 Low Energy (Slow) and Ahead

The fourth simulation case study is shown in Figure 18.9, starting with the aircraft two nautical miles ahead of the nominal profile at a lower initial velocity of 377 kts (i.e. 20 kts slower than the initial nominal velocity). The controller follows the same strategy used in the second case study by initially applying full spoilers with zero throttle, followed by minor throttle corrections.

The results of this verification activity show that the T&E control strategy is effective at correcting minor perturbations to both velocity and along-track distance. The strategy allows the aircraft to rejoin the nominal trajectory to within 0.1 NM of the nominal trajectory (typical RNP approaches specify a limit of within 0.3 NM) in 200 s when the aircraft is behind and 150 seconds when the aircraft is ahead, without overshoot. The strategy also allows the aircraft's velocity to rejoin the nominal velocity profile to within 2 kts in 240 seconds when the aircraft is behind and 200 seconds when it is ahead, again without overshoot.

18.6 Human-Machine Interface Considerations

To provide information on the aircraft energy state and on the required corrections to the pilot in 4D descent operations, a custom 4DT HMI prototype was designed. This prototype

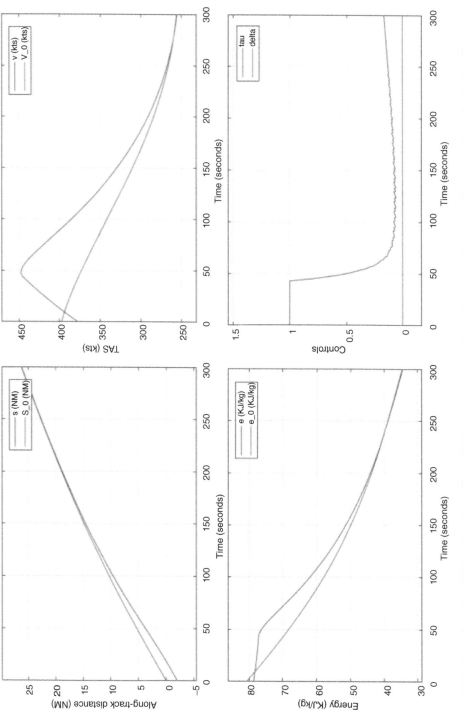

Figure 18.6 Aircraft starting 2 NM behind (top-left) and 20 kts slower (top-right) than the nominal trajectory. The controller varies the throttle (bottom-right) to maintain a constant energy profile (bottom-left) until the aircraft rejoins the along-track profile.

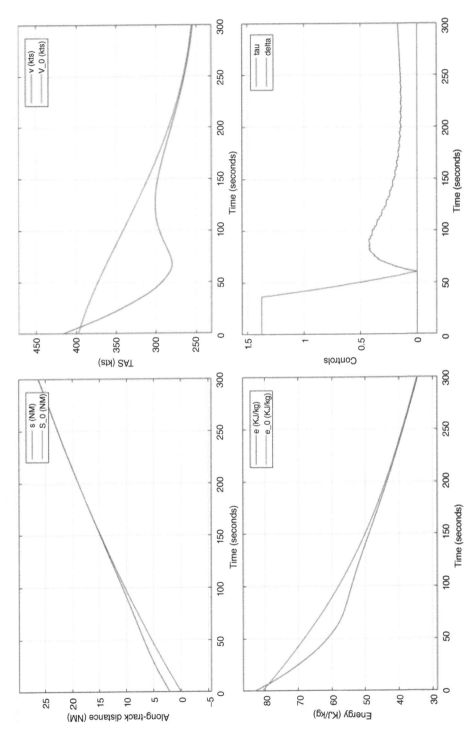

Figure 18.7 Aircraft starting 2 NM ahead (top-left) and 20 kts faster (top-right) than the nominal trajectory. The controller uses the spoiler (bottom-right) to maintain a lower energy profile (bottom-left) until the aircraft rejoins the along-track profile, and then uses the throttle to recover the required energy profile.

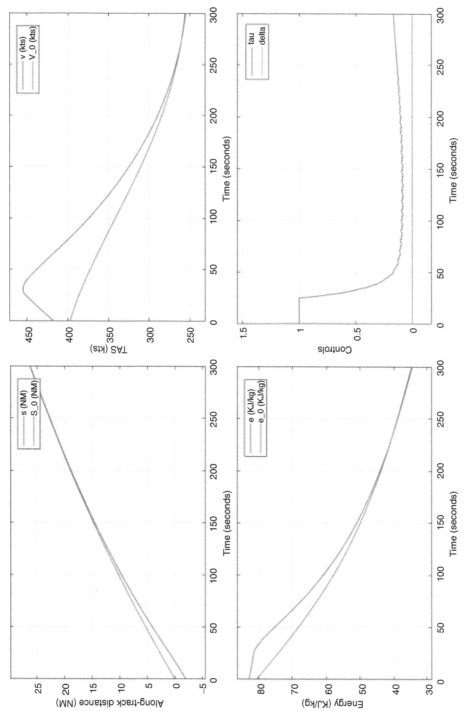

Figure 18.8 Aircraft starting 2 NM behind (top-left) and 20 kts faster (top-right) than the nominal trajectory. The controller uses the throttle (bottom-right) to maintain a constant energy profile (bottom-left) until the aircraft rejoins the along-track profile.

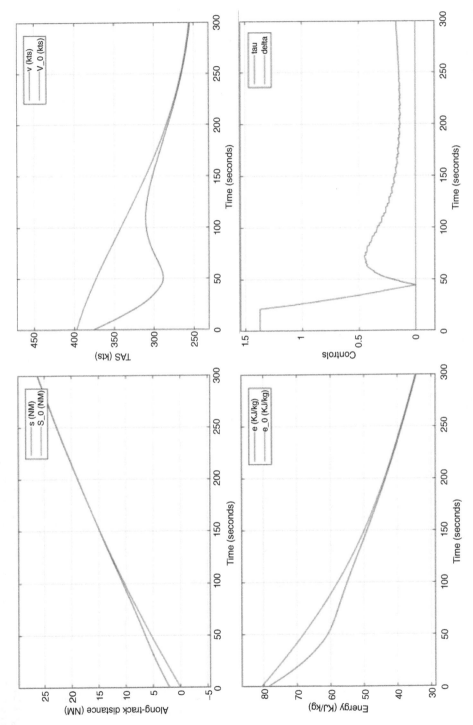

Figure 18.9 Aircraft starting 2 NM ahead (top-left) and 20 kts slower (top-right) than the nominal trajectory. The controller uses the spoiler (bottom-right) to maintain a lower energy profile (bottom-left) until the aircraft rejoins the along-track profile, then uses the throttle to recover the required energy profile.

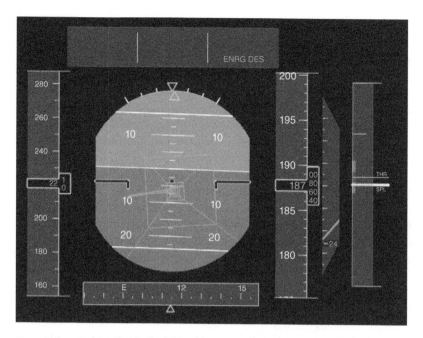

Figure 18.10 VAPS prototype interface supporting energy awareness in 4D descent operations.

is based on the conventional primary flight display (PFD) but also shows the throttle and spoiler corrections as computed by the T&E controller. As illustrated in Figure 18.10, an additional tape indicator was added to the right side of the PFD, depicting both the current and the recommended throttle and spoiler settings. The two control commands are fused into a single throttle-and-spoiler-indicator (TASI), which loosely resembles the controls used in some Airbus-style pedestal panels (where the throttle quadrant is placed directly above the speed brake lever). The recommended throttle/spoiler setting as calculated by the T&E controller is indicated by a single sliding tape with green, yellow, and red bands on the TASI. Conversely, the current throttle and spoiler settings are depicted with separate label indicators moving in opposite directions, with increased throttle indicated by an upward trend and increased spoilers indicated by a downward trend. In addition to the TASI, a highway-in-the-sky (HITS) is overlaid on the electronic attitude director indicator (EADI) to provide attitude guidance to the pilot. HITS have been extensively studied as a means of displaying guidance information in synthetic or enhanced vision systems (SVS/EVS) supporting pilot spatial awareness in 4DT operations [11–14]. Suitable human-in-the-loop studies will be carried out to assess the proposed 4DT HMI prototype in realistic 4D trajectory management scenarios.

18.7 Conclusions

This chapter presented a guidance and control strategy for managing 4D descent operations, which employs an outer guidance loop and a number of inner control loops. The outer guidance loop uses an efficient scheme for computing the active 4D trajectory. The lateral and vertical components are treated separately, with a geometric method used in lateral path planning and an optimal control technique used to compute the vertical profile. In particular, the optimal control technique allows time constraints to be imposed, thereby supporting the generation of 4D waypoints. The active trajectory is updated at regular intervals, or can be triggered to re-compute a new trajectory when excessive deviation from the current 4D trajectory is detected. The control loops comprise the attitude and T&E processes, which respectively correct for deviations in altitude and energy/along-track distance. In particular, this article discusses the T&E control strategy. A proportional-derivative (PD) controller is used to compute suitable throttle and spoiler commands to allow the aircraft to recover from deviations in energy as well as the along-track profile. A number of case studies were conducted, which show that the PD controller was able to compute the necessary throttle and spoiler commands allowing the aircraft to recover from different combinations of initial states (slow/fast vs. ahead/behind). Additionally, a prototype human-machine interface (HMI) was designed to support energy awareness in 4D descent operations. Future work will involve the use of this HMI for human-in-the-loop studies on 4D descent operations.

References

1 Kistan, T., Gardi, A., Sabatini, R. et al. (2017). An evolutionary outlook of air traffic flow management techniques. *Progress in Aerospace Science* 88: 15–42.

2 Liu, J., Gardi, A., Ramasamy, S. et al. (2016). Cognitive pilot-aircraft interface for single-pilot operations. *Knowledge-Based Systems* 112: 37–53.

3 Ramasamy, S., Sabatini, R., Gardi, A., and Liu, J. (2016). LIDAR obstacle warning and avoidance system for unmanned aerial vehicle sense-and-avoid. *Aerospace Science and Technology* 55: 344–358.

4 Lim, Y., Gardi, A., Ramasamy, S. et al. (2017). Cognitive UAS human-machine interfaces and interactions. *Journal of Intelligent & Robotic Systems* 91 (3-4): 755–774.

5 Erzberger, H. (1980). Optimum climb and descent trajectories for airline missions. In: *Theory and applications of optimal control in aerospace systems, AGARDograph AG-251* (ed. NATO Research and Technology Organization (RTO)), 9–1–9–15.

6 Sorensen, J.A. and Waters, M.H. (1981). "Generation of optimum vertical profiles for an advanced flight management system", Analytical Mechanics Associates Inc. and NASA Langley Research Center, Hampton, VA, USA NASA Contract Report CR-165674.

7 Lambregts, A.A. (1983). "Vertical flight path and speed control autopilot design using total energy principles," *AIAA paper,* vol. 83.

8 de Jong, P., Bussink, F., Verhoeven, R. et al. (2016). Time and energy management during approach: a human-in-the-loop study. *Journal of Aircraft*: 1–13.

9 Wu, S.-F. and Guo, S.-F. (1994). Optimum flight trajectory guidance based on total energy control of aircraft. *Journal of Guidance Control and Dynamics* 17: 291–291.

10 Gardi, A., Sabatini, R., and Ramasamy, S. (2016). Multi-objective optimisation of aircraft flight trajectories in the ATM and avionics context. *Progress in Aerospace Science* 83: 1–36.

11 Wickens, C.D., Alexander, A.L., Horrey, W.J. et al. (2004). Traffic and flight guidance depiction on a synthetic vision system display: The effects of clutter on performance and visual attention allocation. In: *Proceedings of the Human Factors and Ergonomics Society Annual Meeting*, 218–222.

12 Williams, K.W. (2002). Impact of aviation highway-in-the-sky displays on pilot situation awareness. *Human Factors* 44: 18–27.

13 Sudesh, K.K., Naidu, P., and Shantha Kumar, N. (2016). "Multi-spectrum-based enhanced synthetic vision system for aircraft DVE operations," .

14 Mulder, M. (1999). "Cybernetics of tunnel-in-the-sky displays," Doctor of Philosophy PhD Thesis, Delft University, The Netherlands.

15 Lim, Y. (2019a). *Cognitive human-machine interfaces and interactions for avionics systems*, PhD Thesis, RMIT University, School of Engineering, Melbourne, VIC 3000, Australia.

16 Lim, Y., Gardi, A., Sabatini, R. et al. (2019b). Optimal energy-based 4D guidance and control for terminal descent operations. *Aerospace Science and Technology* 95: 105436.

19

Contrail Modelling for 4D Trajectory Optimisation

Yixiang Lim[1], Alessandro Gardi[2,3], and Roberto Sabatini[2,3]

[1] *Agency for Science, Technology and Research (ASTAR), Singapore*
[2] *Department of Aerospace Engineering, Khalifa University of Science and Technology, Abu Dhabi, UAE*
[3] *School of Engineering, RMIT University, Melbourne, Victoria, Australia*

19.1 Introduction

With the currently foreseen growth of air traffic globally, the effects of aircraft condensation trails (contrails) are predicted to become significant over large regions well before 2050. Currently, research regarding the persistence or dissipation timescales of contrails is still ongoing, with a particular focus on their evolution into cirrus clouds, since these clouds have been shown to contribute significantly to global warming. The transition into cirrus is a combination of the macro-physical processes such as wind shear (which increases contrail spread), and the microphysics, such as the ageing and growth of the ice particles. Contrail-based trajectory optimisation is a relatively recent field of research, and work done so far has focused mainly on contrail avoidance. A review [1] noted that a general reduction in flight altitudes might not be the most effective solution due to the geographical, seasonal, and diurnal variations in contrail formation regions. For example, contrails are avoided in tropical regions by flying at a lower altitude, while an increase in flight altitudes in temperate regions is required for contrail avoidance. Alternative solutions, such as shifting the air traffic densities towards sunrise and sunset, when the cooling effect of contrails is the strongest, have been suggested in other studies [2, 3]. A 4D-trajectory optimisation model for contrail avoidance has been developed by [4]. This model designates a persistent contrail region as the region that both satisfies the Schmidt-Appleman criterion and is supersaturated with respect to ice, and subsequently a contrail formation cost is associated with such a region. However, not all persistent contrail regions should be penalised, as the radiative impact is also dependent on the time of day and geographic location of the contrail regions. A new scheme is presented in this case study, which was originally published as [17, 18], which uses existing contrail models to compute the associated radiative forcing (RF) based on a set of weather, aircraft, and trajectory inputs and attempts to address any possible implications arising from the adoption of such a comprehensive system. The model is specifically designed to be integrated in a multi-objective trajectory optimisation (MOTO) software framework. The MOTO software is conceived for the implementation in

Sustainable Aviation Technology and Operations: Research and Innovation Perspectives, First Edition.
Edited by Roberto Sabatini and Alessandro Gardi.
© 2024 John Wiley & Sons Ltd. Published 2024 by John Wiley & Sons Ltd.
Companion Website: www.wiley.com/go/sustainableaviation

ground-based and airborne air traffic management (ATM) and avionics systems currently under development as part of this research [5–9], addressing the strategic and tactical online operations. The contrail algorithm will be exploited both as a stand-alone and within the MOTO software to determine and optimise the formation, persistence, and radiative contribution of contrails occurring along a number of trajectories flown under different weather conditions at various times of the day. Some results are included, compared, and discussed here, with a particular focus on evaluating the versatility of the adopted approach.

19.2 Physics of Contrail Formation

Contrails are formed due to the condensation of water vapour in the plume of a jet engine and can be the precursor to the ice clouds called *cirrus*, found at higher altitudes where ice supersaturation is present. The formation of these contrails is thermodynamically triggered by the engine exhaust, and specifically due to the influx of heat, moisture, and soot. The contrail life cycle is subdivided into three phases: the jet, vortex, and dispersion phases, which are characterised by substantially different physical processes. The jet phase lasts for around 20 seconds and encompasses the thermodynamic processes caused by the hot, humid exhaust plume interacting and cooling in the ambient air. The Schmidt-Appleman criterion, developed first by Schmidt (1941) then independently by Appleman (1953), is used to determine the onset of contrail formation [10]. The vortex phase follows from the jet phase and is characterised by a timescale of two minutes. In the vortex phase, the exhaust plume comes under the influence of the wake vortices generated by the aircraft, which cause a downward sinking of the contrail. The initial characteristics of the contrail (its breath, depth, ice properties, etc) can be considered to have stabilised and should be determined. A parametric model developed by [11] is used to model this phase. The dispersion phase follows from the end of the vortex phase and involves the advection, spread, and eventual evaporation of the contrail. Depending on ambient conditions, this can range from a few minutes up to several hours. The integration scheme employed in [11] is adapted for use in this model, with some changes. Instead of a second-order, two-step Runge-Kutta scheme, a forward Euler method is used to determine the advection of the contrail. In addition, while the terminal particle velocity V_T in the former is determined as in [12], this paper uses a simplified model as in [13]. These models are developed and implemented in a Matlab simulation environment, with separate functions to model each phase of the contrail lifecycle.

19.2.1 Weather Data

Weather data is obtained from the Global Forecasting System (GFS). Forecast data is given in a 0.25° resolution, updated four times daily (every six hours), and provides a projection of up to 180 hours in 3-hour intervals. The data is further interpolated to attain a better resolution in space and time. The relative humidity with respect to ice (RHi), obtained as a function of the relative humidity with respect to water RH_w, and temperature T, both obtained from GFS

$$RH_i = RH_w \frac{p_{liq}(T)}{p_{ice}(T)} \tag{19.1}$$

where e_W and e_I are the saturation pressures for water vapour and ice respectively, based on empirical relations by (Sonntag 1996).

$$
p_{liq}(T) = 100 \exp\left[\frac{-6096.9385}{T} + 16.635794 - 0.02711193T \right. \\
\left. + 1.673952e^{-5}T^2 + 2.433502\ln(T)\right]
$$
(19.2)

$$
p_{ice}(T) = 100 \exp\left[\frac{-6024.5282}{T} + 24.7219 + 0.10613868T \right. \\
\left. - 1.3198825e^{-5}T^2 - 0.49382577\ln(T)\right]
$$
(19.3)

with T in degrees Kelvin.

19.2.2 Contrail Model

The contrail model is based on the Contrail Cirrus Prediction Tool (CoCiP) developed by Schumann [11]. The algorithm proceeds as follows. An initial check based on the Schmidt-Appleman criterion is done to flag contrail formation segments along the flight path. A contrail lifetime analysis is then run for flagged segments. The initial size and properties of the plume are computed using a parametric model for the jet and vortex phases. These are then integrated in time using another parametric model for the dispersion phase. A model based on [14] then estimates the contrail-related RF based on the solar zenith angle, the spread, and the lifetime of the contrail. A block diagram of the algorithm is represented in Figure 19.1. In the operational concept depicted in Figure 19.2, an aircraft undertakes a 2-hour journey, departing at 1700 hours and arriving at 1900 hours. It encounters two contrail formation regions, shown in blue. In the first region, the contrail quickly dissipates while persistence is observed in the second region. Midway during the flight, the sun sets. The sunset line is shown as a dashed-line and subsequent contrails after this are greyed out. The model computes the RF of the second contrail and supplies the 4DT optimiser with the information about the red-shaded areas above the threshold RF, marked as a dot-dashed line.

19.2.2.1 Jet Phase

In the jet phase, the Schmidt-Appleman criterion is used to determine the onset of contrail formation. This is given by

$$
G = \frac{EI_{H2O} \cdot c_p \cdot p}{\varepsilon \cdot Q \cdot (1 - \eta)}
$$
(19.4)

where G is the slope of the mixing line, in Pa K^{-1}, and describes the mixing of the exhaust plume with ambient air. EI_{H2O}, Q and η are the emission index of water vapour, heat per mass of fuel, and propulsive efficiency, respectively, and are engine characteristics, p is the ambient pressure, and c_p and ε are the specific heat capacity of air and the ratio of the molecular masses of water and air, respectively.

As the plume cools and mixes, it moves along the mixing line, as shown in Figure 19.3, from the top right to the bottom left, terminating at ambient conditions (Point C). The threshold conditions for contrail formation occur at Point M, where the mixing line touches the water saturation curve. Larger values of G will result in a steeper slope, leading to condensation of the exhaust plume as it crosses the water saturation curve and the formation

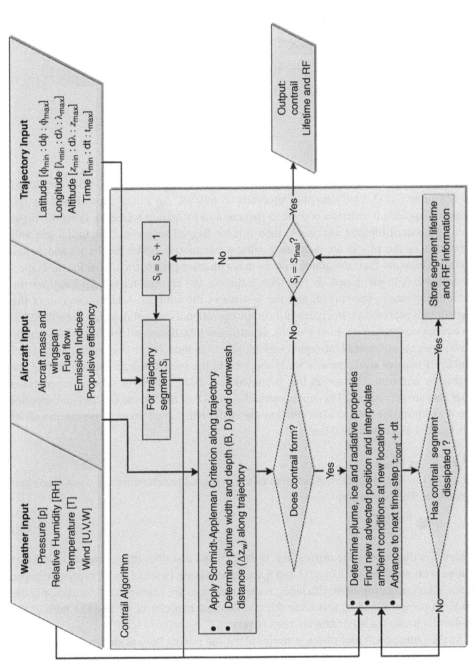

Figure 19.1 Block diagram of the contrail model software algorithm.

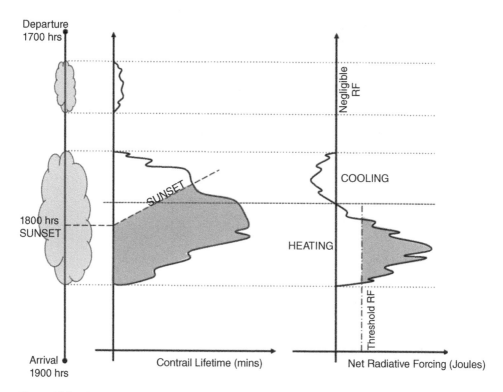

Figure 19.2 Operational concept of the contrail model. The timeline on the left represents the aircraft flight path. The middle graph represents the contrail lifetime and the graph on the right represents the net RF. Depending on the time of day, the RF can be positive or negative.

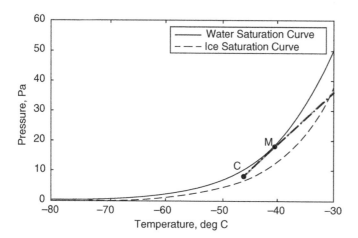

Figure 19.3 Mixing diagram.

of contrails. If ambient conditions are supersaturated with respect to ice, that is, if Point C is above the ice saturation curve, the condensed vapour will freeze into ice crystals, and the contrail might grow into cirrus, given sufficient time. Reference [14] provides an approximation for the temperature at the threshold point in degrees Celsius as:

$$T_{LM} = -46.46 + 9.43 \ln{(G - 0.053)} + 0.72[\ln{(G - 0.053)}]^2 \tag{19.5}$$

where the subscript L denotes the threshold temperature for liquid saturation.

The critical temperature at ambient conditions (Point C) can then be found using the criterion provided by [11] based on curve fitting where T and T_{LM} are given in degrees Celsius:

$$T_{LC} = T_{LM} - (1 - RH_w)\frac{p_{liq}(T_{LM})}{G} - \Delta T_C \tag{19.6}$$

$$\Delta T_C = F_1 RH_w [W - F_2(1 - W)] \tag{19.7}$$

$$W = (1 - RH_w^2)^{x_2} \tag{19.8}$$

$$F_1 = x_1 + x_3 \ln{(G)}, \quad F_2 = \left[\frac{1}{4} - \left(RH_w - \frac{1}{2}\right)^2\right]^4 \tag{19.9}$$

Where $x_1 = 5.686$, $x_2 = 0.3840$, $x_3 = 0.6594$. The criterion for contrail formation is thus matematically formulated as:

$$dT = T - T_{LC} < 0 \tag{19.10}$$

19.2.2.2 Vortex Phase

The parametric model developed by [14] was used to estimate the initial depth and width of the contrail. This is done by first computing the maximum downward displacement Δz_w and then scaling it to obtain the initial depth. The initial width is then parameterized from the depth, dilution, and fuel flow. The parameters that determine the initial size of the contrail are:

Wake vortex separation $b_0 = \pi S_a/4$,
Initial circulation $\Gamma_0 = 4M_a g/(\pi S_a \rho_{air} V_a)$,
Effective time scale $t_0 = 2\pi b_0^2/\Gamma_0$,
Initial velocity scale $w_0 = \Gamma_0/(2\pi b_0)$,
Normalised dissipation rate $\varepsilon^* = (\varepsilon b_0)^{1/3}/w_0$.

These are dependent on the inputs

Wing span S_a,
Aircraft mass M_a,
True air speed V_a,
Air density ρ_{air},
Brunt Vaisaila frequency N_{BV},
Turbulent kinetic energy dissipation rate ε.

Schumann's parameterization distinguishes between strongly and weakly stably stratified conditions. If $N_{BV}t_0 \geq 0.8$:

$$\Delta z_w = 1.49 \frac{w_0}{N_{BV}} \tag{19.11}$$

else, with $\varepsilon^* \leq 0.36$:

$$\frac{\Delta z_w}{b_0} = 7.68 \left(1 - 4.07\, \varepsilon* + 5.67\, \varepsilon*^2\right) \left(0.79 - N_{BV}t_0\right) + 1.88 \tag{19.12}$$

with Δz_w as the maximum sinking.

The initial downward displacement is set to

$$\Delta z_1 = C_{z1} \Delta z_w, \quad C_{z1} = 0.25 \tag{19.13}$$

where the subscript 1 denotes the end of the vortex phase. The initial depth is set to:

$$D_1 = C_{D0} \Delta z_w, \quad C_{D0} = 0.5 \tag{19.14}$$

and the initial width is:

$$B_1 = N_{dil}\left(t_0\right) m_F / \left[(\pi/4)\rho D_1\right] \tag{19.15}$$

with the m_F being the fuel flow in kg m^{-1} and the dilution N_{dil} of the wake vortex formation being

$$N_{dil}(t) \approx 7000\left(t/t_s\right)^{0.8} \tag{19.16}$$

with $t_s = 1$ second. The initial mass mixing ratio of ice I and the ice particle number N are also prescribed for the vortex phase.

$$I_1 = I_0 - \Delta I_{ad} \tag{19.17}$$

$$I_0 = \frac{EI_{H2O}m_F}{(\pi/4)\rho D_1 B_1} + q_0 - q_s\left(p_0, T_0\right) \tag{19.18}$$

$$\Delta I_{ad} = \frac{R_0}{R_1} \left[\frac{p_{ice}\left(T_0 + \Delta T_{ad}\right)}{p_1} - \frac{p_{ice}\left(T_0\right)}{p_0}\right] \tag{19.19}$$

Here, q_0 is the specific humidity of ambient conditions at stage '0' (the start of the vortex phase) and q_s is the specific humidity at saturation point. p_1 is the ambient pressure at stage '1', R_0 and R_1 are the specific heat capacities of air and water respectively, and

$$\Delta T_{ad} = T_0 \left(R_0/c_p\right) \left(p_1 - p_0\right) / p_0 \tag{19.20}$$

The ice number N_1 at the end of the vortex phase is determined by the soot index and a survival factor based on the ice mass ratio.

$$N_1 = f_{surv} N_0 \tag{19.21}$$

$$N_0 = EI_{Soot} m_F, \quad f_{surv} = \frac{I_1}{I_0} \tag{19.22}$$

19.2.2.3 Dispersion Phase

The state of each contrail segment along the flight path evolves with time and is described by the state vector X, denoted as *X(Position, Ambient, Plume, Particle)* – its position in space and time, the relevant ambient conditions, and its plume and ice particle properties. The position at time t is given as:

$$X.Position = (x(t), y(t), z(t), t) \tag{19.23}$$

where x, y, and z give the longitude, latitude, and altitude of the contrail. The advection of the contrail with time is described by:

$$x(t + \Delta t) = x(t) + U(t)\Delta t \tag{19.24}$$

$$y(t + \Delta t) = y(t) + V(t)\Delta t \tag{19.25}$$

$$z(t + \Delta t) = z(t) + W(t)\Delta t \tag{19.26}$$

with the absolute distances converted to geodetic coordinates based on approximations provided by [15]. The ambient conditions include the parameters:

$$X.Ambient = \left(p, T, \rho_{air}, q_a, q_s \right) \tag{19.27}$$

which are pressure, temperature, density, and ambient and saturation specific humidity, respectively, obtained from the GFS, and are functions of *X. Position*. The plume parameters consist of:

$$X.Plume = \left(\sigma, B, D, D_{eff}, A, L, M, I, N, n, M_{H2O}, D_H, D_V \right)$$

Respectively, these are the:

Covariance matrix σ,
Contrail width B,
Contrail depth D,
Effective depth D_{eff},
Contrail area A,
Contrail length L,
Air mass M,
Ice mass mixing ratio I,
Ice particle number N,
Ice concentration n,
Water mass M_{H2O},
Horizontal diffusivity D_H,
Vertical diffusivity D_V.

The terms of the covariance matrix $\sigma = \begin{bmatrix} \sigma_{yy} & \sigma_{yz} \\ \sigma_{yz} & \sigma_{zz} \end{bmatrix}$ are initially given as:

$$\sigma_{yy}(t = t_0) = B^2/8 \tag{19.28}$$

$$\sigma_{zz}(t = t_0) = D^2/8 \tag{19.29}$$

$$\sigma_{yz}\left(t = t_0\right) = 0 \tag{19.30}$$

and evolve with time as:

$$\sigma_{yy}\left(t + \Delta t\right) = \left[\frac{2}{3}S^2 D_V \Delta t^3 + \left(S^2 \sigma_{zz}\left(t\right) + 2D_S S\right)\Delta t^2 + 2\left(D_H + S\sigma_{yz}\left(t\right)\right)\Delta t + \sigma_{yy}\left(t\right)\right] \cdot \left[\frac{L\left(t\right)}{L\left(t + \Delta t\right)}\right]^2 \tag{19.31}$$

$$\sigma_{zz}\left(t + \Delta t\right) = 2D_V \Delta t + \sigma_{zz}\left(t\right) \tag{19.32}$$

$$\sigma_{yz}\left(t + \Delta t\right) = \left[SD_V \Delta t^2 + \left(2D_S + S\sigma_{zz}\left(t\right)\right)\Delta t + \sigma_{yz}\left(t\right)\right] \cdot \left[\frac{L\left(t\right)}{L\left(t + \Delta t\right)}\right] \tag{19.33}$$

with the shear diffusivity D_S set to 0 and the vertical shear of the plume normal velocity S is taken to be the total shear $S_T = \sqrt{\left(\frac{\partial U}{\partial z}\right)^2 + \left(\frac{\partial V}{\partial z}\right)^2}$. The shape and size of the contrail are based on the covariance matrix:

$$B\left(t + \Delta t\right) = \sqrt{8\sigma_{yy}\left(t + \Delta t\right)} \tag{19.34}$$

$$D\left(t + \Delta t\right) = \sqrt{8\sigma_{zz}\left(t + \Delta t\right)} \tag{19.35}$$

$$A\left(t + \Delta t\right) = 2\pi \left\{\frac{1}{3}S^2 D_V^2 (\Delta t)^4 + \frac{2}{3}S^2 D_V \sigma_{zz}\left(t\right)(\Delta t)^3 + \left[2S\sigma_{zz}\left(t\right)\left(D_V - D_S\right)\right. \right.$$
$$\left. - 4\left(D_H D_V - D_S^2\right)\right](\Delta t)^2 + \left[2\sigma_{zz}\left(t\right)\left(D_V + D_H\right) - 4D_S \sigma_{yz}\left(t\right)\right]\Delta t$$
$$\left. + \sigma_{yy}\left(t\right)\sigma_{zz}\left(t\right) - \sigma_{yz}^2\left(t\right)\right\}^{1/2} \tag{19.36}$$

$$D_{eff}\left(t + \Delta t\right) = \frac{A\left(t + \Delta t\right)}{B\left(t + \Delta t\right)} \tag{19.37}$$

and the length $L(t + \Delta t)$ is computed from the positions of the end-points of the contrail $x(t + \Delta t)$ and $y(t + \Delta t)$ in its position vector. The mass properties of the plume can then be calculated:

$$M\left(t + \Delta t\right) = \rho AL\big|_{t+\Delta t} \tag{19.38}$$

$$I\left(t + \Delta t\right) = \frac{M\left(t\right) \cdot \left[I\left(t\right) + q_s\left(t\right) + \Delta M \cdot q_a\right]}{M\left(t + \Delta t\right)} - q_s\left(t + \Delta t\right) \tag{19.39}$$

$$M_{H2O}\left(t + \Delta t\right) = M\left(t + \Delta t\right) \cdot \left(I + q\right)\big|_{t+\Delta t} \tag{19.40}$$

where $\Delta M = M(t + \Delta t) - M(t)$ and $q_a = \frac{q_a(t+\Delta t)+q_a(t)}{2}$

Particle loss due to turbulence and aggregration are modelled to determine the evolution of the ice number, with the adjustable parameters E_A and E_T set to 2:

$$\left(dN/dt\right)_{agg} = -E_A 8\pi r_{eff}^2 V_T N^2 / A \tag{19.41}$$

$$\left(dN/dt\right)_{turb} = -E_T \left(\frac{D_H}{\max\left(B, D\right)^2} + \frac{D_V}{D_{eff}^2}\right)N \tag{19.42}$$

$$\alpha = -\frac{1}{N^2}(\partial N/\partial t)_{agg}, \quad \beta = -\frac{1}{N}(\partial N/\partial t)_{turb} \tag{19.43}$$

$$N(t + \Delta t) = \frac{N(t)\beta \exp(-\beta \Delta t)}{\beta + N(t)\alpha\left[1 - \exp(-\beta \Delta t)\right]} \frac{L(t)}{L(t + \Delta t)} \tag{19.44}$$

where V_T, following Rogers, 1976 (referenced in [13]), is stored in the particle properties:

$$X.Particle = \left(V_T, r_{eff}\right) \tag{19.45}$$

$$V_T = \begin{cases} k_1 r_{eff}^2, & r < 40\mu m \\ k_2 r_{eff}, & 40\mu m < r < 600\mu m \\ k_3\sqrt{r_{eff}}, & r > 600\mu m \end{cases} \tag{19.46}$$

$$k_1 = 1.9e^8, k_2 = 8e^3, k_3 = 2.2e^2\sqrt{\frac{\rho_0}{\rho}} \tag{19.47}$$

and r_{eff} is taken to be the effective particle radius:

$$r_{eff} = \left(\frac{\rho_{air}I}{n\rho_{ice}4\pi/3}\right)^{1/3} \tag{19.48}$$

Finally, the diffusivities are given in [11]:

$$D_H = C_H\left(D^2 S_T\right), \quad C_H = 0.1 \tag{19.49}$$

$$D_V = \frac{C_V}{N_{BV}}w_n' + f_t\left(V_T D_{eff}\right), \quad C_V = 0.2, w_n' = 0.1, f_t = 0.1 \tag{19.50}$$

The time integration ends when the ice concentration n falls below 10^3 m^{-3} (or $1\,l^{-1}$), the ice mass ratio I falls below 10^{-8} (or 10^{-2} mg/kg), or when the time exceeds a given threshold, set to five hours in this case.

19.2.3 Radiative Forcing

The RF model follows [11]. The parameterization scheme is based on a number of parameters that model the particle type (i.e. spherical, hollow, rosette, plate) as well as independent properties. The longwave radiation is positive and dependent on the following independent properties: the outgoing longwave radiation (OLR, W m^{-2}), the atmospheric temperature (T, K), the optical depths of the contrail and its overhead cirrus at 550 nm (τ and τ_c) and the effective particle radius (r_{eff}, μm). The contrail optical depth is computed from [14] as follows:

$$\tau = \beta D_{eff} \tag{19.51}$$

$$\beta = 3Q_{ext}\rho I/\left(4\rho_{ice}r_{eff}\right) \tag{19.52}$$

$$D_{eff} = A/B \tag{19.53}$$

$$Q_{ext} = 2 - (4/\rho_\lambda)\frac{\sin(\rho_\lambda) - \left[1 - \cos(\rho_\lambda)\right]}{\rho_\lambda} \tag{19.54}$$

$$\rho_\lambda = 4\pi r_{eff}(\kappa - 1)/\lambda, \quad \kappa = 1.31, \quad \lambda = 550nm \tag{19.55}$$

While the optical depth of overhead cirrus is assumed to be $\tau_c = 0$. The zenith angle is approximated with ESRL's equations:

$$\gamma = \frac{2\pi}{365} * \left(day\ of\ year - 1 + \frac{hour - 12}{24} \right) \tag{19.56}$$

$$\begin{aligned} eqtime &= 299.18 \left[7.5e^{-5} + 1.868e^{-3} \cos(\gamma) - 3.2077e^{-2} \sin(\gamma) \right. \\ &\left. -1.4615e^{-2} \cos(2\gamma) - 0.40849e^{-2} \sin(2\gamma) \right] \end{aligned} \tag{19.57}$$

$$\begin{aligned} decl &= \left[6.918e^{-3} - 3.99912e^{-1} \cos(\gamma) + 7.026e^{-2} \sin(\gamma) - 6.758e^{-3} \cos(2\gamma) \right. \\ &\left. +9.07e^{-4} \sin(2\gamma) - 2.70e^{-3} \cos(3\gamma) + 1.48e^{-3} \sin(3\gamma) \right] \end{aligned} \tag{19.58}$$

$$time\ offset = eqtime - 4 * long + 60 * timezone \tag{19.59}$$

$$tst = hour * 60 + min + \frac{sec}{60} + time\ offset \tag{19.60}$$

$$ha = \left(\frac{tst}{4} \right) - 180 \tag{19.61}$$

$$\mu = \cos(\theta) = \sin(lat) \sin(decl) + \cos(lat) \cos(decl) \cos(ha) \tag{19.62}$$

The long and short wave RF can then be computed.

$$RF_{LW} = \left[OLR - k_T (T - T_0) \right] \cdot \left\{ 1 - \exp\left[-\delta_\tau F_{LW} (r_{\mathit{eff}}) \tau \right] \right\} E_{LW} (\tau_c) \tag{19.63}$$

$$F_{LW} (r_{\mathit{eff}}) = 1 - \exp\left(-\delta_{lr} r_{\mathit{eff}} \right) \tag{19.64}$$

$$E_{LW} (\tau_c) = \exp\left(-\delta_{lc} \tau_c \right) \tag{19.65}$$

The shortwave radiation is negative and depends on the following independent properties: τ, τ_c, r_{eff}, the cosine of the solar zenith angle $\mu = \cos(\theta)$, and the effective albedo (A_{eff}).

$$RF_{SW} = -SDR(t_A - A_{\mathit{eff}})^2 \alpha_c (\mu, \tau, r_{\mathit{eff}}) E_{SW} (\mu, \tau_c) \tag{19.66}$$

$$\alpha_c (\mu, \tau, r_{\mathit{eff}}) = R_C (\tau_{\mathit{eff}}) \left[C_\mu + A_\mu R_c' (\tau') F_\mu (\mu) \right] \tag{19.67}$$

$$\tau' = \tau F_{SW} (r_{\mathit{eff}}), \quad \tau_{\mathit{eff}} = \tau' / \mu \tag{19.68}$$

$$F_{SW} (r_{\mathit{eff}}) = 1 - F_r \left[1 - \exp\left(-\delta_{sr} r_{\mathit{eff}} \right) \right] \tag{19.69}$$

$$R_C (\tau_{\mathit{eff}}) = 1 - \exp\left(-\Gamma \tau_{\mathit{eff}} \right), \quad R_c (\tau_{\mathit{eff}}) = 1 - \exp\left(-\gamma \tau_{\mathit{eff}} \right), \tag{19.70}$$

$$F_\mu (\mu) = \frac{(1 - \mu)^{B_\mu}}{(1/2)^{B_\mu}} - 1 \tag{19.71}$$

$$E_{SW} (\mu, \tau_c) = \exp\left(-\delta_{sc} \tau_c - \delta_{sc}' \tau_{c,\mathit{eff}} \right) \tag{19.72}$$

$$\tau_{c,\mathit{eff}} = \tau_c / \mu \tag{19.73}$$

The model parameters can be found in [14]; for this chapter, a droxtal habit is assumed.

19.3 Model Verification

For verification, results from the contrail model were compared with [11], which modelled an artificial, long lived contrail (up to 10 hours) in uniform atmospheric conditions. For the purposes of comparison, a uniform RH_i and temperature field of 120% and 217 K was artificially created. The properties of a few contrail segments were captured in Figure 19.4 and are currently in good alignment with the data presented in [11].

Refinement tests were conducted on the algorithm to investigate the impact of varying the time step and grid resolution. The differences in contrail lifetime for a given flight ('Scenario O' in Table 19.2) are presented in Figure 19.5 and the algorithm run time is shown in Table 19.1.

19.3.1 Contrail Mapping for Multi-Objective Trajectory Optimisation

A novel method has been developed to map the contrail region for the purposes of MOTO. Iso-persistence regions are mapped by computing the lifetime of contrails along simulated trajectories that branch out from the flight path. These trajectories are constant in time and can be imagined to be virtual sounding trajectories. The combined virtual sounding trajectories can be interpolated to map out a two-dimensional region of contrail lifetime, like the one depicted in Figure 19.6. The algorithm can be run at different timeframes and at different altitudes to obtain a fully four-dimensional, time-varying volume of iso-persistence. In our simulation case study, the mapping reveals contrail persistence regions along a large part of the flight path from Stockholm to Venice. The region is closely correlated to ambient temperature and RHi variations, and roughly follows the ice super-saturation contour in Figure 19.6.

To check the robustness of the interpolation method, contour maps interpolated with only the 'odd' and 'even' trajectories were compared with the contour map interpolated with all trajectories. These showed good agreement with one another. As the weather data was limited to $45° N < lat < 60° N$, $-10° E < long < 25° E$, the contour plots in Figures 19.7 and 19.8 show a contrail lifetime of 0 minutes outside the given geographical range.

19.3.2 Simulation Case Study – Rerouting

A preliminary simulation case was developed to verify the effectiveness and the computational performances of the contrail model algorithm, run for a flight from Stockholm to Venice; the flight details along with the parameters of the simulated aircraft are presented in Tables 19.2 and 19.3. This simulation case study involves a constant straight and level cruise trajectory between the standard instrument departure (SID) route exit point, which is assumed to be coincident with the top-of-climb (ToC), and the standard terminal arrival route (STAR) entry point, which is assumed to be coincident with the top-of-descent (ToD). The case study examines two different scenarios (A and B) in comparison to the baseline (O) scenario. Scenario A involves a straight level cruise flight at a lower altitude, while Scenario B involves a lateral bypass of the contrail persistent region at the original altitude. The temperature and RH_i fields are shown in Figure 19.9 for the two flight levels. The flight path as specified in scenario B is also shown in the figure. The flight takes place at night so that only the longwave radiative forcing is taken into account (shortwave forcing is zero at night).

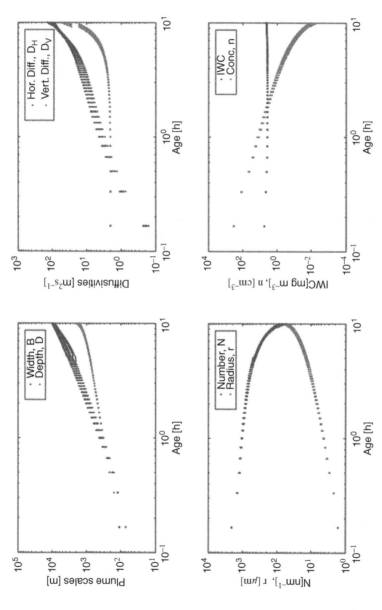

Figure 19.4 Contrail properties versus age, for comparison against Schumann's results. Top-left panel: Width B and depth D. Top-right: Horizontal and vertical diffusivities. Bottom-left: Total number of ice particles per nanometre flight distance N, ice particle volume mean radius r. Bottom-right: Ice particle number concentration per volume n, ice water content per volume ρI.

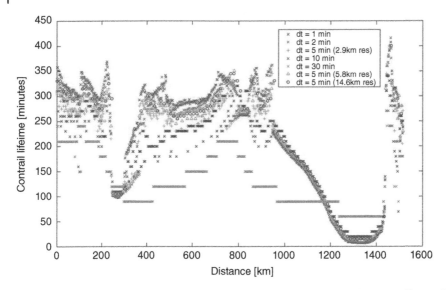

Figure 19.5 Contrail lifetime for the same trajectory and ambient conditions, at different time steps of 1 minutes (blue x's), 2 minutes (red x's), 5 minutes (orange +'s), 10 minutes (purple x's), and 30 minutes (green x's), and different grid resolutions of 5.8 km (blue triangles) and 14.6 km (purple circles) at a time-step of 5 minutes.

Table 19.1 Timed results of refinement test.

Time step (min)	Mean resolution (km)	Algorithm run-time (s)
1	2.9	1142.9
2	2.9	525.7
5	2.9	194.3
10	2.9	89.1
30	2.9	21.7
5	5.8	100.6
5	14.6	41.8

For this simulation, the imported weather data uses a resolution of 0.25° along isobars of 5 kPa, in three-hour intervals. This was interpolated along an altitude with a resolution of 5 ft (converted from pressure isobars) in 10-minute intervals. The trajectory was subdivided into segments of mean resolution 2.9 km for analysis, and the ambient conditions were again interpolated at these points. The time integration of the contrail segments was computed with a time-step of $dt = 2$ minutes. The results are presented in Figure 19.10. The plots show the evolution of contrail properties along the flight path. The contrail formation criterion is based on Eq. (19.4), which measures the difference between ambient and threshold temperature dT; a more negative dT indicates a greater probability of contrail formation. The contrail RF and lifetime are strongly correlated to the RH_i; when the

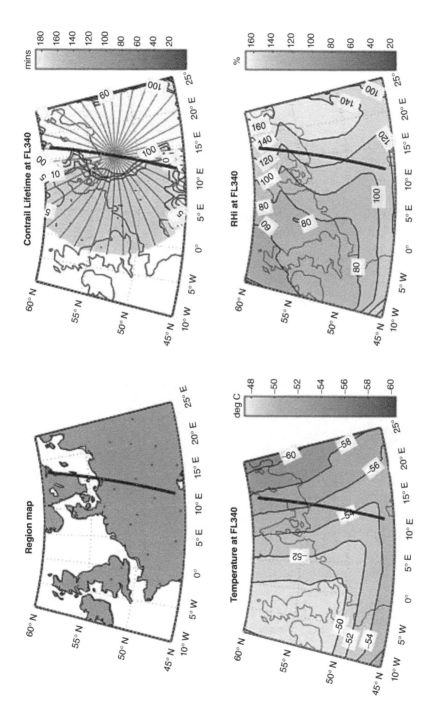

Figure 19.6 Flight from Stockholm to Venice (top left); iso-persistence regions (top right) are mapped along the main flight path (black line) using simulated trajectories (blue lines) at trajectory time $t = 0$, with the contrail lifetime integrated at a time-step of $dt = 2$ minutes. Iso-persistence is closely correlated to ambient temperature (bottom left) and RH_i (bottom right).

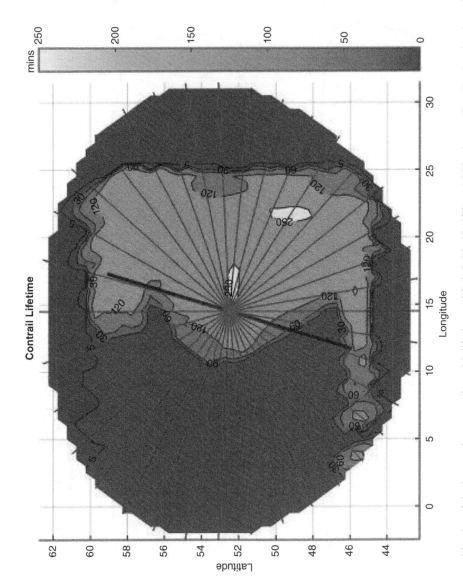

Figure 19.7 A zoomed-in plot of the iso-persistence region with lifetimes of 0, 1, 5, 30, 60, 120, and 250 minutes. The simulated trajectories are spaced in 10° intervals, originating from the midpoint of the flight path.

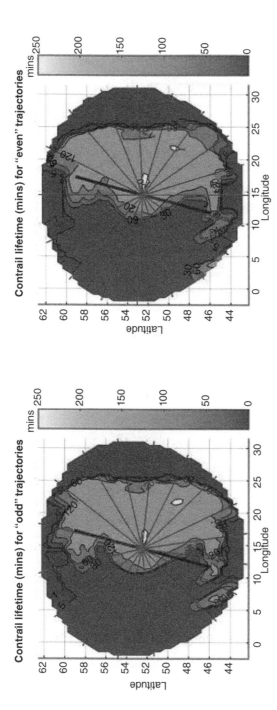

Figure 19.8 Contour maps interpolated from different trajectories for validation purposes. The left plot shows 'odd' trajectories and the right plot shows 'even' trajectories. Both generate similar regions to the 'fully interpolated' map as shown in Figure 19.7.

Table 19.2 Details of the simulated flight.

Flight details			
Origin	Stockholm Arlanda (59.1°N, 17.3°E)		
Destination	Venice Tessera (45.9°N, 11.7°E)		
Scenario	O	A	B
Cruise flight path	NOSLI ROKIB	NOSLI ROKIB	NOSLI Elbe VOR Trasadingen VOR ROKIB
Altitude	FL340	FL320	FL340
Date	11 Apr 2015	11 Apr 2015	11 Apr 2015
Flight time	00:00 to 02:40	00:00 to 02:40	00:00 to 03:05
Net RF	6.8 mW m^{-2}	7.2 mW m^{-2}	4.9 mW m^{-2}

Table 19.3 Aircraft parameters used for the simulation.

Aircraft parameters	
Aircraft model	Airbus A320
Wingspan, S_a	34.1 m
Mass, M_a	64 000 kg
True Airspeed (TAS) in cruise, V_a	236.8 m s^{-1}
Fuel flow, m_F	0.17 kg s^{-1}
Soot number per fuel mass, EI_{soot}	2.8×10^{14} kg^{-1}
Emission index for water, EI_{H2O}	1.23
Fuel combustion heat, Q_{fuel}	43.2 MJ kg^{-1}
Propulsion efficiency, η	0.3

RH_i drops below 100% (ice-subsaturated) the contrail lifetime experiences a sharp drop. Even minor fluctuations in the RH_i can significantly affect the contrail lifetime. This case study indicates that a bypass might be more efficient in contrail avoidance than a general reduction in flight altitude.

19.3.3 Simulation Case Study – Online Trajectory Optimisation

The final verification of the contrail model was achieved upon integration within a MOTO software framework. The aim of this verification case was to highlight the viability of adopting the mapped contrail iso-persistence or iso-radiative forcing region as the basis for the optimisation of the flight trajectory with respect to two or more objectives. The developed contrail model is capable of providing a 4D mapping of the iso-persistence

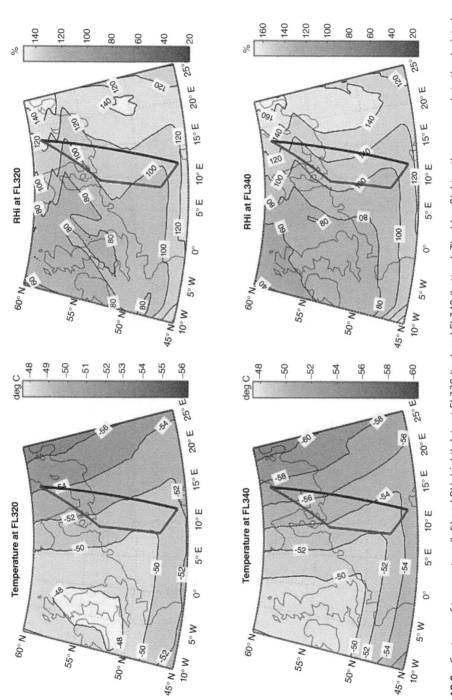

Figure 19.9 Contours of temperature (left) and RH$_i$ (right) data at FL320 (top) and FL340 (bottom). The blue flight path corresponds to the deviated trajectory as per scenario B.

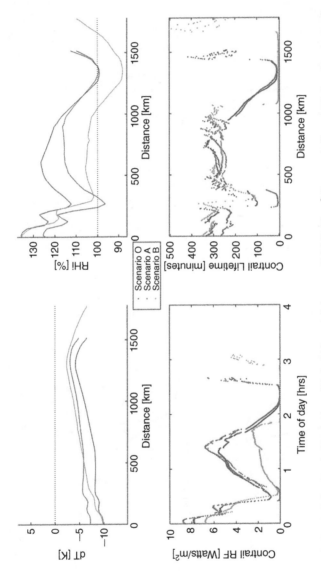

Figure 19.10 Plots of dT in Kelvins (top left), ambient RH$_i$ in % (top right), contrail RF in W m^{-2} (bottom left) and lifetime in minutes (bottom right) for Scenarios O (blue), A (red), and B (orange). The RF from O and A are comparable, while B has significantly reduced RF.

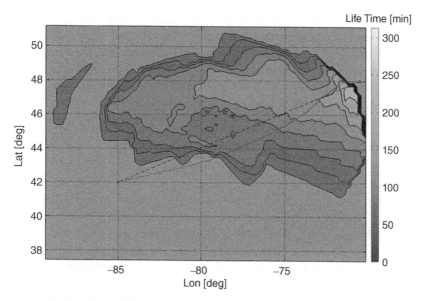

Figure 19.11 Optimal 2D + T trajectory at constant flight level for a given contrail 2D + T iso-persistence mapping region.

and iso-radiative forcing regions, which can be exploited for the optimisation of fully 4D trajectories. Nevertheless, as the current operational approach of ATM restricts aircraft to cruise at discrete flight levels, the solution implemented here adopts a two-dimensional plus time (2D + T) trajectory optimisation approach at a single flight level. The cruise phase of a hypothetical Airbus A320 aircraft flying between Chicago to Quebec City at FL340 was simulated. The 2D + T iso-persistence mapping was generated for a marginally augmented search domain covering parts of the USA and Canada. The origin and destination were amended slightly, such that the trajectory passed through most of the contrail region. The verification case was run with the contrail region generated using the *fmincon* optimisation algorithm provided by Matlab. A cost function that included weighted penalties for both the contrail lifetime and flight time was used. Figure 19.11 shows the results of the 2D + T trajectory optimisation.

19.4 Conclusions

This chapter presented the design and the underlying theoretical framework of a customised contrail model and the related software algorithm specifically conceived for integration in a MOTO software tool. The proposed algorithm allows the selective minimisation of the environmental impacts associated with aircraft-induced contrails and cirrus clouds. In particular, in addition to condensation trail formation, the algorithm determines the evolution of the contrail in terms of lifetime and physical properties, and estimates the net radiative forcing associated with the specific trajectory flown by an aircraft at a specific time. For this purpose, the algorithm entails fully four-dimensional (4D) modelling of the atmospheric thermodynamic state from suitable weather datasets, as well as the solar zenith angle based on the specific time and date. Further research will address the exploitation of the contrail models in a number of MOTO case studies, investigating in particular the possible tradeoffs between the radiative forcing associated with exhaust emissions and that associated with contrails and aircraft-induced cloudiness. These optimisation studies will be performed both in the offline and online operational timeframes, and will target both constant flight level cruise and 4D flight trajectories. For this purpose, the contrail model will be integrated in the ground-based ATM and airborne avionic decision support systems (DSSs) currently being developed [5–9, 16].

References

1 Gierens, K., Lim, L., and Eleftheratos, K. (2008). A review of various strategies for contrail avoidance. *The Open Atmospheric Science Journal* 2: 1–7.

2 Meerkötter, R., Schumann, U., Doelling, D.R. et al. (1999). Radiative forcing by contrails. *Annales Geophysicae* 17: 1080–1094.

3 Myhre, G. and Stordal, F. (2001). On the tradeoff of the solar and thermal infrared radiative impact of contrails. *Geophysical Research Letters* 28: 3119–3122.

4 Soler, M., Zou, B., and Hansen, M. (2014). Flight trajectory design in the presence of contrails: application of a multiphase mixed-integer optimal control approach. *Transportation Research Part C: Emerging Technologies* 48: 172–194.

5 Sabatini, R., Gardi, A., Ramasamy, S. et al. (2015). Modern avionics and ATM systems for green operations. In: *Encyclopedia of Aerospace Engineering* (eds. R. Blockley and W. Shyy). Wiley.

6 Sabatini, R., Gardi, A., Ramasamy, S., Kistan, T., and Marino, M. (2014). Novel ATM and avionic systems for environmentally sustainable aviation, *Practical Responses to Climate Change. Engineers Australia Convention 2014 (PRCC 2014)*, Melbourne, Australia.

7 Gardi, A., Sabatini, R., Kistan, T., Lim, Y., and Ramasamy, S. (2015) 4-Dimensional trajectory functionalities for air traffic management systems, *Integrated Communication, Navigation and Surveillance Conference (ICNS 2015)*, Herndon, VA, USA.

8 Ramasamy, S., Sabatini, R., and Gardi, A. (2015). Novel flight management systems for improved safety and sustainability in the CNS+A context, *Integrated Communication, Navigation and Surveillance Conference (ICNS 2015)*, Herndon, VA, USA.

9 Marino, M., Gardi, A., Sabatini, R., and Kistan, T. (2015). Minimizing the cost of weather cells and persistent contrail formation region avoidance using multi-objective trajectory optimization in air traffic management," *SAE Technical Paper 2015-01-2392.*

10 Appleman, H. (1953). The formation of exhaust condensation trails by jet aircraft. *Bulletin of the American Meteorological Society* 34: 14–20.

11 Schumann, U. (2012). A contrail cirrus prediction model. *Geoscientific Model Development* 5: 543–580.

12 Sölch, I. and Kärcher, B. (2010). A large-eddy model for cirrus clouds with explicit aerosol and ice microphysics and Lagrangian ice particle tracking. *Quarterly Journal of the Royal Meteorological Society* 136: 2074–2093.

13 Khvorostyanov, V.I. and Curry, J.A. (2002). Terminal velocities of droplets and crystals: power laws with continuous parameters over the size spectrum. *Journal of the Atmospheric Sciences* 59: 1872–1884.

14 Schumann, U., Mayer, B., Graf, K., and Mannstein, H. (2012). A parametric radiative forcing model for contrail cirrus. *Journal of Applied Meteorology and Climatology* 51: 1391–1406.

15 Veness, C. (Cited 25 March, 2015). Movable type scripts. Available from: http://www .movable-type.co.uk/scripts/latlong.html

16 Sabatini, R., Gardi, A., Ramasamy, S., Marino, M. and Kistan, T. (2015). Automated ATM system enabling 4DT-based operations, *SAE Technical Paper 2015-01-2539*.

17 Lim, Y. (2015a). *4-Dimensional Trajectory Optimization for Contrail Mitigation*, MSc Thesis, Imperial College London, London, UK, 2015.

18 Lim, Y., Gardi, A., and Sabatini, R. (2015b). Modelling and evaluation of aircraft contrails for 4-dimensional trajectory optimisation. *SAE International Journal of Aerospace* 8: 248–259.

20

Trajectory Optimisation to Minimise the Combined Radiative Forcing Impacts of Contrails and CO_2

Yixiang Lim[1], Alessandro Gardi[2,3], Roberto Sabatini[2,3], and Trevor Kistan[4]

[1] *Agency for Science, Technology and Research (ASTAR), Singapore*
[2] *Department of Aerospace Engineering, Khalifa University of Science and Technology, Abu Dhabi, UAE*
[3] *School of Engineering, RMIT University, Melbourne, Victoria, Australia*
[4] *Thales Australia, Melbourne, Victoria, Australia*

Nomenclature

RF_{LW}	Longwave radiative forcing, Wm^{-2}
RF_{SW}	Shortwave radiative forcing, Wm^{-2}
OLR	Outgoing longwave radiation, Wm^{-2}
SDR	Solar direct irradiance, Wm^{-2}
r_{eff}	Effective particle radius, μm
τ	Contrail optical depth at 550 nm
τ_c	Optical depth at 550 nm of overhead cirrus
T	Atmospheric temperature, K
μ	Cosine of the solar zenith angle
k_T	Longwave RF: model parameter, $Wm^{-2}K^{-1}$
T_0	Longwave RF: model parameter, K
$\delta_\tau, \delta_{lr}, \delta_{lc}$	Longwave RF: Model parameters
t_A, Γ, γ	Shortwave RF: Model parameters
A_μ, B_μ, C_μ, F_r	Shortwave RF: Model parameters
$\delta_{sr}, \delta_{sc}, \delta'_{sc}$	Shortwave RF: Model parameters

20.1 Introduction

Air traffic, in revenue passenger kilometres (RPKs), has been growing at an average of 5% per year. Recent figures indicate an annual growth of 6.1% in the Asia Pacific region, second only to the Middle East, which is growing at 6.9% per annum [1]. This has led to a big push towards more sustainable modes of flight, which means achieving safer, more efficient, and more environmentally-friendly standards in aviation. In the area of avionics and air traffic management (ATM), current operations and procedures are being transformed by new technologies and innovation from such research initiatives as USA's

Sustainable Aviation Technology and Operations: Research and Innovation Perspectives, First Edition.
Edited by Roberto Sabatini and Alessandro Gardi.
© 2024 John Wiley & Sons Ltd. Published 2024 by John Wiley & Sons Ltd.
Companion Website: www.wiley.com/go/sustainableaviation

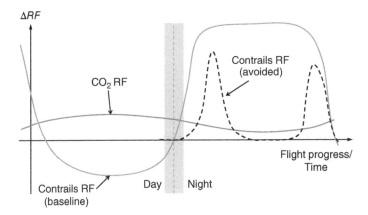

Figure 20.1 Illustrated comparison of the change in RF due to CO_2 and contrails as a function of flight progress.

Next Generation Air Transportation System (NextGen), and the European Single European Sky ATM Research (SESAR).

Multi-objective trajectory optimisation (MOTO) is an area of active research, and has been comprehensively reviewed in [2]. MOTO is part of an initiative to modernise ATM. The decentralisation of flight planning, traditionally done offline from a ground-based decision support system, is now being brought online and into the aircraft, allowing for dynamic re-planning based on user-defined costs and constraints. This will allow for greater savings and more environmentally-friendly trajectories to be flown. The successful implementation of MOTO in next-generation avionics and ATM systems takes into account various objectives, such as fuel consumption, flight time, pollutant emissions, and the associated radiative impact of such pollutants and contrails. These objectives are used to measure the optimality, or cost, of a given trajectory, based on user-defined weights. In particular, the radiative impact of aviation is a cost that can be minimised through MOTO. Studies have shown that the radiative impact is mainly comprised of two main components – contrails and CO_2 (Figure 20.1). The radiative forcing (RF) due to CO_2 is largely a function of fuel usage and remains positive and relatively constant over the course of a flight, while the RF due to contrails is determined by a combination of atmospheric conditions and the time of day. Regions of low temperature (generally $40°$ C) and high relative humidity with respect to ice (generally $RH_i > 0.8$) give rise to contrails, and such regions can be actively avoided (or sought out, if the regions give rise to negative RF) with re-routed trajectories, at the cost of increased fuel and flight time. This case study, originally published in [10], presents a computationally efficient implementation of the contrail models originally proposed in Chapter 2 and their algorithmic integration in a MOTO software for real-time avionics systems similar to the ones covered in other chapters and case studies of this book.

20.2 Contrail Radiative Forcing (RF) Model

Contrails are formed from the condensation of water vapour in the wake of an aircraft engine's exhaust. In favourable environmental conditions, contrails are known to persist for hours, eventually evolving into cirrus clouds. An in-depth review of the contrail life cycle can be found in a recent publication by Paoli and Shariff [3]. Contrail-cirrus are currently known to have a large positive radiative impact, contributing to global warming, yet also have the potential to have a negative radiative impact due to their contribution to the Earth's albedo. The literature has suggested that there are diurnal variations in the radiative impact of contrails, with a greater positive RF at night than in the day [4] (the albedo effect does not have any effect at night). The RF due to contrails comprises two components, the positive longwave radiation and the negative shortwave radiation:

$$RF_{Tot} = RF_{LW} + RF_{SW}$$
$$RF_{LW} \geq 0, \quad RF_{SW} \leq 0 \tag{20.1}$$

The longwave radiation RF_{LW} is emitted from the surface of the Earth. Cirrus-contrails trap some of the outgoing radiation as heat due to the greenhouse effect. The phenomenon is mainly a function of the temperature difference between the contrail layer and the atmospheric temperature T. Based on Schumann's parametric model for contrail RF [5], the RF_{LW} can be obtained as follows:

$$RF_{LW}\left(T, r_{eff}, \tau, \tau_c\right) = \left[OLR - k_T\left(T - T_0\right)\right]$$
$$\cdot \left\{1 - \exp\left[-\delta_\tau F_{LW}\left(r_{eff}\right)\tau\right]\right\} \cdot E_{LW}\left(\tau_c\right) \tag{20.2}$$

The assumed outgoing longwave radiation constant is $OLR = 250 Wm^{-2}$ and $F_{LW}(r_{eff})$ and $E_{LW}(\tau_c)$ are scaling factors to account for the size dependence of the optical properties and the presence of overhead cirrus:

$$F_{LW}\left(r_{eff}\right) = 1 - \exp\left(-\delta_{lr}r_{eff}\right) \tag{20.3}$$

$$E_{LW}\left(\tau_c\right) = \exp\left(-\delta_{lc}\tau_c\right) \tag{20.4}$$

The parametric model accounts for the optical properties of the contrail with eight different ice particles (also termed habits) including spherical, rosette, and plate. The habit depends mainly on atmospheric conditions and a contrail may comprise a combination of different habits, which can be accounted for with a weighted sum of each habit. For the simulation, a rosette habit (which describes particles that form at high RH_i and low temperatures) was assumed, giving values of $T_0 = 152$, $k_T = 1.944$, $\delta_\tau = 0.749$, $\delta_{lr} = 0.170$, and $\delta_{lc} = 0.133$.

The shortwave radiation RF_{SW} comes from the sun and is incident on the Earth. Due to the albedo of the contrail, some of the shortwave radiation is reflected back into space. This reflectance is mainly a function of the optical depth of the contrail, τ (which also contributes to the longwave RF).

$$RF_{SW}\left(T, r_{eff}, \tau, \tau_c\right) = -SDR\left(t_A - A_{eff}\right)^2 \cdot \alpha_c\left(\mu, \tau, r_{eff}\right) \cdot E_{LW}\left(\mu, \tau_c\right) \tag{20.5}$$

The incoming shortwave radiation is assumed to be a constant value of $SDR = 1362\ Wm^{-2}$; an effective albedo of $A_{eff} = 0.2$ is assumed, which is a typical value for grassland areas, consistent with the summer case study presented below. The two functions $\alpha_c(\mu, \tau, r_{eff})$ and $E_{LW}(\mu, \tau_c)$ are scaling factors to account for the contrail albedo and the presence of overhead cirrus.

$$\alpha_c\left(\mu,\tau,r_{eff}\right) = R_C\left(\tau_{eff}\right)\left[C_\mu + A_\mu R'_C\left(\tau'\right)F_\mu\left(\mu\right)\right] \tag{20.6}$$

$$E_{LW}\left(\mu,\tau_c\right) = \exp\left(\delta_{sc}\tau_c - \delta'_{sc}\tau_{c,eff}\right) \tag{20.7}$$

The contrail albedo depends on the effective optical depth, which is a function of the solar zenith angle and particle size,

$$\tau' = \tau F_{SW}\left(r_{eff}\right), \qquad \tau_{eff} = \tau'/\mu \tag{20.8}$$

$$F_{SW}\left(r_{eff}\right) = 1 - F_r\left[1 - \exp\left(-\delta_{sr}r_{eff}\right)\right] \tag{20.9}$$

The reflectances,

$$R_C\left(\tau_{eff}\right) = 1 - \exp\left(-\Gamma\tau_{eff}\right), \qquad R'_C\left(\tau_{eff}\right) = \exp\left(-\gamma\tau_{eff}\right) \tag{20.10}$$

And an empirical function accounting for solar zenith angle dependent sideward scattering

$$F_\mu\left(\mu\right) = \frac{\left(1-\mu\right)^{B_\mu}}{\left(1/2\right)^{B_\mu}} - 1 \tag{20.11}$$

The contrail parameters (length, breadth, ice number, effective ice radius, etc.) were computed based on Schumann's Contrail Cirrus Prediction (CoCiP) tool [6], which evaluates the contrail properties along a linear aircraft trajectory. Contrails are simulated along this trajectory based on ambient weather, with the key parameters being the relative humidity with respect to ice (RH_i) and temperature. A 2-dimensional (4D) mapping algorithm generates and interpolates between multiple linear trajectories along a (latitude, longitude) plane to obtain a 2D contrail field of RF and lifetime. The 2D planes are generated for multiple altitudes and interpolated to obtain a 3D volume, which is then interpolated at different timesteps to obtain a 4D field (Figure 20.2). The detailed explanation and model of the 4D mapping algorithm can be found in Lim et al. [7].

20.3 CO_2 Radiative Forcing (RF) Model

CO_2 emissions are modelled as a function of the aircraft fuel flow, the relations of which can be found in the Base of Aircraft Data (BADA) [8]. The fuel flow (kg/min) is given as the maximum thrust in nominal and idle conditions.

$$FF = \max\left[\tau T_{max}C_{f1}\left(1 + \frac{v_{TAS}}{C_{f2}}\right), \quad C_{f3}\left(1 - \frac{H_P}{C_{f4}}\right)\right] \tag{20.12}$$

The throttle is given by τ, the airspeed by v_{tas} (kts) and the geopotential pressure altitude by H_P(ft). The aircraft-specific parameters T_{max}, C_{f1}, C_{f2}, C_{f3}, C_{f4} can be found in BADA. The CO_2 emissions (kg) can then be approximated as a function of the fuel flow and elapsed time:

$$CO_2 = 3.16\int FF\left(t\right)dt \tag{20.13}$$

Based on the atmospheric mass, a conversion factor of $1\ kg\ CO_2 = 4.69 \times 10^{-12}\ ppm\ CO_2$ is used to obtain CO_2 in parts per million (ppm). This can then be used to calculate the

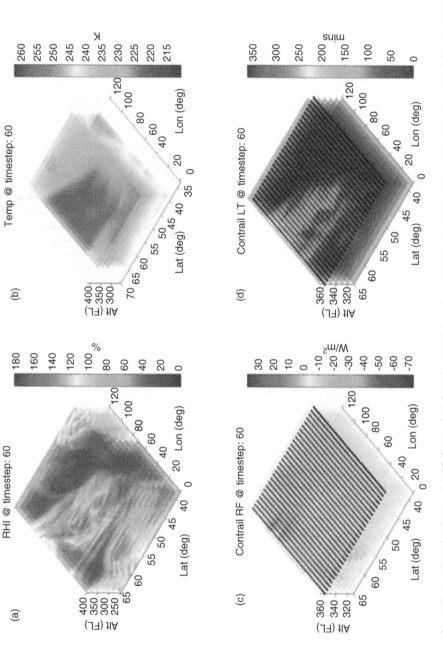

Figure 20.2 Contrail mapping algorithm, with 3D fields of (a) Relative humidity with respect to ice (RH_i), (b) temperature, (c) contrail RF, (d) contrail lifetime. The mapping trajectories in 2D are drawn in (c) and (d).

RF due to CO_2 based on a model from the International Panel of Climate Change (IPCC) [9] as

$$RF_{CO_2} = \alpha_{CO_2} \left[g\left(C\right) - g\left(C_0\right) \right] \tag{20.14}$$

$$g\left(C\right) = \ln\left(1 + 1.2C + 0.005C^2 + 1.4 \times 10^{-6} C^3\right) \tag{20.15}$$

Where $C = C_0 + \Delta C$ is the concentration of CO_2 in ppm and $\alpha_{CO_2} = 3.35$.

20.4 Multi-Objective Trajectory Optimisation

A geometric optimisation was performed based on the *fmincon* function in MATLAB to describe the geometry of the optimal trajectory (in latitude and longitude). The aircraft dynamics were simplified for increased computational efficiency; while the throttle and airspeed were modelled, the aircraft attitude was not considered. Compared to optimal control techniques, geometric optimisation appears to perform better for problems with larger timescales (i.e. in the cruise phase as opposed to within the terminal manoeuvring area) as well as in the presence of stochastic fields (such as wind or contrail regions). The objective was expressed as a weighted product, accounting for the total time, distance, CO_2 emissions and contrail-induced RF:

$$J = Time^{W_{Time}} \cdot Dist^{W_{Dist}} \cdot CO_2^{W_{CO_2}} \cdot RF_{contr}^{W_{contr}} \tag{20.16}$$

20.5 Case Study

The case study examined a direct flight of a Boeing B777–200 from Charles de Gaulle Airport (CDG) in Paris to Beijing Capital International Airport (PEK). Weather data was obtained from the Global Forecast System (GFS) on August 22, 2016. 15 hours of forecast data was obtained in 3-hour intervals, from 1200 hours (GMT + 0) to 0300 (GMT + 0) on August 23, 2016. The weather data includes wind fields, temperature, pressure, and relative humidity with respect to water RH_w. Figure 20.3 shows the RH_i (a function of temperature and RH_w) and temperature. A flight departing CDG at 2 p.m. (GMT + 2) and arriving at Beijing at 5.55 a.m. (GMT + 8) was selected. The flight was flown at 34 000 ft above the mean sea level. As shown in Figure 20.3, the abundance of ice supersaturated regions and low ambient temperatures produce favourable contrail-formation regions along the first two-thirds of the trajectory.

Negative-RF regions appear at the beginning of the flight while positive-RF regions occur in the middle portion of the flight path due to the sunset. Seven cases involving differing weights for w_{CO_2} and w_{contr} were evaluated, with the results shown in Table 20.1 and Figure 20.4. As w_{contr} is increased, the optimised trajectory is observed to deviate more from the original (Great Circle) path due to avoidance of positive-RF contrail regions in the middle of the flight, while favouring areas of negative-RF. This results in increased travel time, distance, and CO_2 emissions.

(a)

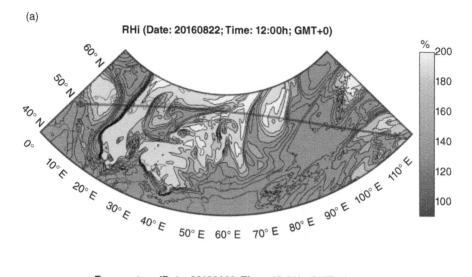

(b)

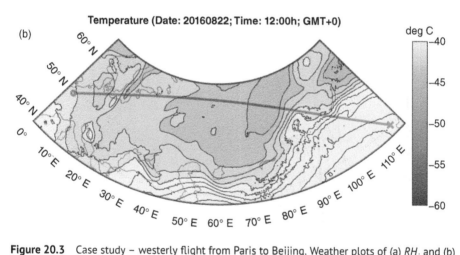

Figure 20.3 Case study – westerly flight from Paris to Beijing. Weather plots of (a) RH_i and (b) temperature.

Table 20.1 Optimisation weights.

Case	w_{CO_2}	$w_{RF_{contr}}$	Total travel time (h)	Total distance travelled (km)	CO_2 emissions (tonnes)	Contrail RF contribution $(W\,m^{-2})$
1	1	1	10.4	8239	131.0	−4.10
2	4	0.667	10.8	8312	140.5	−5.17
3	3.333	1.333	10.5	8272	133.4	−4.05
4	2.667	2	10.7	8423	136.5	−4.24
5	2	2.667	11.0	8617	141.3	−4.24
6	1.333	3.333	11.0	8639	177.3	−4.41
7	0.667	4	12.7	9071	207.4	−4.22

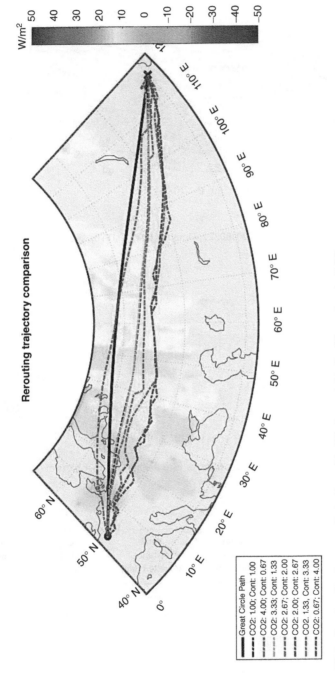

Figure 20.4 Results of trajectory optimisation with differing weights for CO$_2$ and contrail-RF.

20.6 Conclusion

This chapter presents the mathematical models for integrating radiative forcing (RF) of contrails and of carbon dioxide (CO_2) into a MOTO software algorithm. In particular, these models were used to define cost functions in a MOTO case study accounting for an international flight from Paris to Beijing. The case study involves optimal trajectories that minimise the RF of flight trajectories. The results of this study demonstrate the feasibility of MOTO in generating RF-minimal trajectories with respect to contrails and CO_2, and show that not all contrail-persistent regions should be avoided. As contrail-induced RF is dependent on the time of day, with positive RF contributions at night and negative RF contributions during the day, the optimisation process generates trajectories that avoid contrail-persistent regions at night and seeks out contrail-persistent regions by day. This is an important finding, indicating that while more computationally efficient, the simplified contrail avoidance frequently pursued in the literature based on ice super-saturation region models is not adequate in computing trajectories that actually minimise the overall total RF.

References

1 Boeing. (2016). Current Market Outlook 2016-2035, Seattle.

2 Gardi, A., Sabatini, R., and Ramasamy, S. (2016). Multi-objective optimisation of aircraft flight trajectories in the ATM and avionics context. *Progress in Aerospace Science* 83: 1–36.

3 Paoli, R. and Shariff, K. (2016). Contrail modeling and simulation. *Annual Review of Fluid Mechanics* 48: 393–427.

4 Stuber, N., Forster, P., Radel, G., and Shine, K. (2006). The importance of the diurnal and annual cycle of air traffic for contrail radiative forcing. *Nature* 441: 864–867.

5 Schumann, U., Mayer, B., Graf, K., and Mannstein, H. (2012). A parametric radiative forcing model for contrail cirrus. *Journal of Applied Meteorology and Climatology* 51: 1391–1406.

6 Schumann, U. (2012). A contrail cirrus prediction model. *Geoscientific Model Development* 5: 543–580.

7 Lim, Y., Gardi, A., and Sabatini, R. (2015). Modelling and evaluation of aircraft contrails for 4-dimensional trajectory optimisation. *SAE International Journal of Aerospace* 8: 248–259.

8 Eurocontrol. (2013). User Manual for the Base of Aircraft Data (BADA) Revision 3.11, EUROCONTROL, Paris, France.

9 Ramaswamy, V., Boucher, O., Haigh, J. et al. (2001). *Radiative Forcing of Climate Change*. Richland, WA (US): Pacific Northwest National Laboratory (PNNL).

10 Lim, Y., Gardi, A., and Sabatini, R. (2017). Optimal Aircraft Trajectories to Minimize the Radiative Impact of Contrails and CO_2, *Energy Procedia*, 110: 446-452.

21

The W Life Cycle Model – San Francisco Airport Case Study

Anthony Zanetti[1], Alessandro Gardi[1,2], and Roberto Sabatini[1,2]

[1]*RMIT University, Melbourne, Victoria, Australia*
[2]*Department of Aerospace Engineering, Khalifa University of Science and Technology, Abu Dhabi, UAE*

21.1 Introduction

This case study, originally published in [10, 11], discusses the application of the 'W' model introduced in the previous section to a representative case study. The methodology behind this case study has been developed specifically to bring forward a new sustainable life cycle model and to test its validity through a case study application. The discussion is built on a systems engineering approach, embodying sustainable life cycle thinking with the aim of analysing complex systems and their associated problems in a systematic and systemic manner. Sustainability of the aviation industry is a widely researched topic; yet, life cycle management continues to have little presence. Aircraft and air traffic optimisation make up the body of research relating to life cycle concepts; however, a singular focus in addressing externalities is overshadowed if the environmental improvements do not meet the objectives of the overall system. This chapter draws on the critical research gap to build a theoretical framework and qualitative discussion to further its application in aviation operations and to evaluate how systems engineering architectures will help mitigate life cycle emissions. Qualitative research is the most useful method for exploring topics that are new or complex, and to assist in furthering the reader's comprehension of the phenomena at stake [1]. A thorough literature analysis was carried out and presented in the previous section to assess the current flaws in sustainable aviation literature. Onwuegbuzie et al. [2] present a series of derived benefits from reviews, one of them being the ease of distinguishing gaps that need addressing. Meta-synthesis techniques assisted in integrating, evaluating, and interpreting existing data from qualitative papers. These papers were categorised into several themes: supply chain, design, manufacturing, operations, end-of-life, and system models. Relevant information was extracted and applied to aviation. These were then used to develop the 'W' life cycle model. In order to assess the validity of the 'W' life cycle model, an explanatory case study approach is utilised. Explanatory case studies work well when evaluating concepts that are complex in nature and do not warrant the use of experimental methods [3]. The case of San Francisco International Airport is specifically used in this paper, acting as an application of the 'W' model

Sustainable Aviation Technology and Operations: Research and Innovation Perspectives, First Edition.
Edited by Roberto Sabatini and Alessandro Gardi.
© 2024 John Wiley & Sons Ltd. Published 2024 by John Wiley & Sons Ltd.
Companion Website: www.wiley.com/go/sustainableaviation

with the aim of validating how it can be successfully implemented within the aviation context. Data collected was analysed using content analysis, a systematic and objective means of describing and quantifying phenomena [4, 5]. As such, data for this research paper was obtained through a wide range of sources, including; journal articles, scientific textbooks, reports, and company documentation.

21.2   San Francisco International Airport Redevelopment

In order to show the general applicability of the W model, we consider the case of one of the world's most environmentally friendly airports – San Francisco (SFO). San Francisco International has made significant investments in infrastructure revitalization. This is a crucial step towards meeting industry emission targets, as existing infrastructure will be rendered somewhat inadequate in a carbon-neutral future. SFO's efforts are to ensure that airport-wide operations have minimal impact on local, regional, and global environments. This objective goes on to ensure that natural resources are conserved and greenhouse gas emissions are reduced. Current mitigation strategies have been successful in that the airport was presented with an 'Airports Going Green' award [6]. Its sustainable practices go beyond current industry and government standards and, hence, are valuable in showing the general applicability of the W methodology to real life scenarios.

This case study addressed both a specific and holistic view of airport operations at San Francisco International. SFO's overall airport and operational requirements are detailed in both the airport action plan [7] and the sustainability report [6]. Current environmental objectives are to:

- Achieve a 40% reduction in baseline GHG emissions below 1990 by 2025.
- Achieve LEED Gold certification in all new and renovated buildings.
- Increase the solid waste recycling rate to 85% by 2017.
- Reduce energy usage year over year.
- Devise and implement other sustainability initiatives to meet the GHG reduction goals.

Meeting the above targets requires careful consideration into how the airport is operated, the limitations of current infrastructure, and a development plan of refurbishing outdated infrastructure to meet carbon-neutral growth. In the specific case study context, we look at the revitalization of terminal 2, designed and constructed to meet the Leadership in Energy and Environmental Design gold standard (LEED). Haseman [8] writes that terminal 2 was the first to be designed to meet the LEED standard. Environmental metrics gathered before and after terminal refurbishment demonstrate the eco-efficiency of sustainable design. More specifically, the redevelopment is an example of how international airports can collectively rejuvenate existing and outdated terminals to meet world-class environmental standards. The systems process of terminal 2, which can be implemented across a wide spectrum, is depicted against the W model in Figure 21.1. While the terminal refurbishment has already been completed, it is important to highlight that this section endeavours only to show general applicability of the W model, which can then be utilised for new project development.

There are three levels identified for terminal redevelopment. The first (component-level) being the materials used in the reconstruction of existing terminal infrastructure.

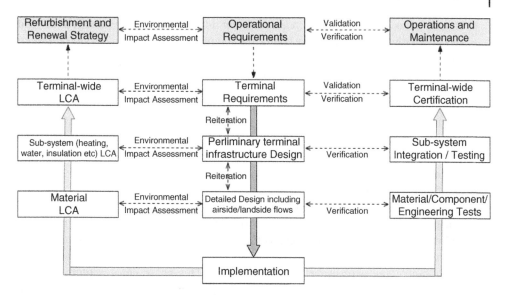

Figure 21.1 Implementation of the W model to San Francisco international airport terminal 2 redevelopment.

The second (sub-system) considers elements that will regulate the requirements of the total terminal. Examples will include heating, water, and insulation systems. Sub-systems will also include the intermodal flow between landside and airside activity. The third element is concerned with the overall terminal precinct as well as the integration of the sub-systems i.e. do they meet expected requirements? Have they been implemented in the most efficient way possible? This broader picture essentially ensures that all the elements within the terminal precinct work together. For example, we are ensuring that a heating system is not overshadowed by poor insulation. These three levels will form the basis of system validation and verification.

San Francisco International has a detailed framework for identifying and analysing customer requirements. These are highlighted in the Sustainable Planning, Design, and Construction Guidelines [9]. SFO utilises a stakeholder engagement process (SEP) to serve as a platform for strategising and establishing goals for any given project [7, 9]. It follows a detailed structure that complements the W methodology and encompasses a complete life cycle assessment of project initiatives. These include:

1. *Project initiation.* Where project requirements and the basis of design are defined.
2. *Planning and design.* Where a preliminary design of infrastructure facilities including measures of emission prevention are completed.
3. *Design of building envelope.* A detailed design of specified infrastructure. This process utilises alternative analysis for aspects such as insulation, windows, and flooring to minimise unwanted heat gain or loss.

   Comprehensive life cycle assessment is undertaken in the SEP process that corresponds to the design and requirement definitions from the first three phases.

4. *Materials procurement and life cycle assessment.* Identifying major materials to be used and alternative choices that may conform to requirements. The choices are ranked on the basis of life cycle environmental impacts and overall economic costs.
5. *Heating, ventilation, and air conditioning systems analysis.* This phase essentially identifies measures to reduce energy consumption and to maintain a self-temperature regulating building. The assessment effectively draws back to the design of the building envelope with an objective to use natural resources for a zero net energy building.
6. *Energy efficiency analysis.* Taking the energy supply issues identified in the previous step into consideration, the energy efficiency analysis should review options for heating, ventilation, and air conditioning system integration. This phase will also consider electrical requirements for lighting.
7. *Indoor air quality analysis.* Another in-depth analysis associated with phase 3. The primary concerns here are to establish the correct methods of ventilation, air purification, and moisture. The SEP should review the performance of air filtration systems from previous projects to aid in system selection.
8. *Water supply and wastewater reduction analysis.* This phase engages plumbing and landscape strategies. Dual plumbing systems are prominent component to SFO terminal infrastructure as they enable the reuse of treated wastewater, thereby reducing the demand for fresh water supplies.
9. *Construction waste management, disposal, and recycling analysis.* A review of past successes to implement a process that disposes of and recycles materials to achieve zero waste.

The results of the terminal 2 refurbishment are threefold. Firstly, the use of preconditioned air and 400 Hz power supply units at aircraft parking stands has reduced the fuel use of auxiliary power units by 1 400 000 gal per year, resulting in an annual reduction of 15 000 tons [8]. Secondly, efficient sub-systems, such as LED lighting, have reduced electrical energy consumption by 2.9 GW hours per year. In addition, efficient machinery has reduced natural gas consumption by 116 000 therms per year. Collectively, a total annual reduction of 750 metric tons has been realised [8]. Perhaps one of the most critical successes in the redevelopment of terminal 2 is the 12 300 tons of CO_2 saved from reusing a substantial portion of existing materials [8]. This provides a good indication of how end-of-life strategies can positively affect the net carbon emissions of a system. Recycling or remanufacturing existing materials extends their overall life and minimises the requirement to procure new resources. Material selection or component level should be considered in all projects going forward for both environmental and economic reasons.

Moving forward to a more holistic assessment of San Francisco airport, we look to uncover the environmental efficiencies gained on a system-wide level. SFO's greenhouse-gas inventory plays a crucial part in our analysis, in that it collectively breaks down total emissions into separate elements, allowing us to address these problems individually while simultaneously addressing the wider airport system. SFO's carbon footprint is defined in the following emission categories and applied in Figure 21.2 [6, 7]:

- *Category 1: SFO controlled emissions*
  Emissions under the control of SFO including airside and landside infrastructure and emissions from all modes of travel on SFO controlled roads.

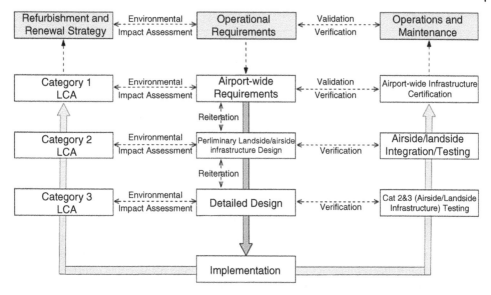

Figure 21.2 A holistic assessment of San Francisco airport development using the W model.

- *Category 2: Airlines, concessionaires, and airline support services emissions*
  Emissions related to the physical boundaries of SFO, including airlines operations, ground service equipment, and other support services.
- *Category 3: Optional emissions*
  Emissions not directly emitted by, but connected with, airport operations. These include passenger travel, delivery trucks, and commuter travel by employees.

The life cycle inventory has been implemented successfully into the W model, highlighting the model's adaptability to assess both the direct and indirect carbon footprint of a system. The idea behind this is to provide a holistic view of airport operations and measure the overall environmental performance of San Francisco International Airport. In practicality, a holistic view of the development process is used to assess the net environmental impact of new or refurbished facilities on the airport system. Analysing the emission trends against each inventory category, we note that SFO controlled emissions have decreased 33.8% below 1990 levels – a clear indication of terminal revitalisation and the use of efficient and cleaner technologies [7, 9]. Subsequently, category 2 and 3 emissions have risen as a result of air traffic and passenger growth. This highlights the need to find an optimal balance between growth and the environment.

The brief discussion below will primarily focus on system level emissions (category 1). These emissions are those which SFO has an influence over and relate directly to efforts of redesigning existing infrastructure. With a closer look at category 1 emissions, we see a significant effort placed in sourcing renewable energy. Table 21.1 shows an increase in energy consumption by 100 000 mW per hour since the year 1990. This would most notably be correlated to the increase in passenger and aircraft traffic. However, while electrical energy consumption continues to increase, we notice a steep decline in greenhouse emissions associated with the use of 100% renewable sources.

Table 21.1 San Francisco electrical consumption and GHG emissions [6, 7, 9].

Emission category	GHG emissions (metric tons)			
	1990	FY2012	FY2013	FY2014
Category 1	49 726	33 328	33 939	32 913
Category 2	838 980	936 841	908 135	936 951
Category 3	7 127 543	9 650 530	8 414 155	10 843 097
Total	8 016 249	10 620 719	9 356 229	11 812 961

Water usage at San Francisco International has shown a significant decline over the past several years. While passenger traffic has increased steadily, annual water usage has dropped by 16.6% from 2008 to 2014 [9]. This decline in consumption is correlated to the renovations of new and existing terminals with dual plumbing allowing for the reuse of reclaimed water. Recycling plays a large part in San Francisco's greening efforts. A total of 8940 tons of solid waste produced at terminals and other facilities were recycled [9]. The solid waste reduction and recycling programmes offset the greenhouse-gas emissions at SFO by 3356 tons in 2013–2014. Construction waste recycling was over 75% in the same year [7]. These end-of-life strategies are an essential component to green life cycle systems. Recycling existing materials into refurbishment projects will offset the need for virgin materials.

Table 21.2 shows the overall significance of utilising the W methodology in airport design and development. A holistic approach to environmental assessment can allow for emission offsets to counter emissions that may be rising in other aspects of the airport's operation. While passenger numbers have increased substantially over the past few years, greenhouse-gas emissions at SFO have decreased. One would assume that an increase in passenger numbers would correlate to increased energy consumption. However, due to efforts aimed at reducing emissions and increasing offsets, GHG is showing trends of gradual decline. The net category 1 greenhouse gas emissions in 2014 were 33.8% less than those recorded in 1990. The collective category 1 GHG emissions at SFO are primarily due to natural gas and fuel consumption. Offset measures, such as solid waste recycling and tree sequestration, contributed to a 9.6% decrease in net category 1 emissions. Airport-wide (category 1–3) mitigation efforts accounted for 89 429 metric tons of carbon mitigation [7]. This analysis has not only given insights to the applicability and flexibility of the W model, but also highlights the critical element of addressing environmental problems in a systemic and systematic manner.

21.3 Conclusion

This chapter addressed the introduction of sustainable life cycle management in aviation, which is a critical gap in current literature. The introduction of life cycle thinking in an air-transportation context has uncovered a new methodology for future system design

Table 21.2 San Francisco international category 1 emissions in 1990 and 2014 [6, 7, 9].

Activity	Category 1 emissions(metric tons)	
	1990	FY 2014
GHG emission levels		
Electric energy and natural gas consumption	29 269	18 803
Fuel consumption	13 138	16 058
Fugitive refrigerant gas emissions	4875	426
Solid waste disposal	2246	753
Wastewater treatment	198	381
Total gross baseline category 1 GHG emissions	49 726	36 420
GHG emission offset levels		
Solid waste recycling offset	0	−3386
Tree sequestration offset	0	−121
Total GHG emission offset	0	−3507
Net category 1 GHG emission	49 726	32 913
GHG emission mitigation level		
Total GHG emission mitigation	0	−89 429

and implementation. It is suggested that air transportation should be considered as a system-of-systems implemented as one to address a common functionality. This also poses the possibility of addressing environmental problems as a whole, moving away from traditional methods of mitigating environmental externalities at only one particular phase of the system life cycle.

The research focussed on two primary questions: (i) what are the limitations in current life cycle models? And (ii) what models and architectures are required to enhance the sustainability of the civil aviation industry? The analysis has uncovered a series of shortfalls in current life cycle methodologies, many of which are not flexible or have the inherent ability to respond to change (e.g. the waterfall model). Perhaps the most common shortfall in current methodologies is the lack of sustainable thinking. System design and implementation have only been validated and verified against economical and performance-based requirements. This leads to the development of the W model that seeks to introduce environmental impact assessment across the entire design and implementation of an aviation system. While it is agreed that previous life cycle models have not been developed in an ecological context, the industry needs to adopt sustainable thinking if it wishes to meet neutral growth targets. A case study analysis of San Francisco International airport shows the application of the W model to real life scenarios. It highlights how current and future projects can be divided into unique sub-levels, which are then addressed independently, but are integrated to achieve system-wide optimality. Its successful implementation in the case of San Francisco highlights how a systemic approach to greenhouse-gas mitigation can be achieved across an entire system. San Francisco International shows evidence of airport-wide mitigation, achievement through declining emissions despite a growth in passenger numbers.

The overall aim of this study has been achieved by a thorough literature analysis, a critical assessment of current life cycle models and their shortfalls, and the development of the W model. The 'W' essentially emphasises that life cycle assessment should be included within the development stage of a system to ensure that emissions during manufacturing, operations, and end-of-life are minimised at their source. It introduces a way forward to sustainable thinking and looks to achieve an optimal balance between economic and environmental requirements.

References

1 Hennink, M., Hutter, I., and Bailey, A. (2010). *Qualitative Research Methods*. Sage.
2 Onwuegbuzie, A., Collins, K., Leech, N. et al. (2010). A meta-framework for conducting mixed research syntheses for stress and coping researchers and beyond. In: *Toward a broader understanding of stress and coping: Mixed methods approaches*, The Research on Stress and Coping in Education Series (Series eds. G.S. Gates, W.H. Gmelch, and M. Wolverton and Vol. eds. K.M.T. Collins, A.J. Onwuegbuzie, and Q.G. Jiao), vol. 5, 169–211.
3 Yin, R.K. (2013). *Case Study Research: Design and Methods*. Sage publications.
4 Elo, S. and Kyngäs, H. (2008). The qualitative content analysis process. *Journal of Advanced Nursing* 62: 107–115.
5 Hsieh, H.-F. and Shannon, S.E. (2005). Three approaches to qualitative content analysis. *Qualitative Health Research* 15: 1277–1288.
6 SFO. (Cited November 8, 2014). San Francisco International Airport 2014 Sustainability Report. Available from: http://media.flysfo.com/media/sfo/community-environment/sfo-2014-sustainability-report.pdf
7 SFO. (Cited November 8, 2015). 2014 SFO Climate Action Plan. Available from: http://media.flysfo.com/media/sfo/community-environment/2014-sfo-climate-action-plan.pdf
8 Haseman, Z. (2013). *Integrating Environmental Sustainability into Airport Contracts*, vol. 42. Transportation Research Board.
9 San Francisco International Airport. (2015). Sustainable Planning, Design and Construction Guidelines - Delivering Healthy, High Performing, and Resilient Facilities, San Francisco International Airport.
10 Zanetti, A. (2015). Introducing Green Life Cycle Management in the Civil Aviation Industry: the State-of-the-Art and the Future, BSc Final Project Report, School of Aerospace, Mechanical, and Manufacturing Engineering, RMIT University, Melbourne, Australia.
11 Zanetti, A., Sabatini, R., and Gardi, A. (2016). Introducing green life cycle management in the civil aviation industry: the state-of-the-art and the future, *International Journal of Sustainable Aviation*, 2: 348-380.

22

Conclusions and Future Research

Roberto Sabatini[1,2] and Alessandro Gardi[1,2]

[1]*Khalifa University of Science and Technology, Abu Dhabi, UAE*
[2]*School of Engineering, RMIT University, Melbourne, Victoria, Australia*

Despite the substantial research advances experienced over the last two decades, the environmental impact of aviation is increasing steadily due to the rapid growth of air traffic and the comparatively slower pace of technological and operational innovations. This disparity is particularly pronounced in key areas such as greenhouse gas and noxious pollutant emissions, low-altitude flight noise, and contrail formation at high-altitude. This unsustainable trend necessitates immediate and resolute action to mitigate its well-documented consequences on climate change, public health and overall ecosystem integrity.

Achieving environmental sustainability in aviation is a major challenge and, despite the ambitious goals set by ICAO and other international/national regulators, the timeframe required for a transition to carbon neutrality appears too stringent, especially if compared to the expected lifespan of present-day aircraft fleets and the insufficient uptake of sustainable aviation technologies. Despite many conflicting forces, the most effective course of action is being identified and a number of major initiatives have been launched in recent years, which are supporting the maturation of critical technology for the "greening" of aircraft, airports and ATM systems of the future.

The chapters of this book have discussed some of the main initiatives currently ongoing for the development and operational uptake of environmentally sustainable technologies and associated lifecycle management processes in aviation. The introduction of new aircraft configurations and materials, fuel efficient/low-emission power plants, sustainable aviation fuels, CNS+A systems and multimodal connectivity in airports, as well as new life-cycle management processes, will contribute substantially to the environmental sustainability of the air transport sector. This transition will also facilitate and accelerate the ongoing modernisation and expansion of ATM operations, which is focussing primarily on safety and efficiency/capacity gains as well as enabling unrestricted access of RPAS to all classes of airspace. Nevertheless, a number of challenges are to be addressed along this evolutionary pathway, including cyber-physical security, certification and trustworthiness of AI-based systems, new through-life support processes and standards, SAF production and logistics/supply chain challenges.

Sustainable Aviation Technology and Operations: Research and Innovation Perspectives, First Edition.
Edited by Roberto Sabatini and Alessandro Gardi.
© 2024 John Wiley & Sons Ltd. Published 2024 by John Wiley & Sons Ltd.
Companion Website: www.wiley.com/go/sustainableaviation

In parallel with air transport developments, progress in spaceflight research has led to the introduction of various manned and unmanned reusable space vehicle concepts, opening up uncharted opportunities for the newborn space transport industry. For future space transport operations to be technically and commercially viable, it is critical that an acceptable level of safety is provided, requiring the development of novel digital tools (e.g., mission planning and decision support systems) that utilize advanced CNS+A technologies, and allowing a seamless integration of space operations in the current ATM network. While the technical maturity of propulsive and vehicle technologies is relatively high, a recent review of emerging platform operational concepts highlighted the challenges (and opportunities) brought in by the adoption of cyber-physical and autonomous systems for integration of point-to-point suborbital spaceflight with conventional atmospheric air transport [1].

Various launch and re-entry methodologies were extensively addressed in the literature, where the physical and computational limitations of these approaches were identified and applicability to future commercial space transport operations was assessed [2–4]. Current research is turning greater attention to the on-orbit phase, where the unique hazards of the space environment are being examined and the necessary elements required for space object de-confliction and collision avoidance modelling are analysed [1, 5]. The regulatory framework evolutions required for spacecraft operations is a conspicuous factor, which requires a holistic approach and extensive government support for the successful development and establishment of sustainable business models, including space debris mitigation strategies, operational risk assessment and liability issues. Within the atmospheric domain, extensions and alternatives to the conventional airspace segregation approaches must be identified including promising ATM and Air Traffic Flow Management (ATFM) techniques to facilitate the integration of new-entrant platforms. Lastly, adequate modelling approaches to meet on-orbit risk criteria must be introduced and evolutionary requirements to improve current operational procedures (and associated regulatory frameworks) must be identified.

While digital innovations advance and transform the Air-and-Space Transport[1] (A&ST) sector, aircraft and spacecraft are becoming more "intelligent" and capable of autonomous decision making. In this rapidly evolving ecosystem, the education sector has a key role in educating the next generation of engineers, scientists and business professionals to embrace the many challenges and opportunities offered by Artificial Intelligence (AI), robotics and autonomous systems. A new multidisciplinary approach to Research and Innovation (R&I) is also needed to accelerate the industrial uptake and maximise the societal benefits of these technologies by addressing (more holistically) the safety, security and ethical implications of A&ST intelligent and autonomous system. Beyond developing and deploying new technologies, the ongoing transformation of the A&ST sector requires significant evolutions of the political, legal and regulatory framework, but also new effective business strategies and new product/service lifecycle management processes focussing on value creation and value capture in digitally transformed and environmentally sustainable A&ST enterprises.

---

1 The terms A&ST is introduced here to include the activities related to DDT&E and operations of aerospace vehicles and ground support systems including both air-and-space passenger and cargo vectors.

To address these new challenges and to realise the commercial opportunities, the research community plays a fundamental role in developing new technological solutions and Sustainable Business Models[2] (SBM), which must rely on effective partnerships between government agencies, industry and research organisations. Key priority areas include:

- development and rapid uptake of fuel-efficient and low-emission technologies (gaseous and noise emissions), introduction of alternative energy sources (including bio-fuels) and cost-effective through-life support solutions for new and ageing aircraft;
- continuous improvements in the efficiency and capacity of airports and increasing integration with other forms of transport to establish successful multi-modal transportation hubs;
- developing an integrated air-and-space traffic management system, which succesfull implement space debris mitigation strategies, operational risk assessment methodologies and liability provisions.

The widespread adoption of digital technologies offers new exciting opportunities for an effective integration of emerging SBM and Digital Business Models (DBM) in the A&ST sector. Current trends include:

- adoption of interconnected CPS and Artificial Intelligence (AI) in DDT&E and, pervasively, in aerospace manufacturing (connected robotics, digital twins, etc.);
- development and progressive deployment of integrated CNS+A systems for enhanced flight operations, also addressing the channeges of highly automated and trusted autonomous UAS operations;
- digital technologies for enhanced airline and airport operations (robotic process automation, intelligent automation, digital ticketing/check-in, intelligent mobility solutions, connected/multimodal transport integration, etc.);
- development of advanced Decision Support Systems (DSS) for ATM and ATFM;
- evolving managerial processes and technical solutions for enhanced A&ST safety and cyber-physical security.

A strategic roadmap for workforce capability development is also a conspicuous factor that the education sector must urgently focus on. The multi-faceted and interconnected A&ST sector requires a more comprehensive and integrated system-of-systems approach that fully encompasses the domains of infrastructure, services and technologies. Furthermore, the integration of all modes of air/space transport within a shared and increasingly crowded airspace will be integral to the safe and sustainable growth of traffic. This is particularly important in low-level airspace operations (including around and within major cities), which will soon include commercial Unmanned Aircraft Systems (UAS) operations. Very importantly, new CPS architectures, different regulatory frameworks and new approaches are required to mitigate safety/security risks (both in the cyber and physical dimensions). Achieving these goals will provide the A&ST sector with a major opportunity to improve its competitiveness, to develop new products/services globally and to increase its contribution to the global economy through profitability and job creation.

---

2 *Sustainable business model* can be comprehenisively defined as a business model that creates, delivers and captures value for all its stakeholders without depleting the natural, economic and social capital it relies on.

Despite the significant traction of Environmental Sustainability (ES) and Digital Transformation (DT) topics in the governance and managerial structures of A&ST enterprises, with various C-suite level executives and director roles emerging in this area, existing Digital and Sustaianble Businesss Models (DBM/SBM) or systematic approaches that could inform the development of such models have not been disclosed or published in the open literature by any of the major industry players in the aerospace sector (Boeing, Airbus, Embraer, Leonardo, Thales, Northrop Grumman, Lockheed Martin, etc.). Additionally, there is no published research on DBM/SBM in aerospace Small-and-Medium Enterprises (SME). It appears, in fact, that this side of the industry is adopting a very cautious approach to the problem, largely due to the direct financial implications associated to the phase-out of certain technologies (e.g., non-compliance with ICAO CAEP standards) and the competitive advantage associated to the availability of innovations (through research) and DDT&E infrastructure supporting the rapid uptake in the market of disruptive digital technologies and eco-solutions. The aviation (flight operations) industry, on the other hand, has developed some useful guidelines and potential business approaches to both environmentally sustainable flight operations [6, 7] and digitally transformed operation/technical services [8]. However, as for the aerospace segment, the scientific literature is mainly focussing on the technological aspects of the problem, while both the scope and significance of published aviation management research is limited at the moment and mainly focussing on either airline, passenger or airport service provider perspectives [9–11]. As of today, there is no published academic or industrial work in the area of integrated digital/sustainable BMI strategies and associated BMD techniques/tools for the A&ST industries. Fusture research should address this research gap and various more specific/derived gaps emerging from a critical review of the current ES/DT literature in the A&ST sector. For completeness, a brief highlight of these gaps is provided below:

- There are large uncertainties over future trends in air transport growth and emissions, depending on the scenarios/assumptions selected to perform projections using current models. Key contributing factors include uncertainties about the pace of introduction of game-changing technologies and the impacts of the current infrastructure constraints ("bottlenecks") in limiting growth both in airport/airspace capacity and demand.
- There is no universally accepted model to quantify the monetary impact of aviation emissions on the environment and the monetary benefits of mitigating those impacts. As already mentioned, different models and different scenarios/assumptions produce different results and there is no consensus on the appropriate level at which any environmental levy should be set.
- As the environmental benefits (reduction of gaseous and noise emissions) achievable with conventional aircraft/power plants configurations have reached a plateau, it is essential to investigate more radical approaches (e.g., rapid uptake of digital technologies and biofuels).
- Current strategies for ensuring aviation sustainability include regulating aircraft design/operations with environmentally-friendly policies (carbon tax/offsetting schemes, noise emission charges, replacing or ruling-out old fleet, etc.). However, in the long term, digital transformation initiatives are essential and will radically transform DDTE&O lifecycle management processes.

- Regulations and standards are currently evolving to support safe operations in Upper Class E airspace above Flight Level (FL) 600 (i.e., 60,000 ft above mean-sea-level). While multiple use cases and business opportunities are emerging, including the possible resumption of supersonic passenger and cargo transport, environmental sustainability aspects must be carefully considered.

- Point-to-point space transport concepts call for research efforts that focus on ATM/STM harmonisation and novel spacecraft-focused Air Traffic Flow Management (ATFM) techniques. Additionally, the environmental sustainability of atmospheric operations (i.e., launch and re-entry) needs to be further verified as pertinent subjects such as spacecraft gaseous emissions and noise are not properly addressed in the scientific literature.

- Current standards and guidelines form an initial basis for the definition of a viable STM infrastructure however, continued efforts are required to delineate a code of conduct, and an equivalent "rules of the air" to define operation norms and enforce standards for space transport operations. Moreover, a lack of certification standards of CNS+A systems above FL600 demands inquest into the actual performance of these systems within and above the near space region as only then can CPS architectures be confidently developed and deployed on a global scale to support a fully integrated (i.e., compatible and interoperable) ATM/STM network.

- An open challenge affecting the A&ST industry uptake of CPS technologies – both autonomous CPS and Cyber-Physical-Human (CPH) systems – is the absence of clearly defined pathways for certification and Lifecycle Management (LCM) of integrated CNS+A systems and, in particular, AI-based technologies used in safety-critical products/services and associated lifecycle management best practices.

- The necessary technology-driven and process-driven regulatory air transport framework evolutions for cyber-physical security (vehicles and infrastructure) must be explored and the associated business implications will have to be considered.

- Despite the significant progress in CPS research, the full economic, social and environmental benefits associated to this technology are far from being fully realized. Major investments are being made worldwide to develop CPS for an increasing number of applications, including aerospace, transport, defence, robotics, communications, security, energy, medical, agriculture, humanitarian, etc. In this context, the TE sector is also facing challenges in order to rapidly and effectively respond to the rising industry demands for quality CPS education. The A&ST disciplines have been relatively proactive in this regard but significant further advancements are required to develop a more holistic approach to industry-focussed research and education.

References

1 Hilton, S., Sabatini, R., Gardi, A. et al. (2019). Space traffic management: towards safe and unsegregated space transport operations. *Progress in Aerospace Sciences* 105: 98–125.

2 Krevor, Z., Howard, R., Mosher, T., and Scott, K. (2011). Dream chaser commercial crewed spacecraft overview. *Proceedings of the 17th AIAA International Space Planes and Hypersonic Systems and Technologies Conference*, San Francisco, CA, USA (April 2011).

3 Longstaff, R. and Bond, A. (2011). The SKYLON project. *Proceedings of the 17th AIAA International Space Planes and Hypersonic Systems and Technologies Conference*, San Francisco, CA, USA (April 2011).

4 Vozoff, M. and Couluris, J. (2008). SpaceX products-advancing the use of space. *Proceeding of the AIAA SPACE 2008 Conference & Exposition*, San Diego, CA, USA (Septemeber 2008).

5 Thangavel, K., Gardi, A., Hilton, S. et al. (2021). Towards multi-domain traffic management. *Proceedings of 72nd International Astronautical Congress, IAC 2021*, Dubai, UAE (October 2021).

6 Abdullah, M.A., Chew, B.C., and Hamid, S.R. (2016). Benchmarking key success factors for the future green airline industry. *Procedia Social and Behavioural Sciences* 224: 246–253.

7 Yan, W., Cui, Z., and Álvarez Gil, M.J. (2016). Assessing the impact of environmental innovation in the airline industry: an empirical study of emerging market economies. *Environmental Innovation and Societal Transitions* 21: 80–94.

8 WEF (2017). Digital transformation initiative – aviation, travel and tourism industry. World Economic Forum (WEF) White paper https://reports.weforum.org/digital-transformation/wp-content/blogs.dir/94/mp/files/pages/files/wef-dti-aviation-travel-and-tourism-white-paper.pdf (accessed 27 February 2023).

9 Alencar Pereira, B. and Caetano, M. (2017). Business model innovation in airlines. *International Journal of Innovation* 5 (2): 184–198.

10 Amadeus (2017). Embracing Airline Digital Transformation: A Spotlight on what Travellers Value. Amadeus IT Group SA – Produced by Travel Weekly Group. https://amadeus.com/en/insights/white-paper/embracing-airline-digital-transformation (accessed 27 February 2023).

11 Halpern, N., Mwesiumo, D., Suau-Sanchez, P. et al. (2021). Ready for digital transformation? The effect of organisational readiness, innovation, airport size and ownership on digital change at airports. *Journal of Air Transport Management* 90: 1–11.

Index

Sustainable Aviation Technology and Operations: Research and Innovation Perspectives, First Edition.
Edited by Roberto Sabatini and Alessandro Gardi.
© 2024 John Wiley & Sons Ltd. Published 2024 by John Wiley & Sons Ltd.
Companion Website: www.wiley.com/go/sustainableaviation